IHDP/Future Earth-Integrated Risk Governance Project Series

This book series, entitled "IHDP/Future Earth-Integrated Risk Governance Project Series" (formerly called IHDP-Integrated Risk Governance Project Series) for the International Human Dimensions Programme on Global Environmental Change — Integrated Risk Governance Project (IHDP/Future Earth -IRG Project), is intended to present in monograph form the most recent scientific achievements in the identification, evaluation and management of emerging global large-scale risks. Future Earth is a flagship initiative of the Science and Technology Alliance for Global Sustainability. It aims to provide critical knowledge required for societies to understand and address challenges posed by global environmental change (GEC) and to seize opportunities for transitions to global sustainability. Future Earth identifies three research themes, i.e., Dynamic Planet, Global Development and Transition toward Sustainability in its plan and adopts a new approach of "co-designing and Co-producing" to incorporate GEC researchers with stakeholders in governments, industry and business, international or intergovernmental organizations, and civil society. Books published in this series are mainly collected research works on theories, methods, models and modeling, and case analyses conducted by scientists from various disciplines and practitioners from various sectors under the IHDP/Future Earth -IRG Project. It includes the IRG Project Science Plan, research on social-ecological system responses, "Entry and Exit Transition" mechanisms, models and modeling, early warning systems, understanding regional dynamics of vulnerability, as well as case comparison studies of large-scale disasters and paradigms for integrated risk governance around the world. This book series, therefore, will be of interest not only to researchers, educators and students working in this field but also to policy-makers and decision-makers in government, industry and civil society around the world. The series will be contributed by the international research teams working on the six scientific themes identified by the IHDP/Future Earth -IRG Project science plan, i.e., Social-Ecological Systems, Entry and Exit Transitions, Early Warning Systems, Models and Modeling, Comparative Case Studies, and Governance and Paradigms, and by six regional offices of the IRG Project around the world.

More information about this series at https://link.springer.com/bookseries/13536

Peijun Shi

Atlas of Global Change Risk of Population and Economic Systems

Springer

Editor-in-Chief
Peijun Shi
Beijing Normal University
Beijing, China

National Key Research and Development Program–Global Change and Mitigation Project Grant No. 2016YFA0602400

ISSN 2363-4979 ISSN 2363-4987 (electronic)
IHDP/Future Earth-Integrated Risk Governance Project Series
ISBN 978-981-16-6693-3 ISBN 978-981-16-6691-9 (eBook)
https://doi.org/10.1007/978-981-16-6691-9

This Springer imprint is published by the registered company Springer Nature Singapore Pte Ltd.
The registered company address is: 152 Beach Road, #21-01/04 Gateway East, Singapore 189721, Singapore

Editorial Committee

Chief Map Designers

Wang, Ying
Wang, Jing'ai

Map Designers

Chen, Shuo
Jing, Yuanyuan
Li, Yichen
Liao, Xinli
Liu, Tian
Liu, Weihang
Ma, Qingyuan
Mu, Qingyang
Shi, Fanya
Sun, Yelin
Zhang, Junlin

Foreword by Mami Mizutori

Reducing disaster risk and ensuring the safety of people and livelihoods are essential for the advancement of sustainable development. Since the Third UN World Conference on Disaster Risk Reduction which resulted in the Sendai Framework for Disaster Risk Reduction 2015–2030, the United Nations system has been committed to supporting the improvement of global understanding of risks toward risk-informed sustainable development. To that end, the United Nations Office for Disaster Risk Reduction has developed the Global Assessment Report and has established the Global Risk Assessment Framework providing tools for countries and regions to better understand and manage risks.

As an amplifier of risk, climate change has brought about stronger storms, intensified coastal flooding, higher temperatures, and longer droughts. As a result, assessing and predicting future risk has become more challenging, especially because of the creation of new patterns of human settlement, land use, livelihoods, and migration in adapting to the emergence of new risks. That is why I was pleased to learn that a research team, led by Beijing Normal University with the support of science and technology partners in China, has generated an important body of work on global future risks across the population and economic dimensions. Its projections provide valuable insights into the potential state of the world if climate action and risk reduction are not accelerated.

I would like to congratulate the research team on creating this Atlas and thank them for their hard work in the interest of global disaster risk reduction. I would also like to thank the Government of the People's Republic of China for supporting this work and for its own efforts toward realizing the goals and targets of the Sendai Framework, especially around disaster risk reduction planning and investing in science and technology.

I sincerely hope that this Atlas will enhance global understanding of current and future risk, and will become a catalyst for action to avoid the creation of new risk while reducing existing risk, and for strengthening resilience so that no one is left behind.

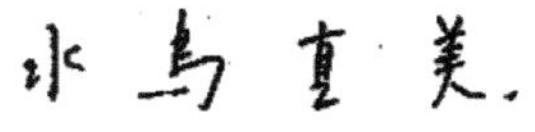

Mami Mizutori
Special Representative of the Secretary-General
for Disaster Risk Reduction, Head of United Nations
Office for Disaster Risk Reduction (UNDRR)

Foreword by Dahe Qin

Climate extremes are an important source of losses from natural hazards and disasters. Global climate change has significantly changed the frequency and intensity of climate extremes, and has significantly increased the possibility of occurrence of disaster chains and disaster compounds. An important content of the Sendai Framework for Disaster Risk Reduction and a critical task for the realization of global sustainable development goals is to fully understand the challenges that climate change brings to disaster risk reduction, and closely integrate disaster risk reduction and climate change response. In 2012, the IPCC issued a special report on "Managing the Risks of Extreme Events and Disasters to Advance Climate Change Adaptation". It focused on the interaction between climate, environment, and human factors that can lead to disasters and examined climate extreme events and disaster risk management at the regional, national, and global scales. It provided guidance for global decision-makers to respond actively to climate extreme events, manage disaster risks, and improve climate change adaption. In 2018, the IPCC issued the "Special Report on Global Warming of 1.5 °C". It estimated the possible risks and impacts of global warming of 1.5 °C, and the potential benefits and opportunities of controlling the temperature rise to 1.5 °C, aiming at providing scientific evidence and motivation for decision-makers to limit global warming to within 1.5 °C.

In order to actively respond to the disaster risk challenges brought about by future climate change, the "Outline of the National Medium- and Long-term Scientific and Technological Development Plan (2006–2020)" released by the Chinese government in 2005 listed "Global Change and Regional Response" as one of the top ten directions of basic research for the country's major strategic needs. In 2010, China launched the National Major Scientific Research Program for global change research. In 2015, China organized a number of institutions to jointly compile the "China National Assessment Report on Risk Management and Adaptation of Climate Extremes and Disasters". The report comprehensively summarized the research outcomes of multiple disciplines related to climate change adaptation and disaster risk management, summarized experiences obtained from past responses to climate extremes and disasters, and put forward policies and practical directions for the future. During the 13th Five-Year Plan period, the National Key Research and Development Program established a set of projects focusing on global change and response, under which a special research direction was established to study the assessment and governance of global change risks.

In this context, the Disaster Risk Science Research Team of Beijing Normal University joined forces with top domestic academic institutions such as the Institute of Atmospheric Physics of the Chinese Academy of Sciences (CAS), the Institute of Geographic Sciences and Natural Resources Research of the CAS, the Northwest Institute of Eco-Environmental Resources of the CAS, and the National Disaster Reduction Research Center of the Ministry of Emergency Management. Together, these institutions have carried out research on the formation mechanism and assessment of global change risks of population and economic systems. The team has projected short-term (2030s) and medium-term (2050s) global-scale climate change and population and economic system changes, developed a set of scenario-based risk assessment models, and assessed the global change risks of population and economic systems at the global scale. Their results have enriched the scientific findings in the field of global change risk research as a token of being pragmatic.

Taking the opportunity of publishing the atlas, we sincerely hope that our peer researchers and practitioners further pay attention to the possible risks in population, economic systems, and the ecological environment brought about by global changes, and further improve the understanding of the risk formation mechanism and improve our models to more accurately assess the risks of exposed population and economic systems. We call for risk-informed integration of climate adaptation and disaster risk governance to promote the realization of sustainable development goals.

Qin Dahe

Dahe Qin
Academician of the Chinese Academy of Sciences
Former Director of China Meteorological Administration

Preface

Global climate change with anthropogenic warming has already posed a severe challenge to the sustainable development of the world and human security. Changes in the frequency and intensity of climate extremes, as well as changes in vulnerability and exposure, brought about by population and economic growth together drive changes in global risks in the future. The Intergovernmental Panel on Climate Change (IPCC) systematically studied the risks of global climate change from the perspectives of climate change projection, impact and risk assessment, and mitigation and adaptation, and specifically focused on the possible risks and impacts of global warming of 1.5 °C. The potential benefits and opportunities of limiting global mean surface temperature rise to within 1.5 °C above the pre-industrial level are estimated. From the three related themes of the dynamic planet, global sustainable development, and the transition to sustainable development, the Future Earth program focuses extensively on preventing systemic risks caused by global changes. The Sendai Framework for Disaster Risk Reduction 2015−2030 closely links disaster risk reduction and climate change response to achieve win–win results.

2019 marks the 30th anniversary of the joint implementation of the United Nation's International Decade for Natural Disaster Reduction. In 1989, in order to actively respond to international actions and national needs of disaster prevention, Beijing Normal University relied on the Cenozoic Paleogeography Laboratory to set up the Natural Disaster Monitoring and Prevention Research Laboratory and began scientific research and disciplinary development for disaster reduction. Over the past three decades, Beijing Normal University has carried out comprehensive research on natural disaster risks from the disciplines such as Geography, Remote Sensing Science and Technology, and Safety Science and Engineering. Research outcomes were summarized in a series of atlases, including the *Atlas of Natural Disaster in China* (Chinese and English editions) published in 1992, the *Atlas of Natural Disaster System of China* (Chinese and English bilingual edition) published in 2003, the *Atlas of Natural Disaster Risk in China* published in 2011 (Chinese and English bilingual editions), and the *World Atlas of Natural Disaster Risk* published in 2015 to welcome the Third World Conference on Disaster Risk Reduction. These atlases systematically present the patterns of spatial and temporal differentiation of natural hazards, disasters, and risks in China and the world.

The future risk challenges brought about by global changes have attracted the attention of the whole world. With the financial support from China's 12th Five-Year Plan National Major Scientific Research Plan "Global Change Environmental Risk Formation Mechanism and Adaptation Research" (2012CB955400) and 13th Five-Year Plan National Key R&D Program Project "Global Change Risk of Population and Economic Systems: Mechanisms and Assessments" (2016YFA0602400), we have moved from a traditional risk study by using loss uncertainty estimation based on historical data to a scenario-based global change risk study integrating future climate and population and economic system changes. Our latest project carried out the projection of changes in the climate, and in the population and economic system exposures (population, crops, industrial added value, road transportation system, and gross domestic product) for the near (2030s, 2016–2035) and medium (2050s, 2046–2065) future periods, under the combined scenarios of Representative Concentration Pathways (RCP2.6, RCP4.5, and RCP8.5) and Shared Socioeconomic Pathways (SSP1, SSP2, and SSP3), and at global-scale medium and high resolution (50 km × 50 km or 30 km × 30 km). The project also conducted a risk assessment for population mortality, crop yield loss, and economic loss. The most representative results have been collected to enter this atlas for hard-copy publication, while more systematic and comprehensive results and multi-modal uncertainty analysis results are publicly accessible via a digital map platform (http://gcr.grisk.info).

Academic work behind this atlas was mainly conducted by the Disaster Risk Science Research Team of Beijing Normal University, relying on the State Key Laboratory of Earth Surface Processes and Resource Ecology, and the Key Laboratory of Environmental Change and Natural Disaster of the Ministry of Education. Joint efforts have been devoted by the Institute of Atmospheric Physics of the Chinese Academy of Sciences (CAS), the Institute of Geographic Sciences and Natural Resources of the CAS, the Northwest Institute of Eco-Environmental Resources of the CAS, and the National Disaster Reduction Center of the Ministry of Emergency Management.

In the process of compiling this atlas, as well as in the development of disaster risk science research of Beijing Normal University, we have received strong support from many institutions at home and abroad. Here, we especially express our sincere gratitude to academic institutions and other organizations that provided support in literature, data, and technology. Domestic institutions and organizations include the National Climate Center of China Meteorological Administration, Xinjiang Institute of Ecology and Geography of the CAS, Institute of Mountain Hazards and Environment of the CAS, Institute of Tibetan Plateau Research of the CAS, School of Urban and Environmental Sciences of Peking University, Department of Earth System Sciences of Tsinghua University, School of Geographical Sciences of East China Normal University, School of Environmental and Geographical Sciences of Shanghai Normal University, School of Geographical Sciences of Qinghai Normal University, Academy of Plateau Science and Sustainability of the People's Government of Qinghai Province and Beijing Normal University, and Academy of Disaster Reduction and Emergency Management of the Ministry of Emergency Management and Ministry of Education. Internationally renowned universities and research institutions that have helped us include: University of Maryland, Boston University, Columbia University, Old Dominion University, Nanyang Technological University, International Institute for Applied Systems Analysis, University of Leuven, Max Planck Institute of Biogeochemistry, University of Stuttgart, University of Trier, Australian Commonwealth Scientific and Industrial Research Organization, Australian National University, Kyoto University, the United Nations Office for Disaster Risk Reduction (UNDRR), and UNDRR Asia-Pacific Technology and Academic Advisory Committee.

We express our sincere thanks to experts and scholars who generously offered valuable advice and comments on the compilation of this atlas, including Academicians Dahe Qin, Jifan Chou, Guanhua Xu, Guodong Cheng, Zhisheng An, Guoxiong Wu, Tandong Yao, Weijian Zhou, Bojie Fu, Huadong Guo, Peng Cui, Fahu Chen, Chenghu Zhou, Jianbing Peng, and Jun Chen, and Professors Hui Lin, Jianguo Wu, Quansheng Ge, Panmao Zhai, Shirong Liu, Peng Gong, Yongjian Ding, Ye Qi, Dabang Jiang, Jianqi Sun, Qiuhong Tang, Changqing Song, Baoyuan Liu, Qian Ye, Guoyi Han, Ming Wang, Qiang Zhang, and Jianjun Wu and Ms. Mami Mizutori! We are also very grateful to Dr. Ying Li and Dr. Juan Du for their contribution to the language editing and proofreading of the entire atlas.

We sincerely hope that the course of global climate change response and comprehensive disaster risk reduction will attain greater achievements in the future and make greater contributions to the building of a community with a shared future for mankind. Limited by the data and academic research and technical capacities of the research team, there are inevitably inadequacies in this atlas. Readers are very welcome to offer their criticisms and help us to make improvements.

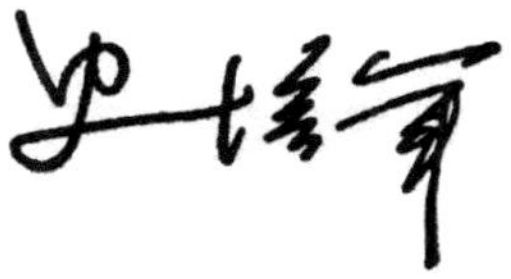

Peijun Shi
State Key Laboratory of Earth
Surface Processes and Resource Ecology
Key Laboratory of Environmental Change and Natural Disaster, MOE
Academy of Disaster Reduction and Emergency Management
MOEM and MOE
Beijing Normal University
Beijing, China

Institutions

Leading Institutions

State Key Laboratory of Earth Surface Processes and Resource Ecology, Beijing Normal University (BNU), China
Key Laboratory of Environmental Change and Natural Disaster, Ministry of Education of China, BNU, China
Academy of Disaster Reduction and Emergency Management, Ministry of Emergency Management (MEM) and Ministry of Education (MOE), China
Faculty of Geographical Science, BNU, China

Participating Institutions

Institute of Atmospheric Physics, Chinese Academy of Sciences (CAS), China
Institute of Geographical Sciences and Natural Resources Research, CAS, China
Cold and Arid Regions Environmental and Engineering Research Institute, CAS, China
National Disaster Reduction Center of China, MEM, China

Collaborating Institutions/Organizations

IHDP/Future Earth-Integrated Risk Governance Project
Ministry of Education, P. R. China
State Administration of Foreign Experts Affairs, P. R. China
The United Nations Office for Disaster Risk Reduction (UNDRR)
Asia Science Technology Academia Advisory Group (ASTAAG)

Data Sources

Food and Agriculture Organization (FAO), UN
International Institute for Applied Systems Analysis (IIASA), Austria
United States Geological Survey (USGS), USA
National Aeronautics and Space Administration (NASA), USA
The International Disaster Database (EM-DAT), Centre for Research on the Epidemiology of Disasters (CRED), Belgium

Sponsors

National Key Research and Development Program Project (2016YFA0602400), Ministry of Science and Technology, P. R. China
Expertise-Introduction Project for Disciplinary Innovation of Universities (BP0820003), Ministry of Education and State Administration of Foreign Experts Affairs, China
Ministry of Education, P. R. China
National Natural Science Foundation of China
National Disaster Reduction Commission of China

Contents

World Political Map

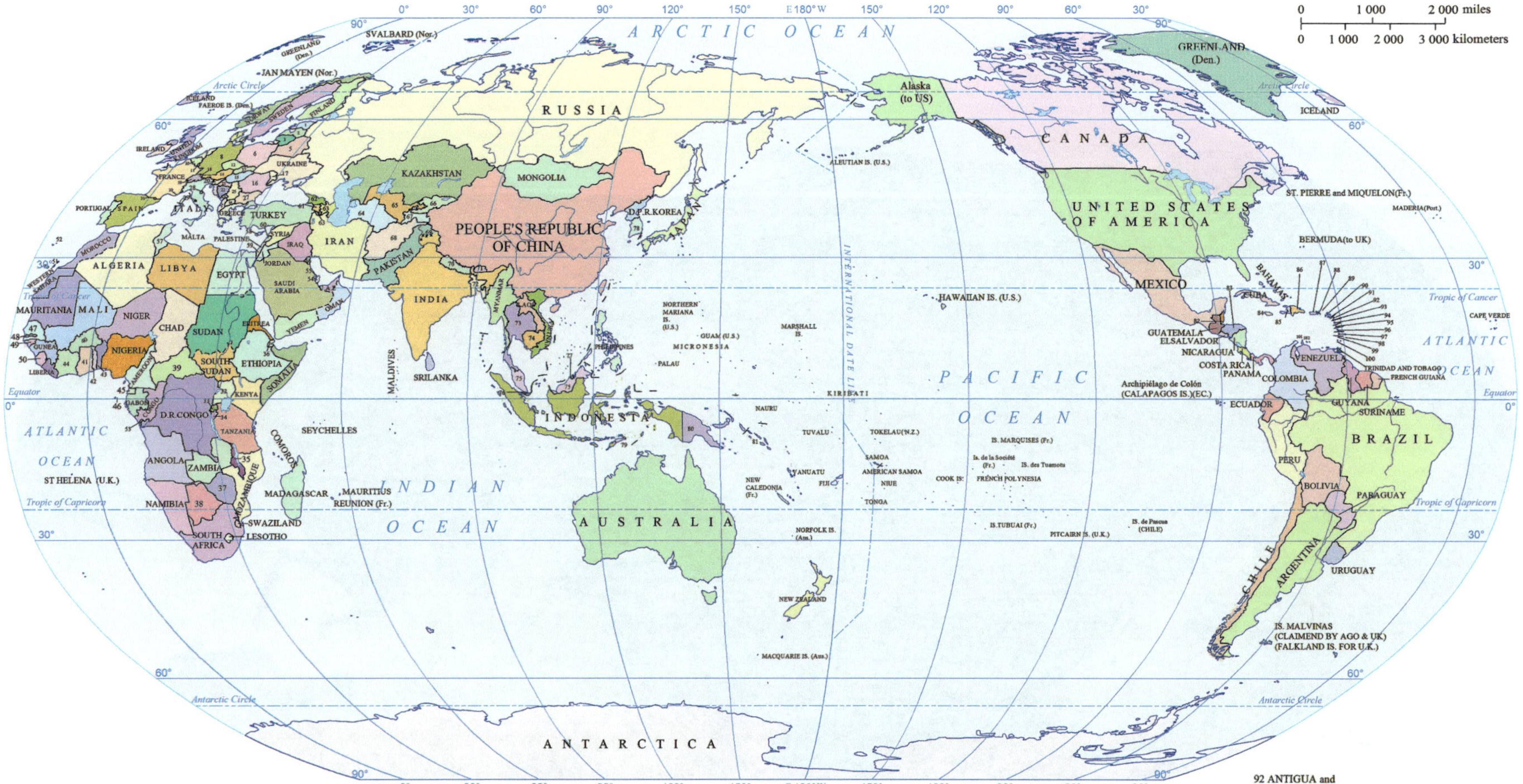

1 ESTONIA
2 LATVIA
3 LITHUANIA
4 RUSSIA
5 BELARUS
6 POLAND
7 DENMARK
8 GERMANY
9 NETHERLANDS
10 BELGIUM
11 LUXEMBOURG
12 CZECH REPUBLIC
13 SLOVAKIA
14 AUSTRIA
15 HUNGARY
16 ROMANIA
17 MOLDOVA
18 LIECHTENSTEIN
19 SWITZERLAND
20 SLOVENIA
21 CROATIA
22 BOSNIA AND HERZEGOVINA
23 MONTENEGRO
24 ALBANIA
25 SERBIA
26 MACEDONIA
27 BULGARIA
28 SAN MARINO
29 VATICAN CITY
30 MONACO
31 ANDORRA
32 UGANDA
33 RWANDA
34 BURUNDI
35 MALAWI
36 DJIBOUTI
37 ZIMBABWE
38 BOTS WANA
39 CENTRAL AFRICAN
40 BURKINA FASO
41 GHANA
42 TOGO
43 BENIN
44 CôTE D'IVOIRE
45 EQUAT.GUINEA
46 SAO TOME AND PRINCIPE
47 SENEGAL
48 GAMBIA
49 GUINEA-BISSAU
50 SIERRA LEONE
51 CANARY IS. (Sp.)
52 MADEIRA (Porl.)
53 ANGOLA
54 QATAR
55 BAHRAIN
56 KUWAIT
57 TUNISIA
58 CYPRUS
59 ISRAEL
60 LEBANON
61 ARMENIA
62 GEORGIA
63 AZERBAIJAN
64 TURKMENISTAN
65 UZBEKISTAN
66 KYRGYZSTAN
67 TAJIKISTAN
68 AFGHANISTAN
69 KASHMIR
70 NEPAL
71 BHUTAN
72 BANGLADESH
73 THAILANO
74 CAMBODIA
75 MALAYSIA
76 SINGAPORE
77 BRUNEI
78 R.O.KOREA
79 EAST TIMOR
80 PAPUA NEW GUINEA
81 SOLOMON IS.
82 BELIZE
83 HONDURAS
84 JAMAICA
85 HAIN
86 DOMINICAN REP
87 PUERTO RICO(U.S.)
88 VIRGIN IS.(U.K.)
89 VIRGIN IS.(U.S.)
90 ANGUILLA(U.K.)
91 ST.KITIS AND NEVIS
92 ANTIGUA and BARBUDA
93 MONSERRAT(U.K.)
94 GUADELOUPE(FR.)
95 DOMINICA
96 MARTINIQUE(FR.)
97 ST.LUCIA
98 BARBADOS
99 ST.VINCENT AND THE GRENADINES
100 GRENADA
101 ARUBA(NETH.)
102 GURACAO(NETH.)

Data Source: World Atlas of Natural Disaster Risk (2015)

Global Satellite Image

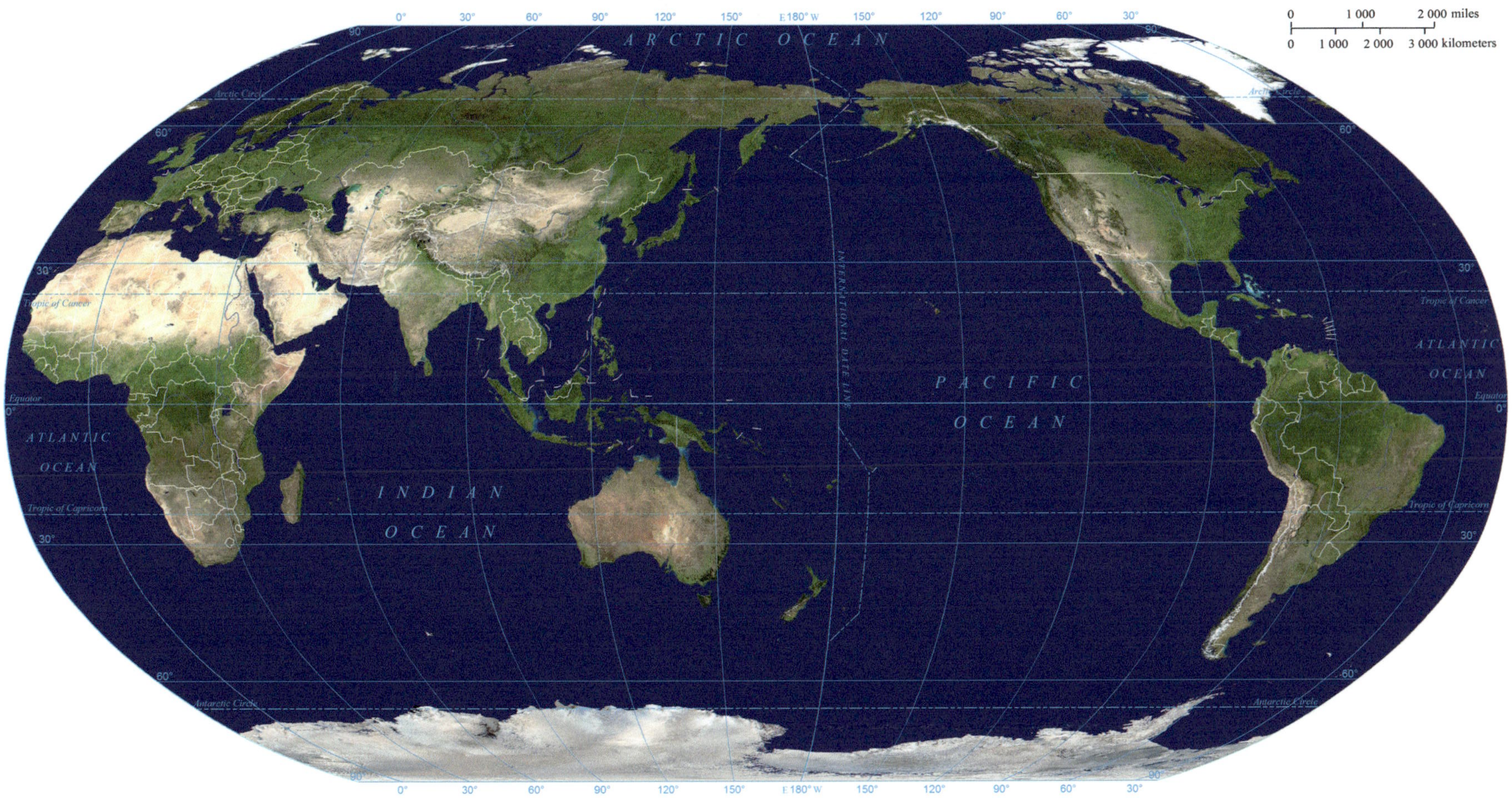

Source: National Aeronautics and Space Administration(NASA), http://visibleearth.nasa.gov/view.php?id=57752

Environments

Mapping Environments of the World

Peijun Shi, Jing'ai Wang, Ying Wang, and Tian Liu

1 Introduction

A regional disaster system is essentially the product of the interaction of humans and nature, which is composed of the disaster-formative environment, exposure, hazard, and disaster losses. It is a dynamic system with complex characteristics on the surface of the earth (Fig. 1) (Shi 1991).

Disaster-formative environment (E): Broadly defined, it is the natural and human environments. The regional differences of disaster-formative environments not only have a profound impact on the generation of hazards but also have an obvious influence on the human casualties and property losses caused by the hazards. Hazard (H): A hazard is a process or phenomenon that may pose negative impacts on the economy, society, and ecology, including both natural factors and human factors that are associated with the natural ones. Exposure (S): It covers humanity itself and lifeline systems, production systems, cultural and social systems, and various natural resources and ecological systems. Disaster loss and effect (Ds): It includes casualties and psychological impacts, direct and indirect economic losses, building (structural) destruction, social network (non-structural) disorder, ecosystem degradation, environmental pollution, resource damages, and so on (Shi 1991, 1996, 2002, 2005, 2009, 2019).

Authors: Peijun Shi, Jing'ai Wang, Ying Wang, Tian Liu.
Map Designers: Tian Liu, Yuanyuan Jing, Yelin Sun, Fanya Shi, Jing'ai Wang, Ying Wang.
Language Editor: Ying Wang.

P. Shi (✉) · J. Wang · Y. Wang
State Key Laboratory of Earth Surface Processes and Resource Ecology, Beijing Normal University, Beijing, 100875, China
e-mail: spj@bnu.edu.cn

P. Shi · J. Wang · Y. Wang · T. Liu
Faculty of Geographical Science, Beijing Normal University, Beijing, 100875, China

2 Environments

The disaster-formative environment shown in this part mainly refers to the natural physical environments, namely geology, landform, climate, hydrology, vegetation, and soil.

Land elevation and slope of the terrain will affect the spatial distribution of disasters. River systems will affect the occurrence and scope of floods. Land cover and soil will directly or indirectly influence the severity of floods and droughts. In addition, climate zones directly or indirectly reflect the distribution of extreme climatic events.

3 Maps Based on Reference Data

The maps based on reference data contain World Political Map, Global Land Cover, Global Soil, Global Climate Zones, Global River Systems, Global Digital Elevation Models, Global Terrain Slope, and Global Satellite Image, and their sources are shown in Table 1.

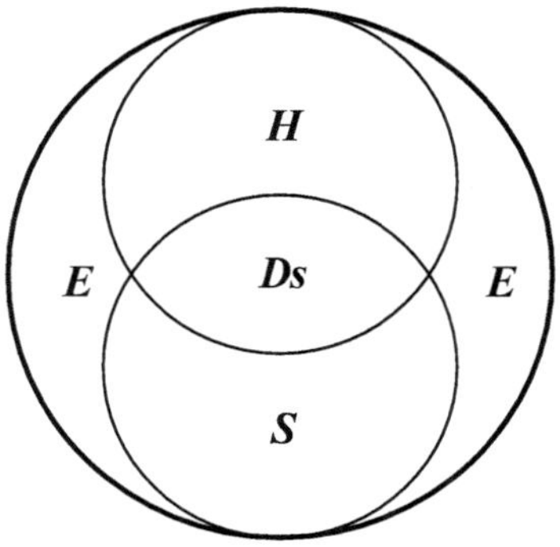

E: Disaster-formative environment, **H**: Hazard,
S: Exposure, **Ds**: Loss and effects

Fig. 1 Disaster structural system

P. Shi, *Atlas of Global Change Risk of Population and Economic Systems*, IHDP/Future Earth-Integrated Risk Governance Project Series, https://doi.org/10.1007/978-981-16-6691-9_1

Table 1 Maps included in the Environments section

Maps	Source
World Political Map	World Atlas of Natural Disaster Risk (2015)
Global Land Cover	Copernicus Global Land Service https://lcviewer.vito.be/download
Global Soil	Food and Agriculture Organization (FAO), United Nations Environment Programme (UNEP), Land Degradation Assessment in Drylands (LDAD) Project http://www.fao.org/geonetwork/srv/en/metadata.show?id=37139&currTab=distribution
Global Climate Zones	Food and Agriculture Organization (FAO) https://data.apps.fao.org/map/catalog/srv/eng/catalog.search?uuid=7538cb25-7b2e-4030-8454-7197a49af48a#/home
Global River Systems	Food and Agriculture Organization (FAO) http://www.fao.org/geonetwork/srv/en/main.home?uuid=9264483f-ca14-496b-aeae-fe1b8aebf520
Global Digital Elevation Models	NOAA's National Centers for Environmental Information (NCEI) https://data.nodc.noaa.gov/cgi-bin/iso?id=gov.noaa.ngdc.mgg.dem:316
Global Terrain Slope	Calculated based on Global Digital Elevation Models of the NOAA National Centers for Environmental Information
Global Satellite Image	National Aeronautics and Space Administration (NASA) http://visibleearth.nasa.gov/view.php?id=57752

4 Maps

Global Land Cover

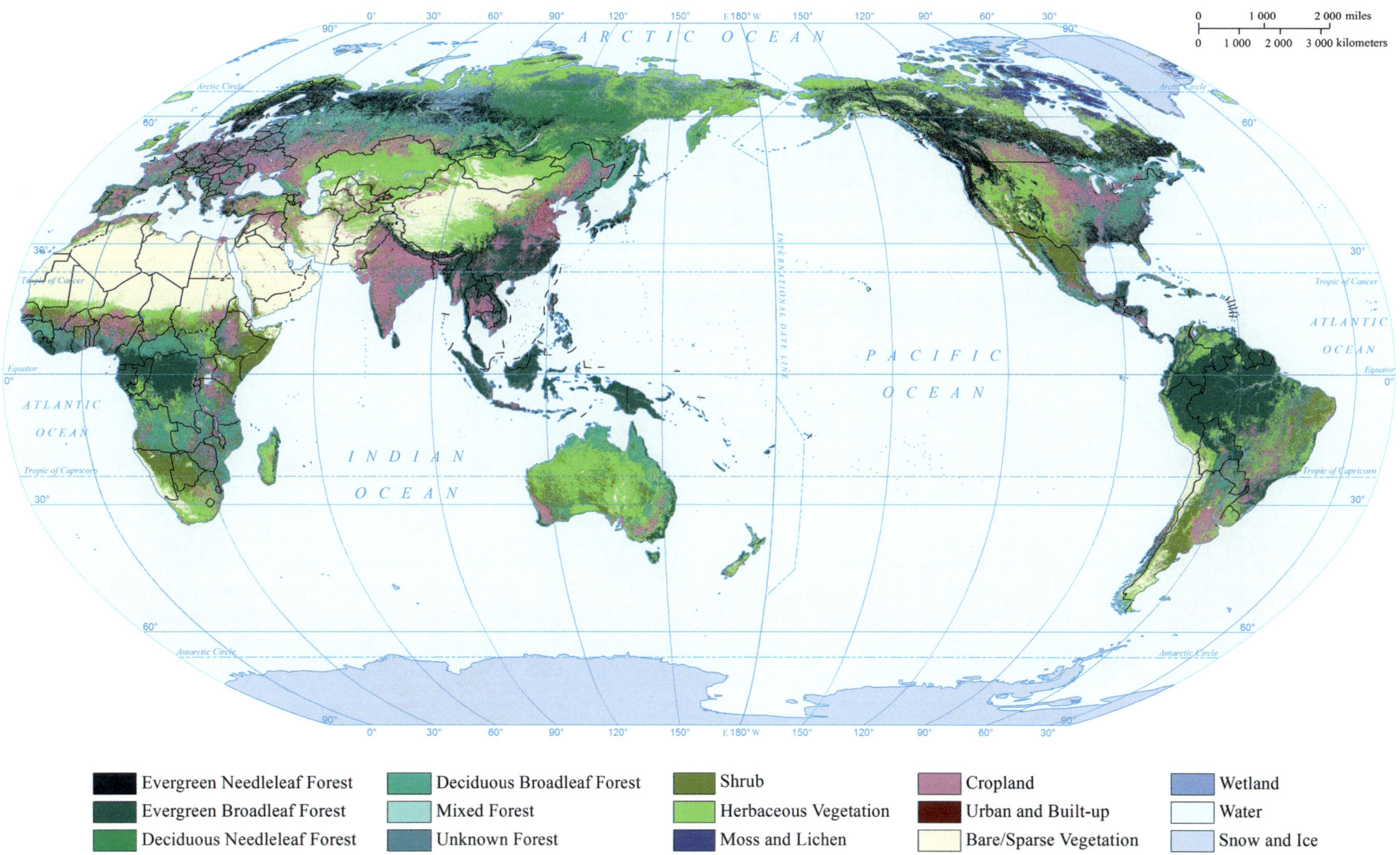

Data Source: Copernicus Global Land Service, © Copernicus Service Information year of publication
https://lcviewer.vito.be/download

Global Soil

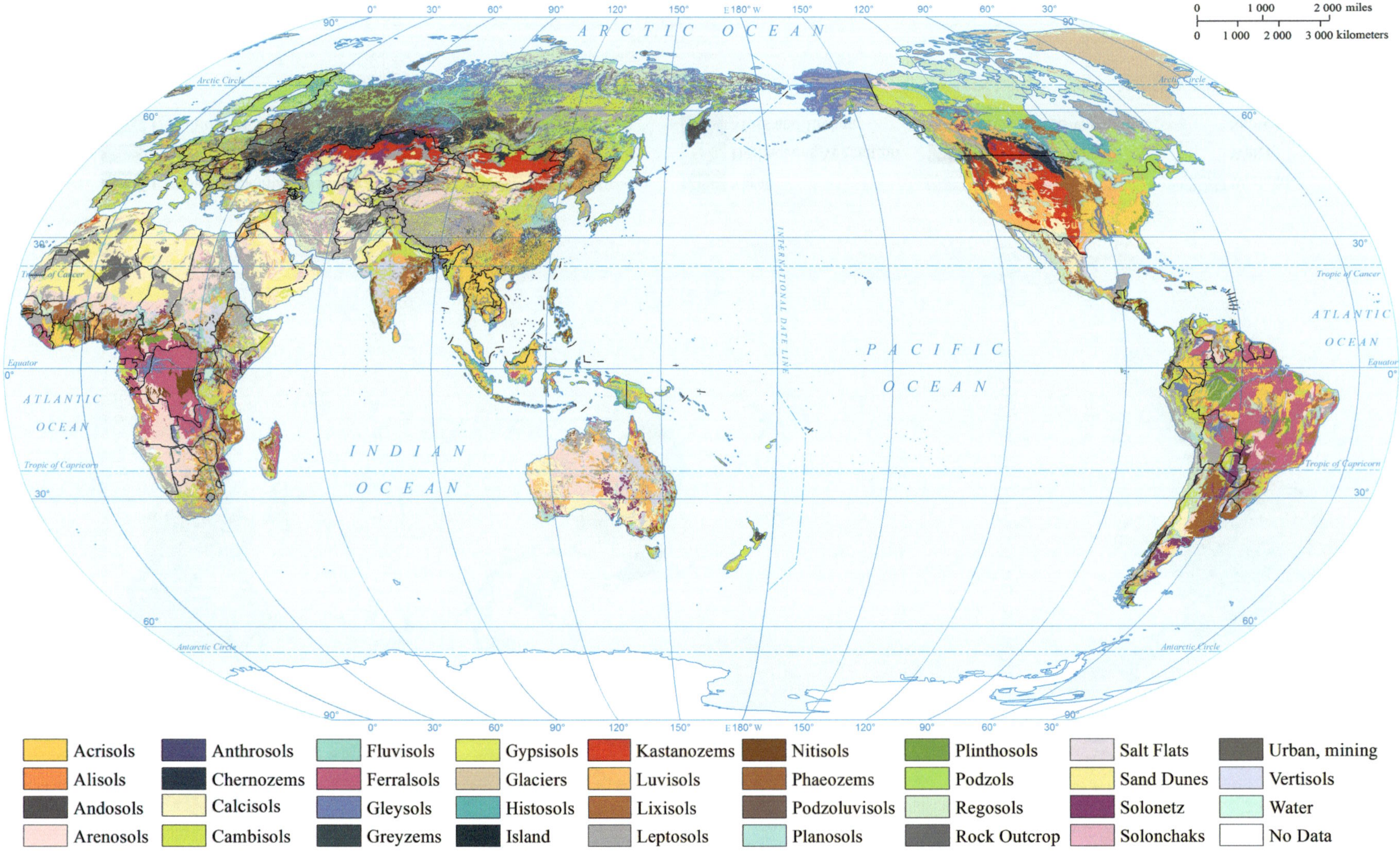

Data Source: Food and Agriculture Organization(FAO) United Nations Environment Programme(UNEP) Land Degradation Assessment in Drylands(LDAD) Project
http://www.fao.org/geonetwork/srv/en/metadata.show?id=37139&currTab=distribution

Global Climate Zone

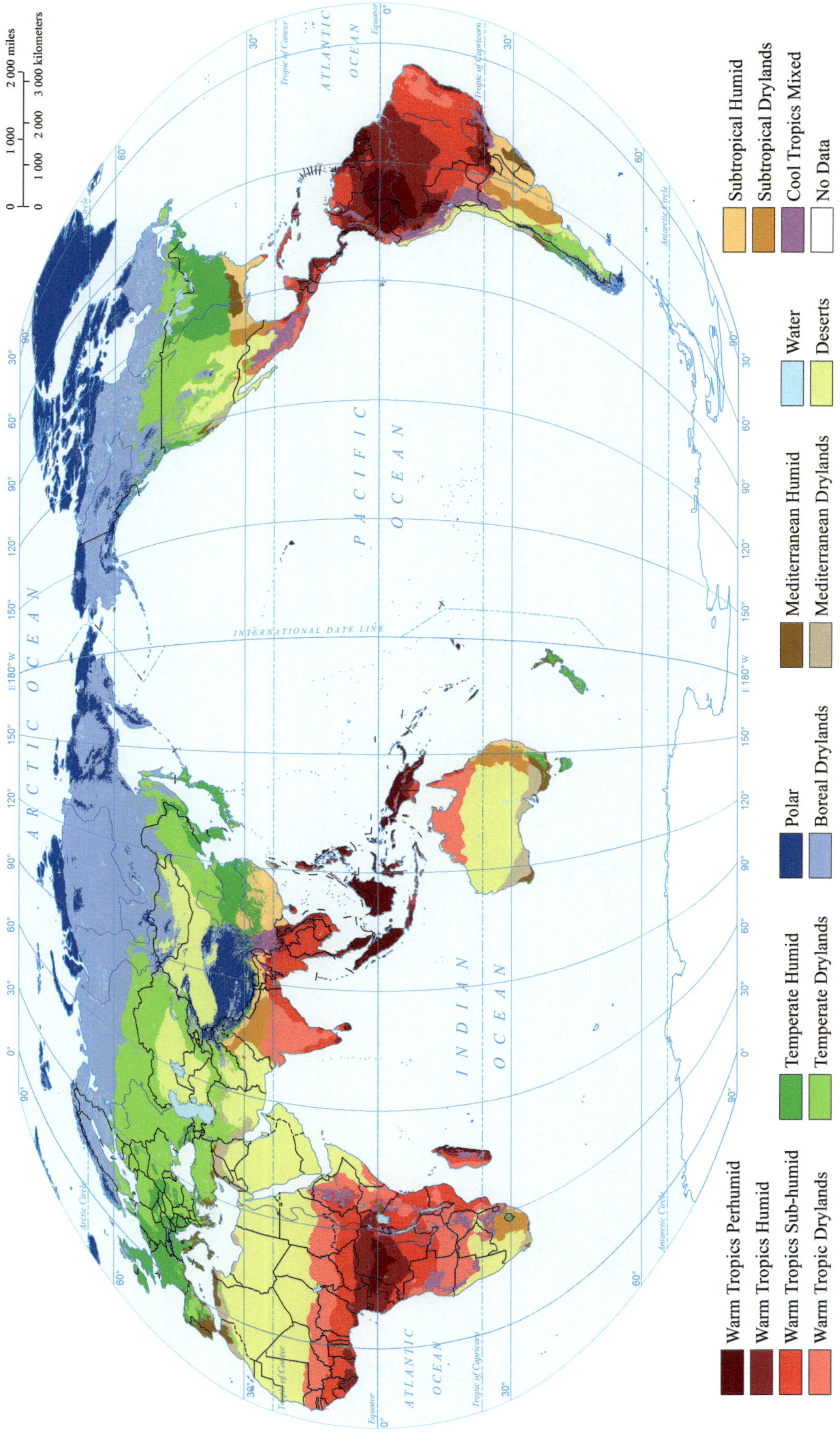

Data Source: Food and Agriculture Organization(FAO)
http://www.fao.org/geonetwork/srv/en/metadata.show?id=37139&currTab=distribution

Global River Systems

Data Source: Food and Agriculture Organization(FAO) AQUASTAT - programme of the Land and Water Division Montana
http://www.fao.org/geonetwork?uuid=7538cb25-7b2e-4030-8454-7197a49af48a

Global Terrain Slope

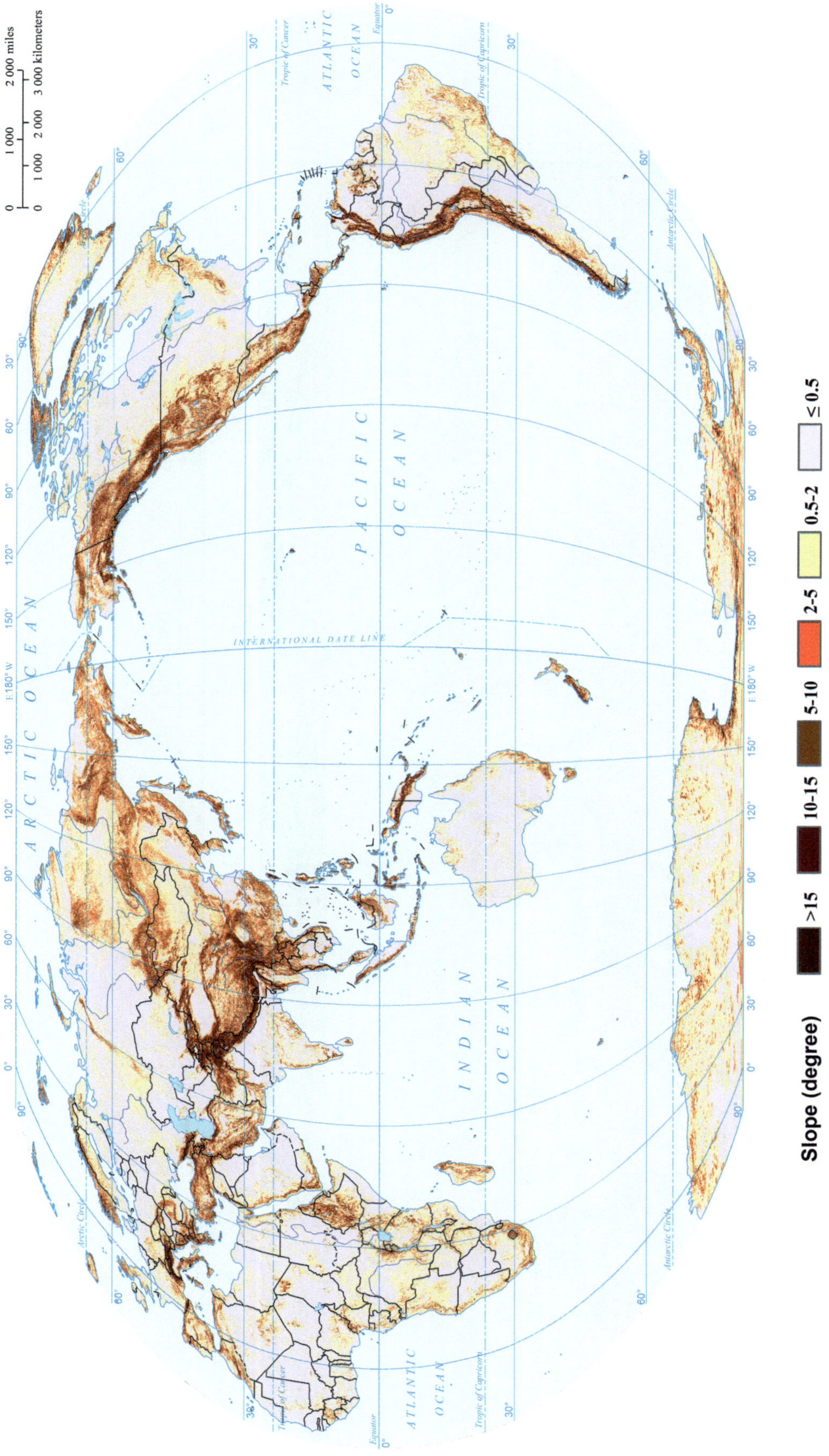

Global Land Elevation

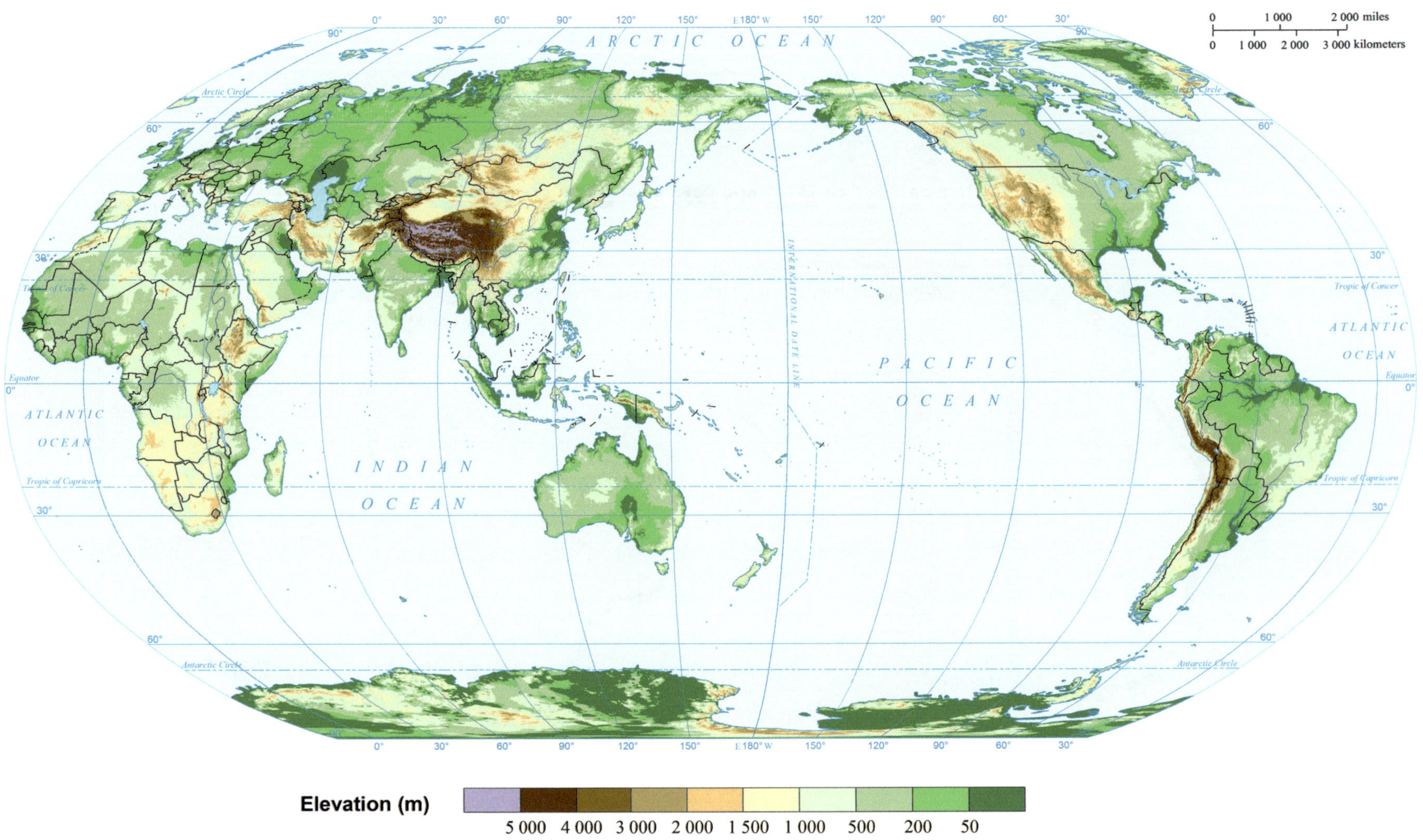

Data Source: NOAA's National Centers for Environmental Information (NCEI)
https://data.nodc.noaa.gov/cgi-bin/iso?id=gov.noaa.ngdc.mgg.dem:316

References

Shi, P.J. 1991. Study on the theory of disaster research and its practice. *Journal of Nanjing University (Natural Sciences)* 11 (Supplement): 37–42 (in Chinese).

Shi, P.J. 1996. Theory and practice of disaster study. *Journal of Natural Disasters* 5 (4): 6–17 (in Chinese).

Shi, P.J. 2002. Theory on disaster science and disaster dynamics. *Journal of Natural Disasters* 11 (3): 1–9 (in Chinese).

Shi, P.J. 2005. Theory and practice on disaster system research—The fourth discussion. *Journal of Natural Disasters* 14 (6): 1–7 (in Chinese).

Shi, P.J. 2009. Theory and practice on disaster system research—The fifth discussion. *Journal of Natural Disasters* 18 (5): 1–9 (in Chinese).

Shi, P.J. 2019. *Disaster Risk Science*, 2nd ed. Beijing Normal University Press and Springer Nature.

Climate Changes

Mapping Temperature Changes

Xin Qi, Miaoni Gao, Tao Zhu, Siyu Li, Sicheng He, and Jing Yang

1 Introduction

Under the background of global warming, extreme temperature events have significantly increased and hit various parts of the globe (Alexander et al. 2006; Piao et al. 2010; Fischer and Knutti 2014; Gao et al. 2019; Qi et al. 2019)—for example, extreme high temperature occurred during the summer of 2010 over Central Europe-Russia (Grumm 2011) and the super cold surge swept across China at the end of 2020 (Zheng et al. 2021). As a serious worldwide challenge, extreme temperature events bring severe damages to public health, agricultural production, and socioeconomic systems (Easterling et al. 2000; Sun et al. 2018; Wang et al. 2019). Therefore, assessing future global temperature changes is crucial for tackling climate change and disaster mitigation and prevention.

Several studies have attempted to project future changes in temperature at the global scale or with a focus on certain regions through the coarse global climate models (GCMs) or high-resolution regional climate models (Zobel et al. 2017; Dosio et al. 2018; Nangombe et al. 2018). However, the global temperature changes including the mean state, variance, and extreme temperature in the future based on fine-resolution multiple GCM outputs are rarely reported in the previous literature. A high-resolution (0.25° × 0.25°) dataset named NASA Earth Exchange Global Daily Downscaled Projections (NEX-GDDP) has been released (Thrasher et al. 2012, 2013), which enables the temperature assessment from a global perspective. NEX-GDDP is a statistical downscaling dataset using the bias correction and spatial disaggregation method based on the simulations of the GCMs from the Coupled Model Intercomparison Project Phase 5 (CMIP5) and historical observation (Wood et al. 2004; Maurer et al. 2010). Compared with the original GCM outputs, the historical fidelity of climatology and extreme temperature derived from the downscaled NEX-GDDP has been improved (Bao and Wen 2017; Luo et al. 2020), which provides us a new opportunity to perform a comprehensive assessment of future changes in temperature.

Therefore, this section initiatively investigates the prospective changes in the mean state, variance, and extreme values of global temperature under three greenhouse gas emissions scenarios, including Representative Concentration Pathway (RCP) 2.6, RCP4.5, and RCP8.5, for two target periods (the 2030s and 2050s). The present results provide a fundamental reference for the relevant climate risk identification and assessment.

Authors: Xin Qi, Miaoni Gao, Tao Zhu, Siyu Li, Sicheng He, Jing Yang.
Map Designers: Tian Liu, Yelin Sun, Fanya Shi, Jing'ai Wang, Ying Wang.
Language Editor: Jing Yang.

X. Qi · T. Zhu · S. Li · S. He · J. Yang (✉)
Faculty of Geographical Science, Beijing Normal University, Beijing, 100875, China
e-mail: yangjing@bnu.edu.cn

J. Yang
State Key Laboratory of Earth Surface Processes and Resource Ecology, Faculty of Geographical Science, Beijing Normal University, Beijing, 100875, China

M. Gao
School of Geographical Science, Nanjing University of Information Science and Technology, Nanjing, 210044, China

P. Shi, *Atlas of Global Change Risk of Population and Economic Systems*, IHDP/Future Earth-Integrated Risk Governance Project Series, https://doi.org/10.1007/978-981-16-6691-9_2

2 Data

The global daily maximum and minimum temperature for the period from 1950 to 2100 were retrieved from the NEX-GDDP dataset, including downscaled projections from the 21 models under RCP4.5 and RCP8.5 greenhouse gas emissions scenarios for which daily datasets were produced and distributed under CMIP5. The spatial resolution of the data is 0.25° (~25 km × 25 km). In addition, the projection with the same resolution for RCP2.6 from the 13 models was assimilated by the Institute of Atmospheric Physics (IAP) and Chinese Academy of Sciences (CAS), which covers the region between 60°S and 90°N (Xu and Wang 2019). The NEX-GDDP dataset can be freely accessed on the following website: https://www.nccs.nasa.gov/services/data-collections/land-based-products/nex-gddp.

3 Method

Here the temporal range is divided into three periods: the historical period (2000s) defined as 1986–2005, the future period 2030s (2016–2035), and the future period 2050s (2046–2065). Note that summer and winter refer to June–July–August (JJA) and December–January–February (DJF), respectively. Furthermore, the extreme temperature is measured by both an absolute index (*TXx*) and two percentile indices (*TX90p* and *TN10p*) according to the definitions of extremes indices recommended by the Expert Team on Climate Change Detection and Indices (ETCCDI) (Karl and Easterling 1999; Zhang et al. 2011; Fan et al. 2020). *T*mean is defined as the average daily minimum and maximum temperature. *Tstd* is defined as the standard deviation of the daily mean temperature. *TXx* refers to the multi-model ensemble of annual maximum near-surface air temperature during the historical period or future time periods based on the NEX-GDDP models. *TX90p* (*TN10p*) refers to the percentage of the days with the daily maximum (minimum) temperature exceeding (below) the local calendar day 90th (10th) percentile centered on a 5-day window for the base period 1961–1990. In addition, The *T*mean change is calculated by subtracting the *T*mean during the historical period (2000s) from the *T*mean under the RCP scenarios in the future. The model uncertainty of the *T*mean is represented by the standard deviation of the *T*mean under the RCP scenarios during the different periods based on all selected models. Changes and model uncertainties of *Tstd*, *TXx*, *TX90p*, and *TN10p* are calculated in the same way as *T*mean.

4 Major Findings

Figure 1 shows the daily *T*mean over nine regions under different greenhouse gas emissions scenarios, respectively, for the 2030s and 2050s. Compared to the historical period (gray bar), the nine selected regions in all RCP scenarios are expected to experience warming in the future. Under the same RCP scenarios, *T*mean in the 2050s is higher than in the 2030s in all regions.

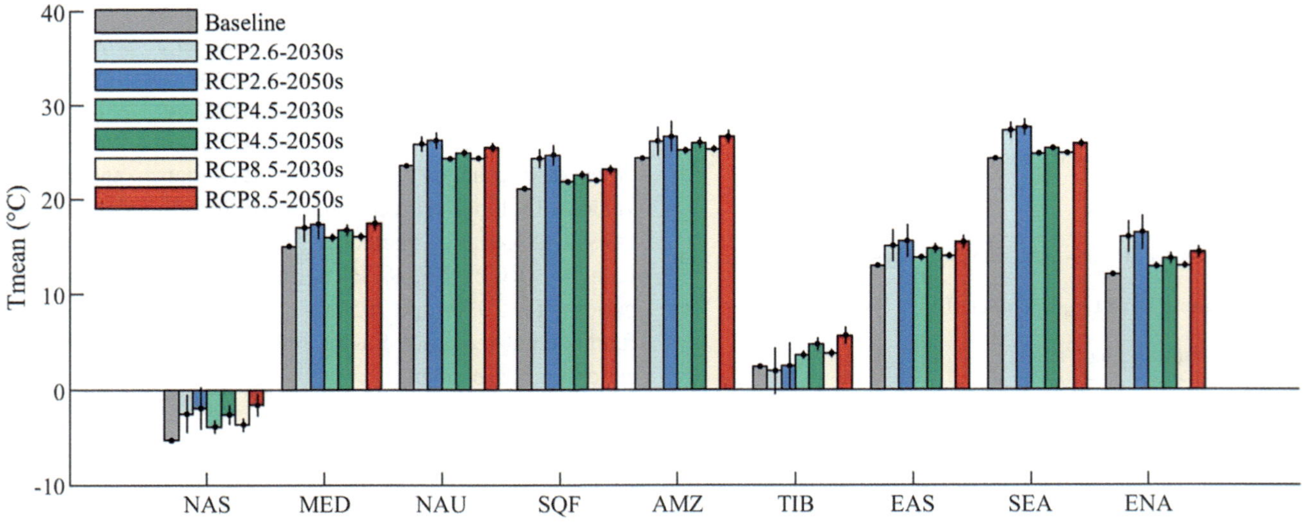

Fig. 1 Area-averaged *T*mean in nine regions under different Representative Concentration Pathway (RCP) scenarios. The error bar represents the one standard deviation across all selected models—13 general circulation models (GCMs) (RCP2.6) and 21 GCMs (RCP4.5 and RCP8.5). NAS, MED, NAU, SQF, AMZ, TIB, EAS, SEA, and ENA represent North Asia (47–70°N, 60.5–180.5°E), Mediterranean Basin (30–47°N, 10.5°W–37.5°E), Northern Australia (28–10°S, 109.5–155.5°E), South Equatorial Africa (26–0°S, 0.5–55.5°E), Amazon Basin (20°S–10°N, 78.5–34.5°W), Tibet (30–47°N, 80.5–104.5°E), East Asia (20–47°N, 104.5–140.5°E), Southeast Asia (10°S–20°N, 100.5–150.5°E), and Eastern North America (25–50°N, 85.5–60.5°W), respectively. The regional division follows Giorgi and Bi (2005)

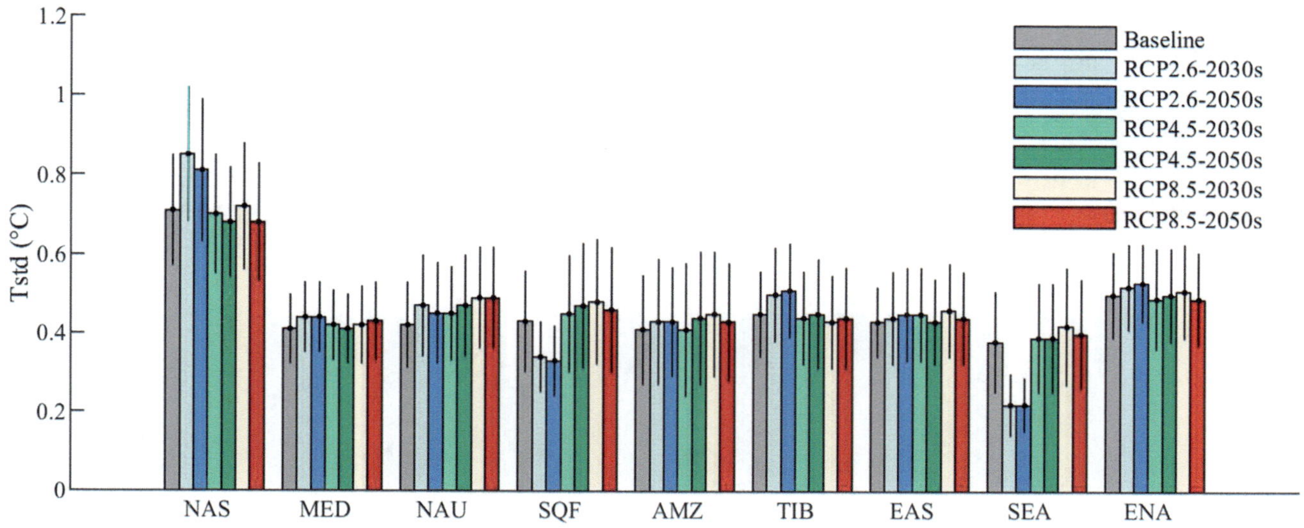

Fig. 2 Standard deviation of the daily mean temperature (*Tstd*) in nine regions under different Representative Concentration Pathway (RCP) scenarios. The error bar represents the one standard deviation across all selected models. Region abbreviations are the same as in Fig. 1

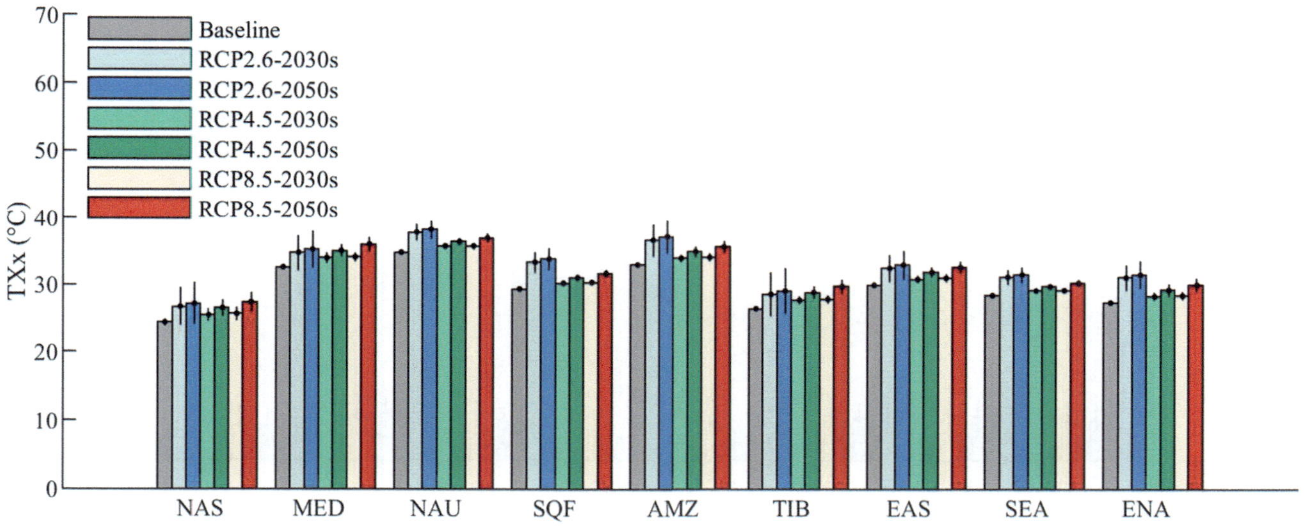

Fig. 3 Annual maximum near-surface air temperature (***TXx***) in nine regions under different Representative Concentration Pathway (RCP) scenarios. The error bar represents the one standard deviation across all selected models. Region abbreviations are the same as in Fig. 1

The standard deviations of the daily mean temperature for the nine regions under different greenhouse gas emissions scenarios are shown in Fig. 2. The temperature deviations in most areas are about 0.4 °C in all scenarios. In comparison, the North Asia region exhibits the largest temperature deviation with high uncertainty, while the lowest deviations appear in South Equatorial Africa and Southeast Asia under the RCP2.6 scenarios.

The annual maximum near-surface air temperature for the nine regions under different RCP scenarios is shown for three epochs in Fig. 3. Similar to the *T*mean, the rising *TXx* occurs in all regions under all scenarios in comparison with the historical period (gray bar). Regions with high *TXx* are mainly located in the Mediterranean Basin, Northern Australia, and the Amazon Basin. Additionally, the *TXx* difference between the 2050s and the 2030s is larger under the RCP8.5 scenarios.

The percentage of the days with a daily maximum temperature greater than the 90th percentile of the base period for the nine regions under different RCP scenarios is shown

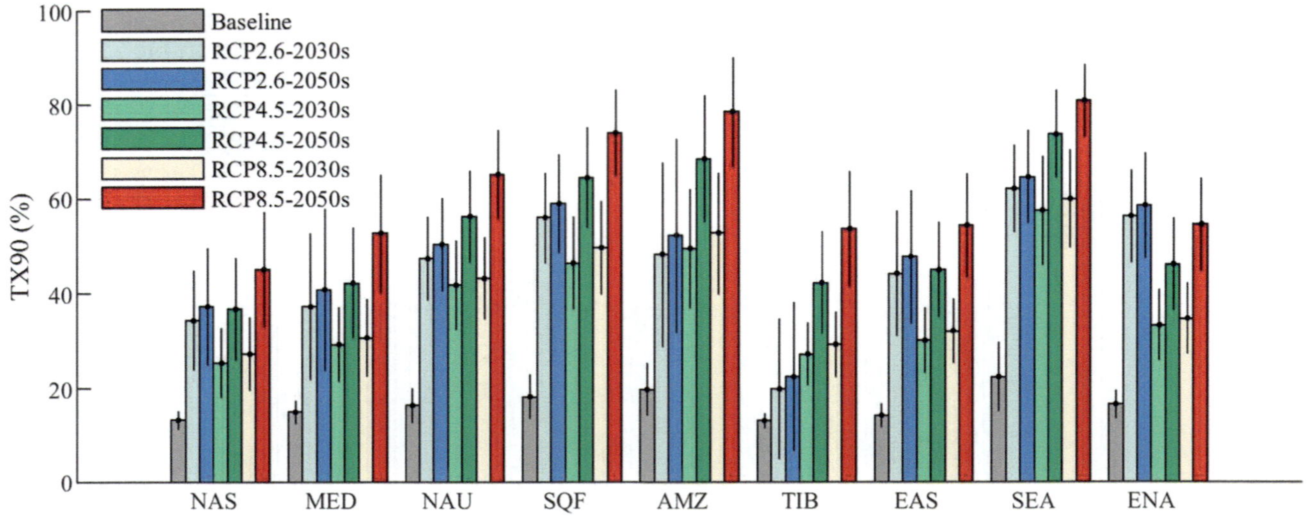

Fig. 4 Percentage of the days with the daily maximum temperature greater than the 90th percentile of the base period (*TX90p*) in nine regions under different Representative Concentration Pathway (RCP) scenarios. The error bar represents the one standard deviation across all selected models. Region abbreviations are the same as in Fig. 1

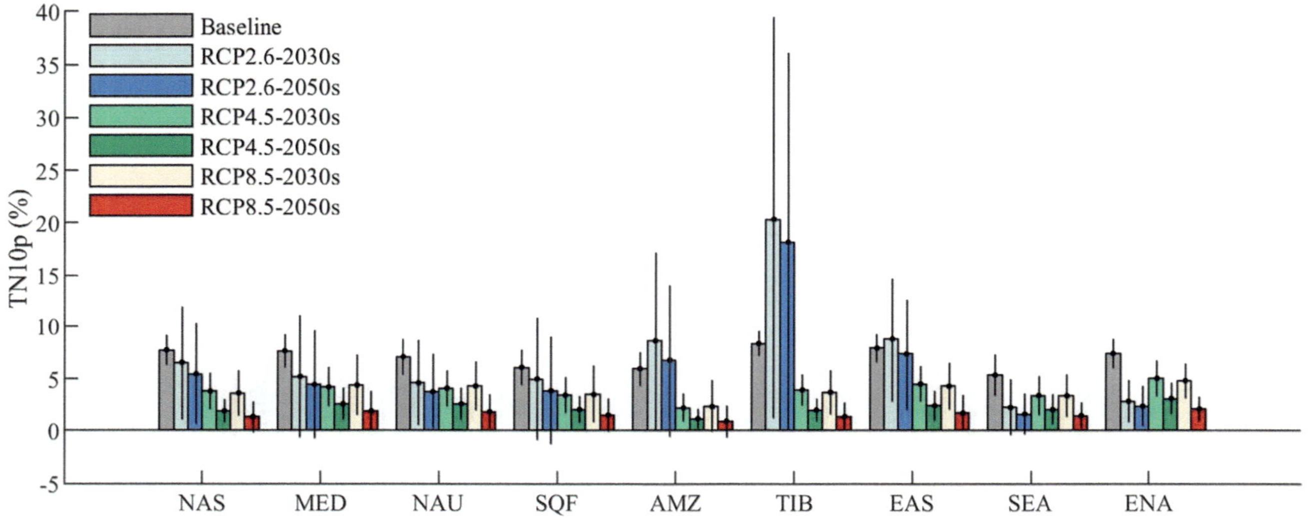

Fig. 5 Percentage of the days with the daily minimum temperature less than the 10th percentile of the base period (*TN10p*) in nine regions under three Representative Concentration Pathway (RCP) scenarios. The error bar represents the one standard deviation across all selected models. Region abbreviations are the same as in Fig. 1

in Fig. 4. Invariably, *TX90p* will increase significantly in all regions in the future, regardless of the scenario. In particular, in the 2050s, *TX90p* is expected to even exceed the historical period by a factor of four under the RCP8.5 scenario.

The percentage of the days with the daily minimum temperature less than the 10th percentile of the base period is shown for the nine regions under the three RCP scenarios in three epochs in Fig. 5. Compared to the historical period (gray bar), six of the nine selected regions—North Asia, the Mediterranean Basin, Northern Australia, South Equatorial Africa, Southeast Asia, and Eastern North America—are expected to have a decreased *TN10p* in the future in all RCP scenarios. In the 2030s, *TN10p* in the other three regions (the Amazon Basin, Tibet, and East Asia) increases under RCP2.6.

Annual mean temperature (Baseline)

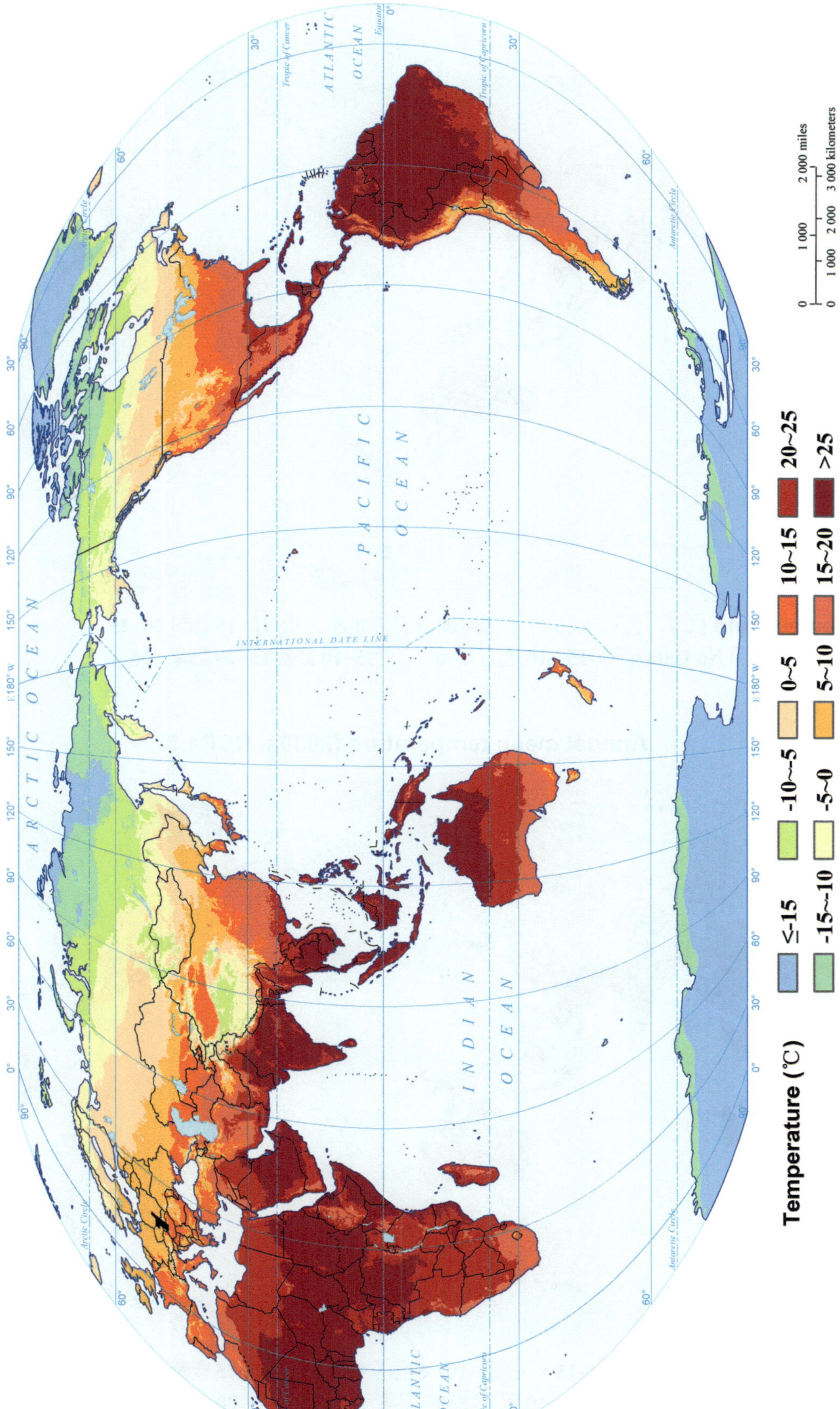

Annual mean temperature (2030s, RCP2.6)

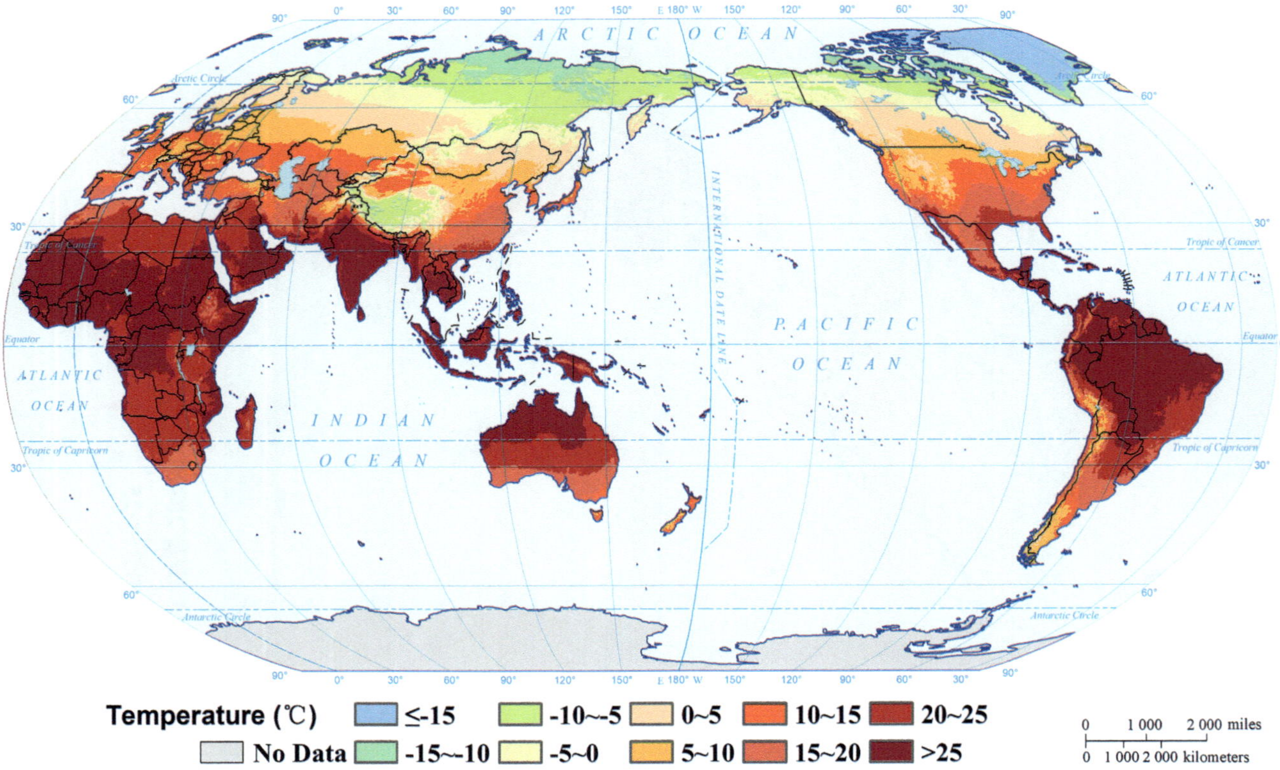

Annual mean temperature (2030s, RCP4.5)

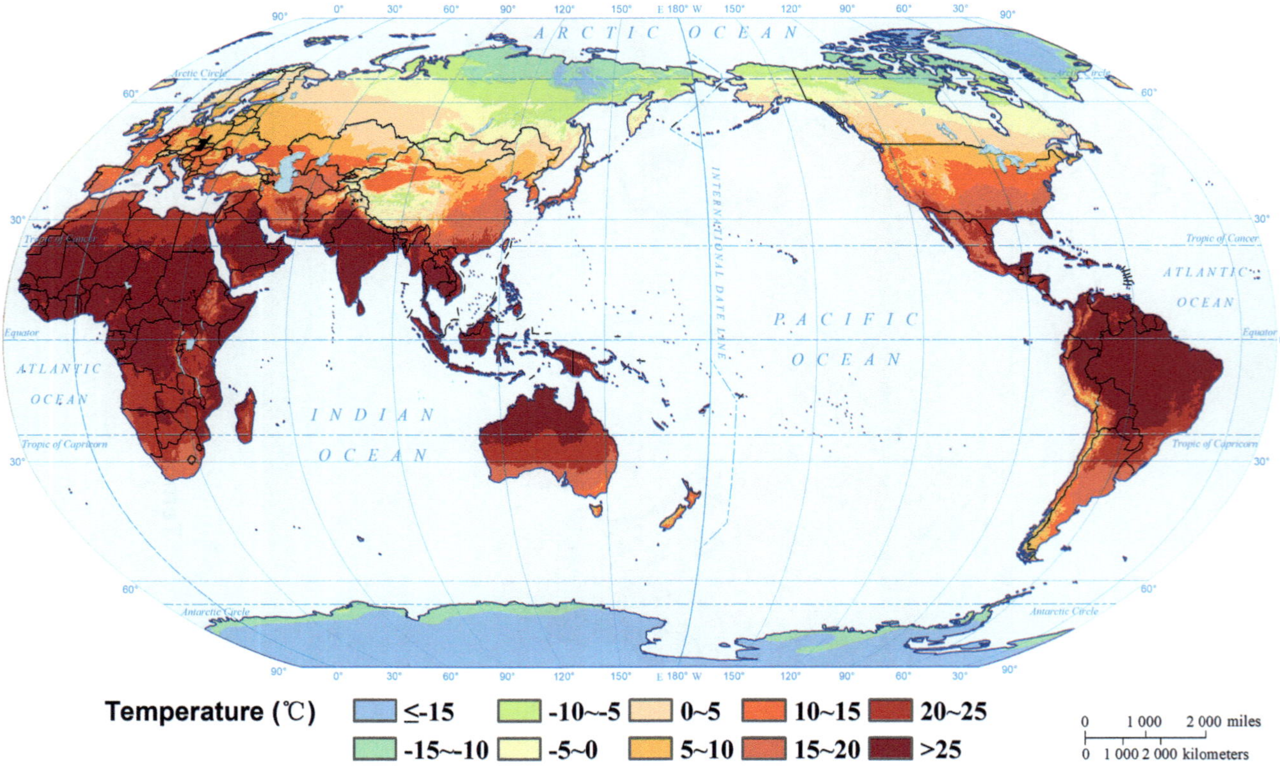

Annual mean temperature (2030s, RCP8.5)

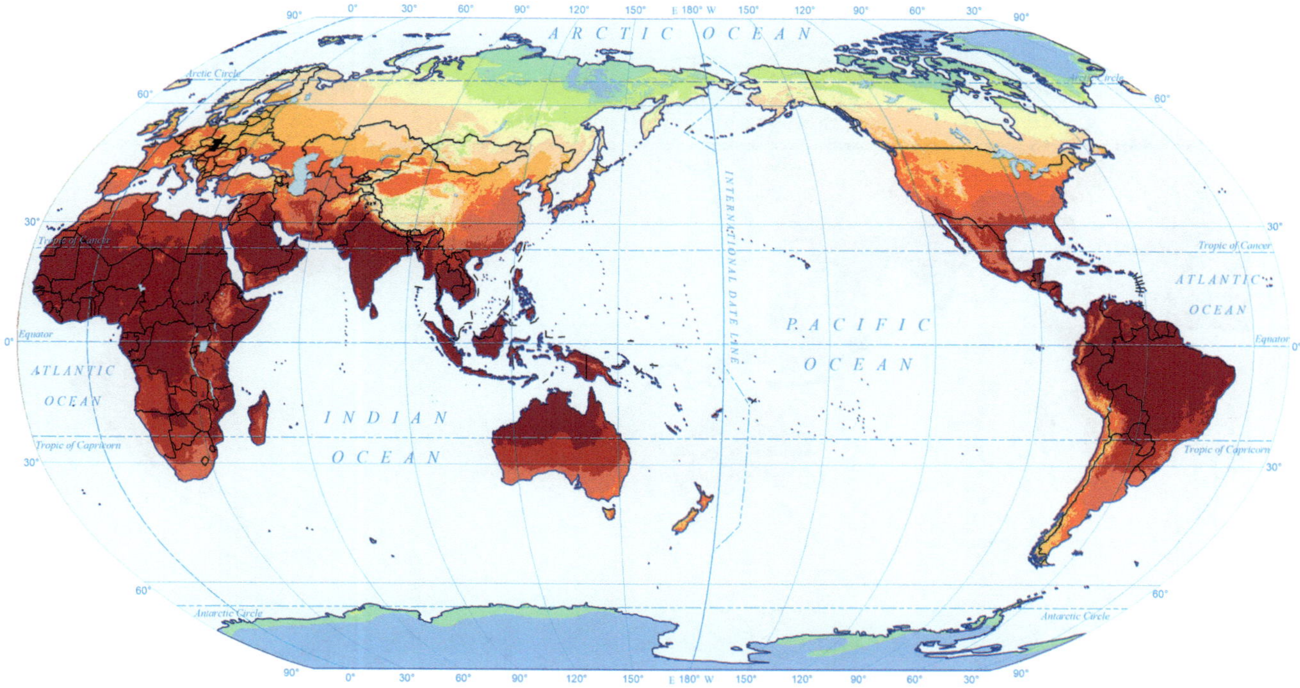

Temperature (℃) ≤-15 | -15~-10 | -10~-5 | -5~0 | 0~5 | 5~10 | 10~15 | 15~20 | 20~25 | >25

0 1 000 2 000 miles
0 1 000 2 000 kilometers

Annual mean temperature (2050s, RCP2.6)

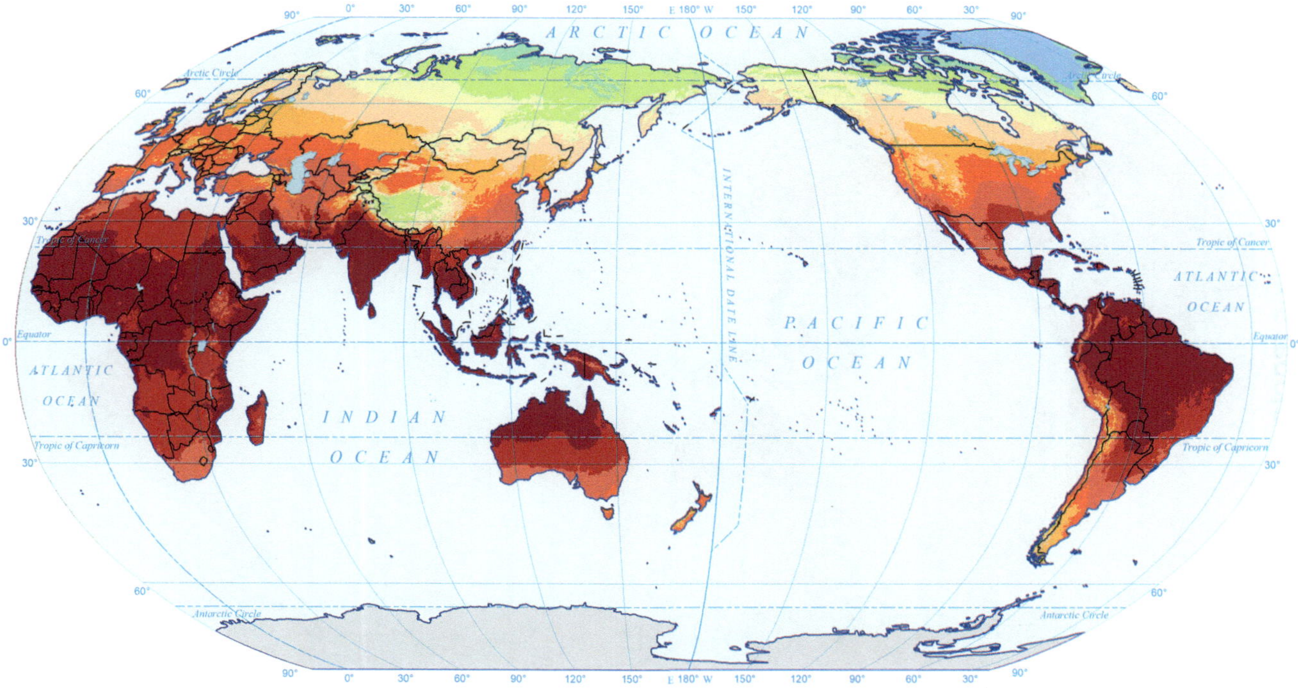

Temperature (℃) No Data | ≤-15 | -15~-10 | -10~-5 | -5~0 | 0~5 | 5~10 | 10~15 | 15~20 | 20~25 | >25

0 1 000 2 000 miles
0 1 000 2 000 kilometers

Annual mean temperature (2050s, RCP4.5)

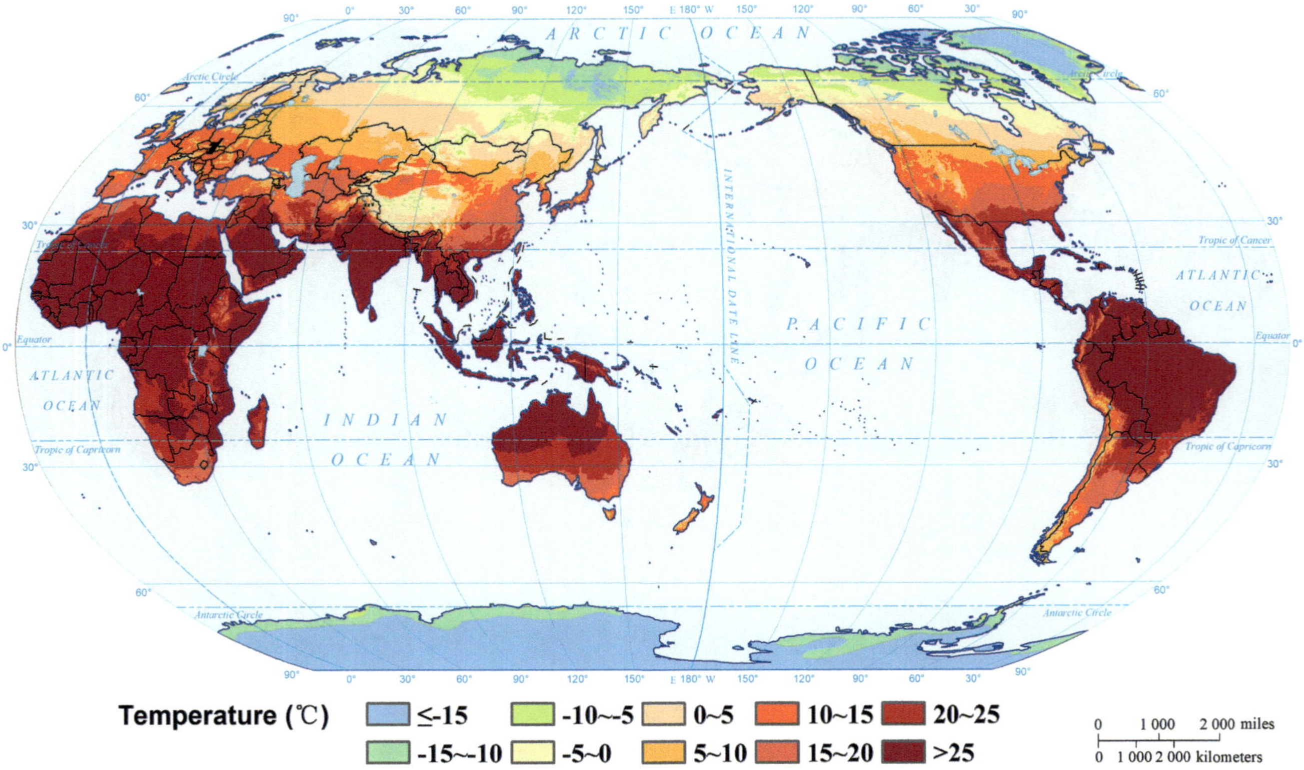

Annual mean temperature (2050s, RCP8.5)

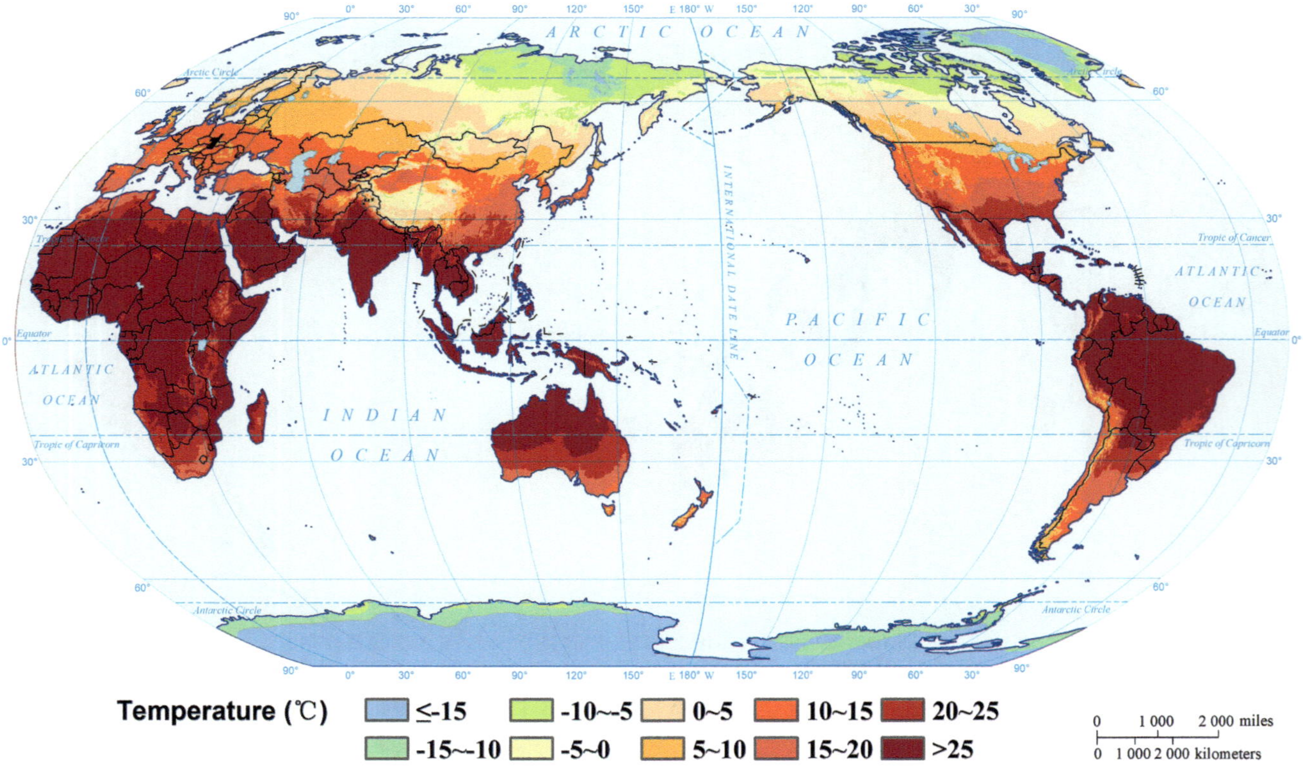

Annual mean temperature variability (Baseline)

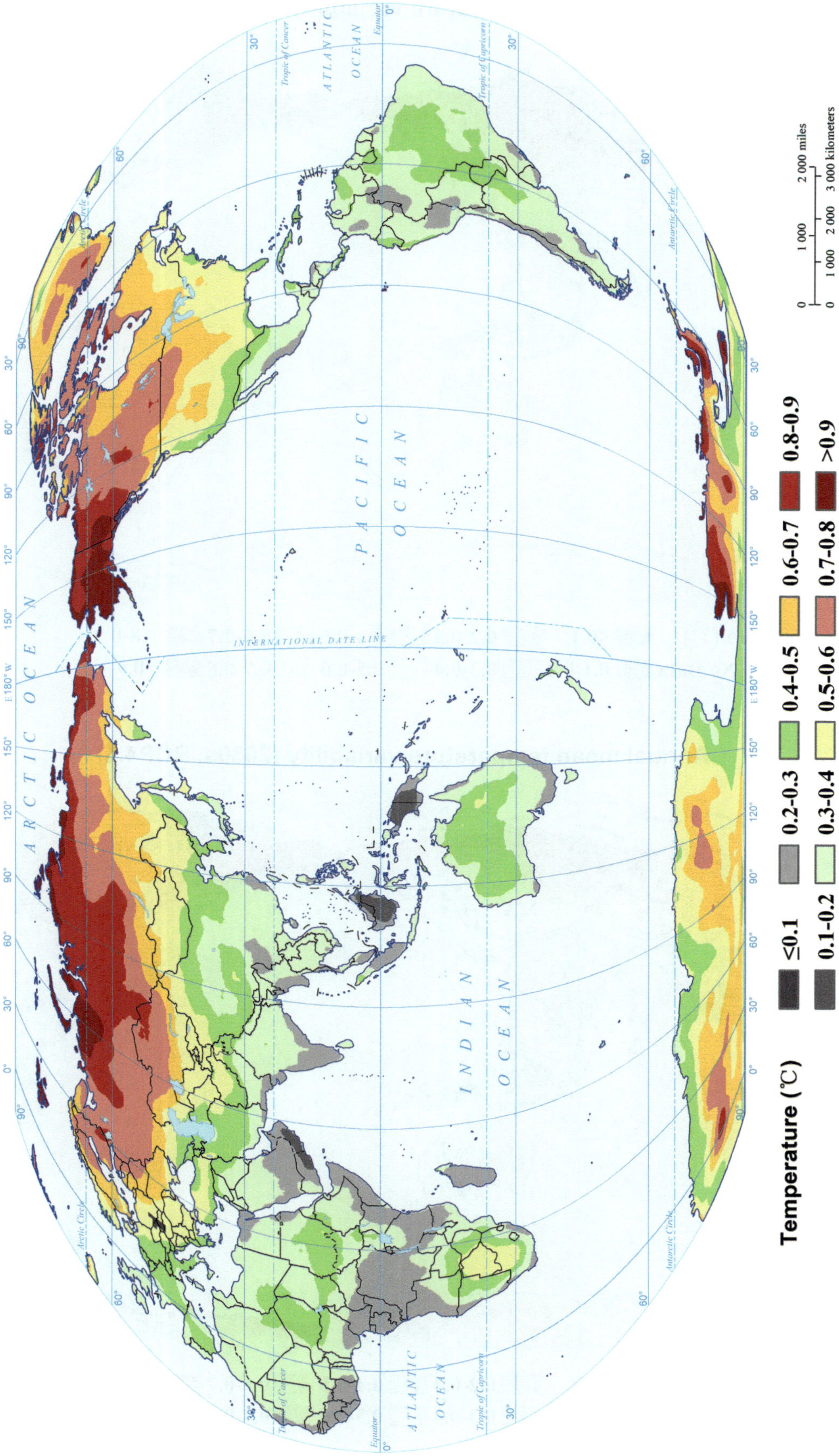

Annual mean temperature variability (2030s, RCP2.6)

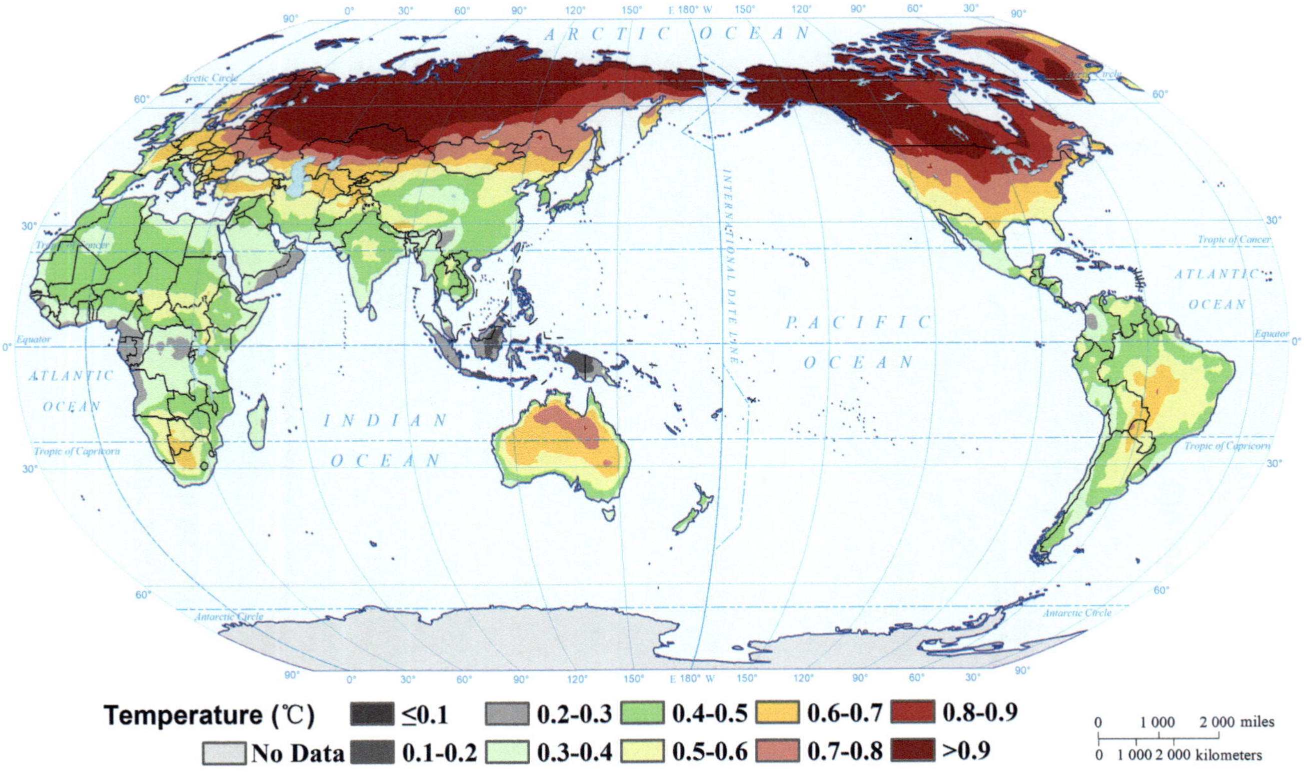

Annual mean temperature variability (2030s, RCP4.5)

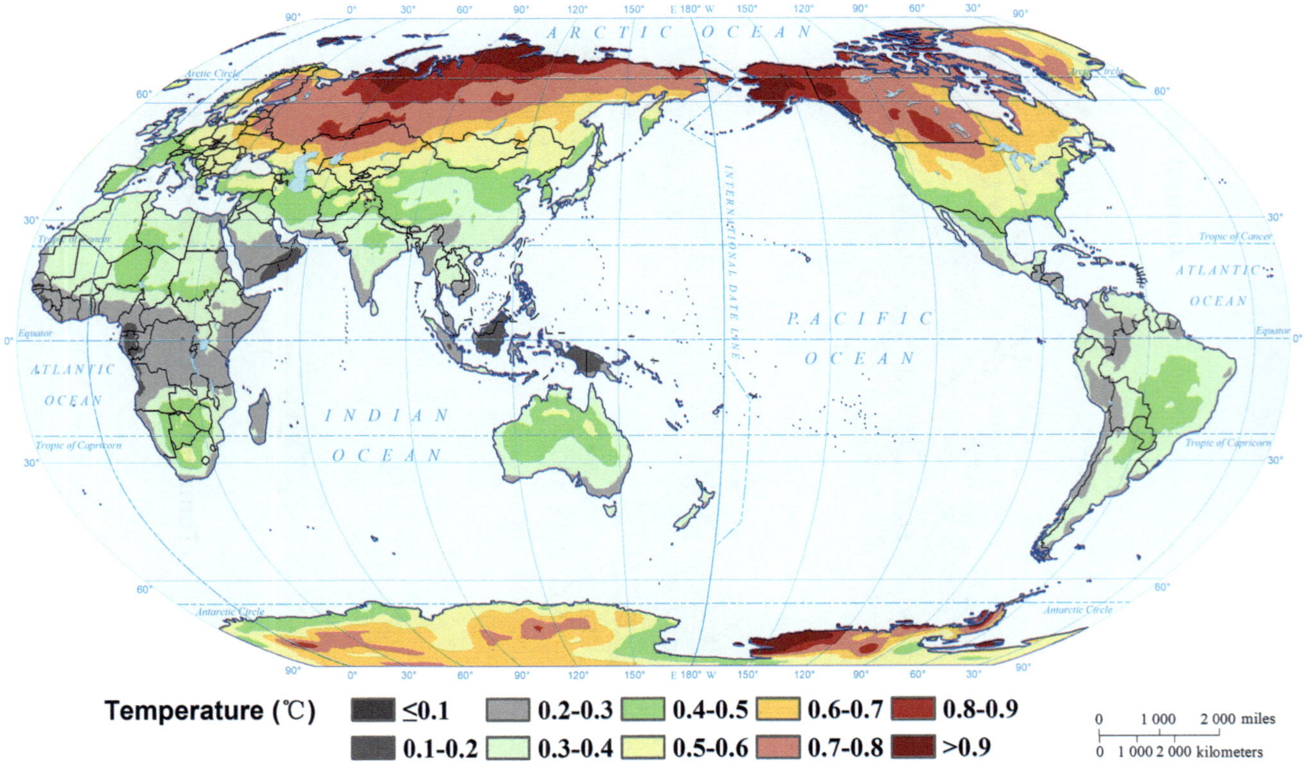

Annual mean temperature variability (2030s, RCP8.5)

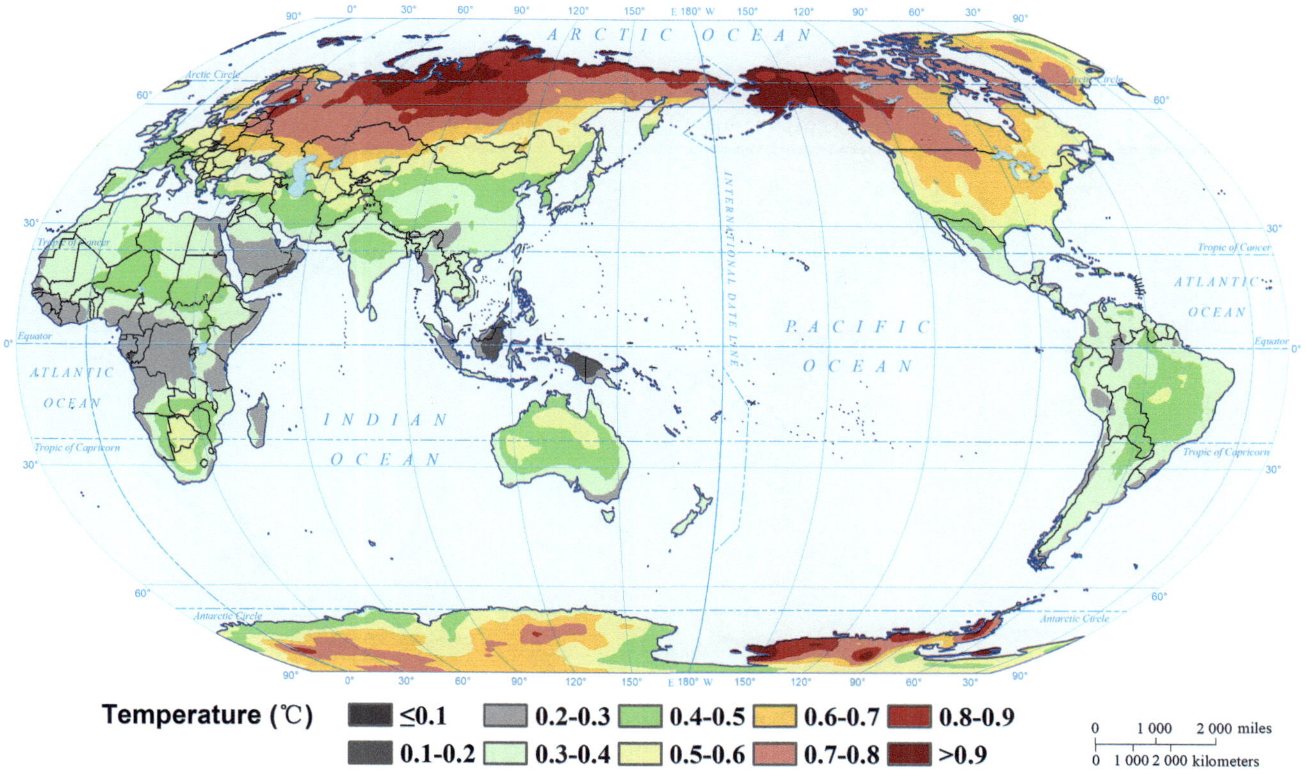

Annual mean temperature variability (2050s, RCP2.6)

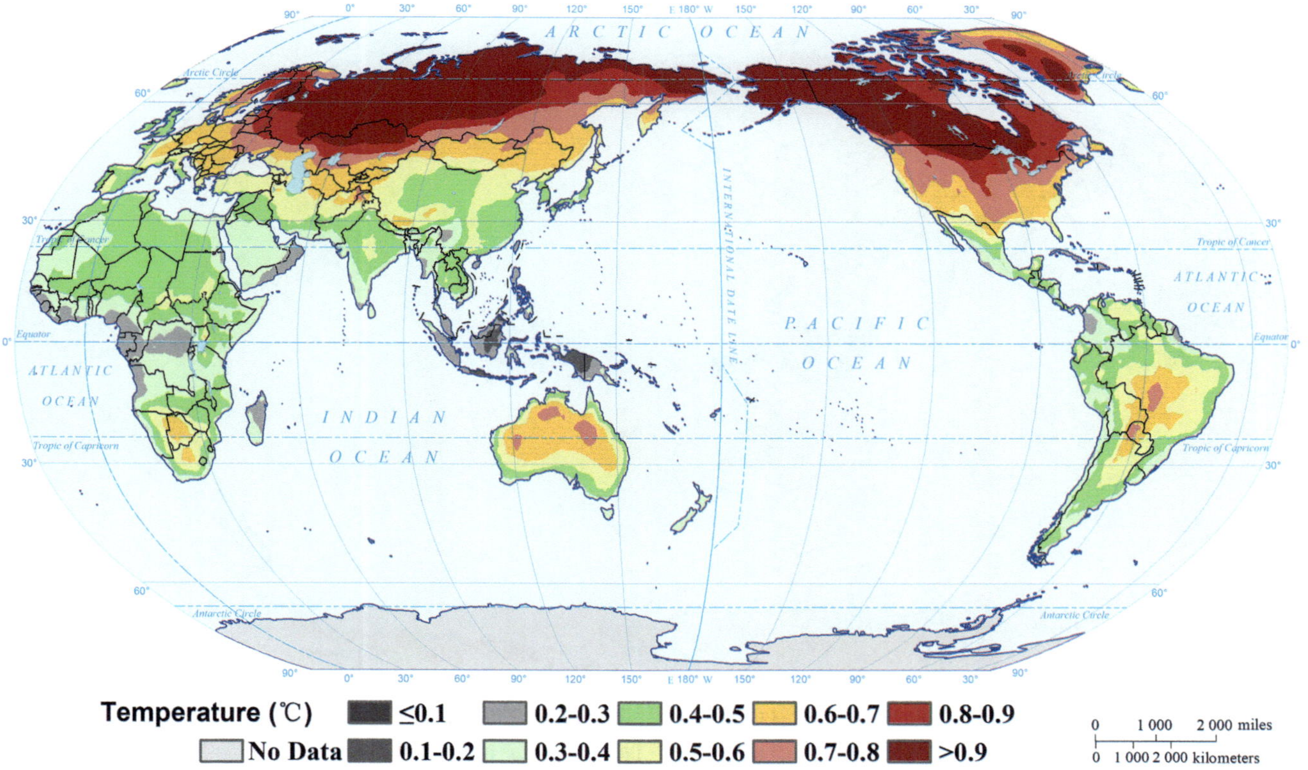

Annual mean temperature variability (2050s, RCP4.5)

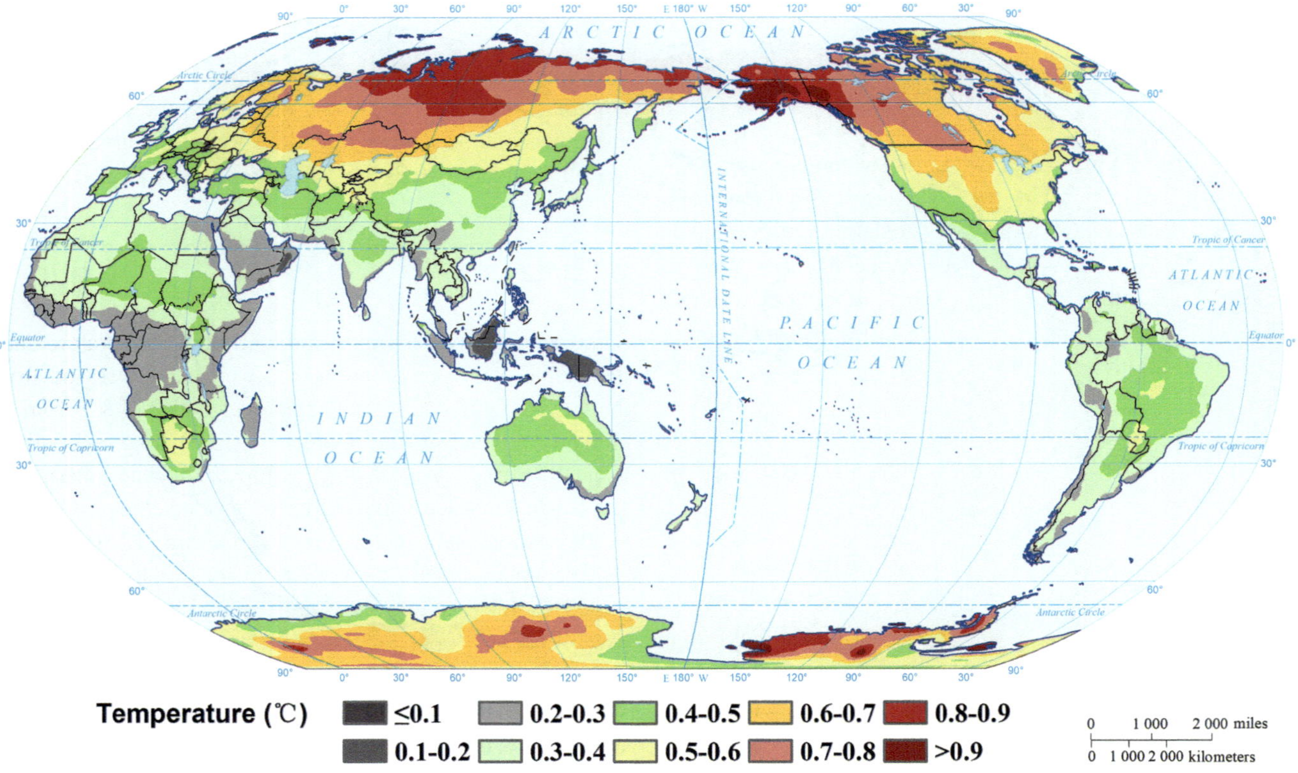

Annual mean temperature variability (2050s, RCP8.5)

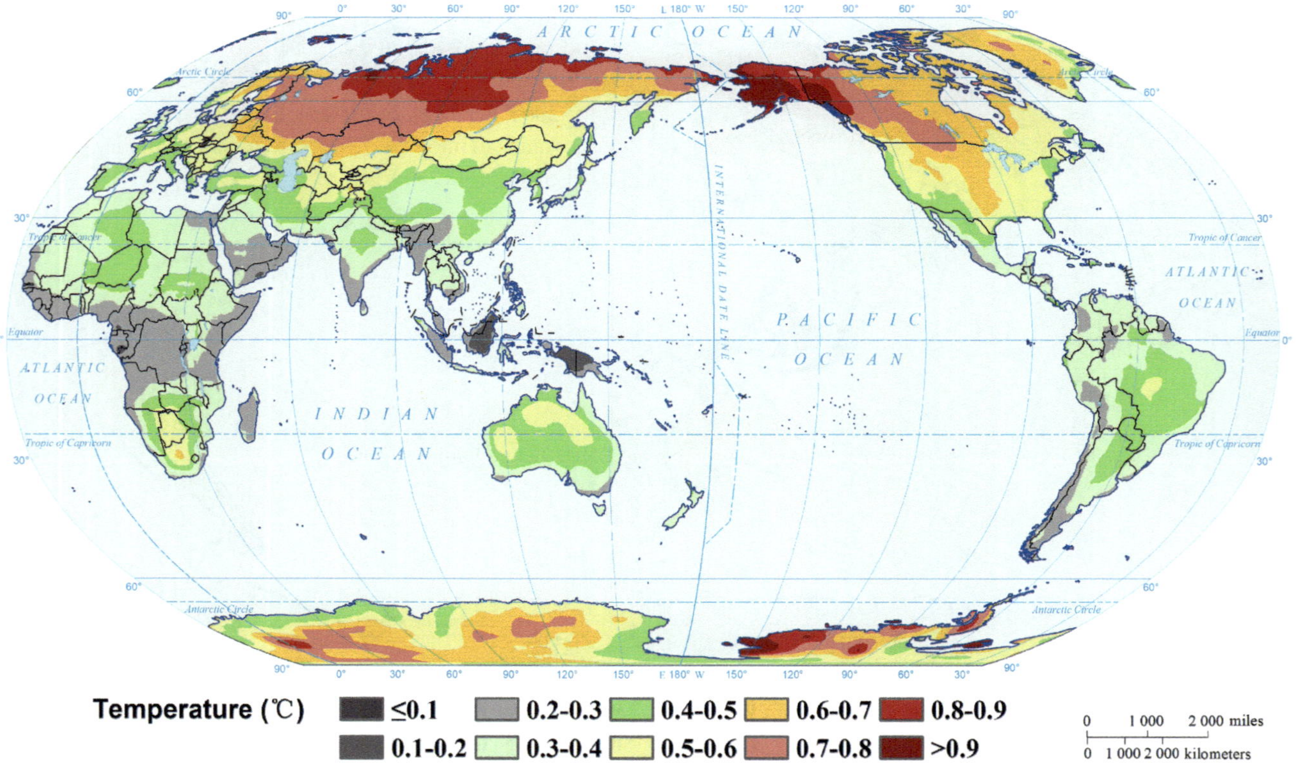

Annual maximum air temperature (Baseline)

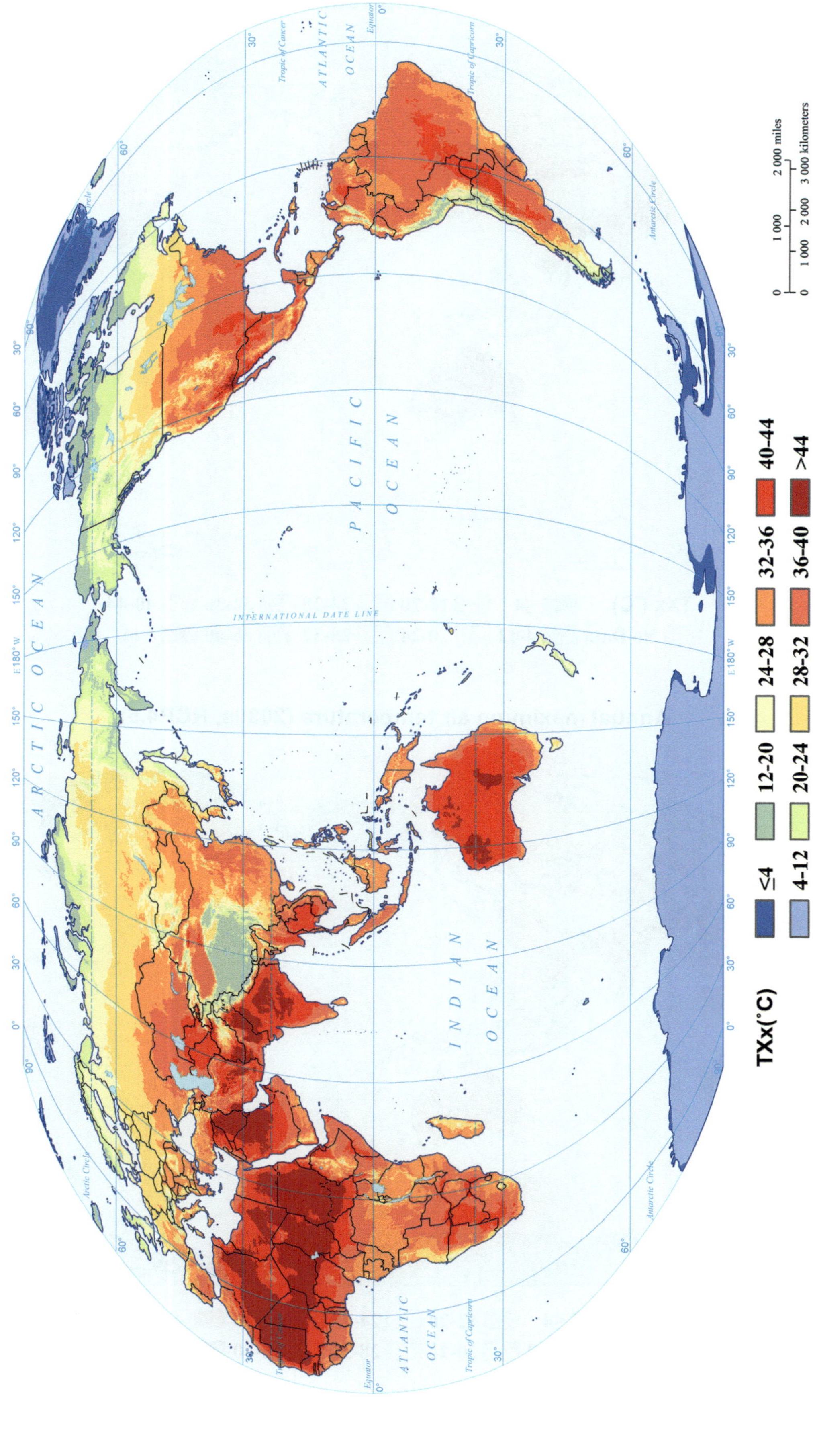

Annual maximum air temperature (2030s, RCP2.6)

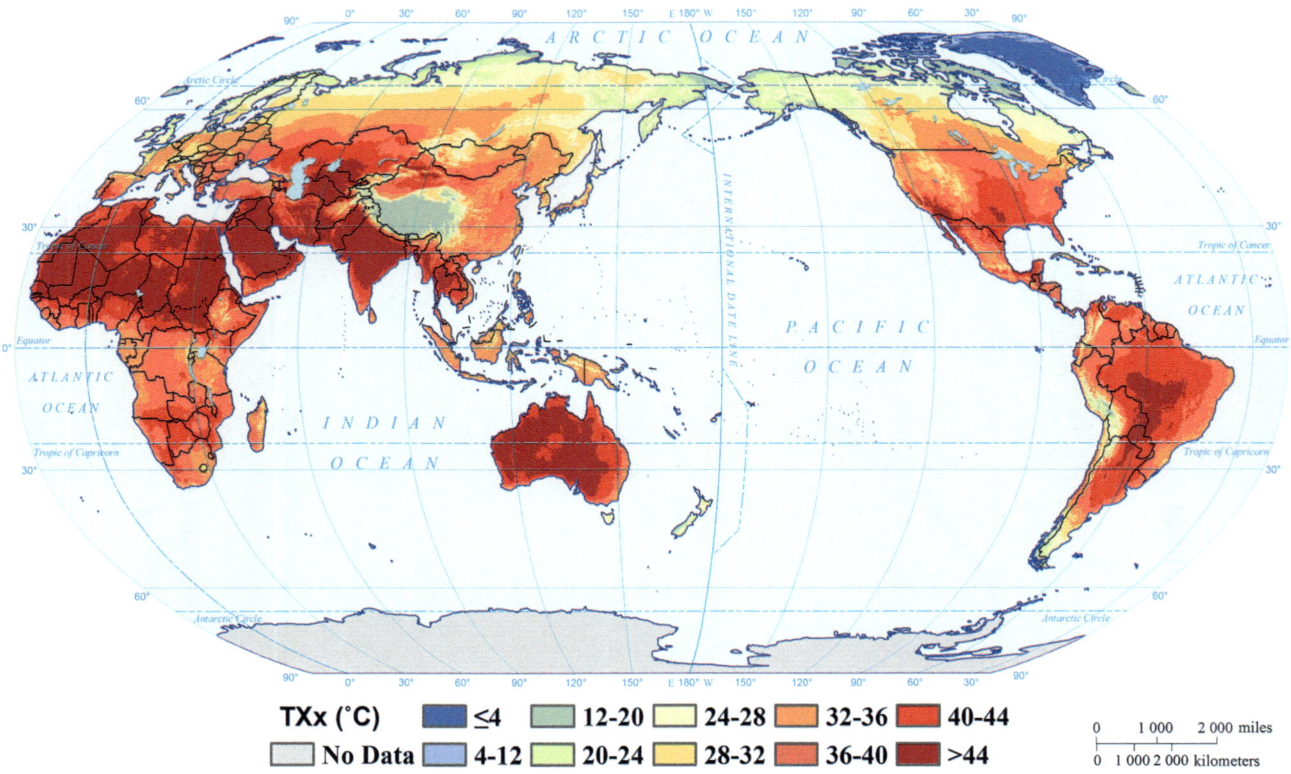

Annual maximum air temperature (2030s, RCP4.5)

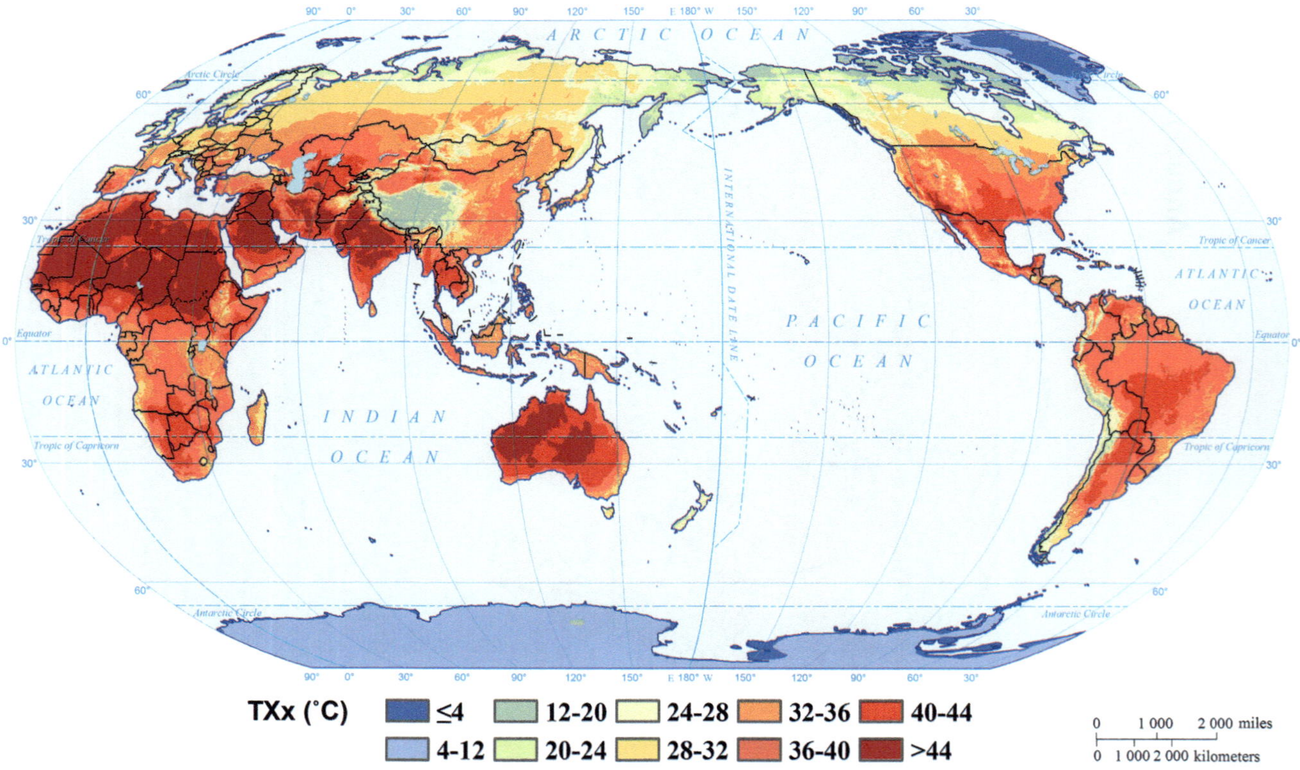

Annual maximum air temperature (2030s, RCP8.5)

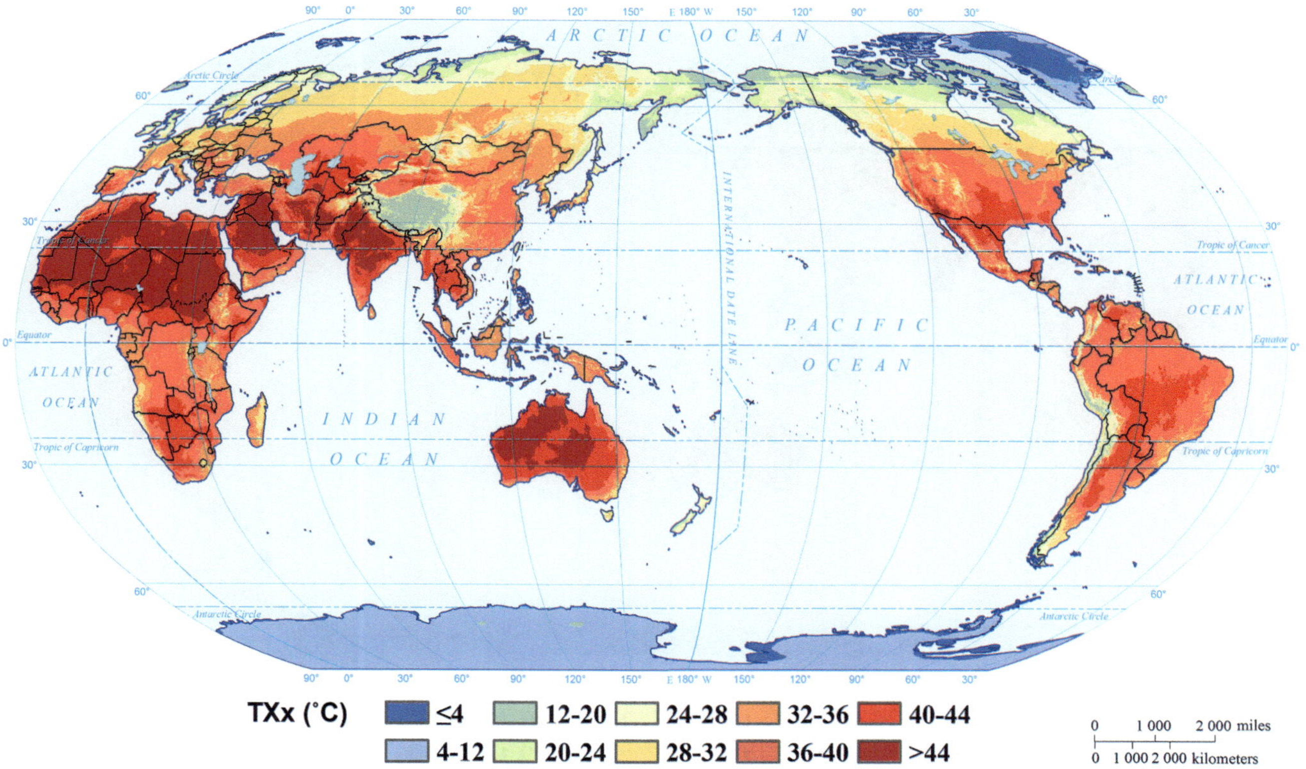

Annual maximum air temperature (2050s, RCP2.6)

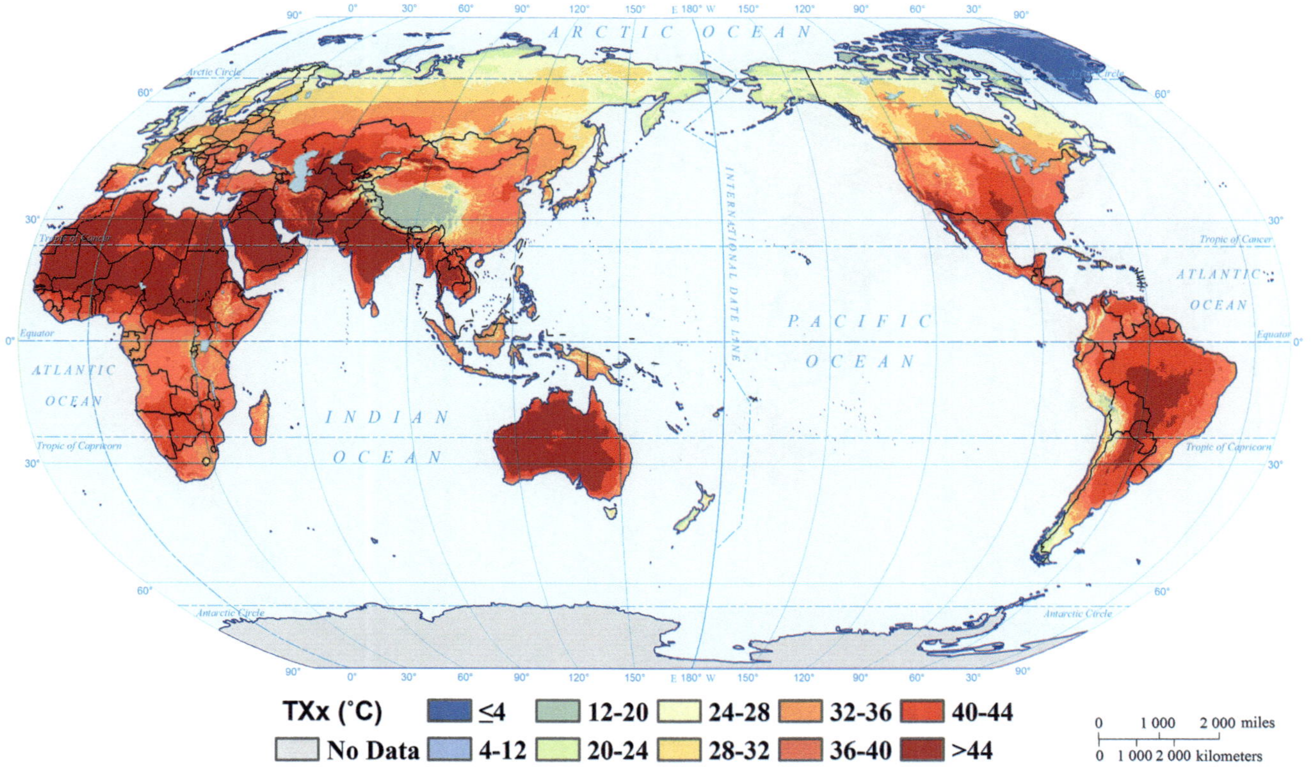

Annual maximum air temperature (2050s, RCP4.5)

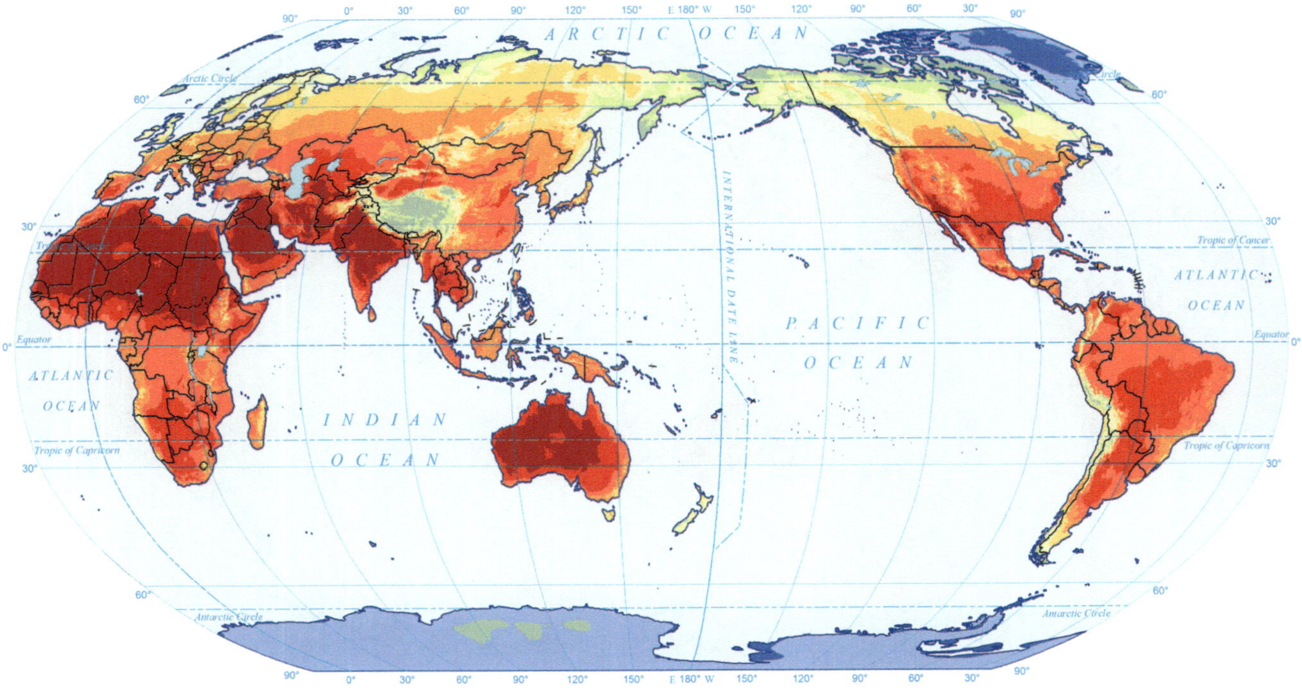

TXx (°C) ≤4 4-12 12-20 20-24 24-28 28-32 32-36 36-40 40-44 >44

0 1 000 2 000 miles
0 1 000 2 000 kilometers

Annual maximum air temperature (2050s, RCP8.5)

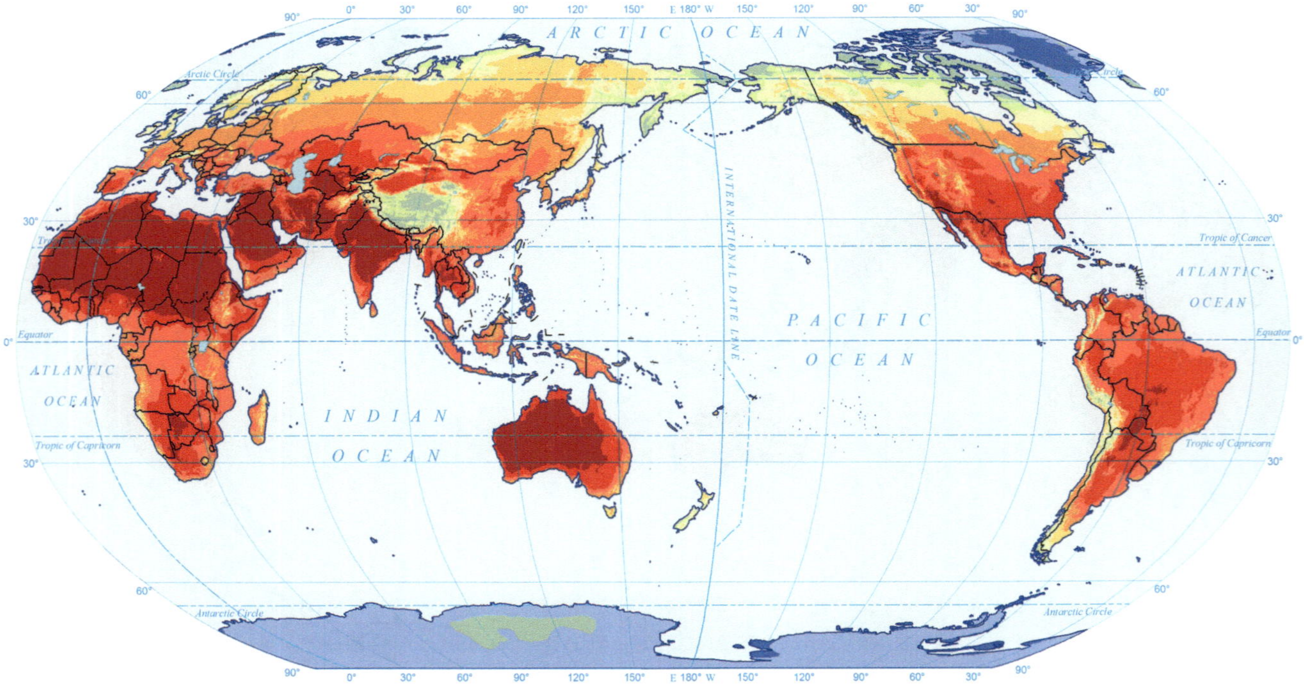

TXx (°C) ≤4 4-12 12-20 20-24 24-28 28-32 32-36 36-40 40-44 >44

0 1 000 2 000 miles
0 1 000 2 000 kilometers

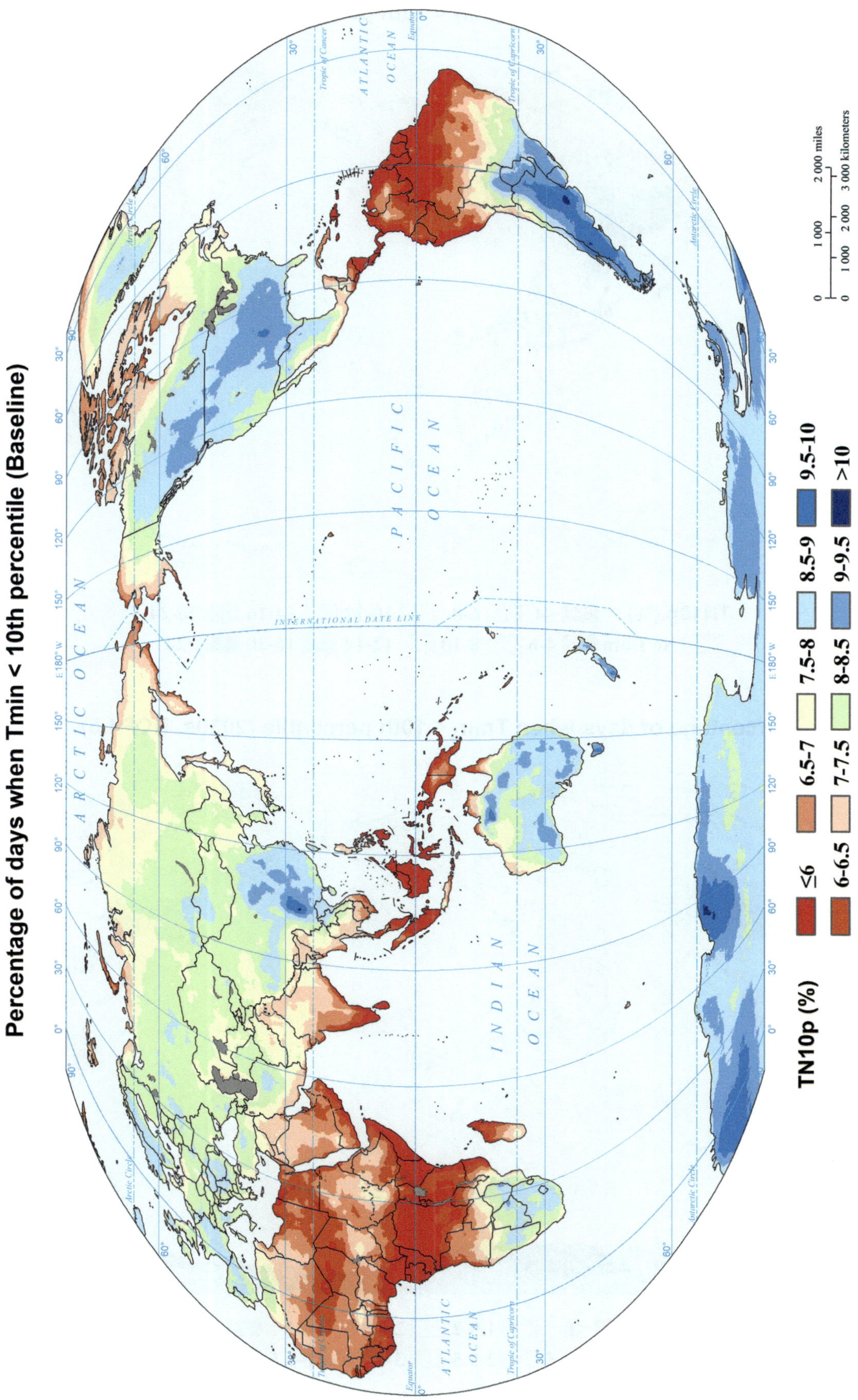
Percentage of days when Tmin < 10th percentile (Baseline)
TN10p (%)
≤6
6-6.5
6.5-7
7-7.5
7.5-8
8-8.5
8.5-9
9-9.5
9.5-10
>10
0 1 000 2 000 miles
0 1 000 2 000 3 000 kilometers
ARCTIC OCEAN
ATLANTIC OCEAN
PACIFIC OCEAN
INDIAN OCEAN
INTERNATIONAL DATE LINE
Arctic Circle
Tropic of Cancer
Equator
Tropic of Capricorn
Antarctic Circle

Percentage of days when Tmin < 10th percentile (2030s, RCP2.6)

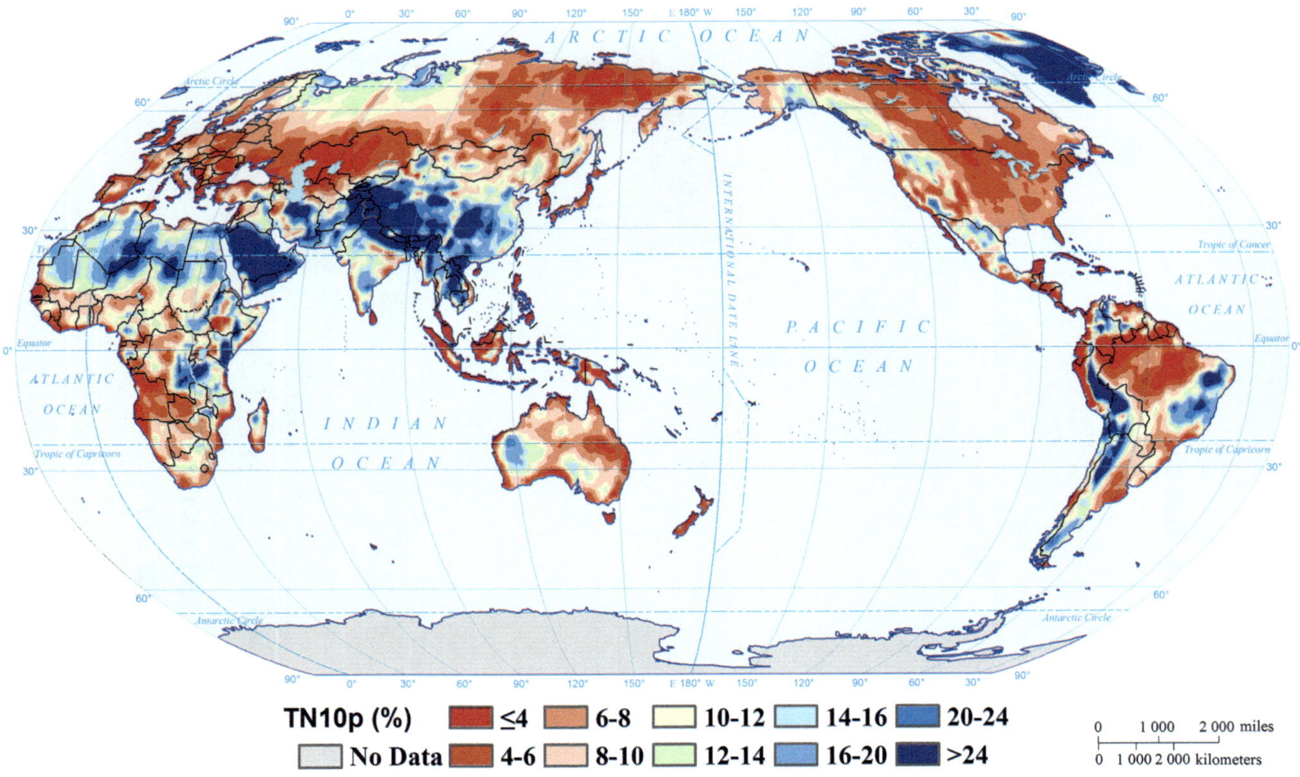

Percentage of days when Tmin < 10th percentile (2030s, RCP4.5)

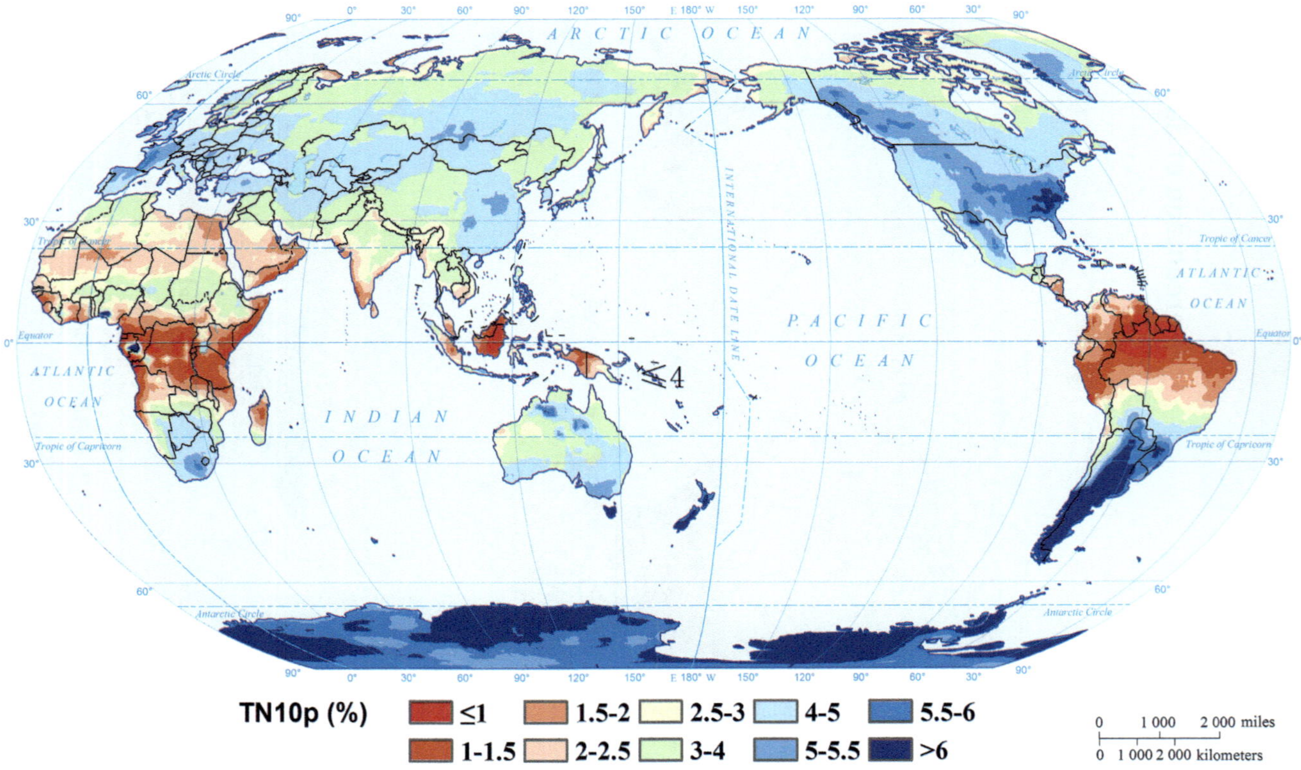

Percentage of days when Tmin < 10th percentile (2030s, RCP8.5)

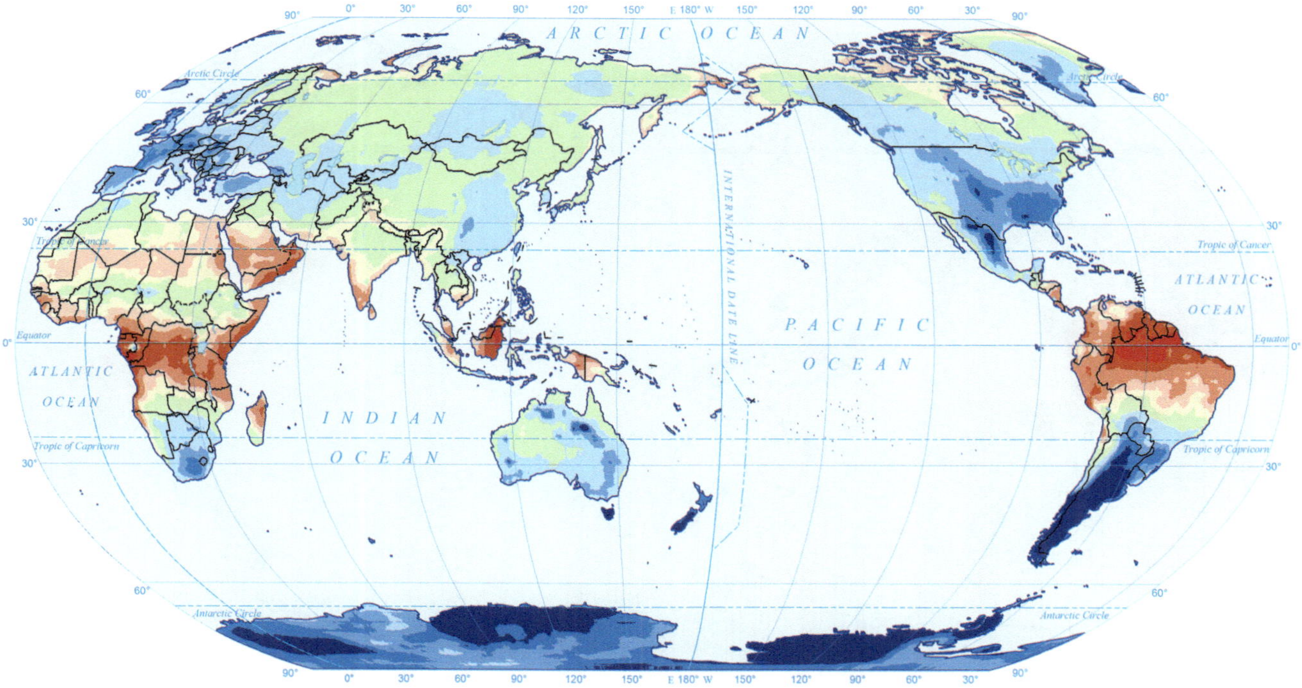

TN10p (%) ≤1 | 1-1.5 | 1.5-2 | 2-2.5 | 2.5-3 | 3-4 | 4-5 | 5-5.5 | 5.5-6 | >6

0 1 000 2 000 miles
0 1 000 2 000 kilometers

Percentage of days when Tmin < 10th percentile (2050s, RCP2.6)

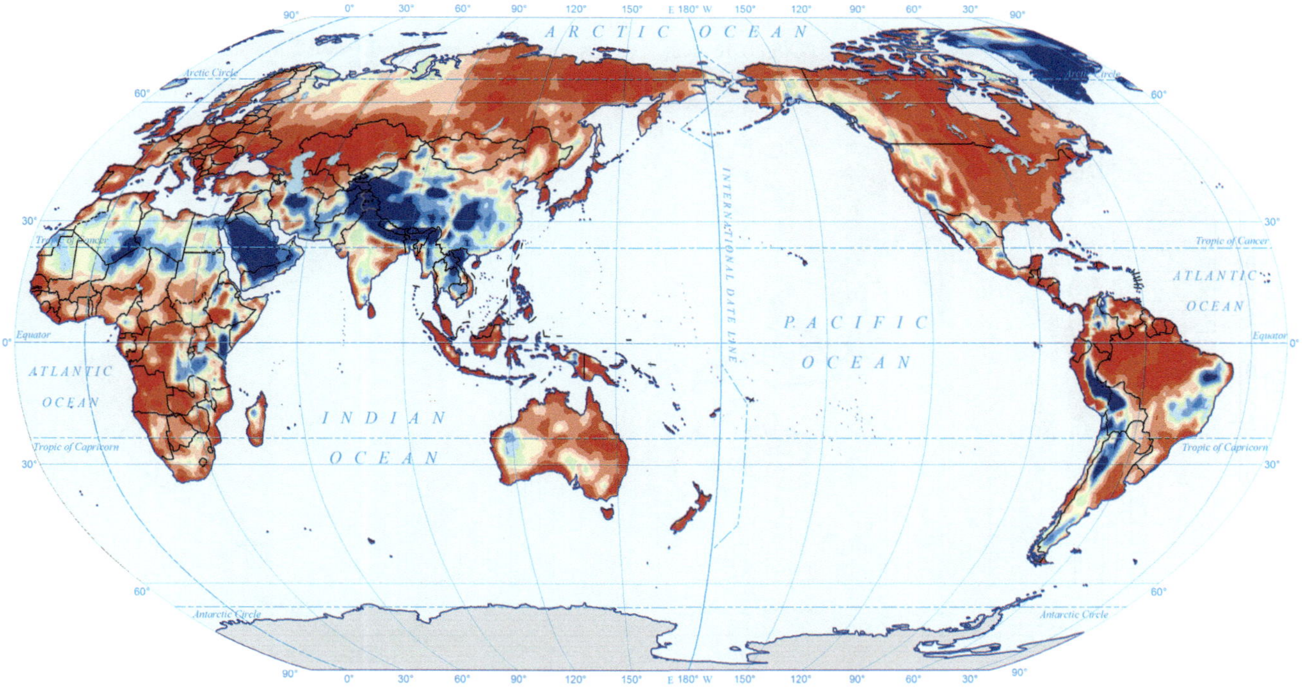

TN10p (%) No Data | ≤4 | 4-6 | 6-8 | 8-10 | 10-12 | 12-14 | 14-16 | 16-20 | 20-24 | >24

0 1 000 2 000 miles
0 1 000 2 000 kilometers

Percentage of days when Tmin < 10th percentile (2050s, RCP4.5)

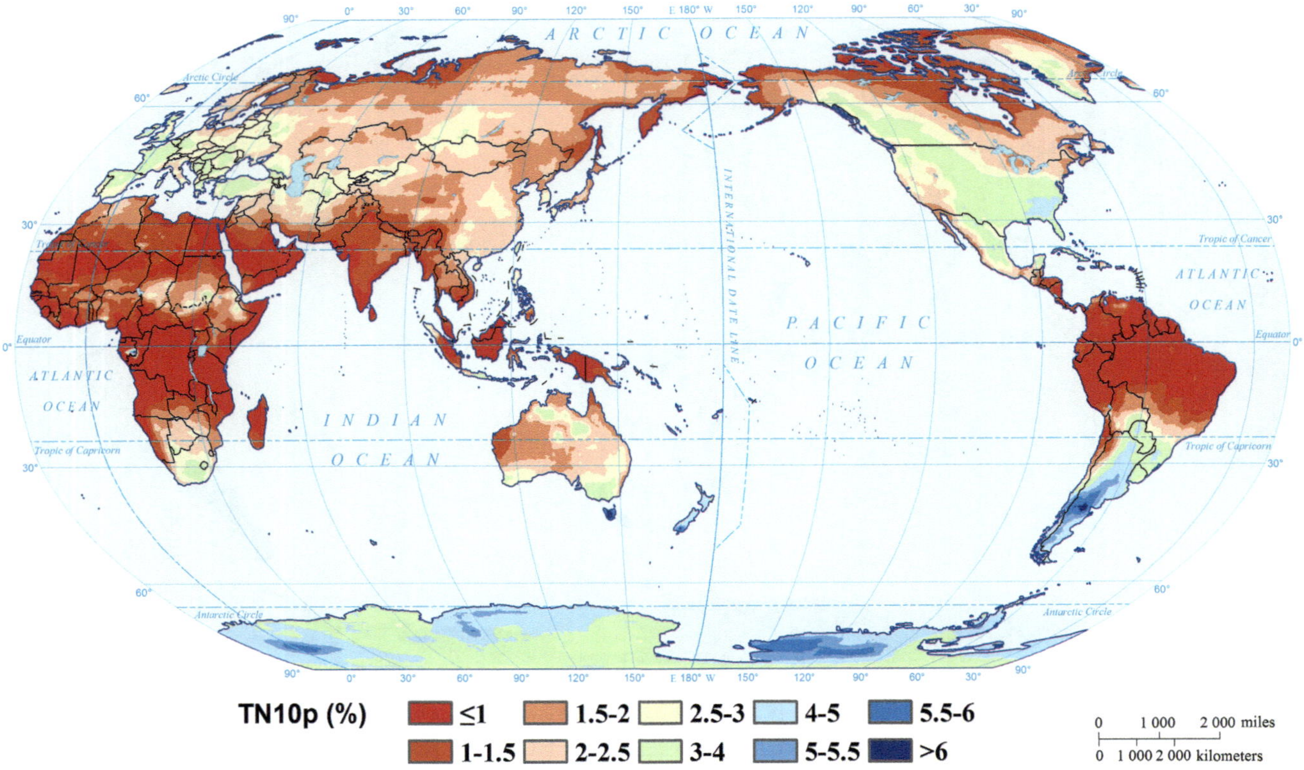

Percentage of days when Tmin < 10th percentile (2050s, RCP8.5)

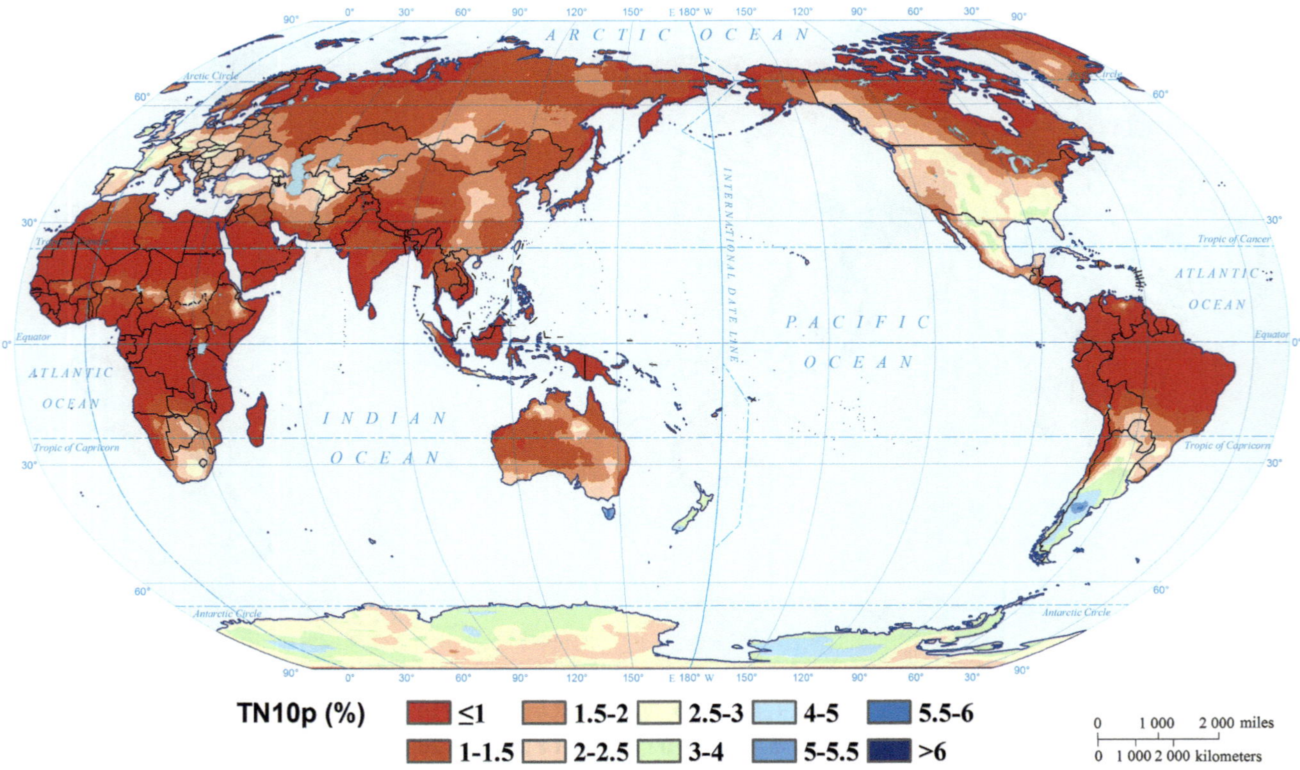

Percentage of days when Tmax > 90th percentile (Baseline)

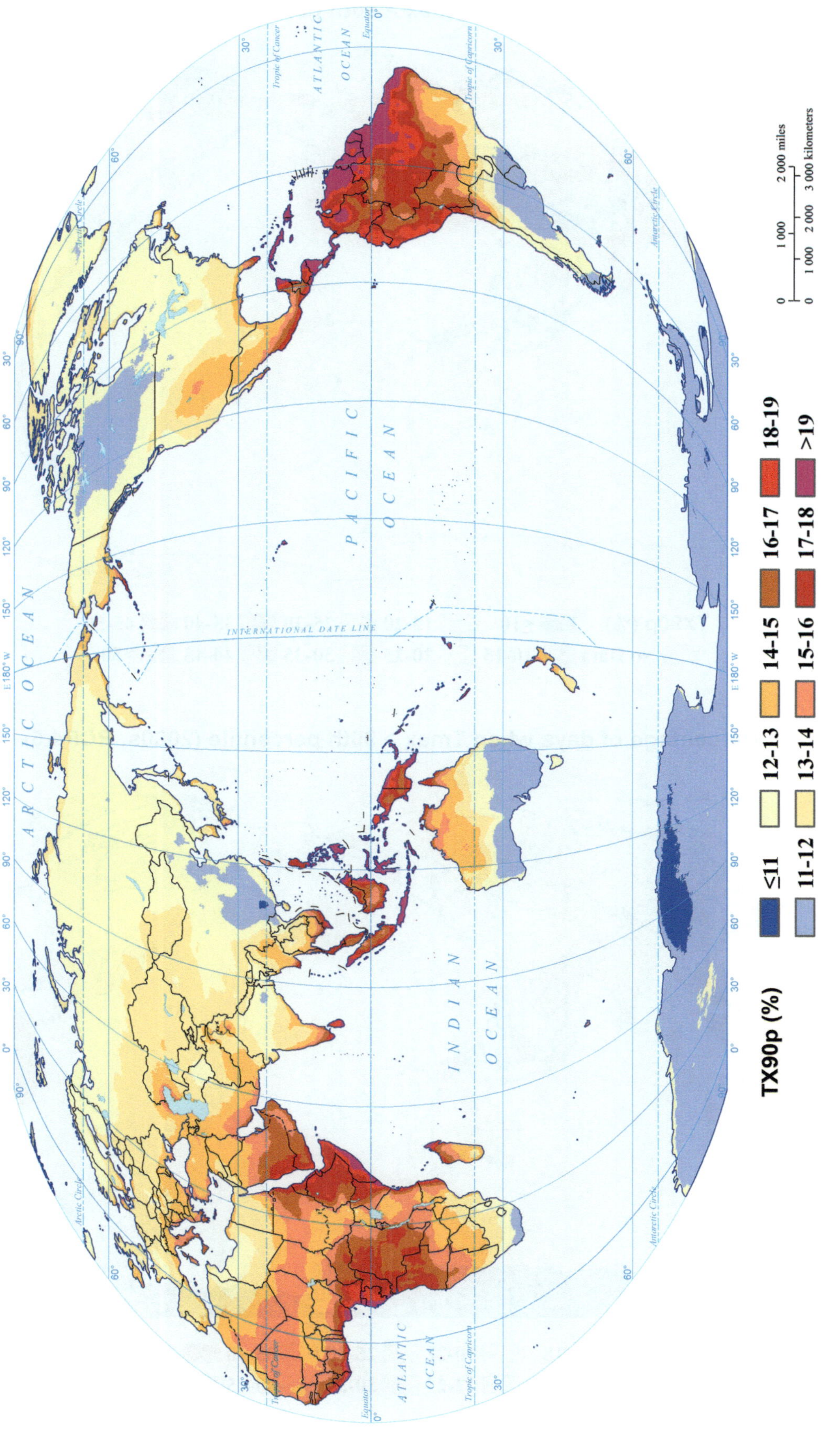

Percentage of days when Tmax > 90th percentile (2030s, RCP2.6)

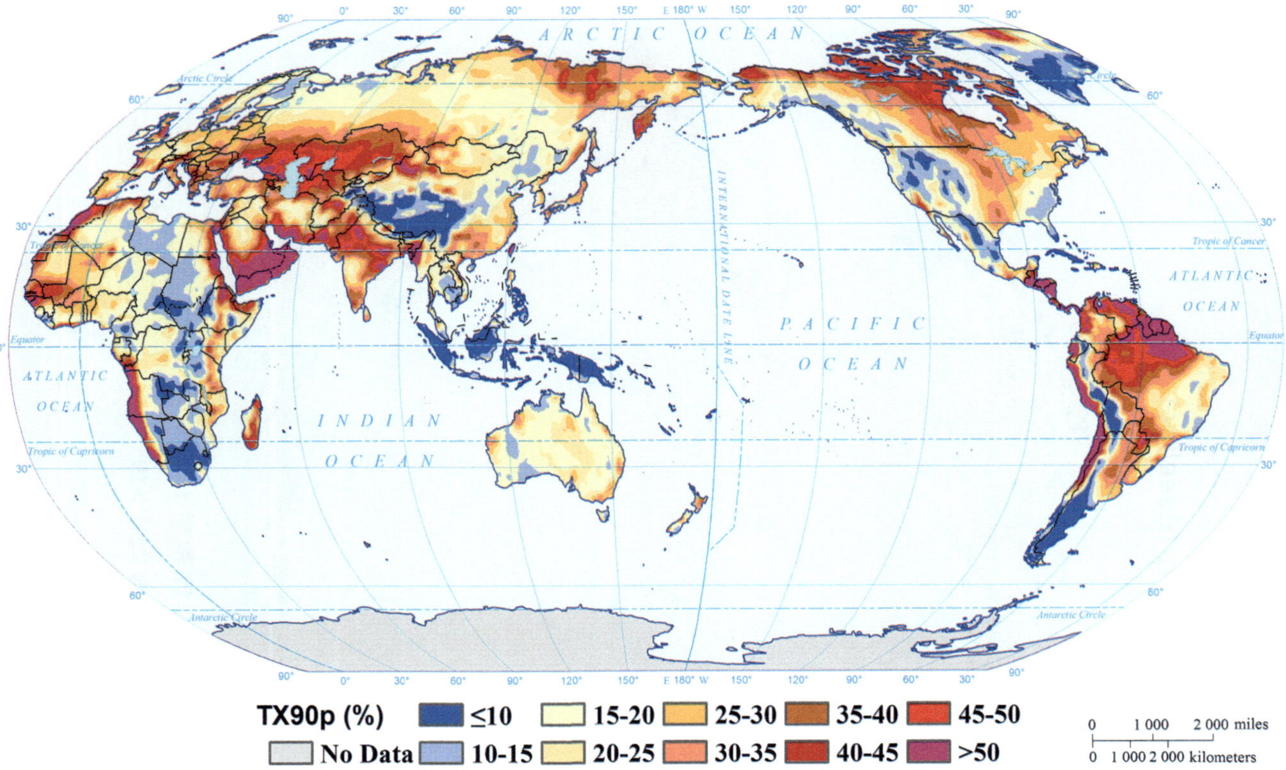

Percentage of days when Tmax > 90th percentile (2030s, RCP4.5)

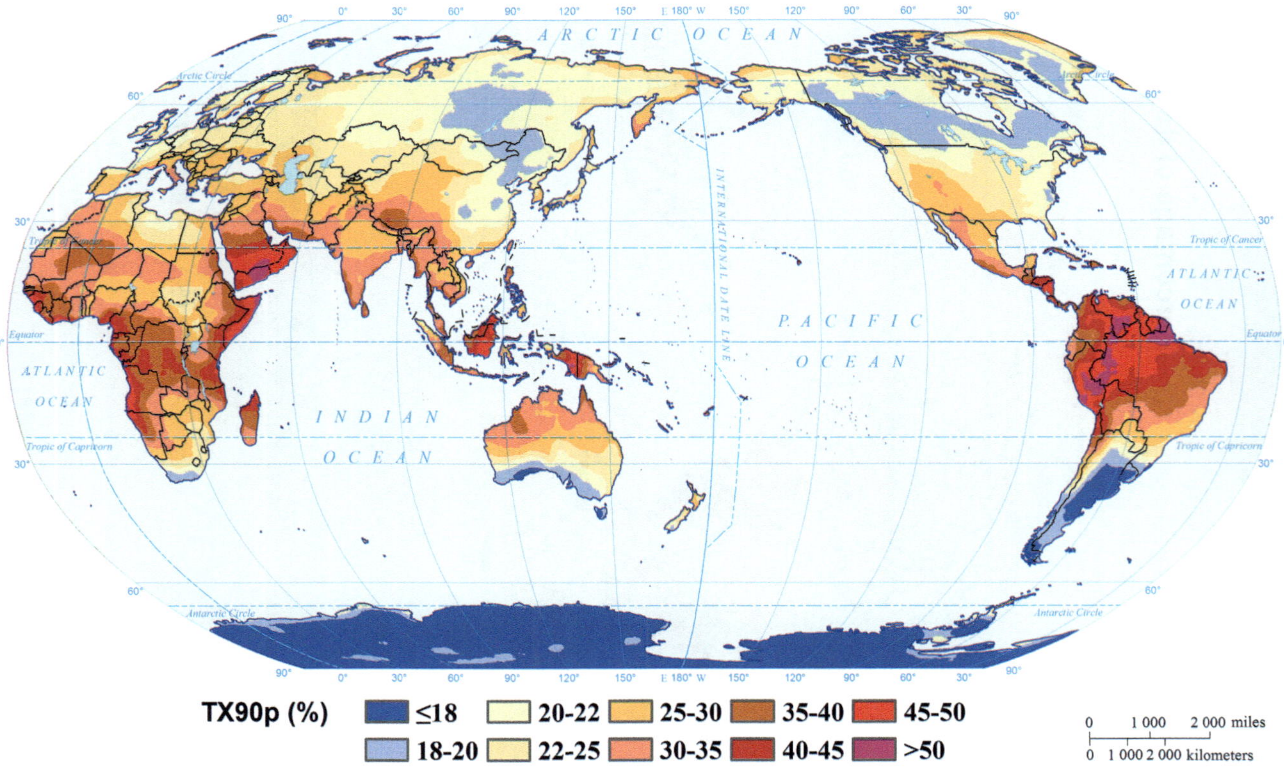

Percentage of days when Tmax > 90th percentile (2030s, RCP8.5)

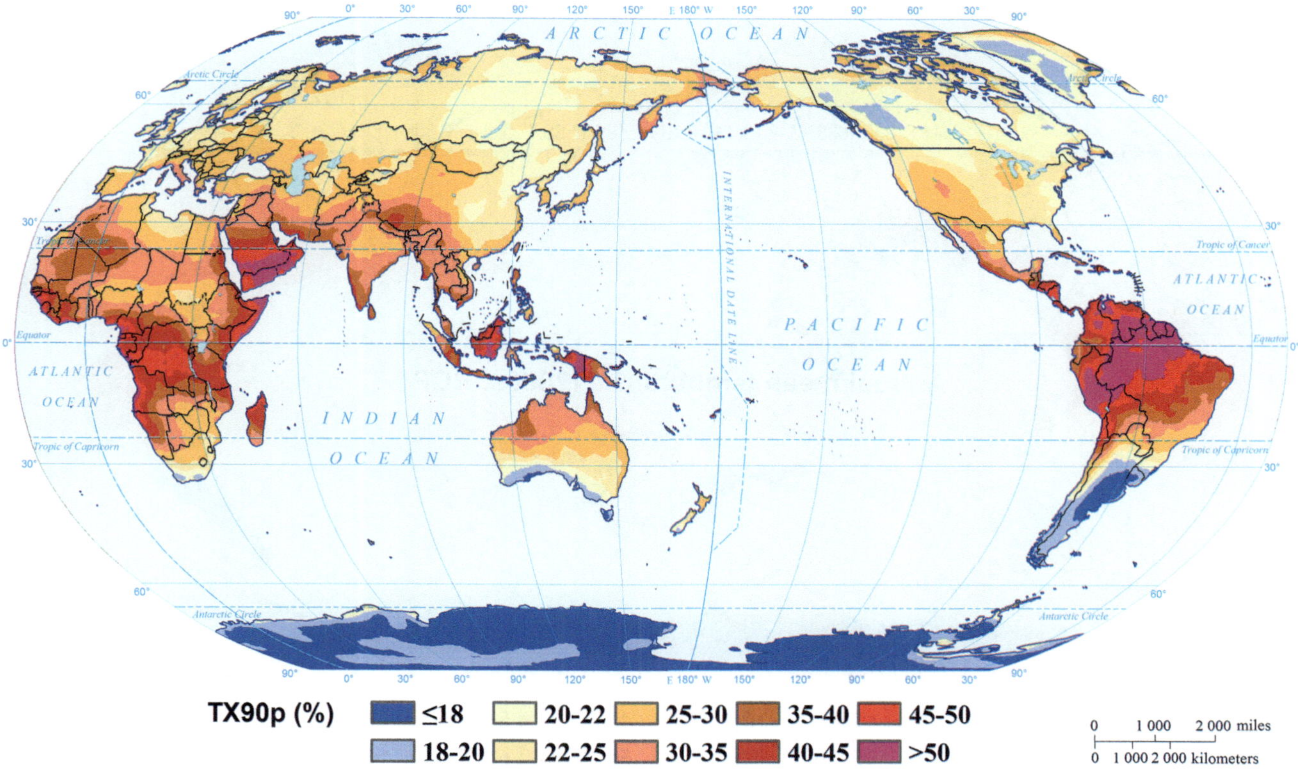

Percentage of days when Tmax > 90th percentile (2050s, RCP2.6)

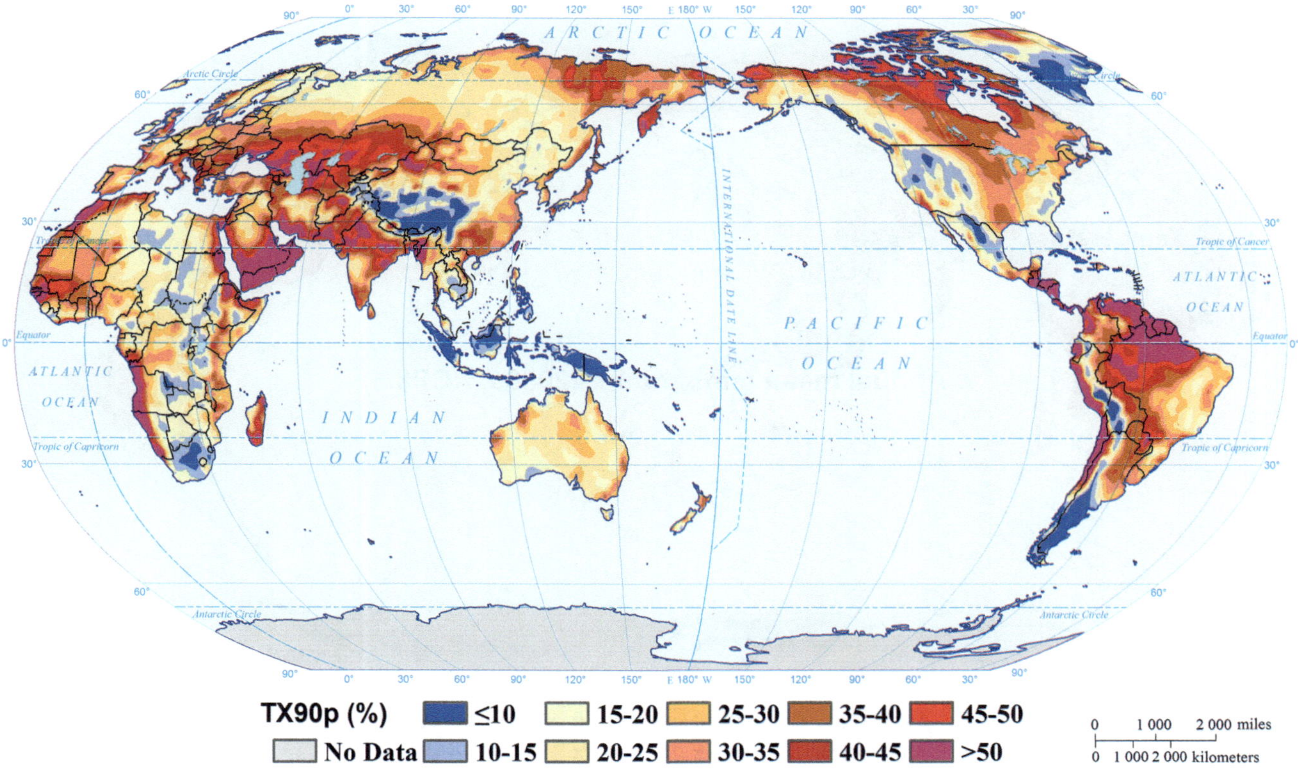

Percentage of days when Tmax > 90th percentile (2050s, RCP4.5)

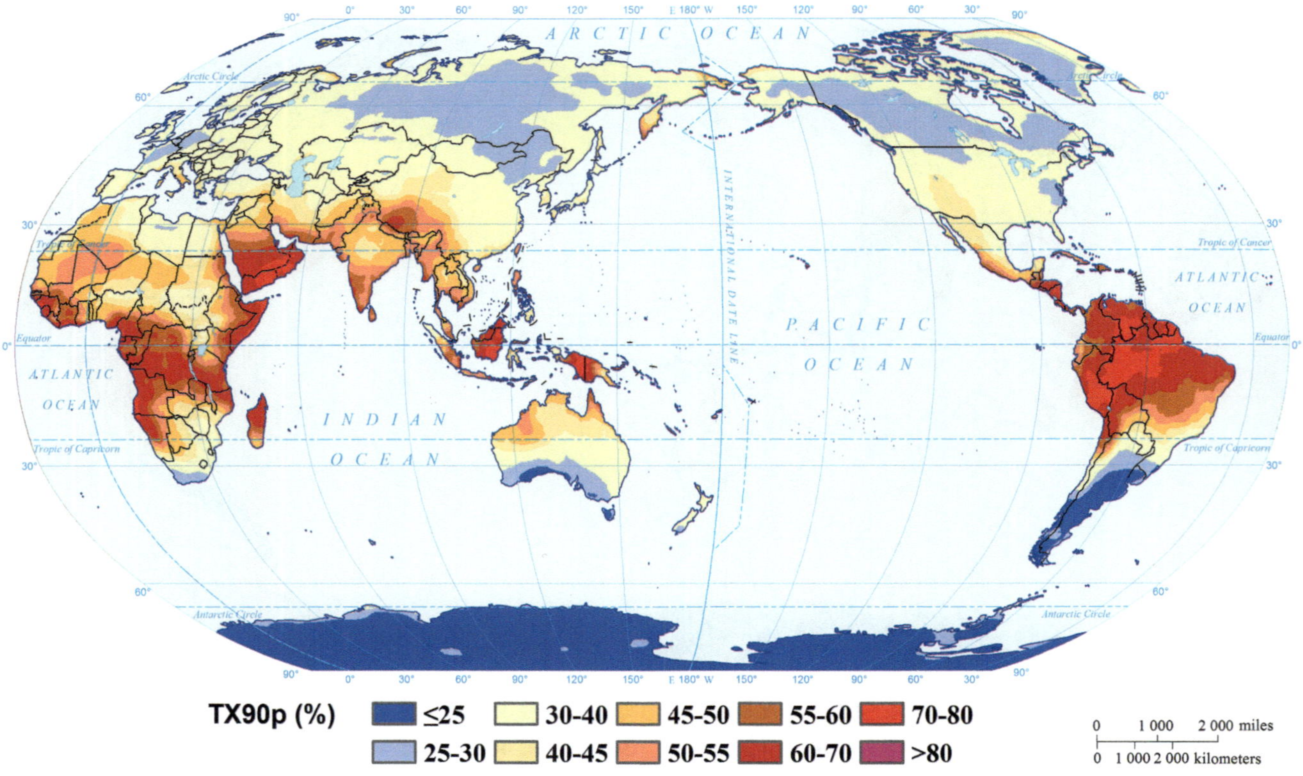

Percentage of days when Tmax > 90th percentile (2050s, RCP8.5)

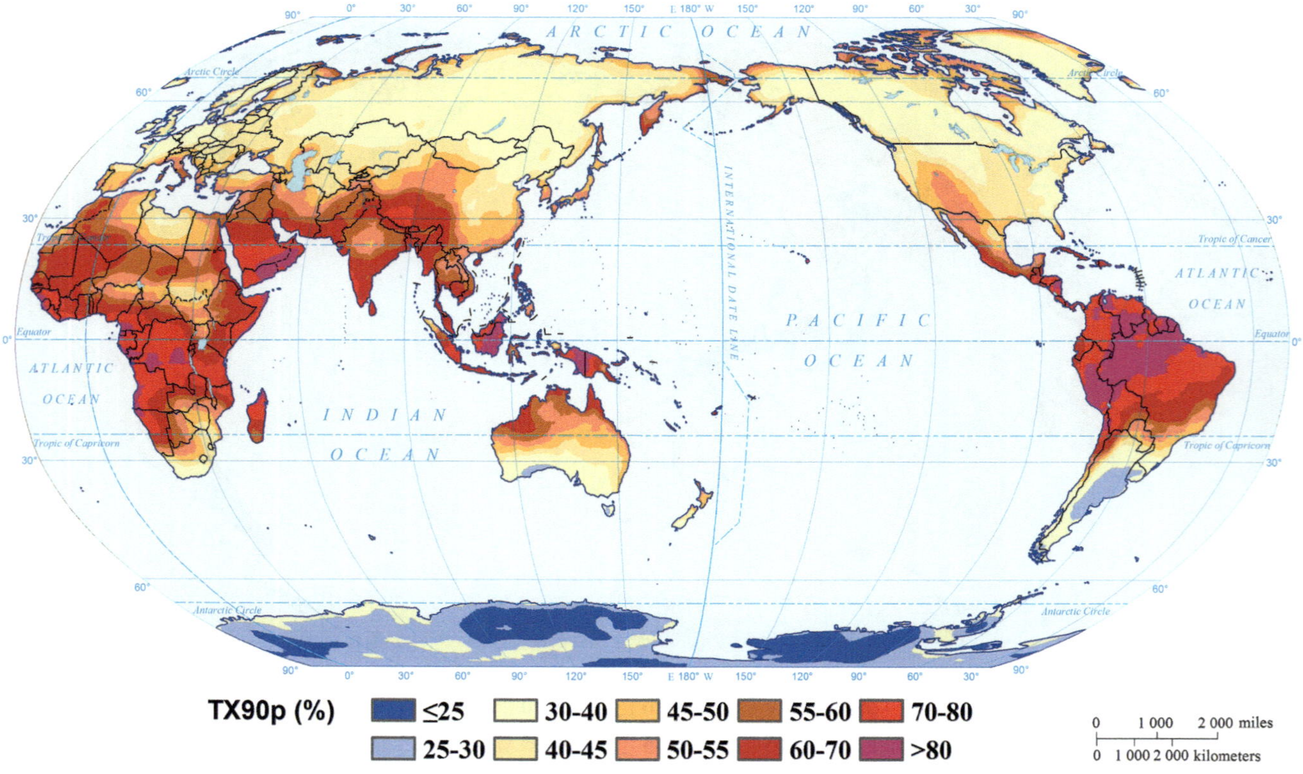

References

Alexander, L.V., X. Zhang, T.C. Peterson, J. Caesar, B. Gleason, A.M. G. Klein Tank, M. Haylock, D. Collins, et al. 2006. Global observed changes in daily climate extremes of temperature and precipitation. *Journal of Geophysical Research: Atmospheres* 111 (D5): D05109.

Bao, Y., and X.Y. Wen. 2017. Projection of China's near- and long-term climate in a new high-resolution daily downscaled dataset NEX-GDDP. *Journal of Meteorological Research* 31 (1): 236–249.

Dosio, A., L. Mentaschi, E.M. Fischer, and K. Wyser. 2018. Extreme heat waves under 1.5 °C and 2 °C global warming. *Environmental Research Letter* 13(5): 54006.

Easterling, D.R., G.A. Meehl, C. Parmesan, S.A. Changnon, T.R. Karl, and L.O. Mearns. 2000. Climate extremes: Observations, modeling, and impacts. *Science* 289 (5487): 2068–2074.

Fan, X., C. Miao, Q. Duan, C. Shen, and Y. Wu. 2020. The performance of CMIP6 versus CMIP5 in simulating temperature extremes over the global land surface. *Journal of Geophysical Research: Atmospheres* 125: e2020JD033031.

Fischer, E.M., and R. Knutti. 2014. Detection of spatially aggregated changes in temperature and precipitation extremes. *Geophysical Research Letters* 41 (2): 547–554.

Gao, M.N., J. Yang, D.Y. Gong, P.J. Shi, Z.G. Han, and S.J. Kim. 2019. Footprints of atlantic multi-decadal oscillation on the low-frequency variation of extreme high temperature in the northern hemisphere. *Journal of Climate* 32: 791–802.

Giorgi, F., and X. Bi. 2005. Regional changes in surface climate interannual variability for the 21st century from ensembles of global model simulations. *Geophysical Research Letters* 32: L13701.

Grumm, R.H. 2011. The Central European and Russian heat event of July–August 2010. *Bulletin of the American Meteorological Society* 92 (10): 1285–1296.

Karl, T.R., and D.R. Easterling. 1999. Climate extremes: Selected review and future research directions. *Climatic Change* 42 (1): 309–325.

Luo, Z.Q., J. Yang, M.N. Gao, and D.L. Chen. 2020. Extreme hot days over three global megaregions: Historical fidelity and future projection. *Atmospheric Science Letters* 21: e1003.

Maurer, E.P., H.G. Hidalgo, T. Das, M.D. Dettinger, and D.R. Cayan. 2010. The utility of daily large-scale climate data in the assessment of climate change impacts on daily streamflow in California. *Hydrology and Earth System Sciences* 14 (6): 1125–1138.

Nangombe, S., T.J. Zhou, X.W. Zhang, B. Wu, S. Hu, L.W. Zou, and D.H. Li. 2018. Record-breaking climate extremes in Africa under stabilized 1.5 °C and 2 °C global warming scenarios. *Nature Climate Change* 8(5): 375–380.

Piao, S., P. Ciais, Y. Huang, Z. Shen, S. Peng, J. Li, L. Zhou, H. Liu, et al. 2010. The impacts of climate change on water resources and agriculture in China. *Nature* 467 (7311): 43–51.

Qi, X., J. Yang, M. Gao, H. Yang, and H. Liu. 2019. Roles of the tropical/extratropical intraseasonal oscillations on generating the heat wave over Yangtze River valley: A numerical study. *Journal of Geophysical Research: Atmospheres* 124: 3110–3123.

Sun, Y., T. Hu, X.B. Zhang, H. Wan, P. Stott, and C.H. Lu. 2018. Anthropogenic influence on the eastern China 2016 super cold surge. *Bulletin of the American Meteorological Society* 99 (1): S123–127.

Thrasher, B., E.P. Maurer, C. McKellar, and P.B. Duffy. 2012. Technical note: Bias correcting climate model simulated daily temperature extremes with quantile mapping. *Hydrology and Earth System Sciences* 16 (9): 3309–3314.

Thrasher, B., J. Xiong, W. Wang, F. Melton, A. Michaelis, and R. Nemani. 2013. Downscaled climate projections suitable for resource management. *Eos, Transactions American Geophysical Union* 94 (37): 321–323.

Wang, Y.J., A.Q. Wang, J.Q. Zhai, H. Tao, T. Jiang, B.D. Su, J. Yang, G.J. Wang, et al. 2019. Tens of thousands additional deaths annually in cities of China between 1.5 °C and 2.0 °C warming. *Nature Communications* 10(1): 3376.

Wood, A.W., L.R. Leung, V. Sridhar, and D.P. Lettenmaier. 2004. Hydrologic implications of dynamical and statistical approaches to downscaling climate model outputs. *Climatic Change* 62 (1–3): 189–216.

Xu, L., and A. Wang. 2019. Application of the bias correction and spatial downscaling algorithm on the temperature extremes from CMIP5 multi-model ensembles in China. *Earth and Space Science* 6 (12): 2508–2524.

Zhang, X., L. Alexander, G.C. Hegerl, P. Jones, A.K. Tank, T.C. Peterson, B. Trewin, and F.W. Zwiers. 2011. Indices for monitoring changes in extremes based on daily temperature and precipitation data. *Wires Climate Change* 2 (6): 851–870.

Zheng, F., Y. Yuan, Y.H. Ding, K.X. Li, X.H. Fang, Y.H. Zhao, Y. Sun, J. Zhu, et al. 2021. The 2020/21 extremely cold winter in China influenced by the synergistic effect of La Niña and warm Arctic. Advances in Atmospheric Sciences https://doi.org/10.1007/s00376-021-1033-y.

Zobel, Z., J.L. Wang, D.J. Wuebbles, and V.R. Kotamarthi. 2017. High-resolution dynamical downscaling ensemble projections of future extreme temperature distributions for the United States. *Earth's Future* 5 (12): 1234–1251.

Mapping Precipitation Changes

Xianghui Kong, Xiaoxin Wang, Huopo Chen, Aihui Wang, Dan Wan, Lianlian Xu, Yue Miao, Ju Huang, Yang Liu, Ruiheng Xie, Yue Chen, and Xianmei Lang

1 Introduction

Compared to the observed changes in temperature, the changes in precipitation show more uncertainty (Hartmann et al. 2013). The IPCC AR5 indicated that anthropogenic forcing has contributed to a global-scale intensification of heavy precipitation since the second half of the twentieth century (IPCC 2013) and the intensity of daily precipitation increases more under the higher warming scenarios (Weber et al. 2018).

To achieve a comprehensive understanding of changes in precipitation in the future, this section initiatively assesses the change of precipitation characteristics, such as mean amount, variability, and extremes under three greenhouse gas emissions scenarios, including Representative Concentration Pathway (RCP) 2.6, RCP4.5, and RCP8.5.

2 Data

Same as the daily maximum and minimum temperature data, the global daily precipitation data were also retrieved from the NEX-GDDP dataset under RCP4.5 and RCP8.5 from 21 climate models in the Coupled Model Intercomparison Project Phase 5 (CMIP5) (https://www.nccs.nasa.gov/services/data-collections/land-based-products/nex-gddp). Furthermore, the precipitation data from 13 models in CMIP5 under the RCP2.6 scenario have also been downscaled by the Institute of Atmospheric Physics (IAP) Chinese Academy of Sciences (CAS) (Xu and Wang 2019). This dataset covers all grids between 60°S and 90°N global land area. The spatial resolution of the data for all maps is 0.25° (~25 km × 25 km).

3 Method

The precipitation extremes cover three time periods, including the historical period (1986–2005, denoted as the 2000s), and two future periods 2016–2035 (2030s) and 2046–2065 (2050s). Summer represents June–July–August (JJA), and winter is December and January–February (DJF) of the following year.

Authors: Xianghui Kong, Xiaoxin Wang, Huopo Chen, Aihui Wang, Dan Wan, Lianlian Xu, Yue Miao, Ju Huang, Yang Liu, Ruiheng Xie, Yue Chen, Xianmei Lang.
Map Designers: Yelin Sun, Tian Liu, Fanya Shi, Jing'ai Wang, Ying Wang.
Language Editor: Aihui Wang.

X. Kong · H. Chen · A. Wang (✉) · D. Wan · L. Xu · Y. Miao · J. Huang · Y. Liu · R. Xie
Institute of Atmospheric Physics, Nansen-Zhu International Research Centre, Chinese Academy of Sciences, Beijing, 100029, China
e-mail: wangaihui@mail.iap.ac.cn

X. Wang
Climate Change Research Center, Institute of Atmospheric Physics, Chinese Academy of Sciences, Beijing, 100029, China

L. Xu
School of Atmospheric Sciences, Sun Yat-Sen University, Zhuhai, 519082, China

Y. Chen
College of Atmospheric Sciences, Lanzhou University, Lanzhou, 730000, China

X. Lang
International Center for Climate and Environment Sciences, Institute of Atmospheric Physics, Chinese Academy of Sciences, Beijing, 100029, China

P. Shi, *Atlas of Global Change Risk of Population and Economic Systems*, IHDP/Future Earth-Integrated Risk Governance Project Series, https://doi.org/10.1007/978-981-16-6691-9_3

Total precipitation in wet days (*Pr*) is defined as the 20-year mean of summation of all daily precipitation amount ≥ 1 mm d^{-1} during the 2000s, 2030s, and 2050s. The *Pr* change (%) is defined as:

$$Pr_{\text{change}} = 100\% \times (Pr_{2030s} - Pr_{2000s})/Pr_{2000s} \quad (1)$$

$$Pr_{\text{change}} = 100\% \times (Pr_{2050s} - Pr_{2000s})/Pr_{2000s} \quad (2)$$

The inter-model uncertainty of *Pr* (ensemble spread) is defined as the standard deviation of the Pr_{change} across all models.

Precipitation variability is defined as the standard deviation of *Pr* during three different periods. The change of precipitation variability during the 2030s and 2050s is calculated similarly as Eqs. (1) and (2), respectively. The inter-model uncertainty is the standard deviation of the precipitation variability across all models.

Precipitation extreme indices, including *RX1day*, *RX5-day*, and *R10mm*, are adopted from the Expert Team on Climate Change Detection and Indices (ETCCDI, see Klein Tank et al. (2009) and Zhang et al. (2011)).

RX1day is the maximum 1-day precipitation amount (mm/day). *R10mm* is the number of days when daily precipitation amount ≥ 10 mm. *RX5day* is the maximum consecutive 5-day precipitation. The definition of *RX1day* change, *RX5day* change, and *R10mm* change and their inter-model uncertainty are similarly defined as those in Eqs. (1) and (2).

4 Major Findings

Nine regions were selected following Giorgi and Bi (2005) to quantitatively compare the changes of precipitation under the three greenhouse gas emissions scenarios. These regions are sensitive to global warming (Xu et al. 2019). Figures 1, 2, 3, 4 and 5 show the area-weighted average annual total precipitation in wet days, precipitation variability, annual maximum 1-day precipitation (RX1D), annual days of daily precipitation equal to or greater than 10 mm (R10mm), and annual maximum consecutive 5-day precipitation (RX5D), respectively. Generally, the changes in precipitation depend on the greenhouse gas emissions scenario and the region.

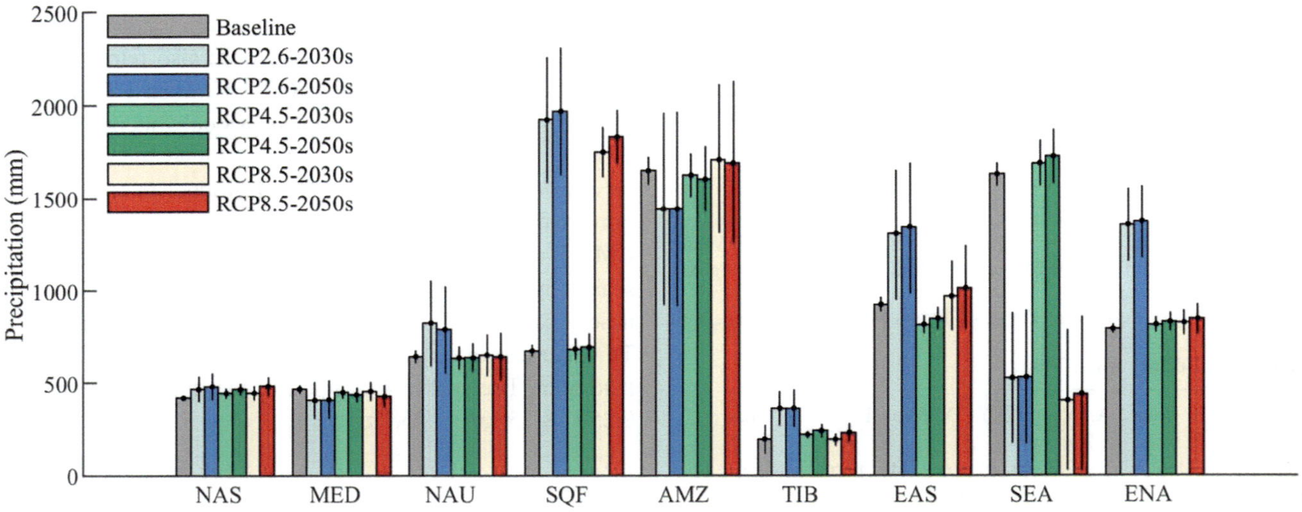

Fig. 1 Annual total precipitation in wet days (unit: mm) in nine regions under different Representative Concentration Pathway (RCP) scenarios. The error bar represents the one standard deviation across all selected models—13 general circulation models (GCMs) (RCP2.6) and 21 GCMs (RCP4.5 and RCP8.5). NAS, MED, NAU, SQF, AMZ, TIB, EAS, SEA, and ENA represent North Asia (47–70°N, 60.5–180.5°E), Mediterranean Basin (30–47°N, 10.5°W–37.5°E), Northern Australia (28–10°S, 109.5–155.5°E), South Equatorial Africa (26–0°S, 0.5–55.5°E), Amazon Basin (20°S–10°N, 78.5–34.5°W), Tibet (30–47°N, 80.5–104.5°E), East Asia (20–47°N, 104.5–140.5°E), Southeast Asia (10°S–20°N, 100.5–150.5°E), and Eastern North America (25–50°N, 85.5–60.5°W), respectively. The regional division follows Giorgi and Bi (2005)

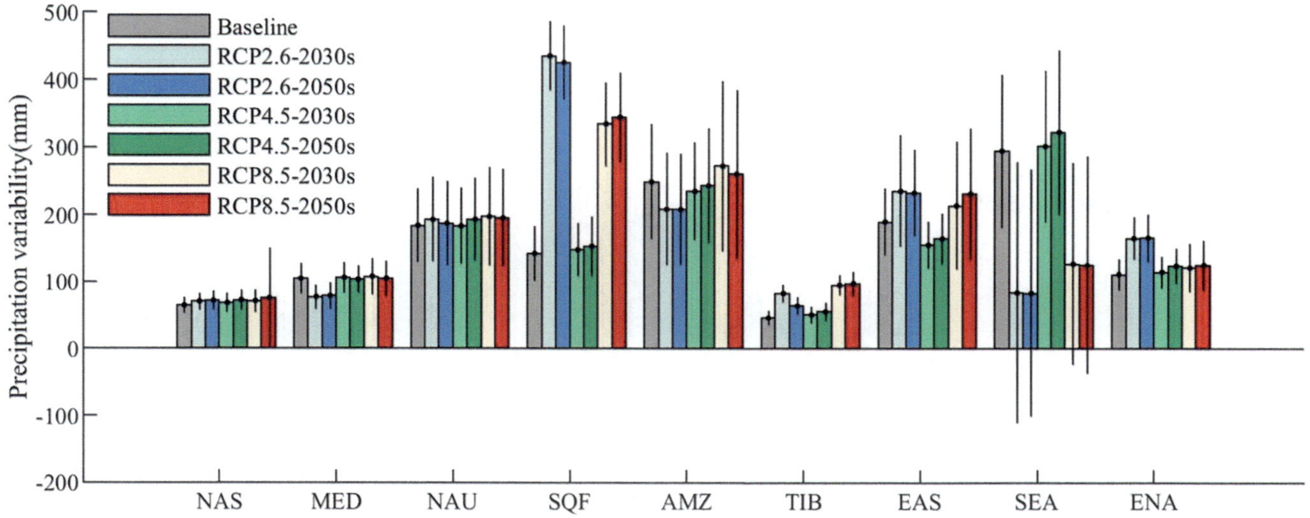

Fig. 2 Variability of annual total precipitation in wet days (unit: mm) in nine regions under different Representative Concentration Pathway (RCP) scenarios. The error bar represents the one standard deviation across all selected models. Region abbreviations are the same as in Fig. 1

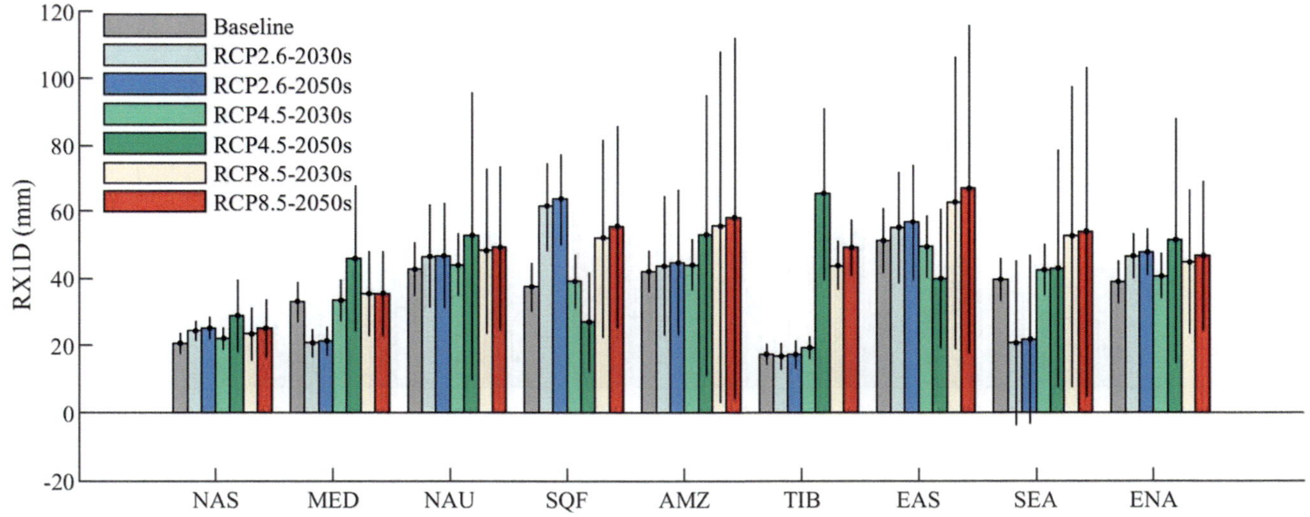

Fig. 3 Annual maximum 1-day precipitation (unit: mm) in nine regions under different Representative Concentration Pathway (RCP) scenarios. The error bar represents the one standard deviation across all selected models. Region abbreviations are the same as in Fig. 1

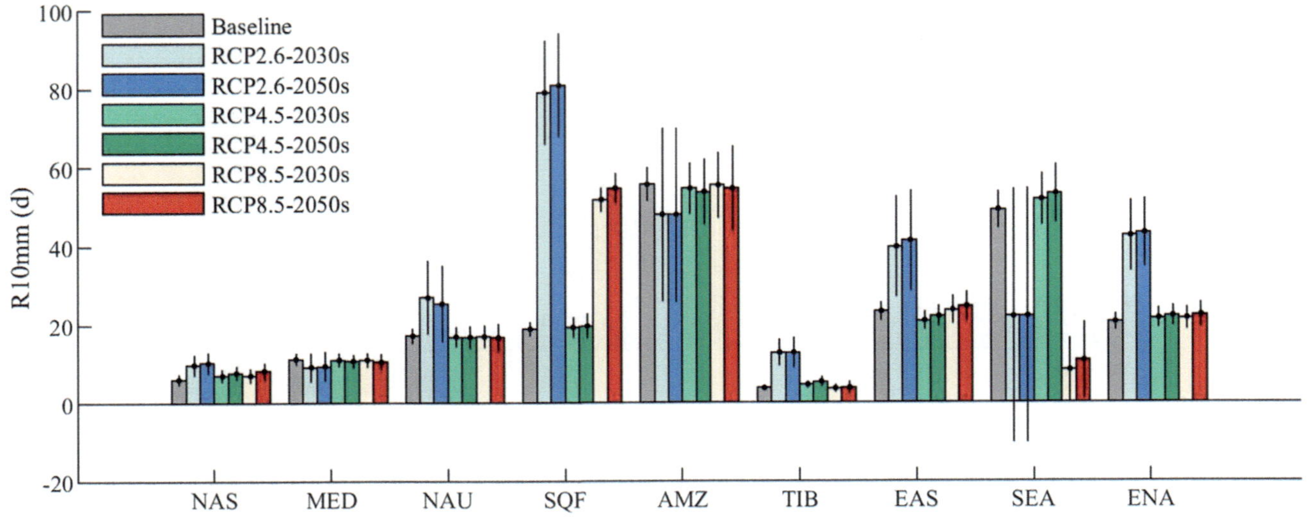

Fig. 4 Annual days of daily precipitation equal to or greater than 10 mm (unit: d) in nine regions under different Representative Concentration Pathway (RCP) scenarios. The error bar represents the one standard deviation across all selected models. Region abbreviations are the same as in Fig. 1

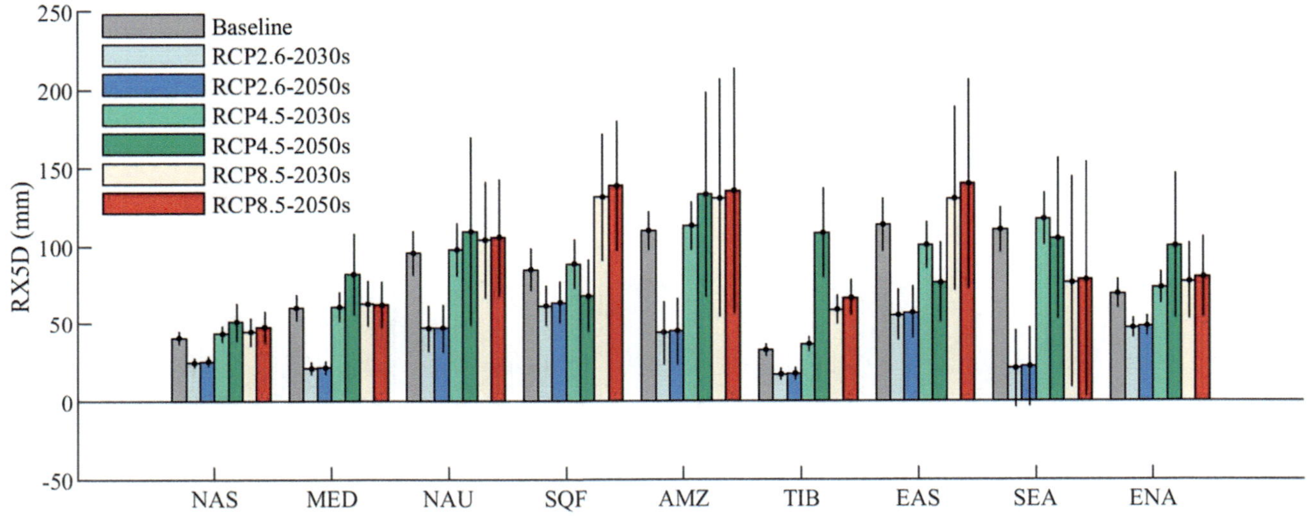

Fig. 5 Annual maximum consecutive 5-day precipitation (unit: mm) in nine regions under different Representative Concentration Pathway (RCP) scenarios. The error bar represents the one standard deviation across all selected models. Region abbreviations are the same as in Fig. 1

Annual total precipitation in wet days (Baseline)

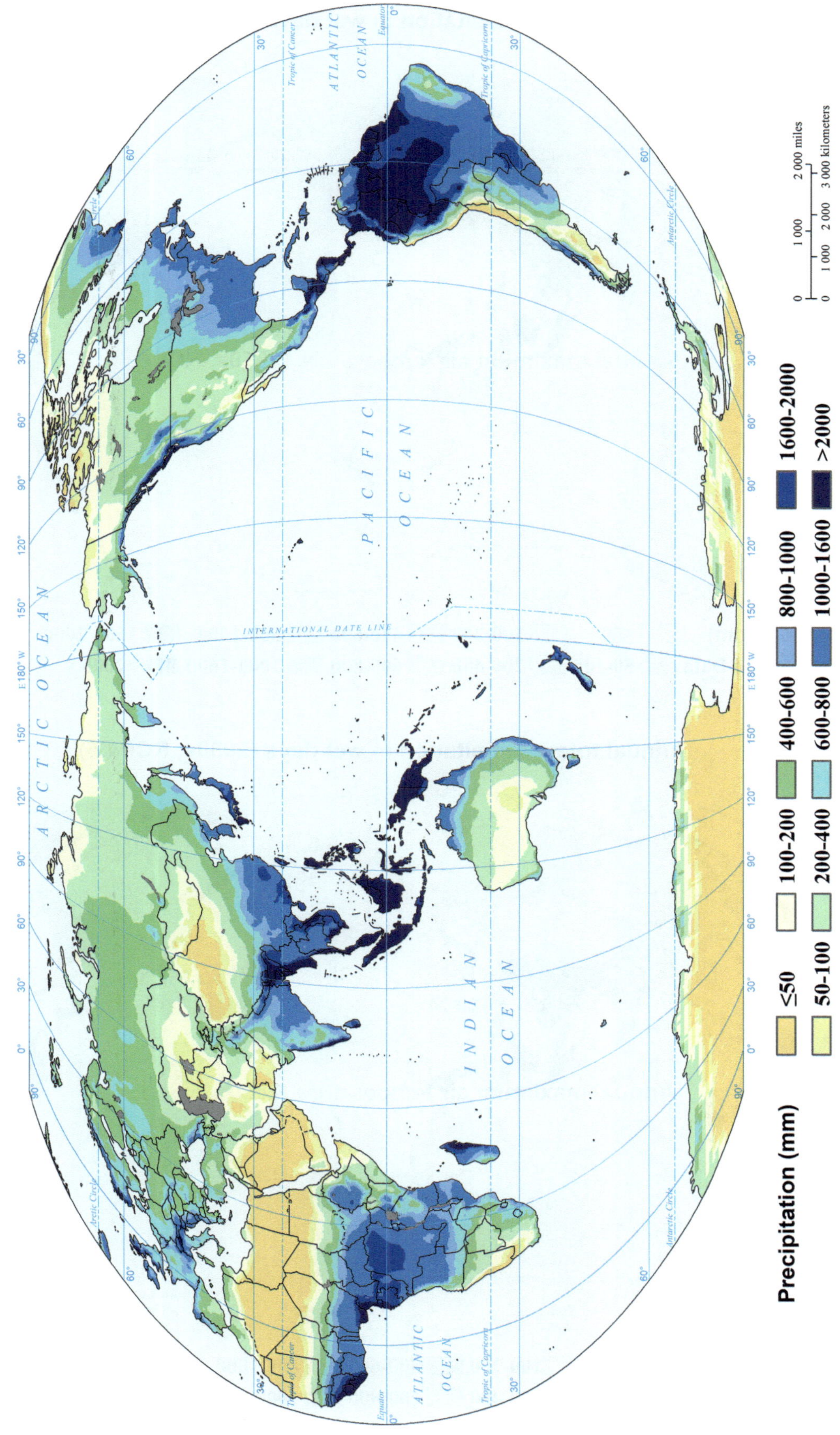

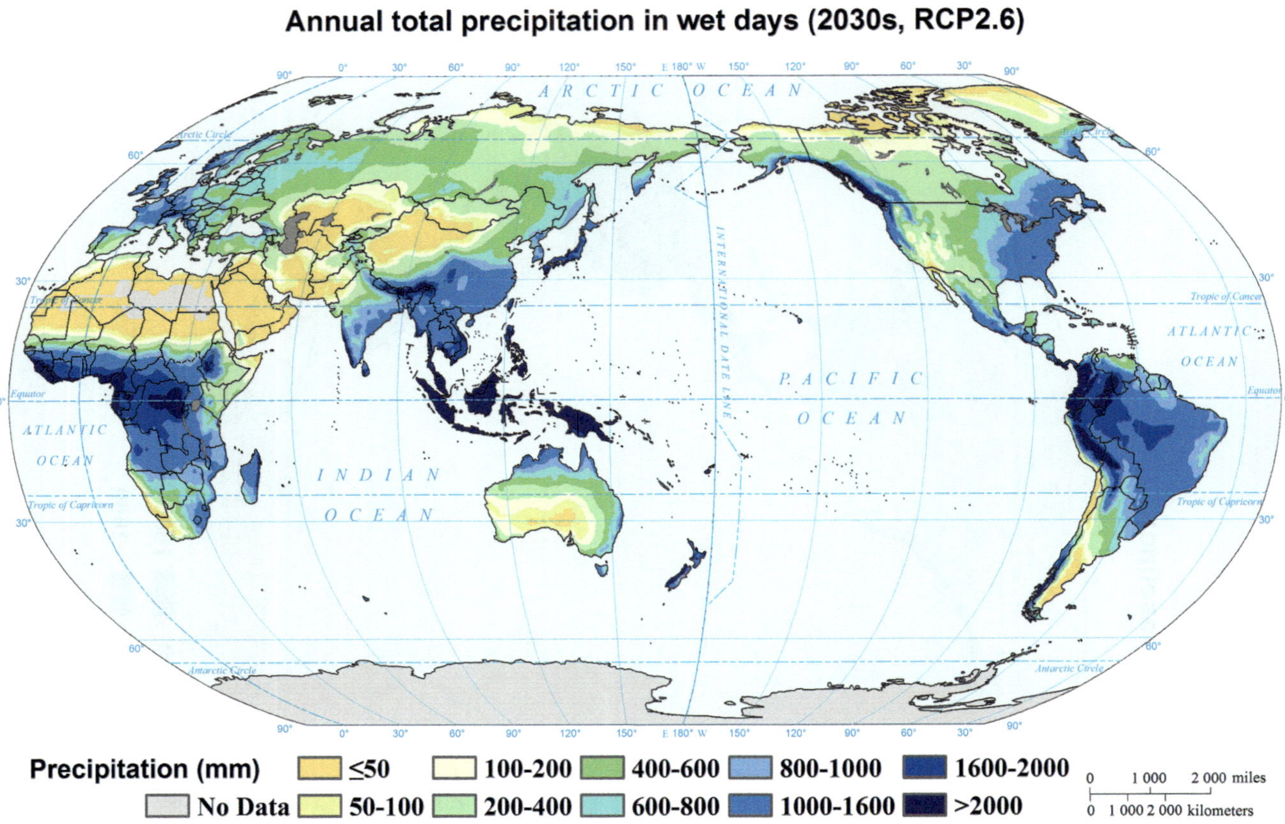
Annual total precipitation in wet days (2030s, RCP2.6)
ARCTIC OCEAN
PACIFIC OCEAN
ATLANTIC OCEAN
INDIAN OCEAN
INTERNATIONAL DATE LINE
Tropic of Cancer
Equator
Tropic of Capricorn
Arctic Circle
Antarctic Circle
Precipitation (mm)
No Data
≤50
50-100
100-200
200-400
400-600
600-800
800-1000
1000-1600
1600-2000
>2000
0 1 000 2 000 miles
0 1 000 2 000 kilometers

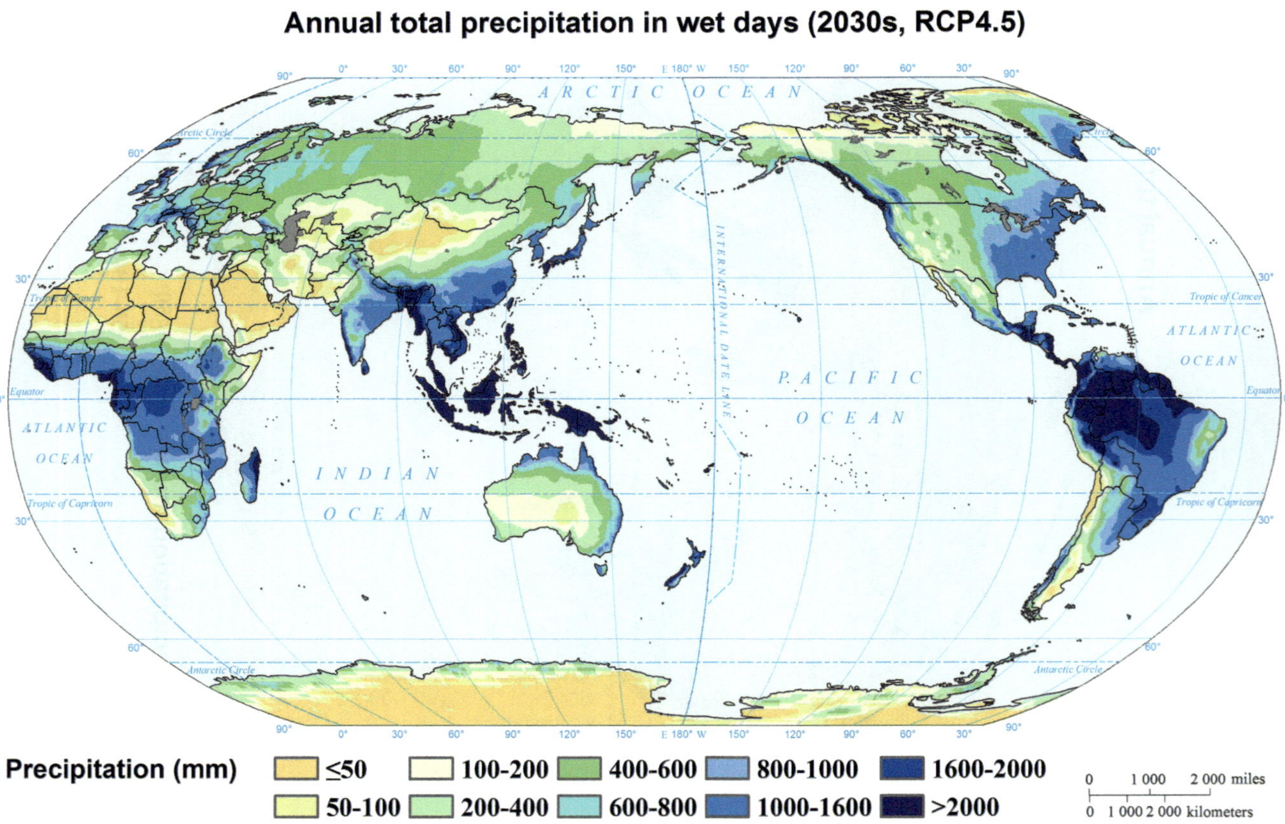
Annual total precipitation in wet days (2030s, RCP4.5)
ARCTIC OCEAN
PACIFIC OCEAN
ATLANTIC OCEAN
INDIAN OCEAN
INTERNATIONAL DATE LINE
Tropic of Cancer
Equator
Tropic of Capricorn
Arctic Circle
Antarctic Circle
Precipitation (mm)
≤50
50-100
100-200
200-400
400-600
600-800
800-1000
1000-1600
1600-2000
>2000
0 1 000 2 000 miles
0 1 000 2 000 kilometers

Annual total precipitation in wet days (2030s, RCP8.5)

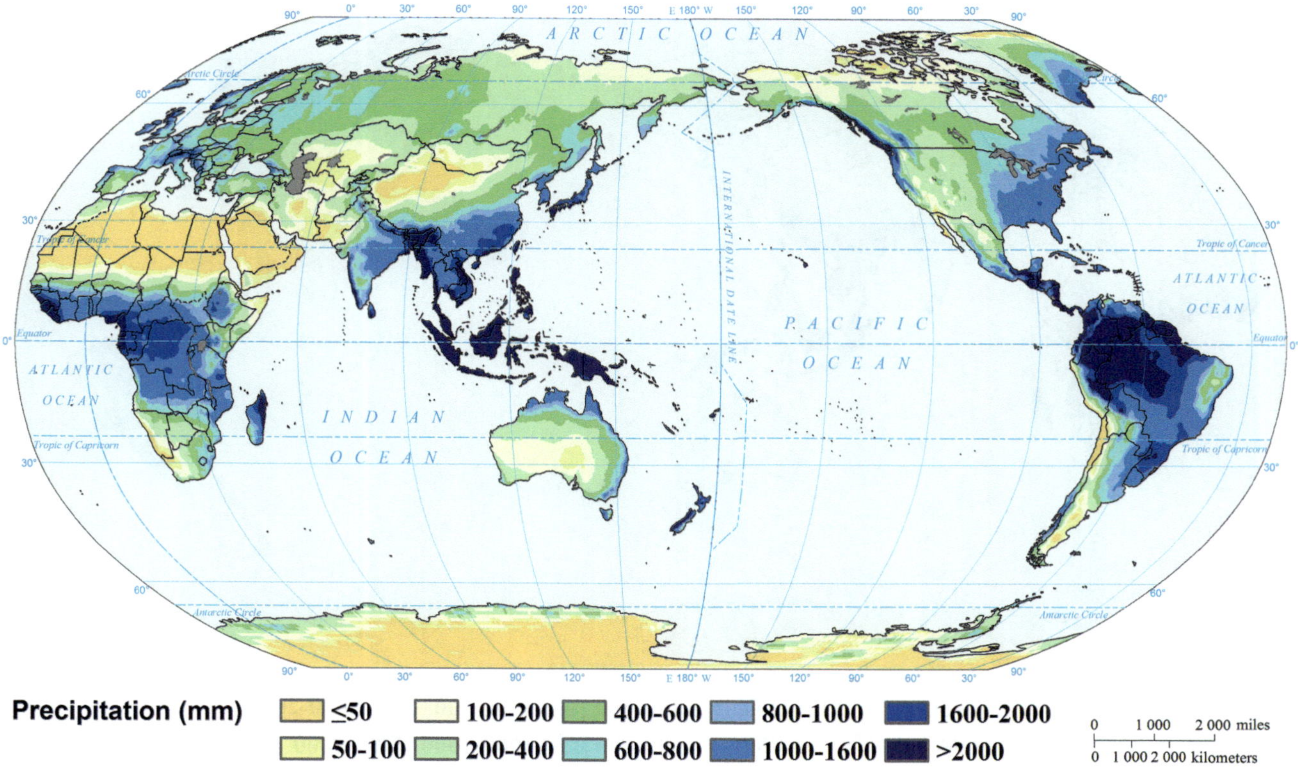

Annual total precipitation in wet days (2050s, RCP2.6)

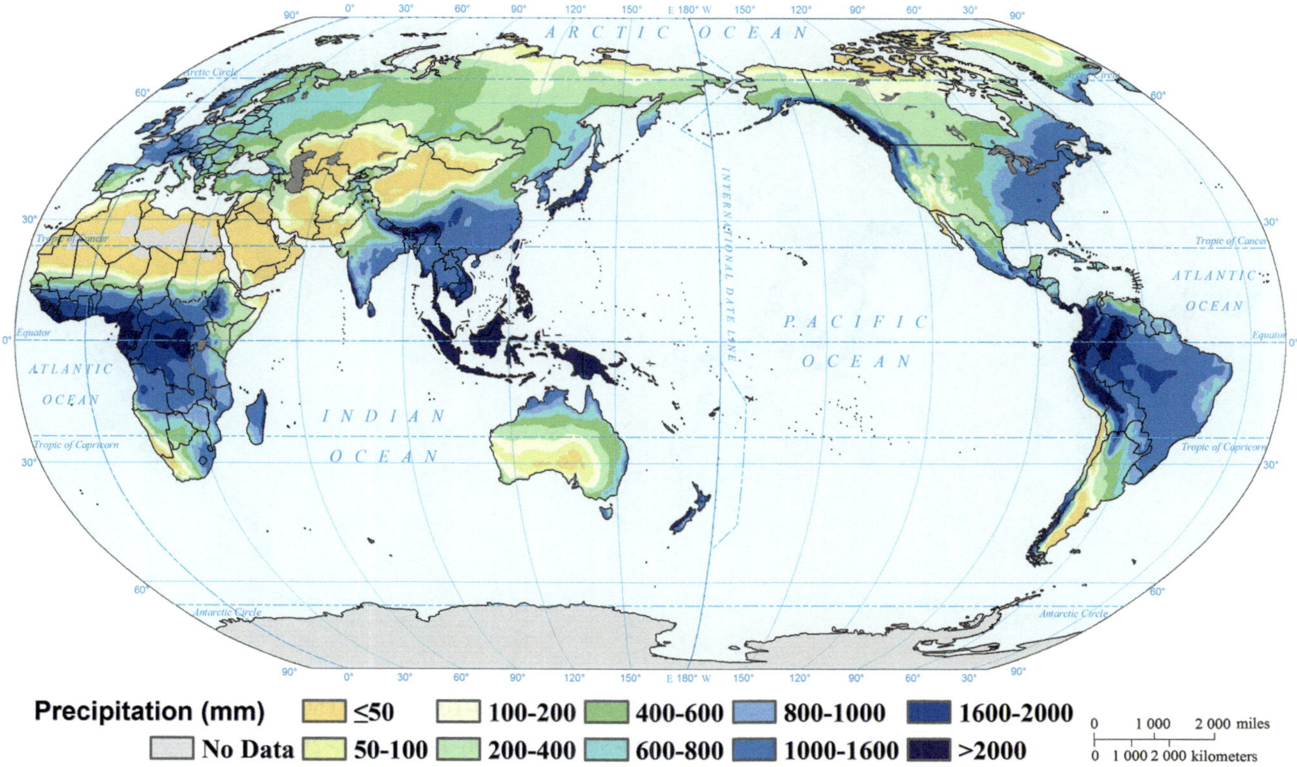

Annual total precipitation in wet days (2050s, RCP4.5)

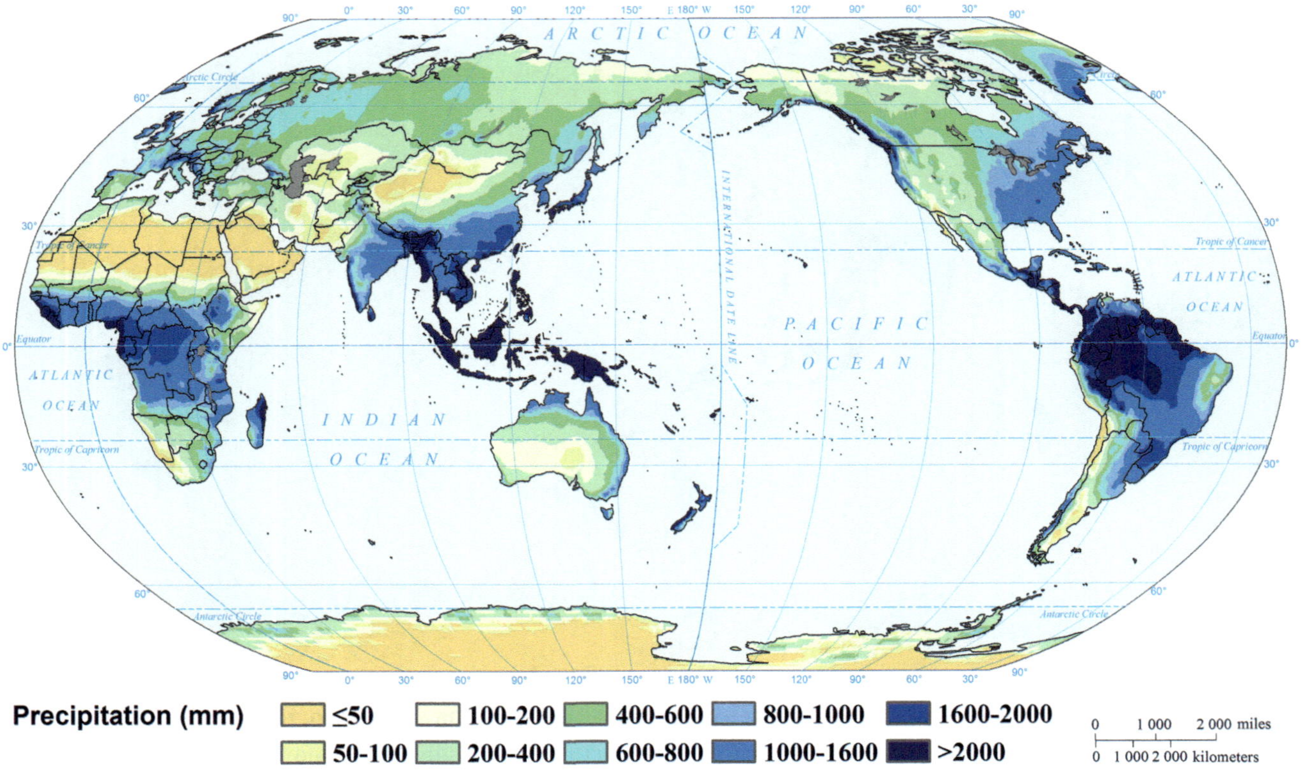

Annual total precipitation in wet days (2050s, RCP8.5)

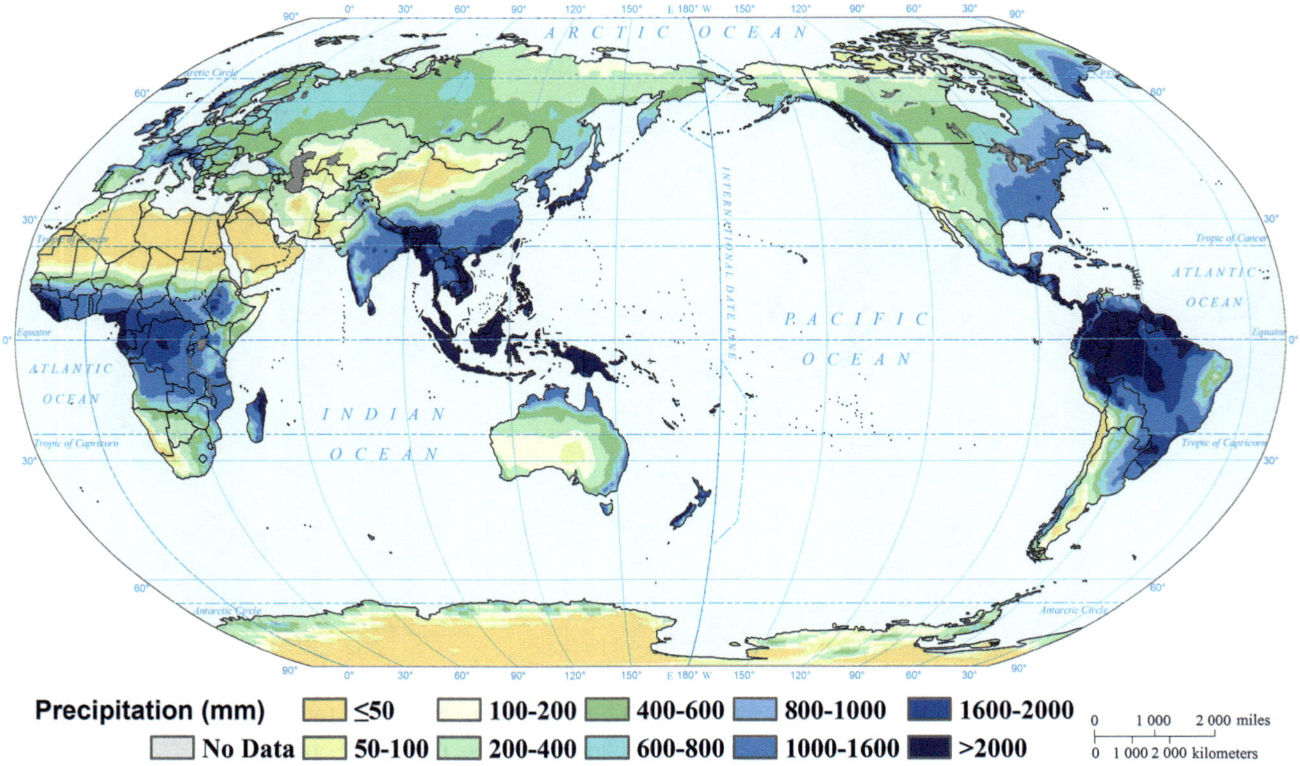

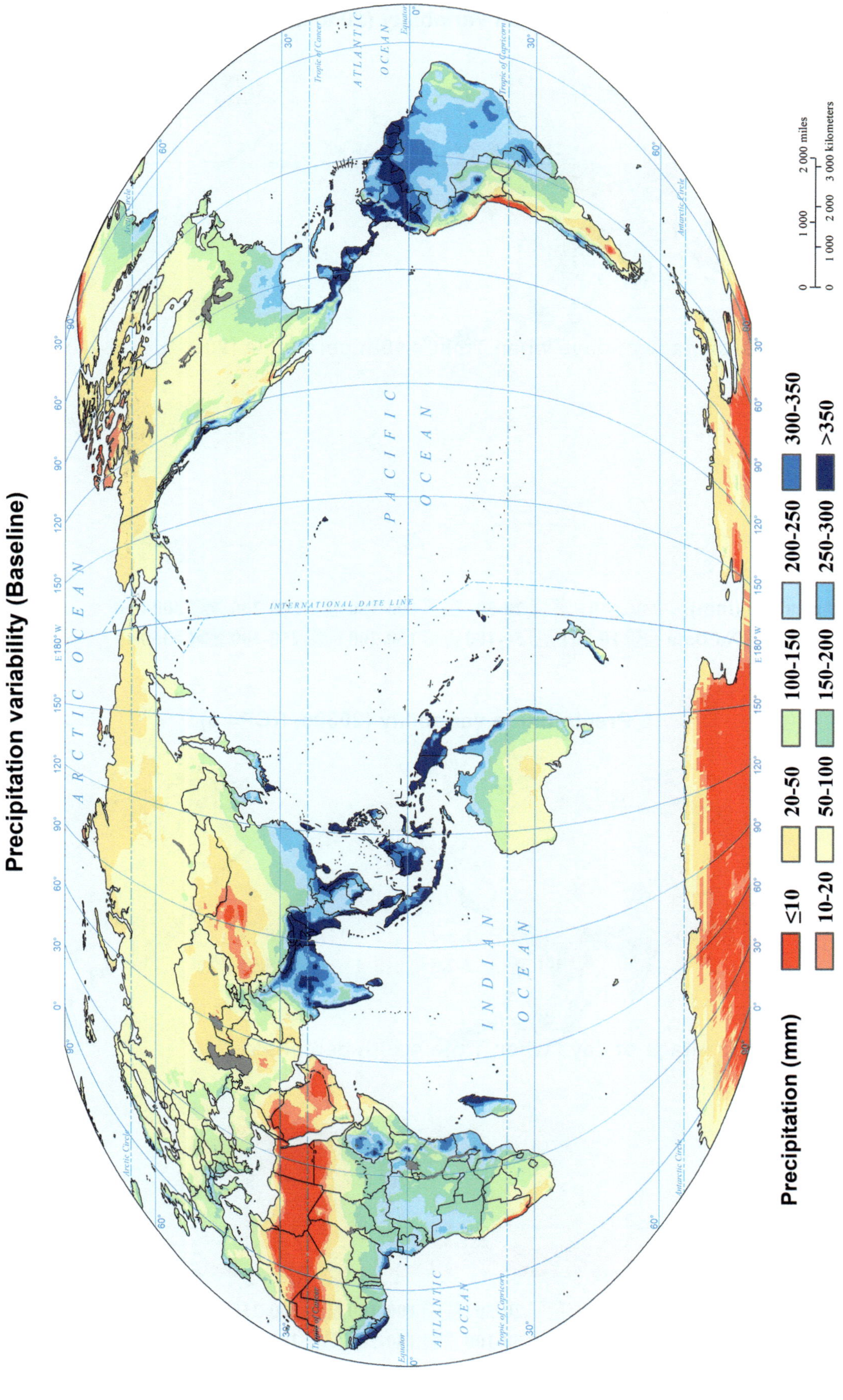
Precipitation variability (Baseline)
ARCTIC OCEAN
PACIFIC OCEAN
ATLANTIC OCEAN
INDIAN OCEAN
INTERNATIONAL DATE LINE
Tropic of Cancer
Tropic of Capricorn
Equator
Arctic Circle
Antarctic Circle
Precipitation (mm)
≤10
10-20
20-50
50-100
100-150
150-200
200-250
250-300
300-350
>350
0 1 000 2 000 miles
0 1 000 2 000 3 000 kilometers

Precipitation variability (2030s, RCP2.6)

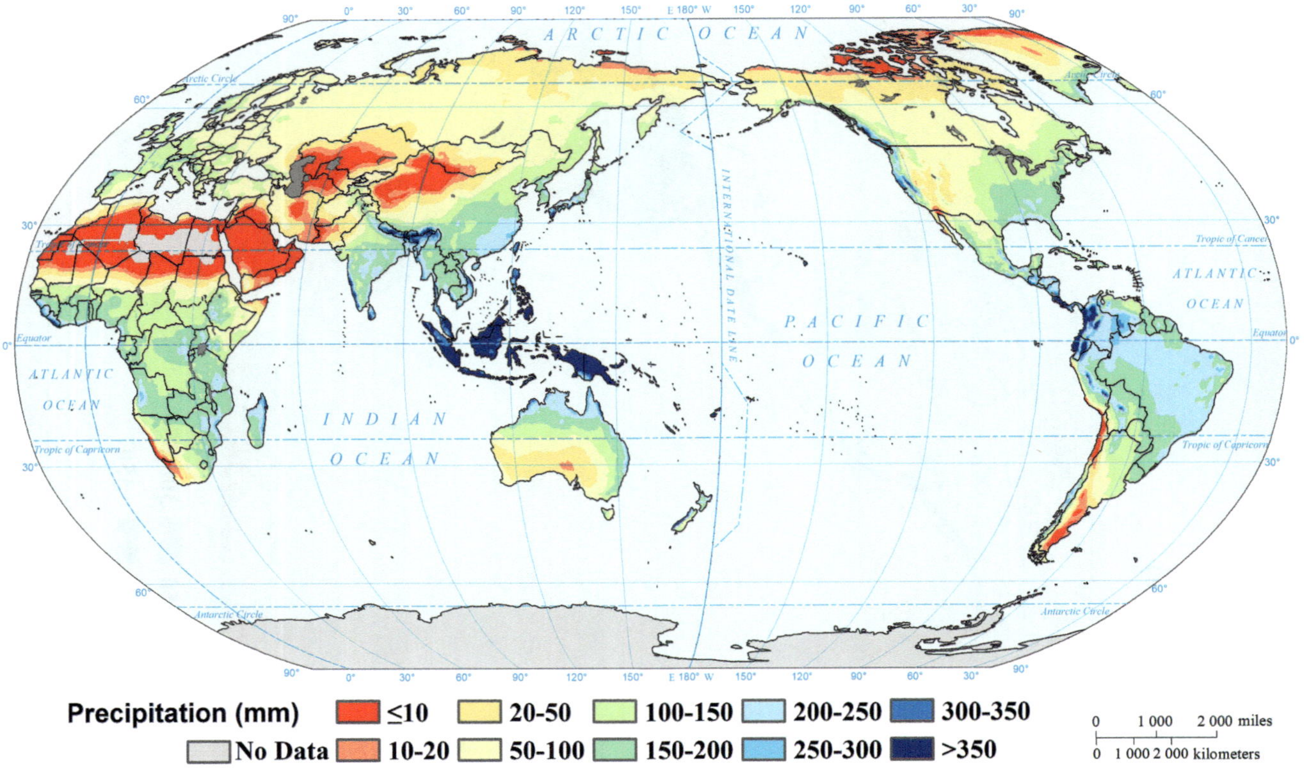

Precipitation variability (2030s, RCP4.5)

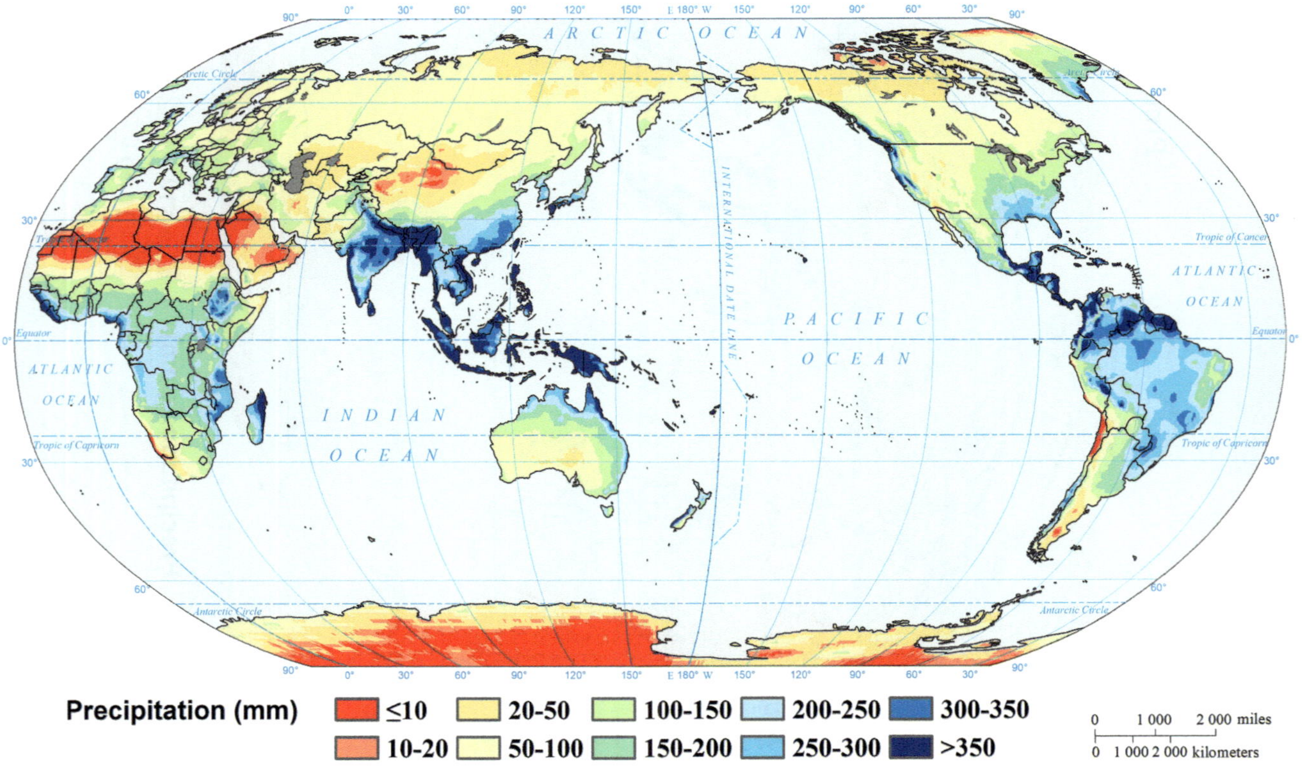

Precipitation variability (2030s, RCP8.5)

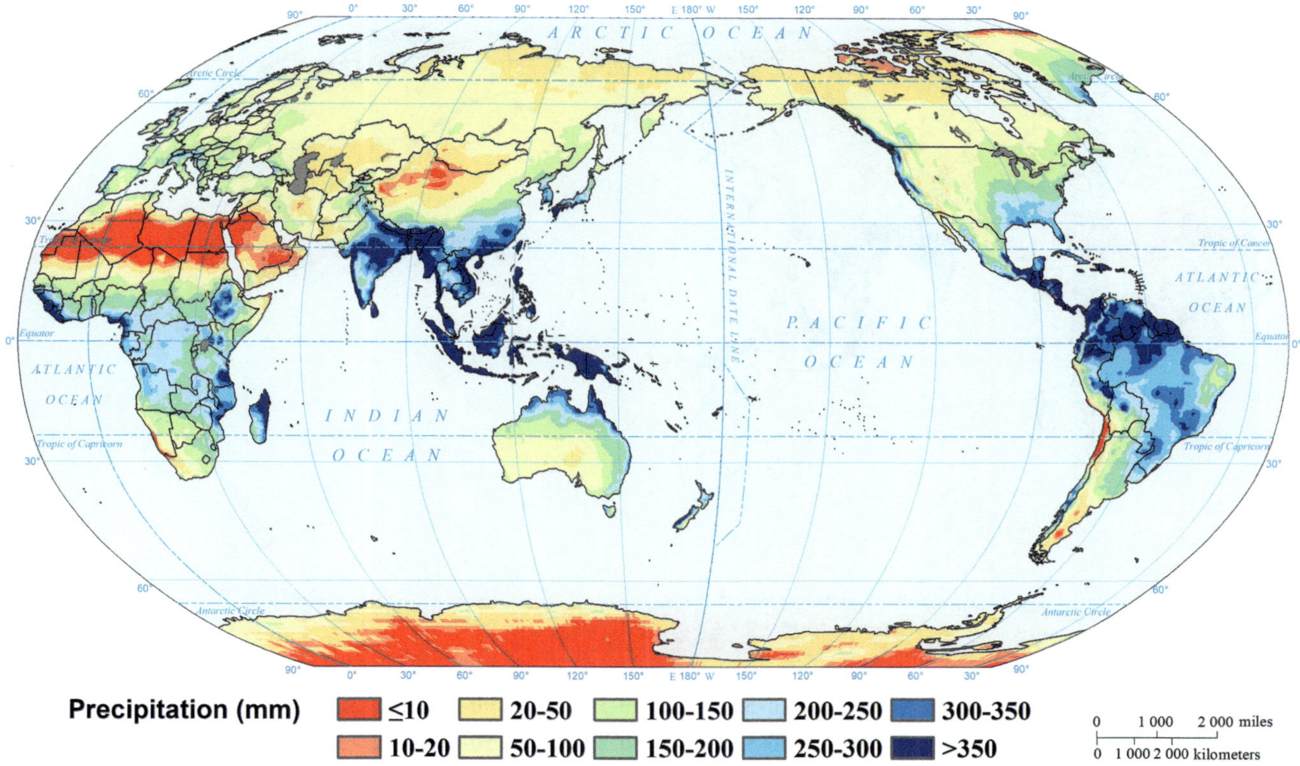

Precipitation variability (2050s, RCP2.6)

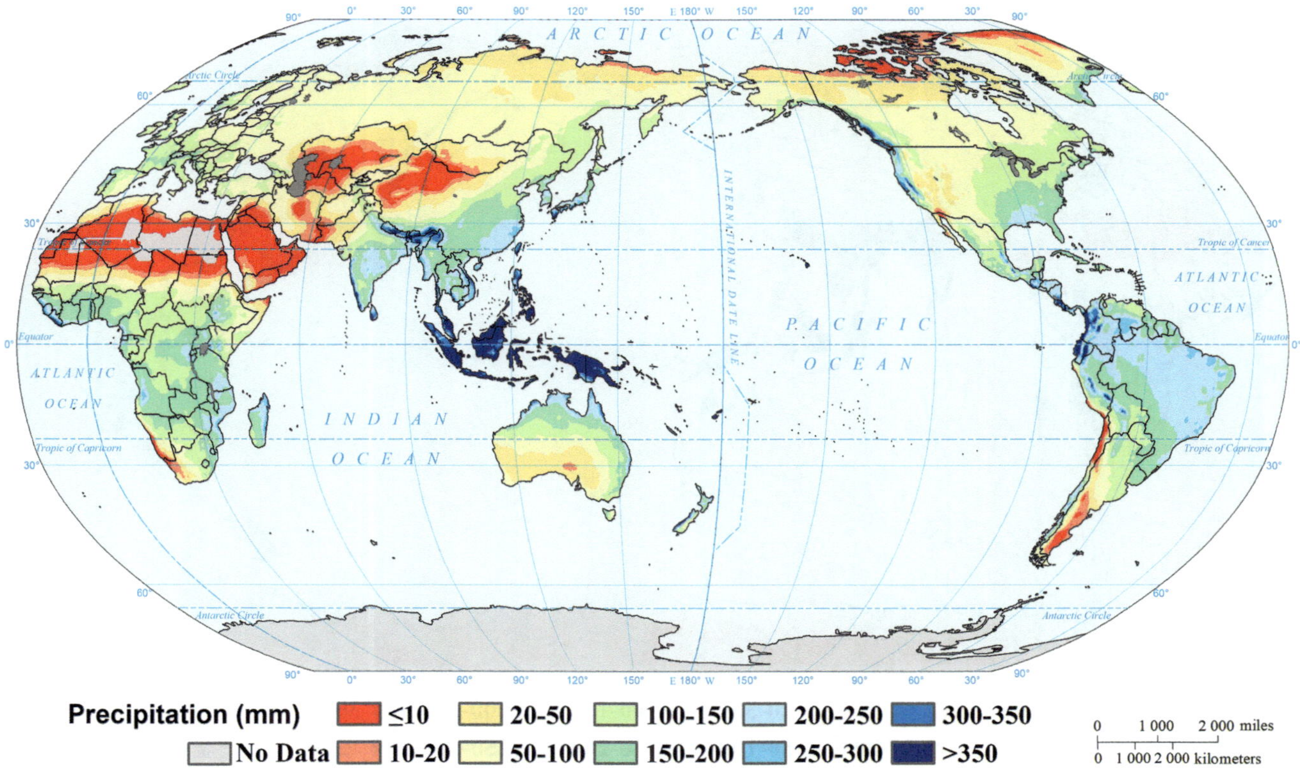

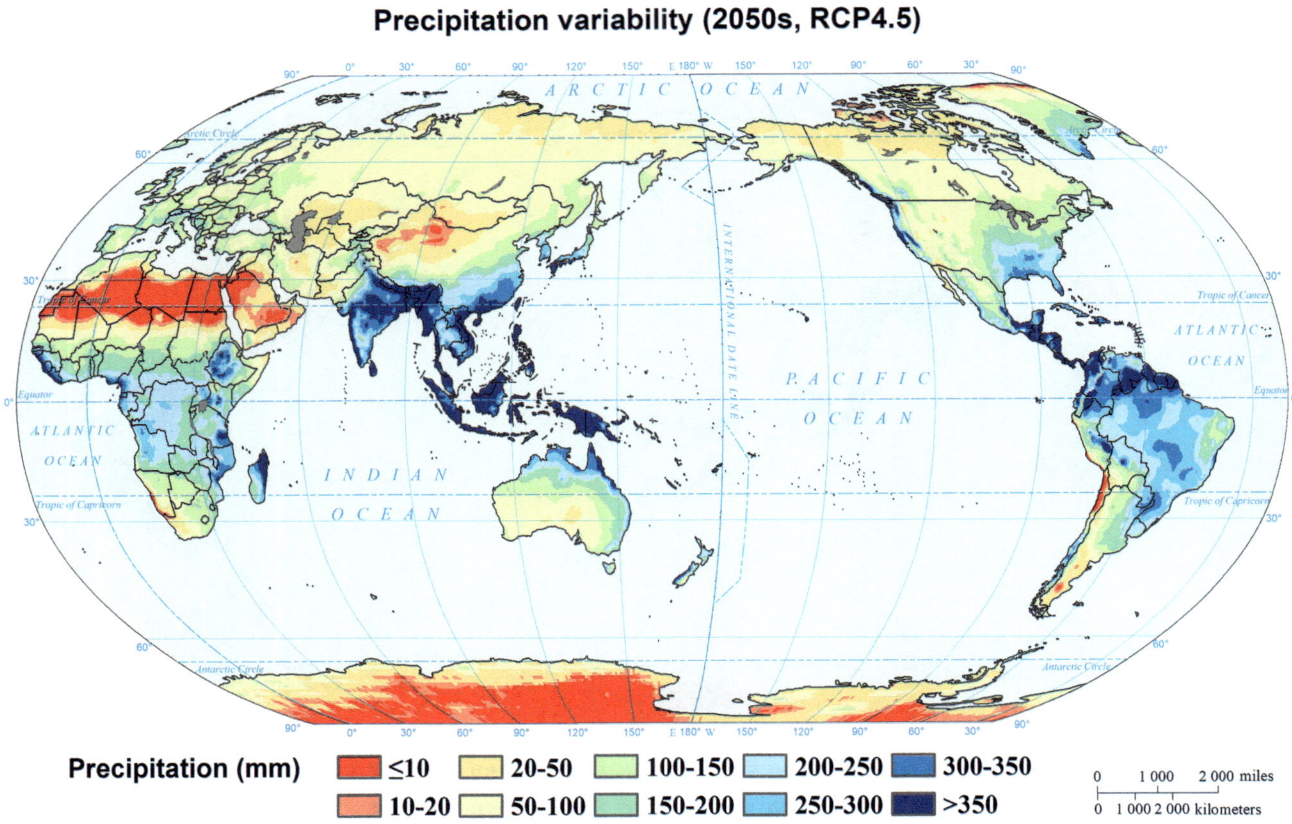
Precipitation variability (2050s, RCP4.5)
ARCTIC OCEAN
PACIFIC OCEAN
INDIAN OCEAN
ATLANTIC OCEAN
INTERNATIONAL DATE LINE
Arctic Circle
Tropic of Cancer
Equator
Tropic of Capricorn
Antarctic Circle
Precipitation (mm)
≤10
10-20
20-50
50-100
100-150
150-200
200-250
250-300
300-350
>350
0 1 000 2 000 miles
0 1 000 2 000 kilometers

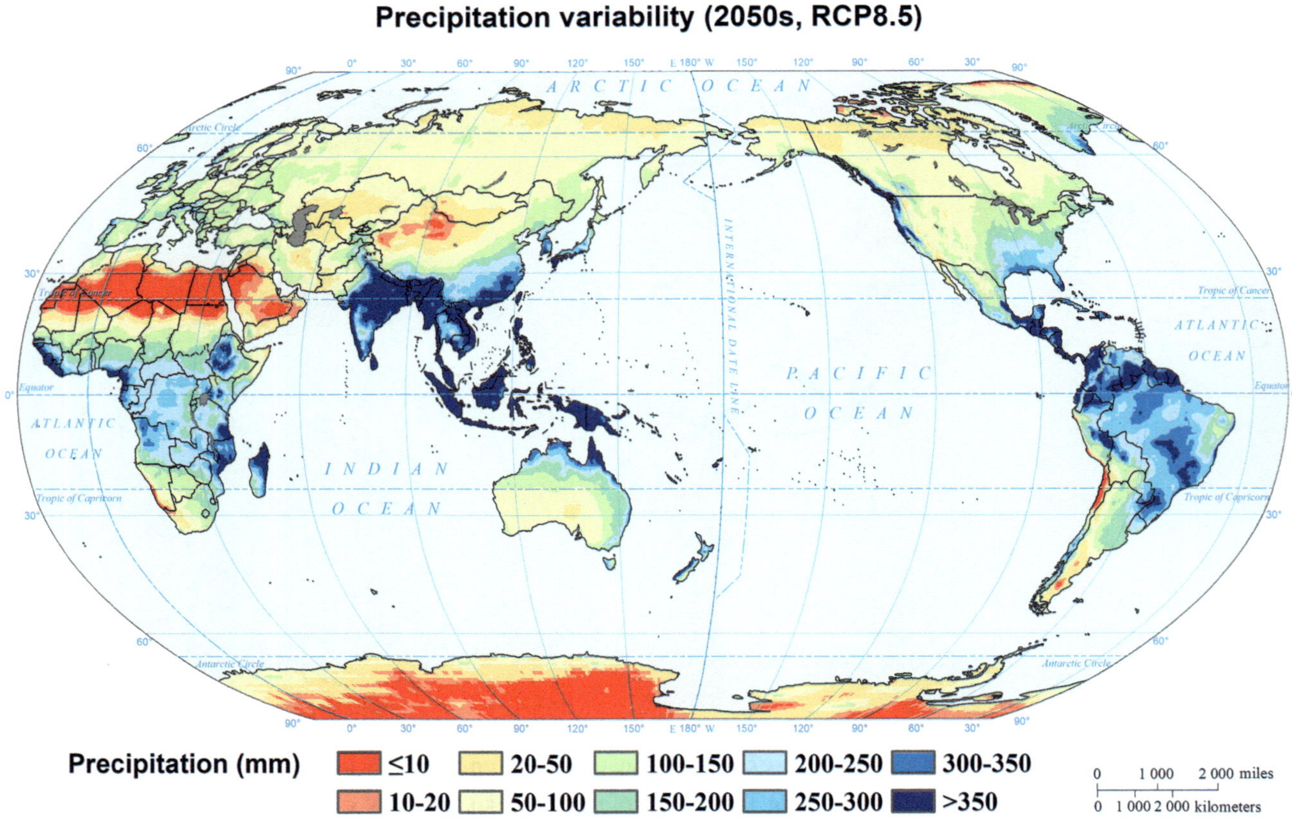
Precipitation variability (2050s, RCP8.5)
ARCTIC OCEAN
PACIFIC OCEAN
INDIAN OCEAN
ATLANTIC OCEAN
INTERNATIONAL DATE LINE
Arctic Circle
Tropic of Cancer
Equator
Tropic of Capricorn
Antarctic Circle
Precipitation (mm)
≤10
10-20
20-50
50-100
100-150
150-200
200-250
250-300
300-350
>350
0 1 000 2 000 miles
0 1 000 2 000 kilometers

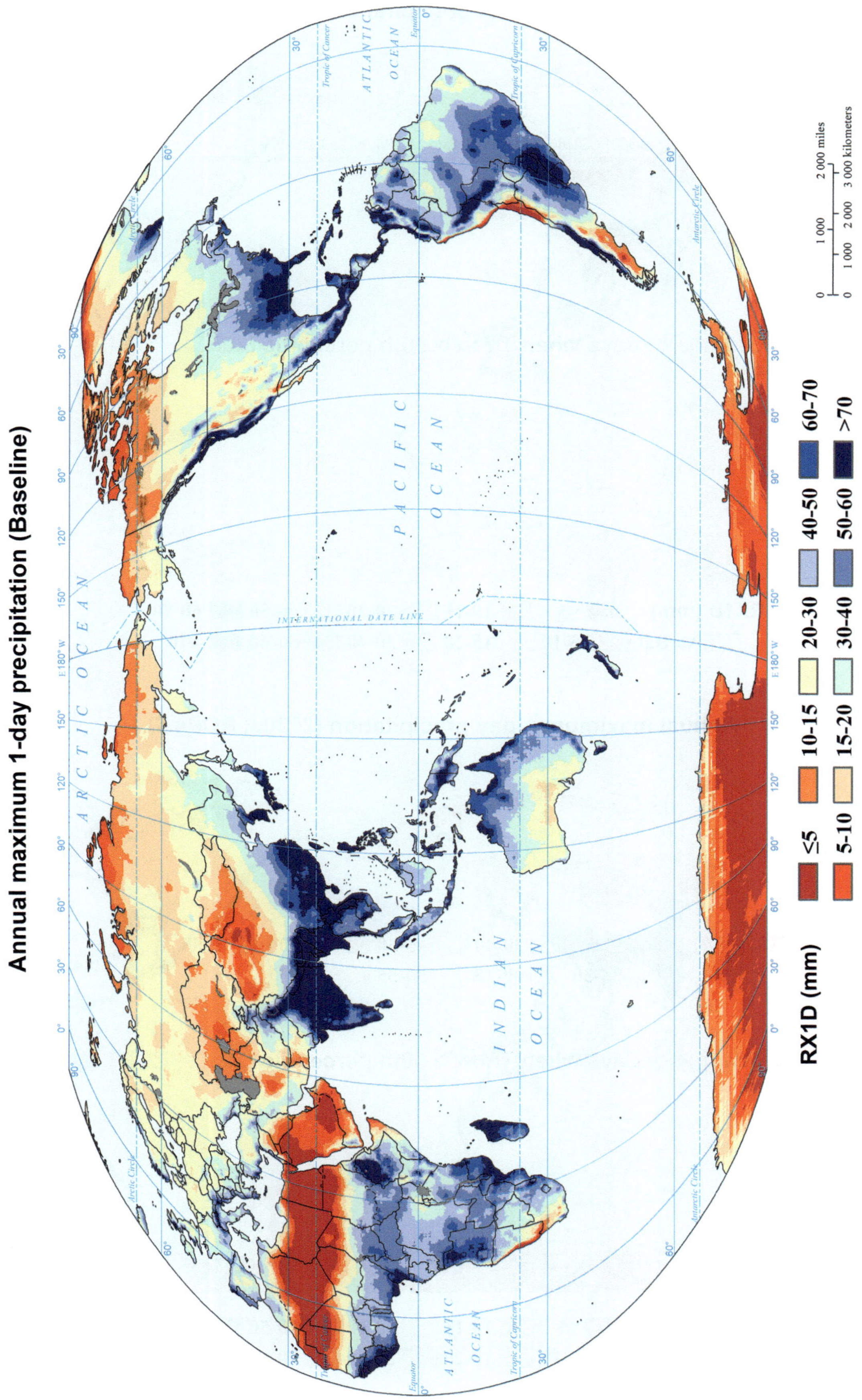
Annual maximum 1-day precipitation (Baseline)
RX1D (mm)
≤5
5-10
10-15
15-20
20-30
30-40
40-50
50-60
60-70
>70
ARCTIC OCEAN
ATLANTIC OCEAN
PACIFIC OCEAN
INDIAN OCEAN
INTERNATIONAL DATE LINE
Arctic Circle
Tropic of Cancer
Equator
Tropic of Capricorn
Antarctic Circle
0 1 000 2 000 miles
0 1 000 2 000 3 000 kilometers

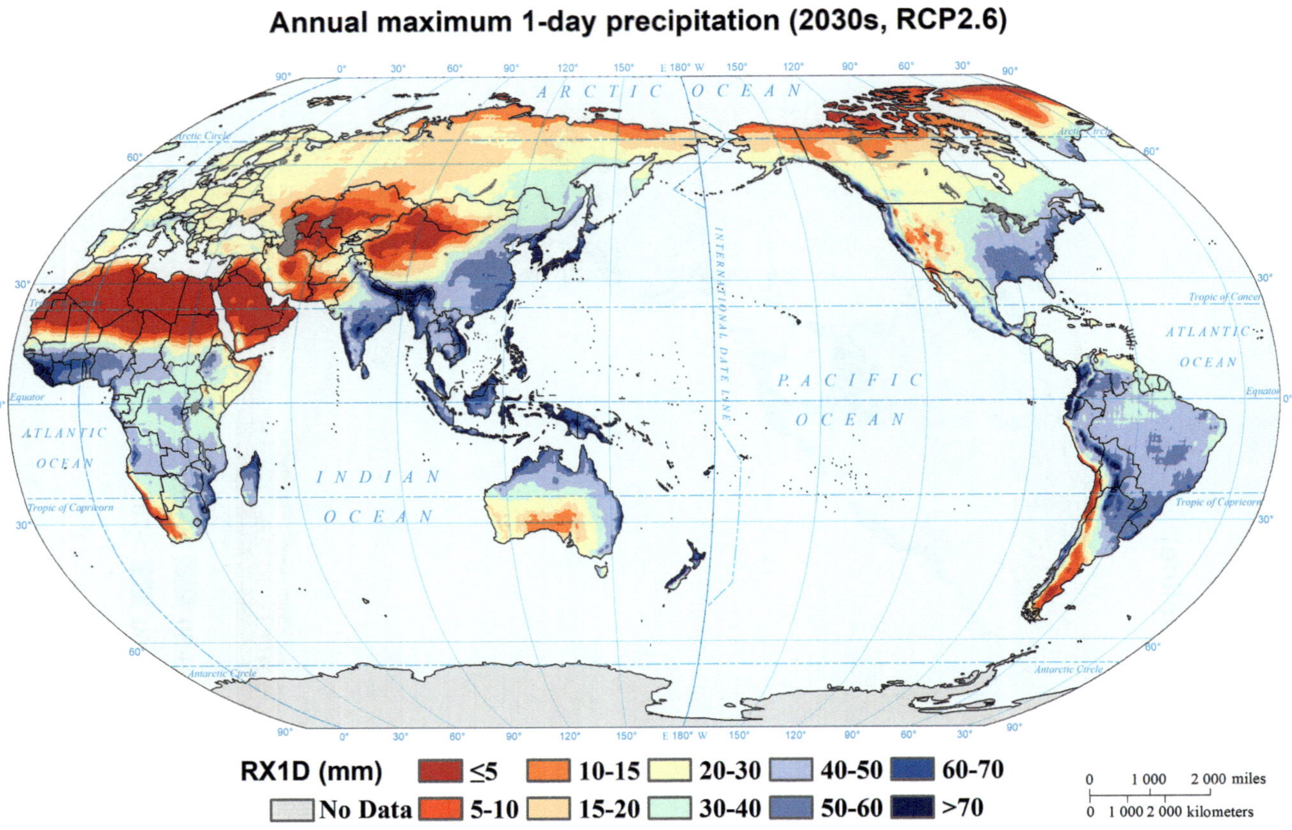
Annual maximum 1-day precipitation (2030s, RCP2.6)
ARCTIC OCEAN
PACIFIC OCEAN
ATLANTIC OCEAN
INDIAN OCEAN
INTERNATIONAL DATE LINE
Arctic Circle
Tropic of Cancer
Equator
Tropic of Capricorn
Antarctic Circle
RX1D (mm)
≤5
5-10
10-15
15-20
20-30
30-40
40-50
50-60
60-70
>70
No Data
0 1 000 2 000 miles
0 1 000 2 000 kilometers

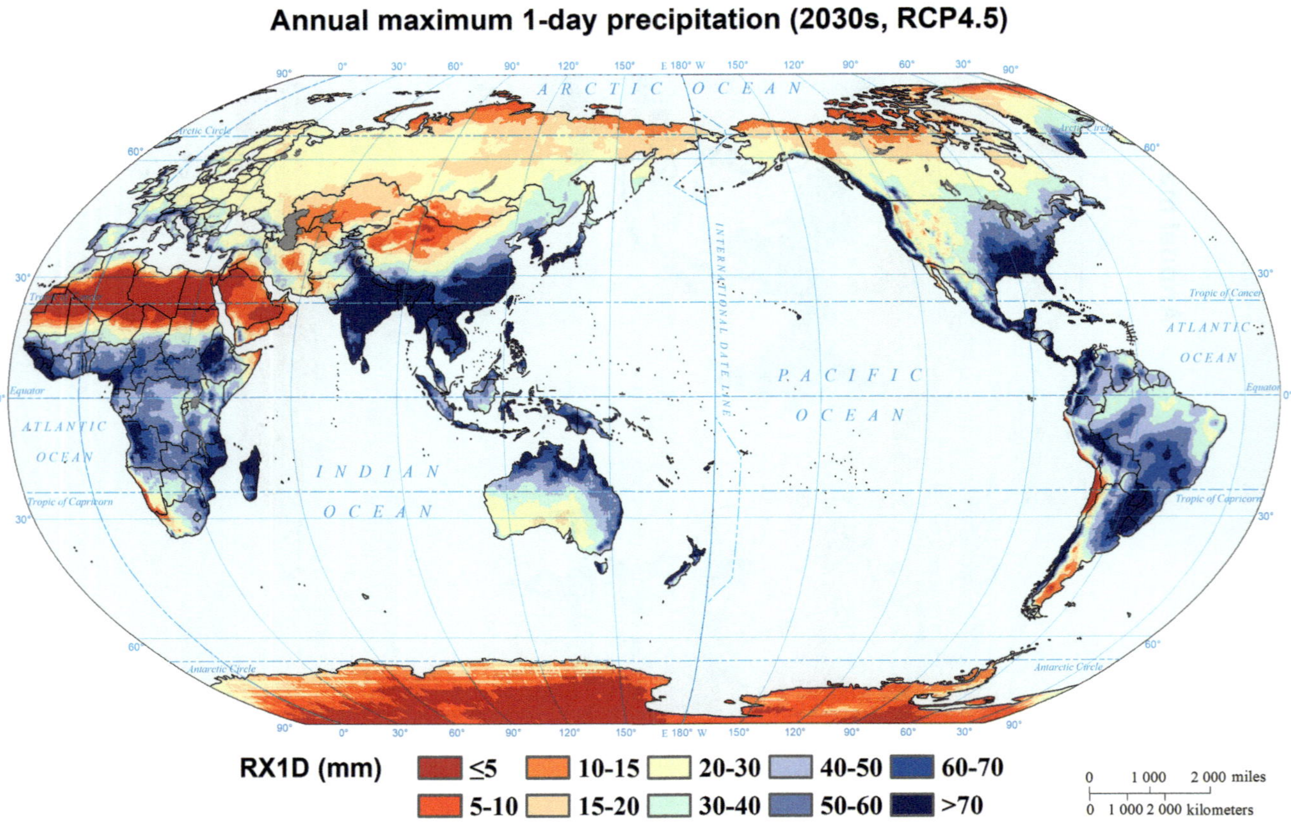
Annual maximum 1-day precipitation (2030s, RCP4.5)
ARCTIC OCEAN
PACIFIC OCEAN
ATLANTIC OCEAN
INDIAN OCEAN
INTERNATIONAL DATE LINE
Arctic Circle
Tropic of Cancer
Equator
Tropic of Capricorn
Antarctic Circle
RX1D (mm)
≤5
5-10
10-15
15-20
20-30
30-40
40-50
50-60
60-70
>70
0 1 000 2 000 miles
0 1 000 2 000 kilometers

Annual maximum 1-day precipitation (2030s, RCP8.5)

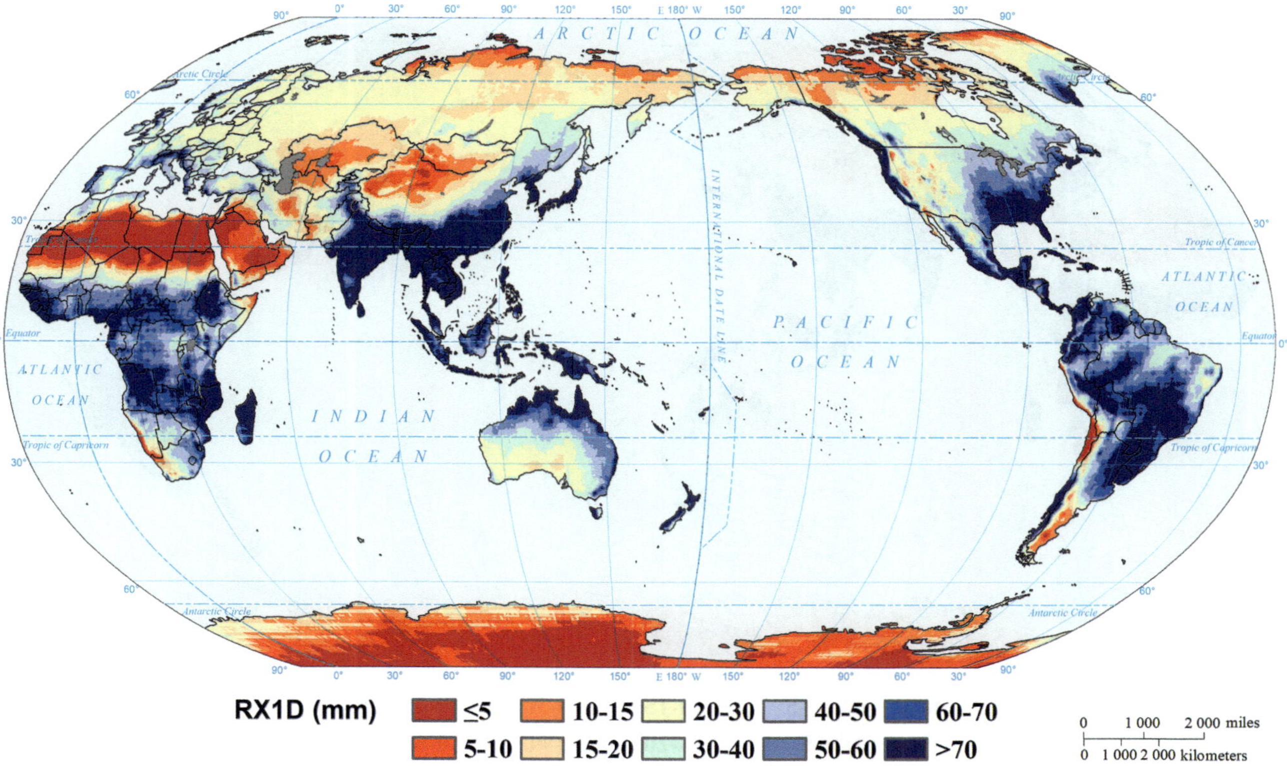

Annual maximum 1-day precipitation (2050s, RCP2.6)

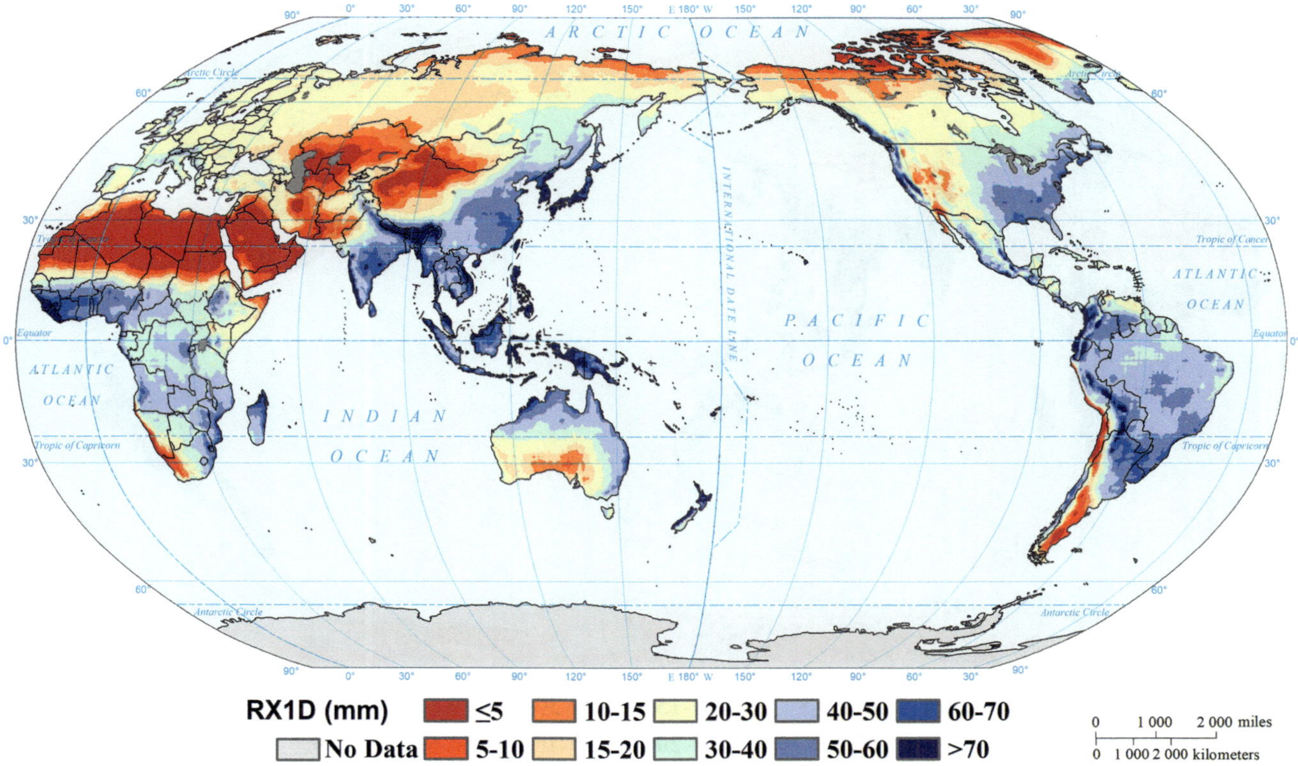

Annual maximum 1-day precipitation (2050s, RCP4.5)

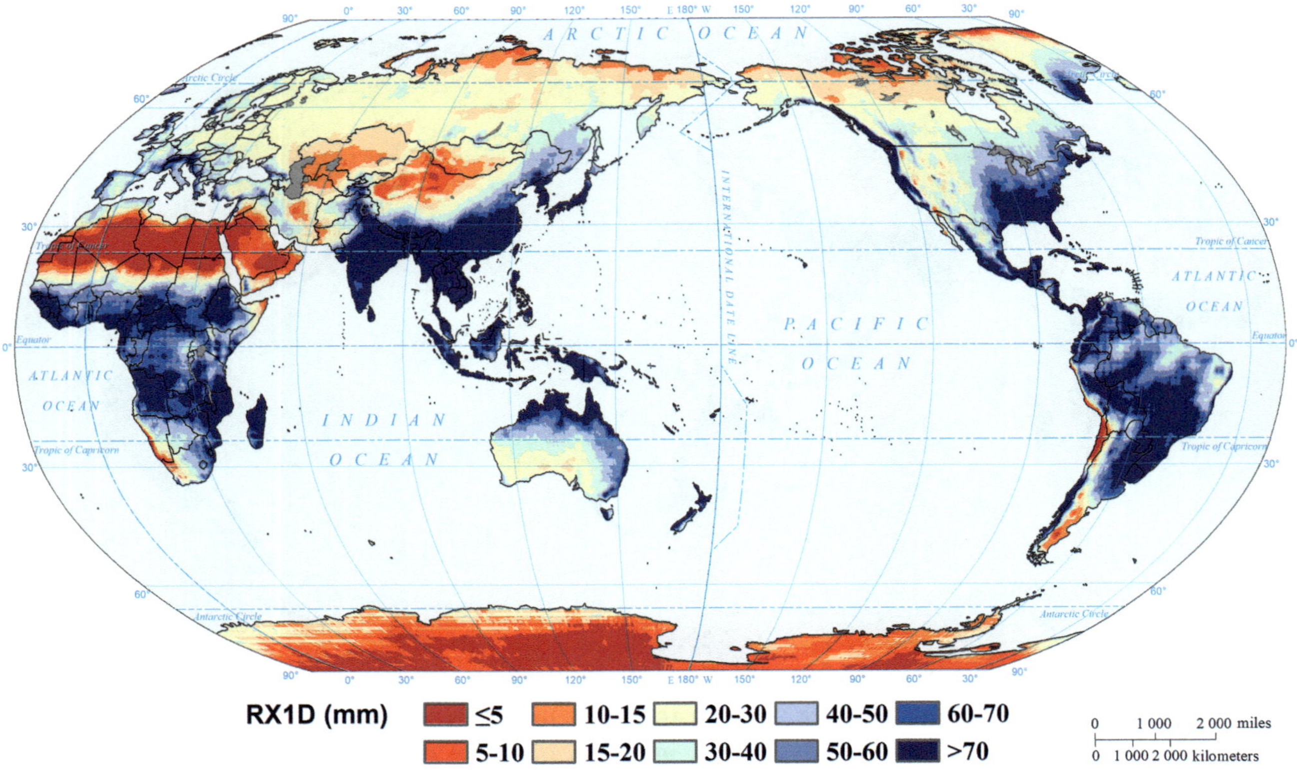

Annual maximum 1-day precipitation (2050s, RCP8.5)

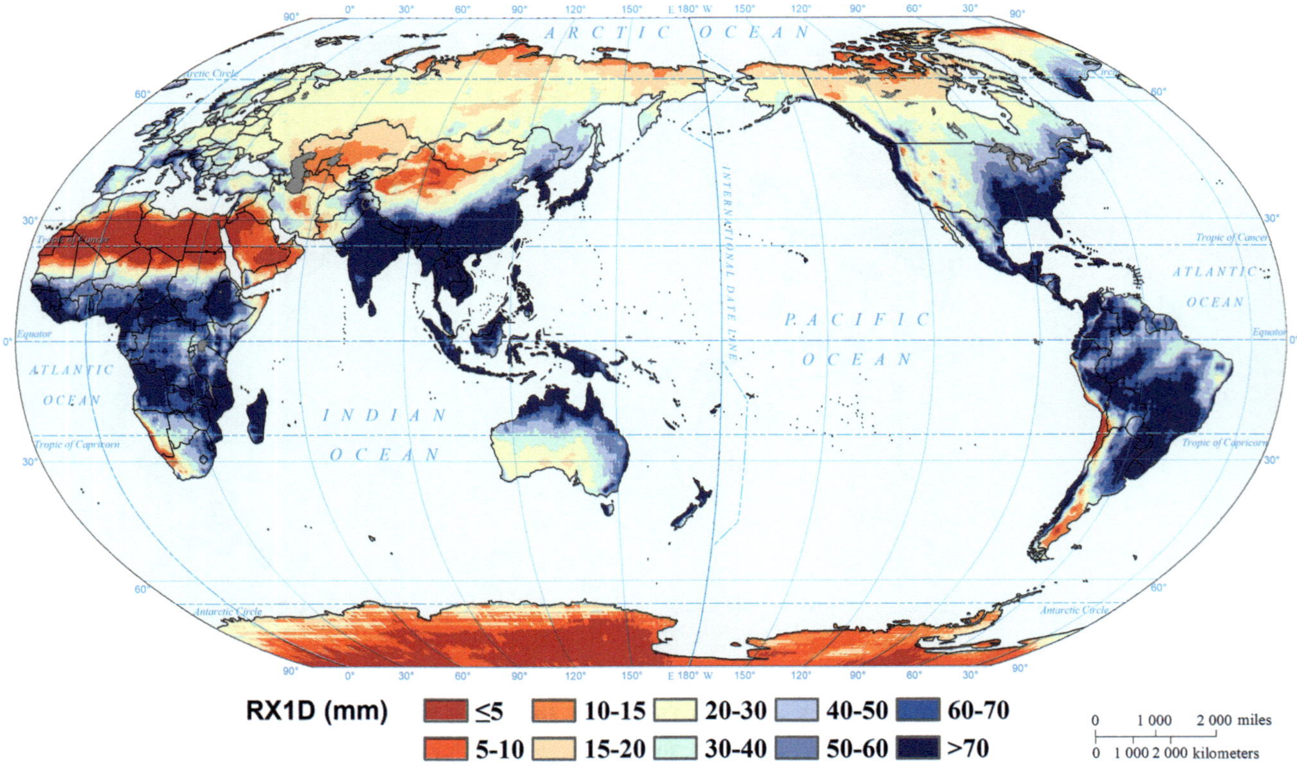

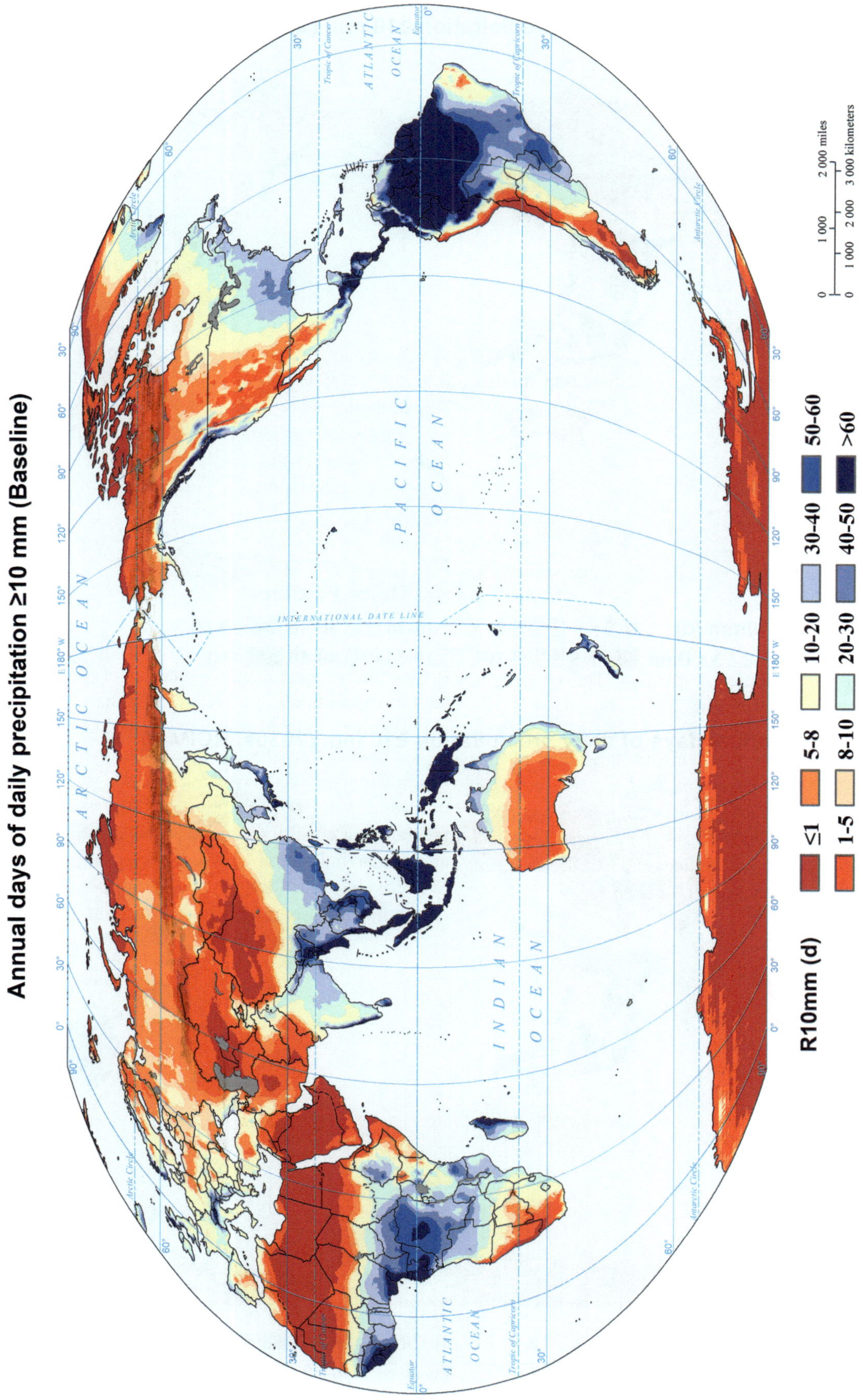
Annual days of daily precipitation ≥10 mm (Baseline)
R10mm (d)
≤1
1-5
5-8
8-10
10-20
20-30
30-40
40-50
50-60
>60
0 1 000 2 000 miles
0 1 000 2 000 3 000 kilometers
ARCTIC OCEAN
PACIFIC OCEAN
ATLANTIC OCEAN
INDIAN OCEAN
INTERNATIONAL DATE LINE
Tropic of Cancer
Tropic of Capricorn
Equator
Arctic Circle
Antarctic Circle

Annual days of daily precipitation ≥10 mm (2030s, RCP2.6)

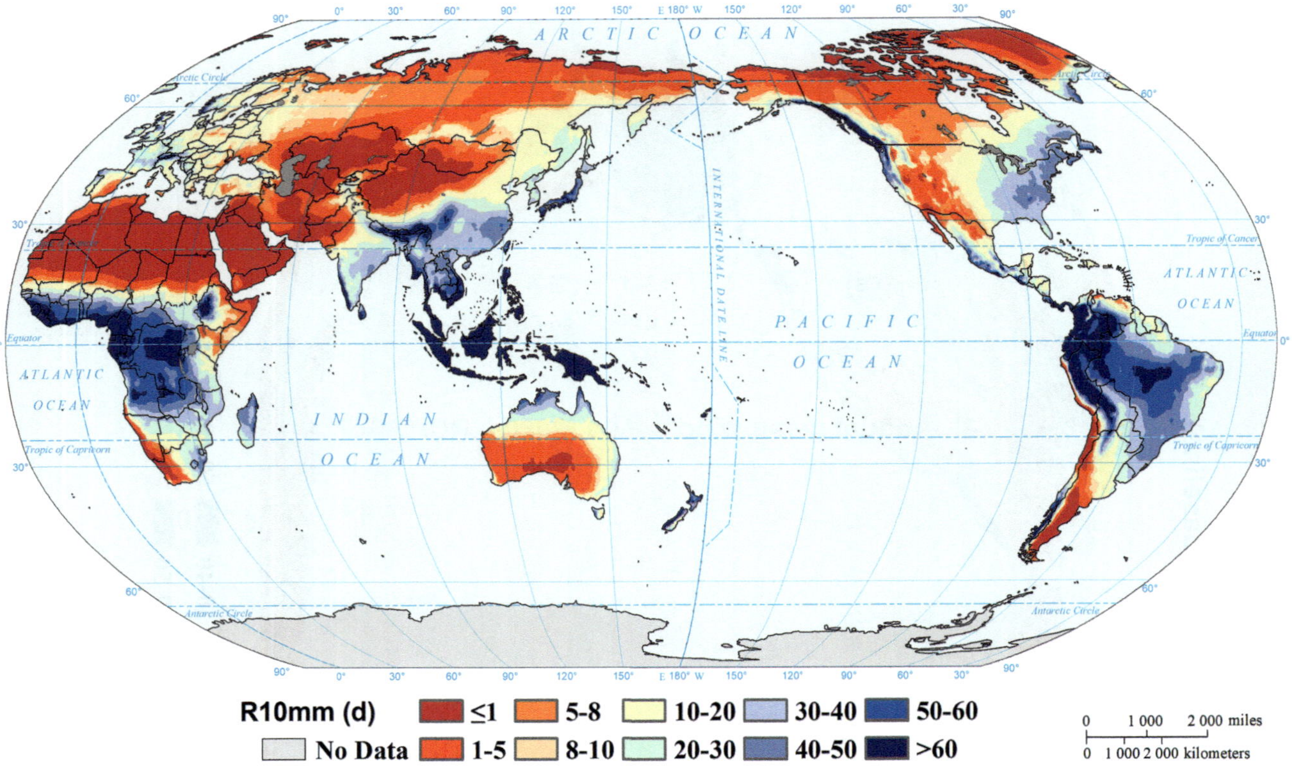

Annual days of daily precipitation ≥10 mm (2030s, RCP4.5)

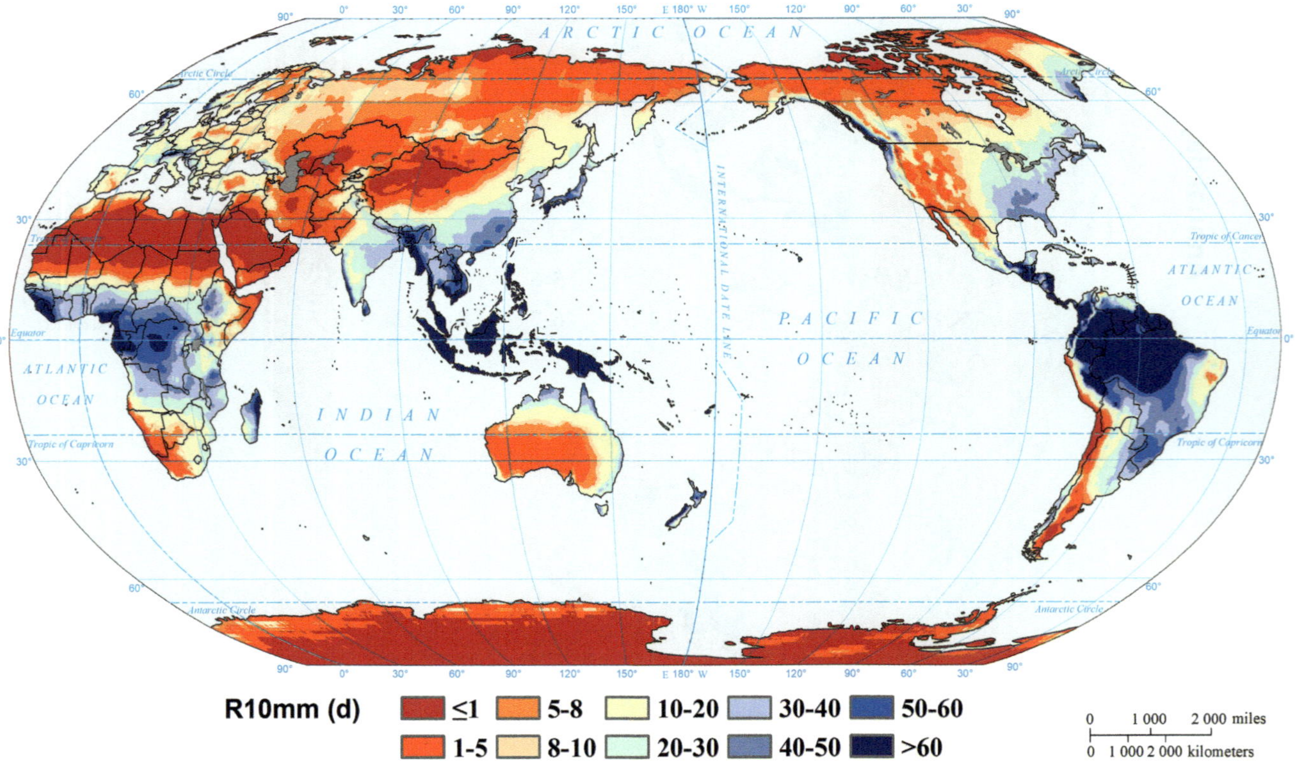

Annual days of daily precipitation ≥10 mm (2030s, RCP8.5)

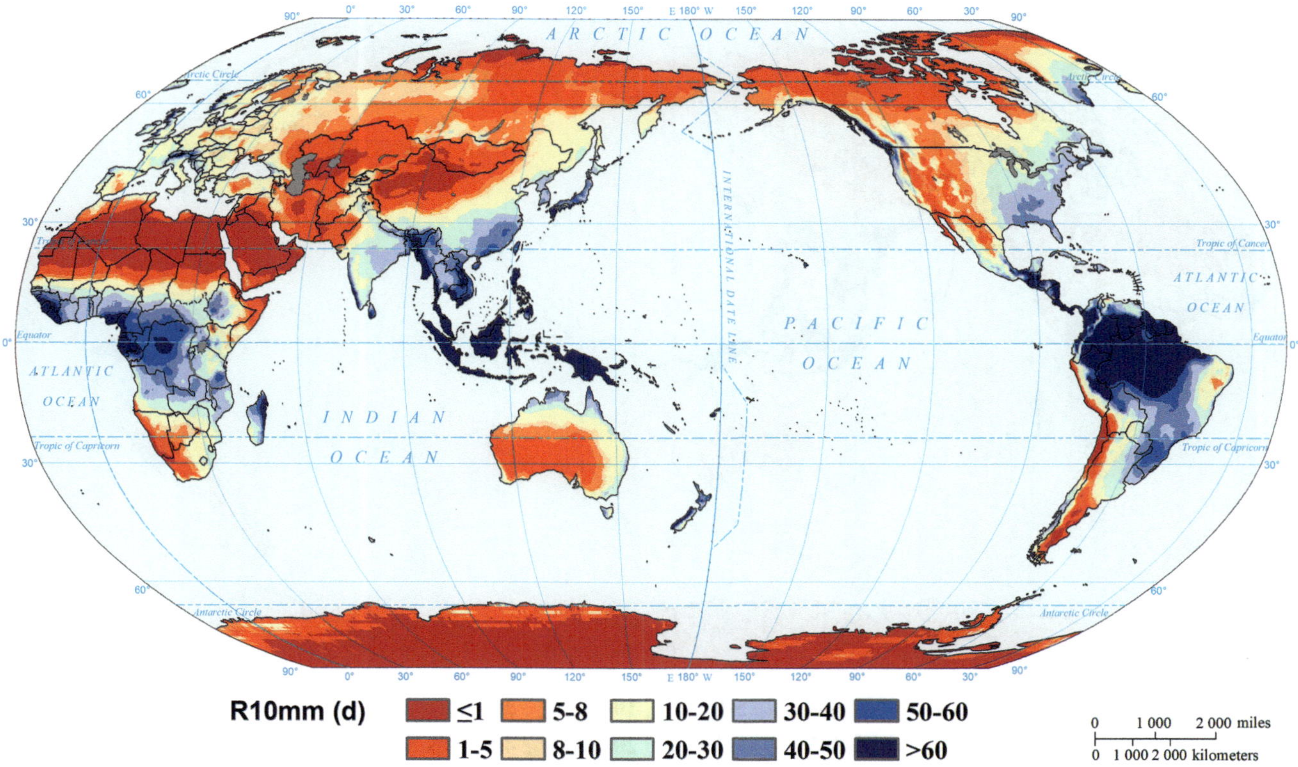

Annual days of daily precipitation ≥10 mm (2050s, RCP2.6)

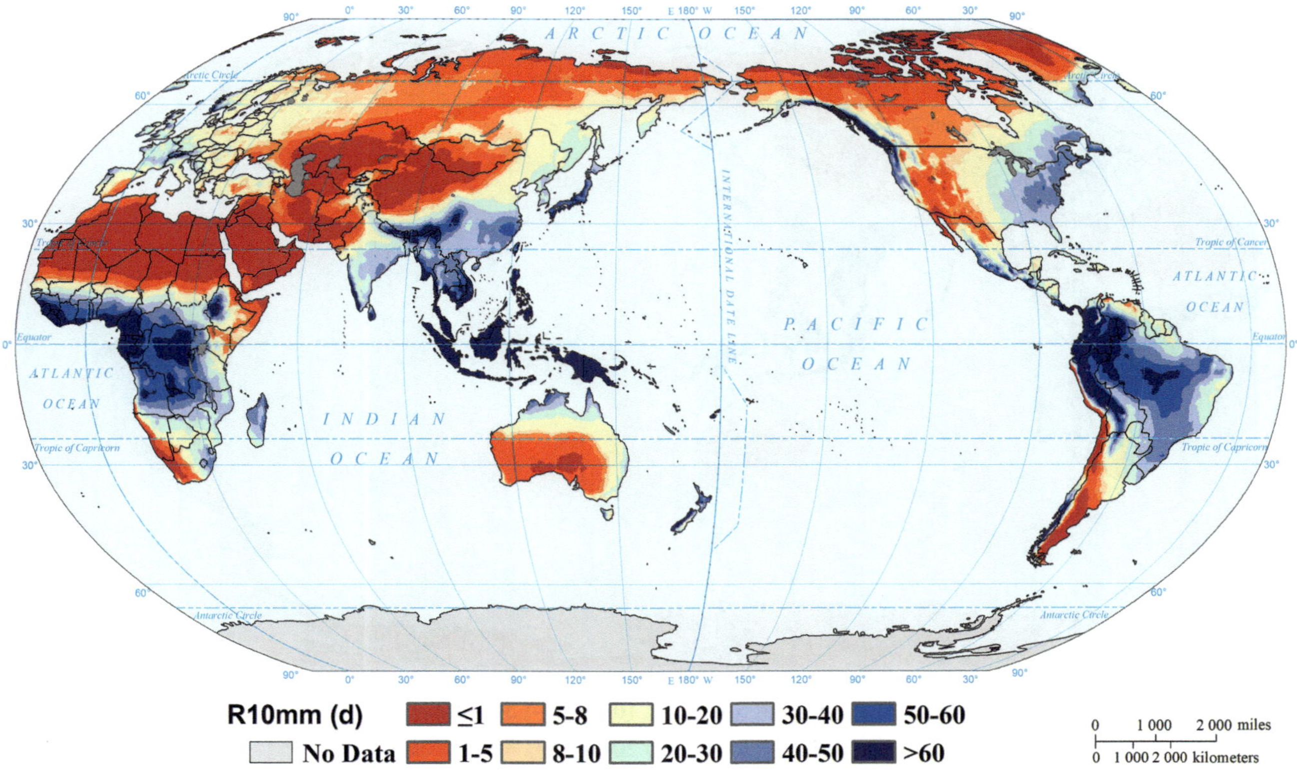

Annual days of daily precipitation ≥10 mm (2050s, RCP4.5)

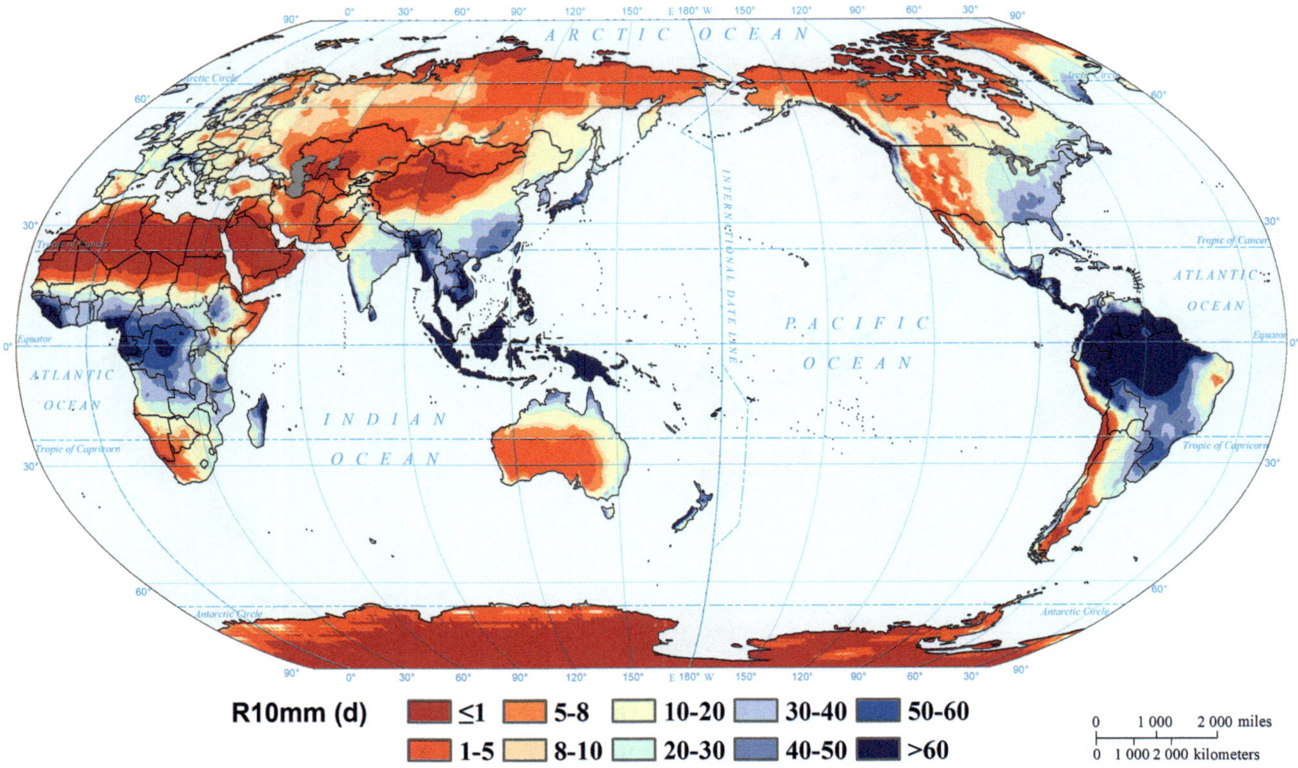

Annual days of daily precipitation ≥10 mm (2050s, RCP8.5)

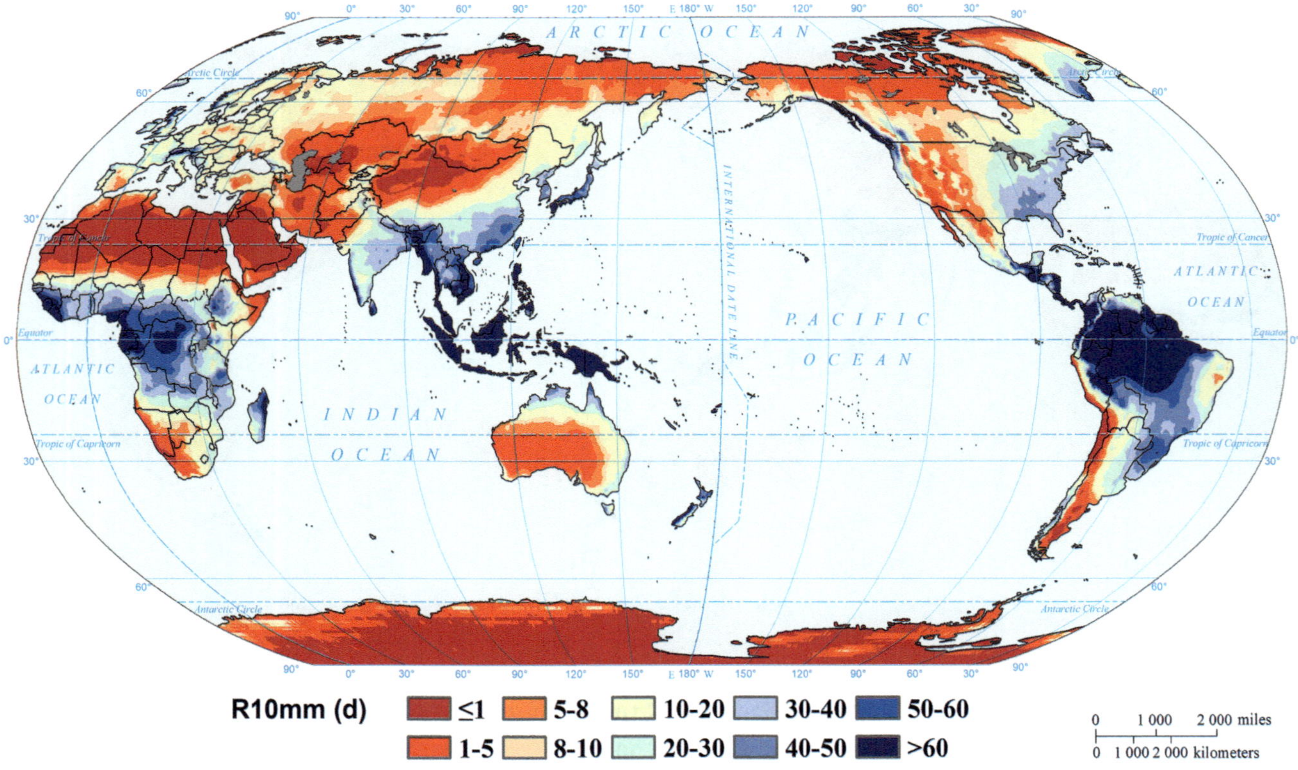

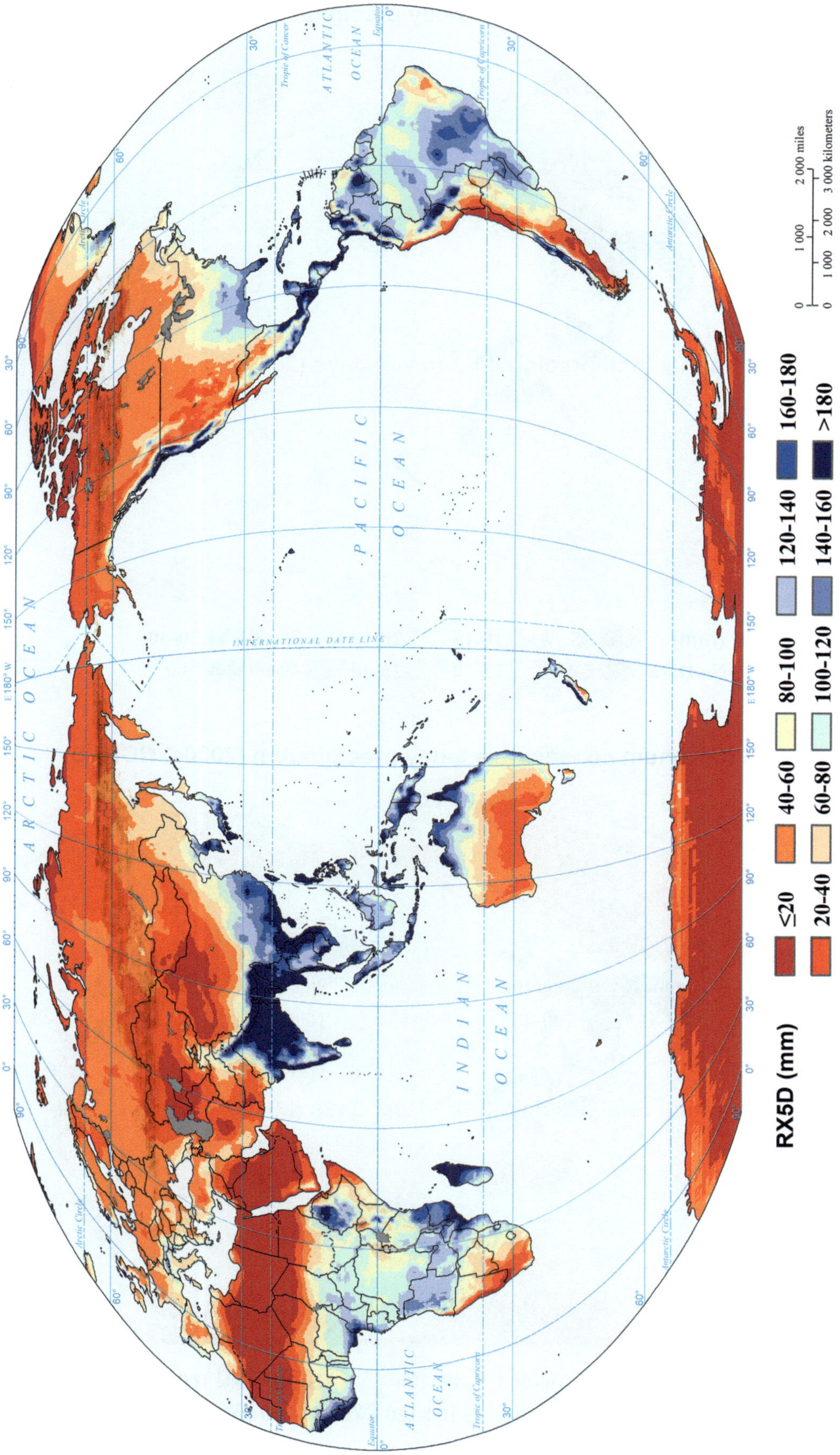

Annual maximum consecutive 5-day precipitation (Baseline)

Annual maximum consecutive 5-day precipitation (2030s, RCP2.6)

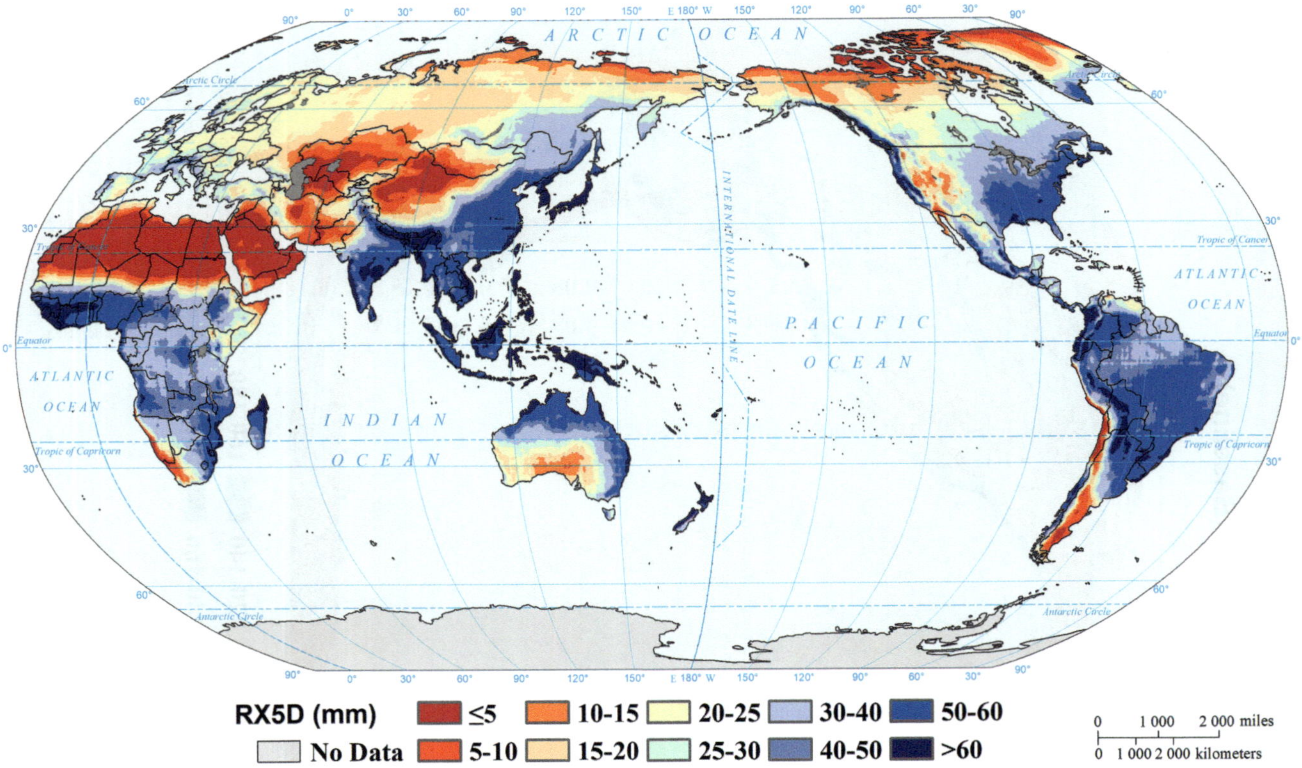

Annual maximum consecutive 5-day precipitation (2030s, RCP4.5)

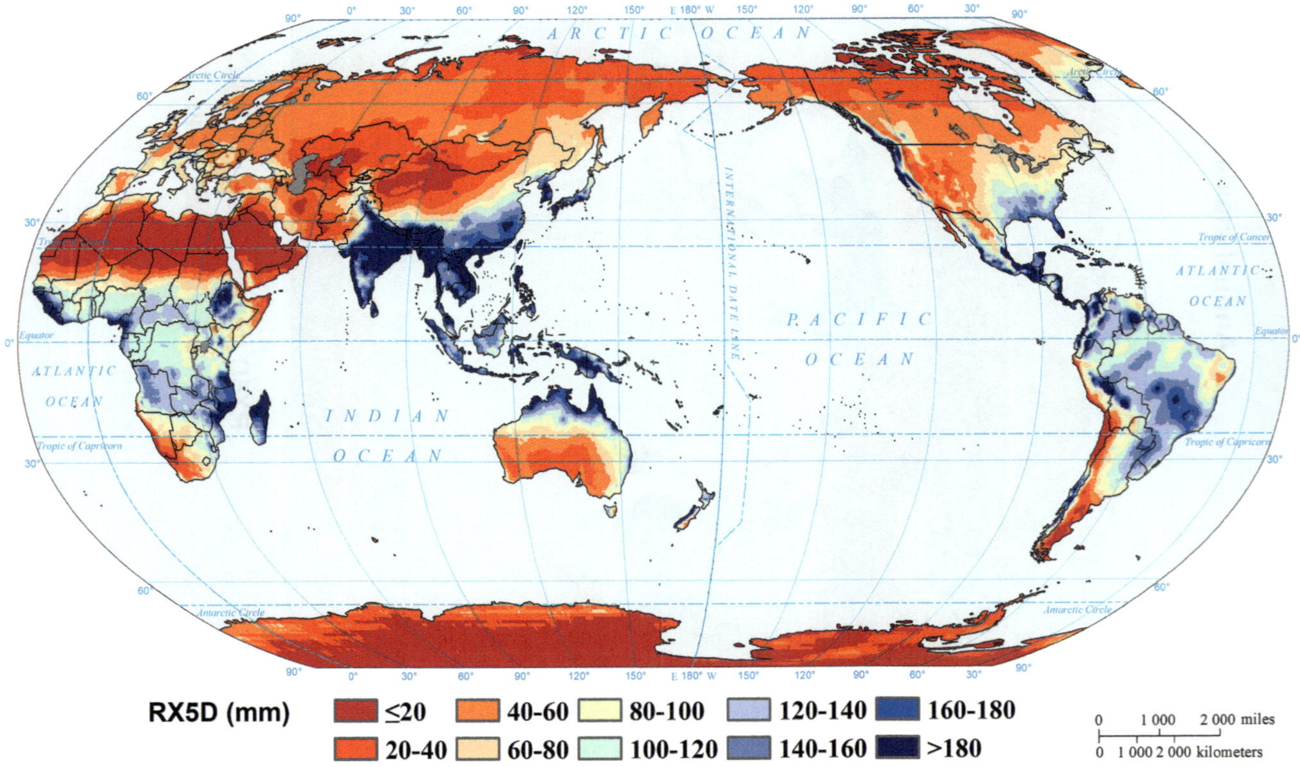

Annual maximum consecutive 5-day precipitation(2030s, RCP8.5)

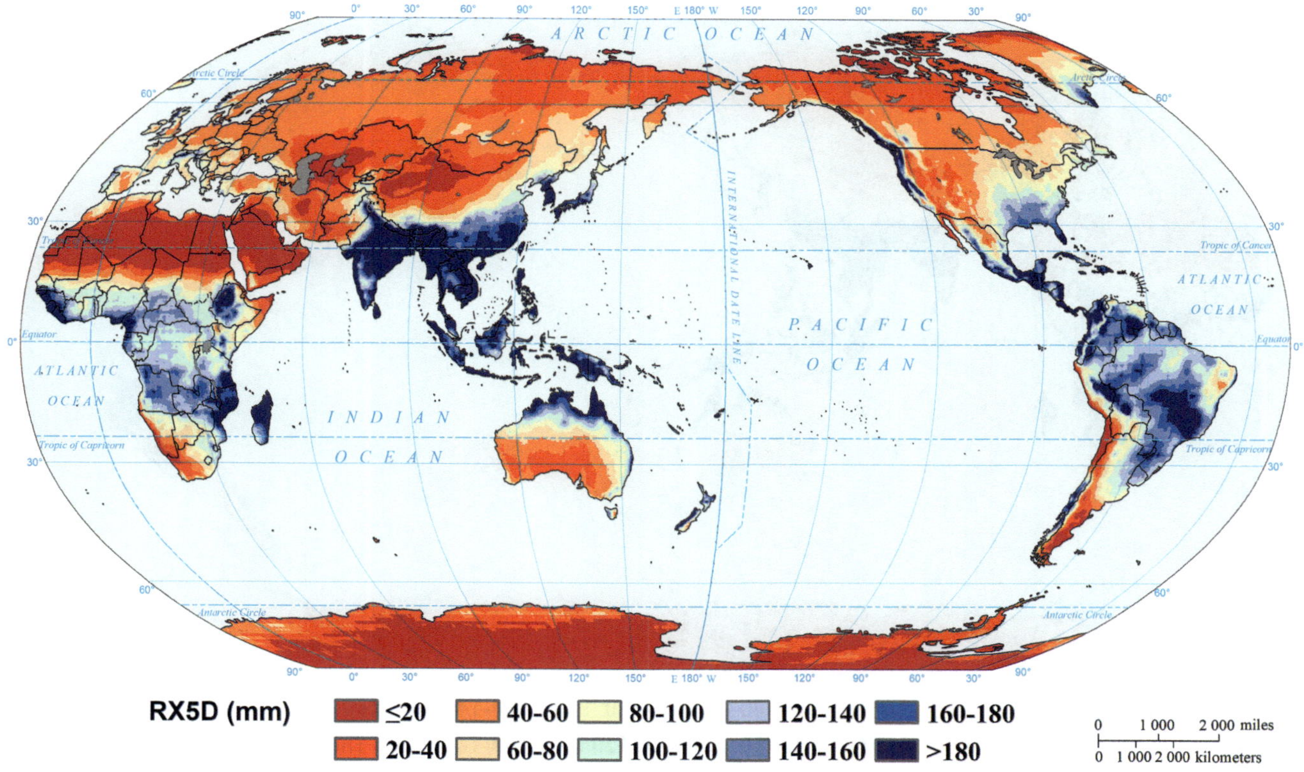

Annual maximum consecutive 5-day precipitation (2050s, RCP2.6)

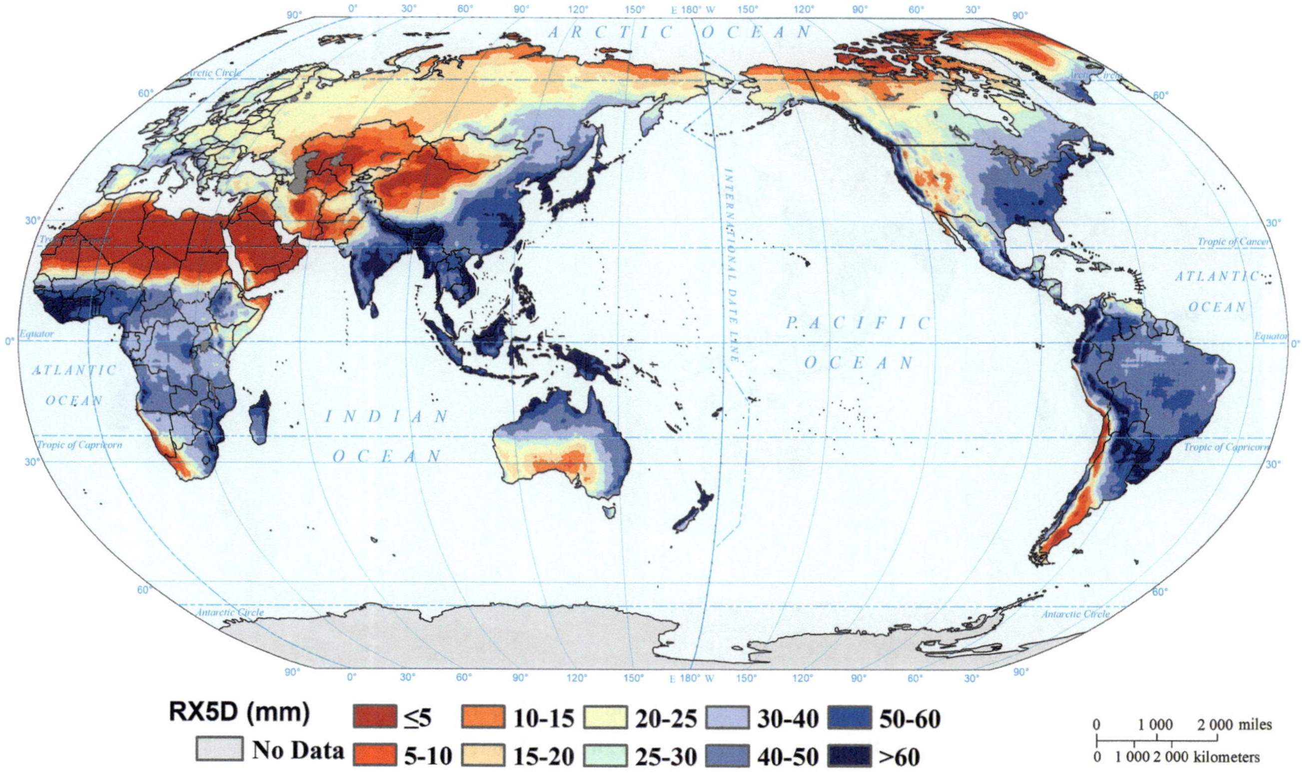

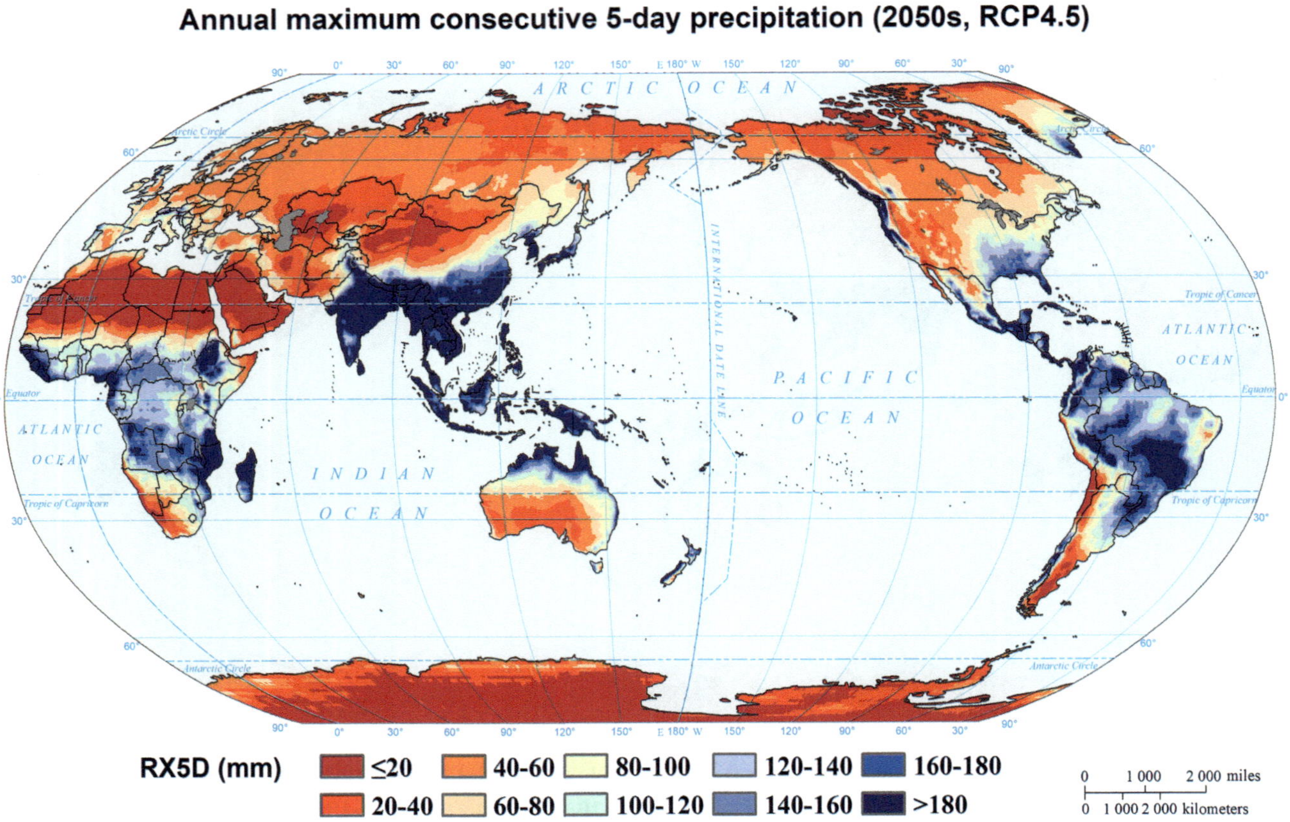
Annual maximum consecutive 5-day precipitation (2050s, RCP4.5)
ARCTIC OCEAN
PACIFIC OCEAN
INDIAN OCEAN
ATLANTIC OCEAN
INTERNATIONAL DATE LINE
Arctic Circle
Tropic of Cancer
Equator
Tropic of Capricorn
Antarctic Circle
RX5D (mm)
≤20
20-40
40-60
60-80
80-100
100-120
120-140
140-160
160-180
>180
0 1 000 2 000 miles
0 1 000 2 000 kilometers

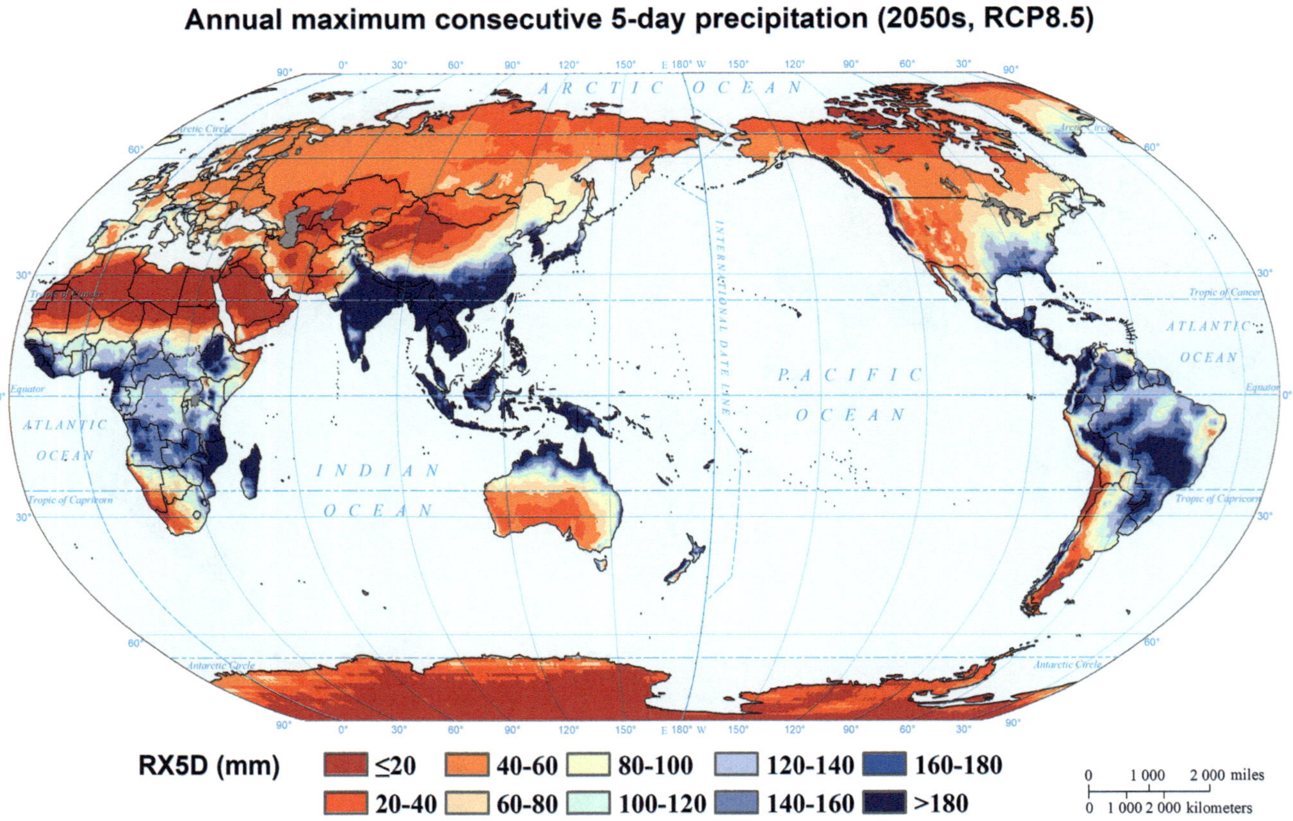
Annual maximum consecutive 5-day precipitation (2050s, RCP8.5)
ARCTIC OCEAN
PACIFIC OCEAN
INDIAN OCEAN
ATLANTIC OCEAN
INTERNATIONAL DATE LINE
Arctic Circle
Tropic of Cancer
Equator
Tropic of Capricorn
Antarctic Circle
RX5D (mm)
≤20
20-40
40-60
60-80
80-100
100-120
120-140
140-160
160-180
>180
0 1 000 2 000 miles
0 1 000 2 000 kilometers

References

Giorgi, F., and X. Bi. 2005. Regional changes in surface climate interannual variability for the 21st century from ensembles of global model simulations. *Geophysical Research Letters* 32: L13701.

Hartmann, D.L., and Coauthors. 2013. Observations: Atmosphere and surface. In *Climate Change 2013: The Physical Science Basis. Contribution of Working Group I to the Fifth Assessment Report of the Intergovernmental Panel on Climate Change*, ed. T.F. Stocker et al., 159–254. Cambridge, UK: Cambridge University Press.

IPCC. 2013. Climate change 2013: The physical science basis. In *Contribution of Working Group I to the Fifth Assessment Report of the Intergovernmental Panel on Climate Change*. Cambridge, UK, New York, USA: Cambridge University Press.

Klein Tank, A.M.G., F.W. Zwiers, and X. Zhang. 2009. Guidelines on analysis of extremes in a changing climate in support of informed decisions for adaptation. Climate data and monitoring WCDMP-No. 72, WMO-TD No. 1500.

Weber, T., A. Haensler, D. Rechid, S. Pfeifer, B. Eggert, and D. Jacob. 2018. Analyzing regional climate change in Africa in a 1.5, 2, and 3 degrees C global warming world. *Earth's Future* 6(4): 643–655.

Xu, L., and A. Wang. 2019. Application of the bias correction and spatial downscaling algorithm on the temperature extremes from CMIP5 multi-model ensembles in China. *Earth and Space Science* 6 (12): 2508–2524.

Xu, L., A. Wang, D. Wang, and H. Wang. 2019. Hot spots of climate extremes in the future. *Journal of Geophysical Research: Atmospheres* 124: 3035–3049.

Zhang, X., L. Alexander, G.C. Hegerl, P. Jones, A.K. Tank, T.C. Peterson, B. Trewin, and F.W. Zwiers. 2011. Indices for monitoring changes in extremes based on daily temperature and precipitation data. *Wires Climate Change* 2: 851–870.

Mapping Wind Speed Changes

Rui Mao, Cuicui Shi, Qi Zong, Xingya Feng, Yijie Sun, Yufei Wang, and Guohao Liang

1 Introduction

Wind variability has a major impact on water cycles, wind energy, and natural hazards and disasters such as hurricanes and typhoons. In the past decades, the global and regional mean near-surface wind speed (sfcWind) has shown a significantly downward trend, especially in the mid-latitudes in the Northern Hemisphere (Wu et al. 2018). The global mean terrestrial sfcWind has decreased linearly at a rate of 0.08 m s^{-1} per decade during 1981–2011. Vautard et al. (2010) analyzed changes in sfcWind at stations across the globe and found that 73% of the total stations presented a decrease in the annual mean sfcWind, with linear trends of −0.09, −0.16, −0.12, and −0.07 m s^{-1} per decade in Europe, Central Asia, East Asia, and North America, respectively. In addition, a pronounced reduction in extreme sfcWind has also been observed in Europe and the United States (Yan et al. 2002; Pryor et al. 2012). However, because long-term observational data are lacking in most land areas and oceans, the uncertainty in the long-term trend of sfcWind is high, particularly in the Southern Hemisphere and over oceans.

The Coupled Model Intercomparison Project Phase 5 (CMIP5) aims to understand the past climate changes, make projections, and estimate future uncertainties. For example, with respect to the historical period (1979–2005), the multi-model ensemble mean of simulations in the CMIP5 projects a decrease in the annual maximum wind speeds at the end of this century (2074–2100) over the contiguous United States and most regions of Asia in the high latitudes and an increase in the annual maximum wind speeds over Amazonia, India, Africa, and Southeast Asia, whereas few changes are projected to occur over Europe (Kumar et al. 2015). It is worth noting that studies on extreme wind speed in Asia, Africa, and South America as well as those on mean sfcWind over the Southern Hemisphere by using the CMIP5 simulations are lacking. Moreover, the uncertainty analysis in regional mean sfcWind and extreme wind speed in the CMIP5 results is ignored in previous studies. For example, the CMIP5 models fail to represent the current wind climate over the Bay of Bengal (Krishnan and Bhaskaran 2019) and show relatively poor skill for the long-term temporal trends over the Northern Hemisphere (Tian et al. 2019).

To achieve a comprehensive understanding of the changes in global sfcWind and extreme wind speed in the future and their uncertainties in the results of CMIP5 models, this section initiatively assesses the changes of the mean state, variance, and extreme conditions of sfcWind under three greenhouse gas emission scenarios (Representative Concentration Pathway (RCP2.6, RCP4.5, RCP8.5).

Authors: Rui Mao, Cuicui Shi, Qi Zong, Xingya Feng, Yijie Sun, Yufei Wang, Guohao Liang.
Map Designers: Fanya Shi, Yelin Sun, Tian Liu, Jing'ai Wang, Ying Wang.
Language Editor: Rui Mao.

R. Mao (✉) · C. Shi · Q. Zong · X. Feng · Y. Sun · Y. Wang · G. Liang
Faculty of Geographical Science, Beijing Normal University, Beijing, 100875, China
e-mail: mr@bnu.edu.cn

R. Mao · C. Shi · Q. Zong · X. Feng · Y. Sun · Y. Wang · G. Liang
State Key Laboratory of Earth Surface Processes and Resource Ecology, Faculty of Geographical Science, Beijing Normal University, Beijing, 100875, China

R. Mao · C. Shi · Q. Zong · X. Feng · Y. Sun · Y. Wang · G. Liang
Key Laboratory of Environmental Change and Natural Disaster, Ministry of Education, Beijing Normal University, Beijing, 100875, China

P. Shi, *Atlas of Global Change Risk of Population and Economic Systems*, IHDP/Future Earth-Integrated Risk Governance Project Series, https://doi.org/10.1007/978-981-16-6691-9_4

2 Data

The global average daily sfcWind and daily maximum sfcWind data are provided by the Coupled Model Intercomparison Project Phase 5 (CMIP5) from the World Climate Research Programme (WRCP), where daily sfcWind includes 24 models for the historical period and RCP4.5 and RCP8.5 scenarios and 16 models for the RCP2.6 scenario, and daily maximum sfcWind includes 19 models for historical period and RCP4.5 and RCP8.5 scenarios and 11 models for RCP2.6 scenario. The spatial resolution of the data is 0.5 degrees ($\sim$50 km $\times$ 50 km). The CMIP5 data can be freely accessed on this website: https://esgf-node.llnl.gov/search/cmip5/.

3 Method

We divided the temporal range into three periods, including the historical period (1986–2005, denoted as the 2000s), and two future time periods: 2016–2035 (2030s) and 2046–2065 (2050s). In this study, summer is defined as June–July–August (JJA), and winter is defined as December and January–February of the following year (DJF).

The mean near-surface wind speed (sfcWind) is defined as the 20-year mean of all daily near-surface wind speed for the historical period and the two future time periods under the three RCP scenarios. The sfcWind change was calculated by subtracting the sfcWind during the historical period (2000s) from that under the RCP scenarios in the future temporal ranges. The model uncertainty of the mean sfcWind is represented by the standard deviation of the mean sfcWind across the models for the three periods.

The near-surface wind speed variability (sfcWind std) is defined as the standard deviation of the daily mean near-surface wind speed during the three periods. The change in sfcWind variability was calculated by subtracting the sfcWind std during the historical period (2000s) from the sfcWind std under the RCP scenarios in the future temporal ranges. The model uncertainty of the sfcWind std is represented by the standard deviation of the sfcWind std for the three periods across the models.

The maximum near-surface wind speed (sfcWindmax) mean is defined as the 20-year mean of all daily maximum near-surface wind speed for the historical period and the two future time periods under the three RCP scenarios. The sfcWindmax change was calculated by subtracting the sfcWindmax during the historical period (2000s) from the sfcWindmax under the RCP scenarios in the future temporal ranges. The model uncertainty of the mean sfcWindmax is represented by the standard deviation of the mean sfcWindmax for the three periods across the models.

The maximum near-surface wind speed variability (sfcWindmax std) is defined as the standard deviation of the daily mean maximum near-surface wind speed during the three periods. The change in sfcWindmax variability was calculated by subtracting the sfcWindmax std during the historical period (2000s) from the sfcWindmax std under the RCP scenarios in the future temporal ranges. The model uncertainty of the sfcWindmax std is represented by the standard deviation of the sfcWindmax std for the three periods across the models.

Because of the varying spatial resolutions of the simulations in the CMIP5, the variables are interpolated by using a bilinear interpolation with a high spatial resolution of $0.5° \times 0.5°$.

4 Major Findings

Surface wind speed, surface wind speed variability, and surface maximum wind speed in nine global regions under climate change are shown in Figs. 1, 2 and 3. In general, under the climate warming scenarios, the surface wind speed increases in North Asia, the Mediterranean Basin, Tibet, and Eastern North America, and decreases in low latitude areas, such as Northern Australia and the Amazon Basin. Compared with the historical period, the interannual variability of wind speed will increase. However, with climate warming, the interannual variation of surface wind speed will be different in different regions—it is projected to decrease in the Mediterranean Basin and Tibet, and increase in the Amazon Basin and Southeast Asia. The overall change of the surface maximum wind speed is not obvious, and only in Tibet and the Mediterranean Basin it will decrease with the climate change.

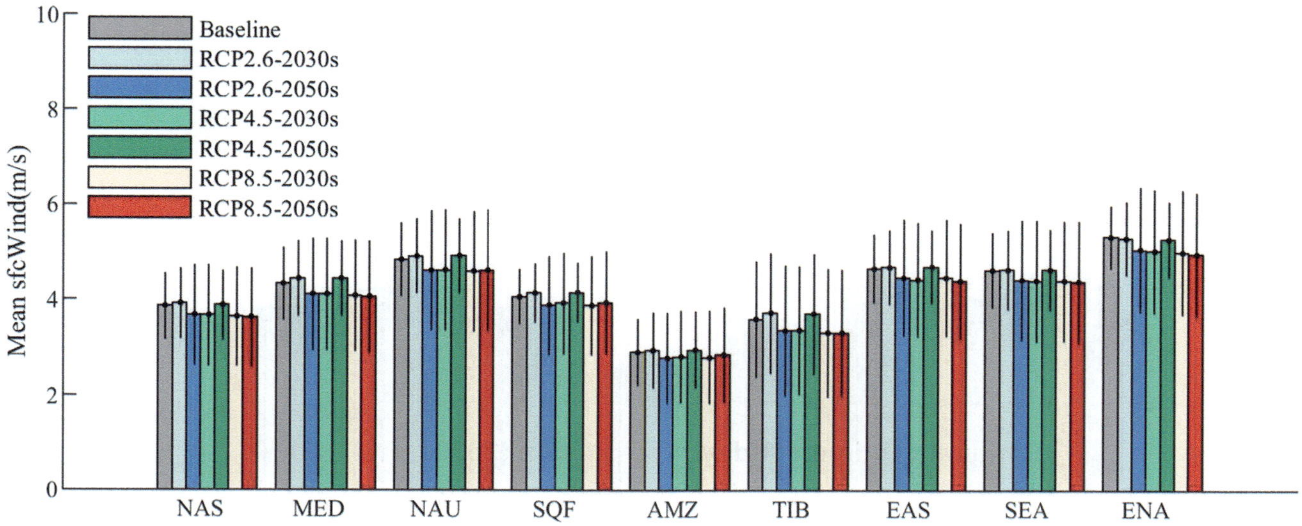

Fig. 1 Mean surface wind speed (unit: m/s) in nine regions under different Representative Concentration Pathway (RCP) scenarios. The error bar represents the one standard deviation across 16 general circulation models (GCMs) (RCP2.6) and 24 GCMs (RCP4.5 and RCP8.5). NAS, MED, NAU, SQF, AMZ, TIB, EAS, SEA, and ENA represent North Asia (47–70°N, 60.5–180.5°E), Mediterranean Basin (30–47°N, 10.5°W–37.5°E), Northern Australia (28–10°S, 109.5–155.5°E), South Equatorial Africa (26–0°S, 0.5–55.5°E), Amazon Basin (20°S–10°N, 78.5–34.5°W), Tibet (30–47°N, 80.5–104.5°E), East Asia (20–47°N, 104.5–140.5°E), Southeast Asia (10°S–20°N, 100.5–150.5°E), and Eastern North America (25–50°N, 85.5–60.5°W), respectively. Regional division follows Giorgi and Bi (2005)

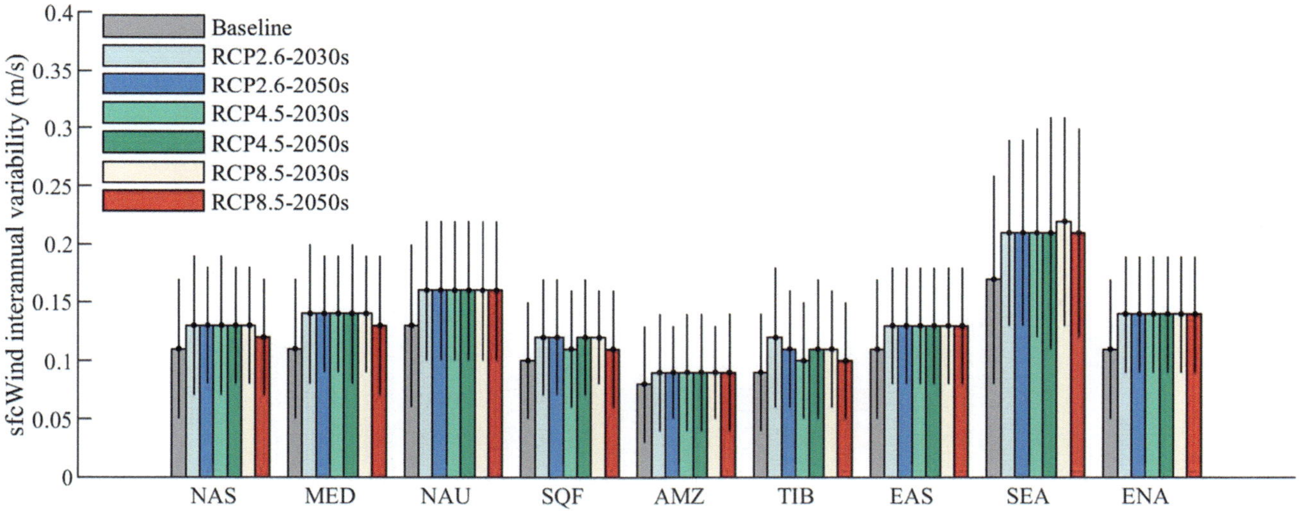

Fig. 2 Variability of mean surface wind speed (unit: m/s) in nine regions under different Representative Concentration Pathway (RCP) scenarios. The error bar represents the one standard deviation across 16 general circulation models (GCMs) (RCP2.6) and 24 GCMs (RCP4.5 and RCP8.5). Region abbreviations are the same as in Fig. 1

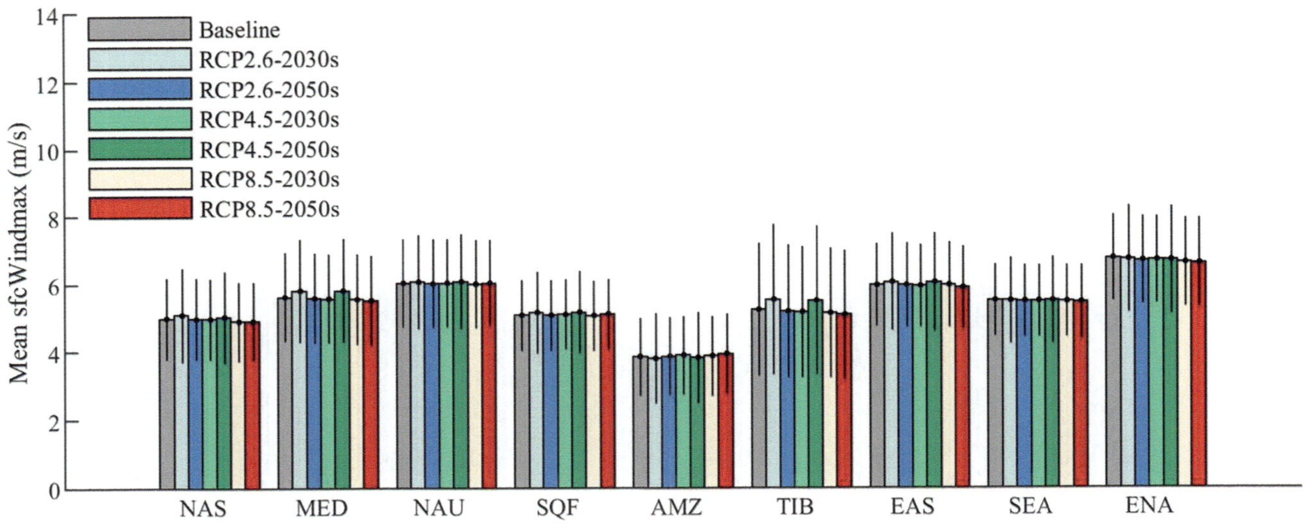

Fig. 3 Surface maximum wind speed (unit: m/s) in nine regions under different Representative Concentration Pathway (RCP) scenarios. The error bar represents the one standard deviation across 11 general circulation models (GCMs) (RCP2.6) and 19 GCMs (RCP4.5 and RCP8.5). Region abbreviations are the same as in Fig. 1

Annual mean wind speed (Baseline)

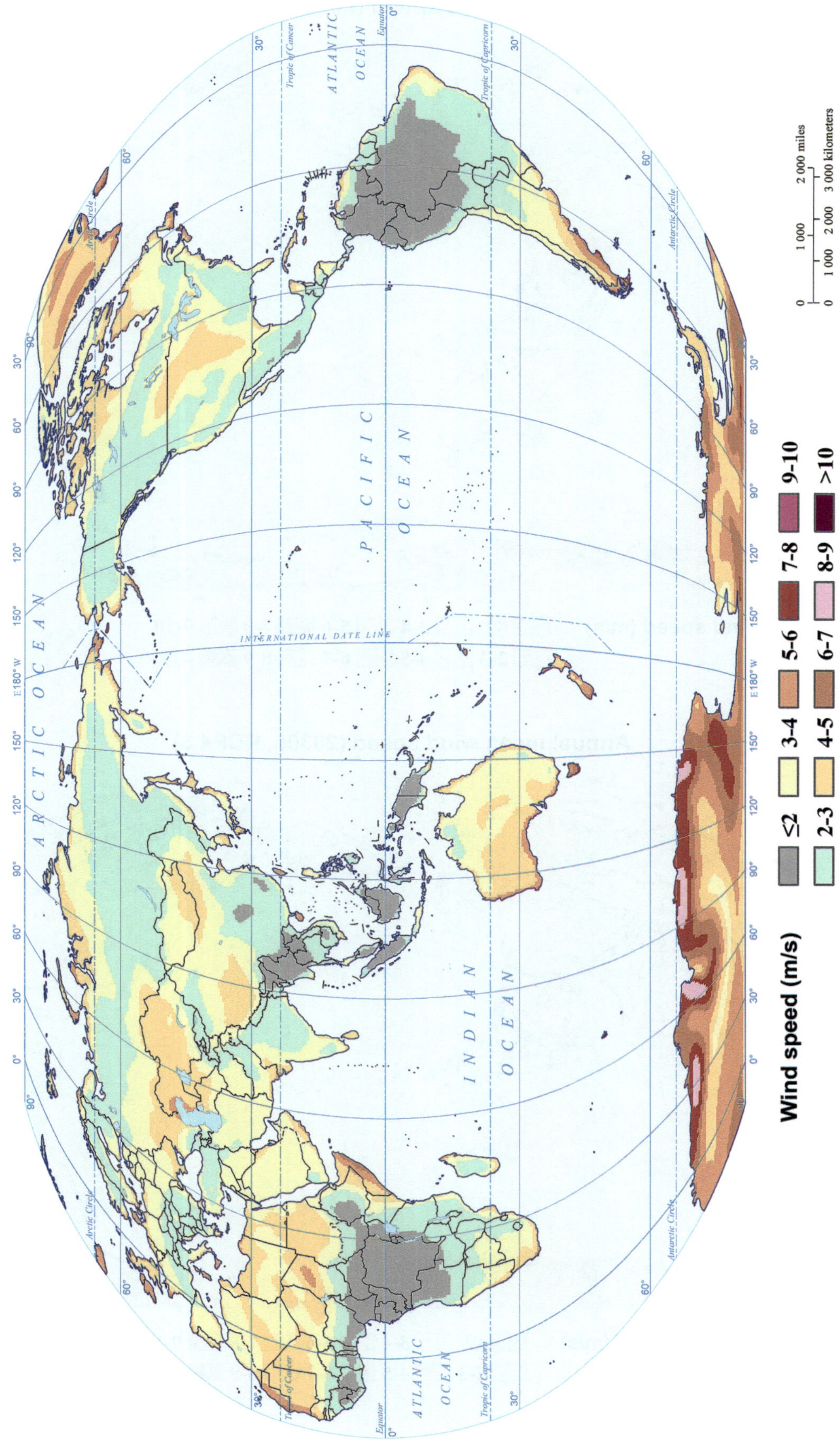

Annual mean wind speed (2030s, RCP2.6)

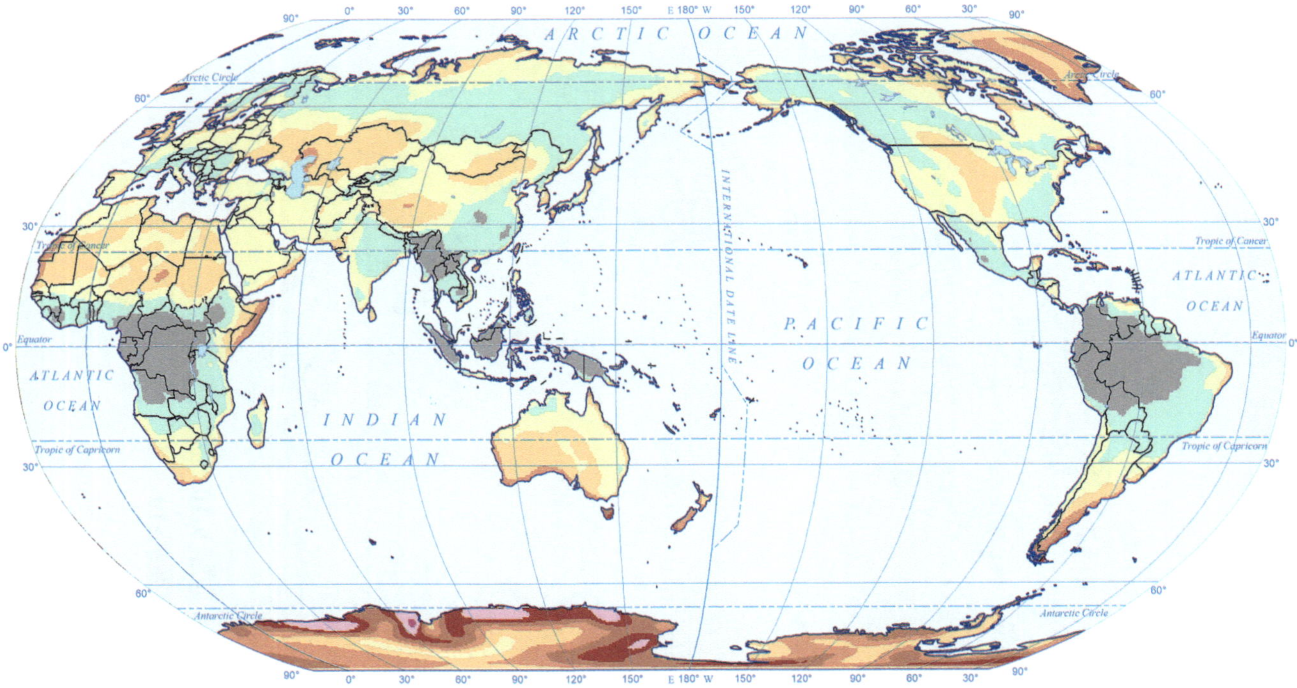

Wind speed (m/s) ≤2 2-3 3-4 4-5 5-6 6-7 7-8 8-9 9-10 >10

0 1 000 2 000 miles
0 1 000 2 000 kilometers

Annual mean wind speed (2030s, RCP4.5)

f

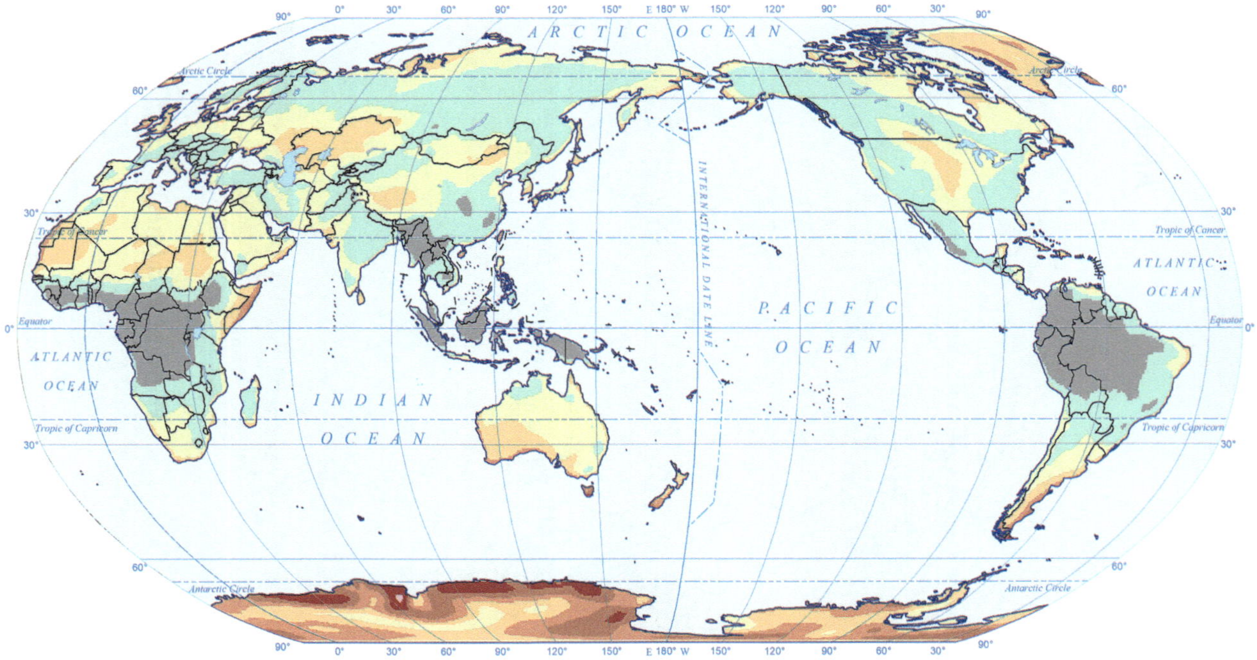

Wind speed (m/s) ≤2 2-3 3-4 4-5 5-6 6-7 7-8 8-9 9-10 >10

0 1 000 2 000 miles
0 1 000 2 000 kilometers

Annual mean wind speed (2030s, RCP8.5)

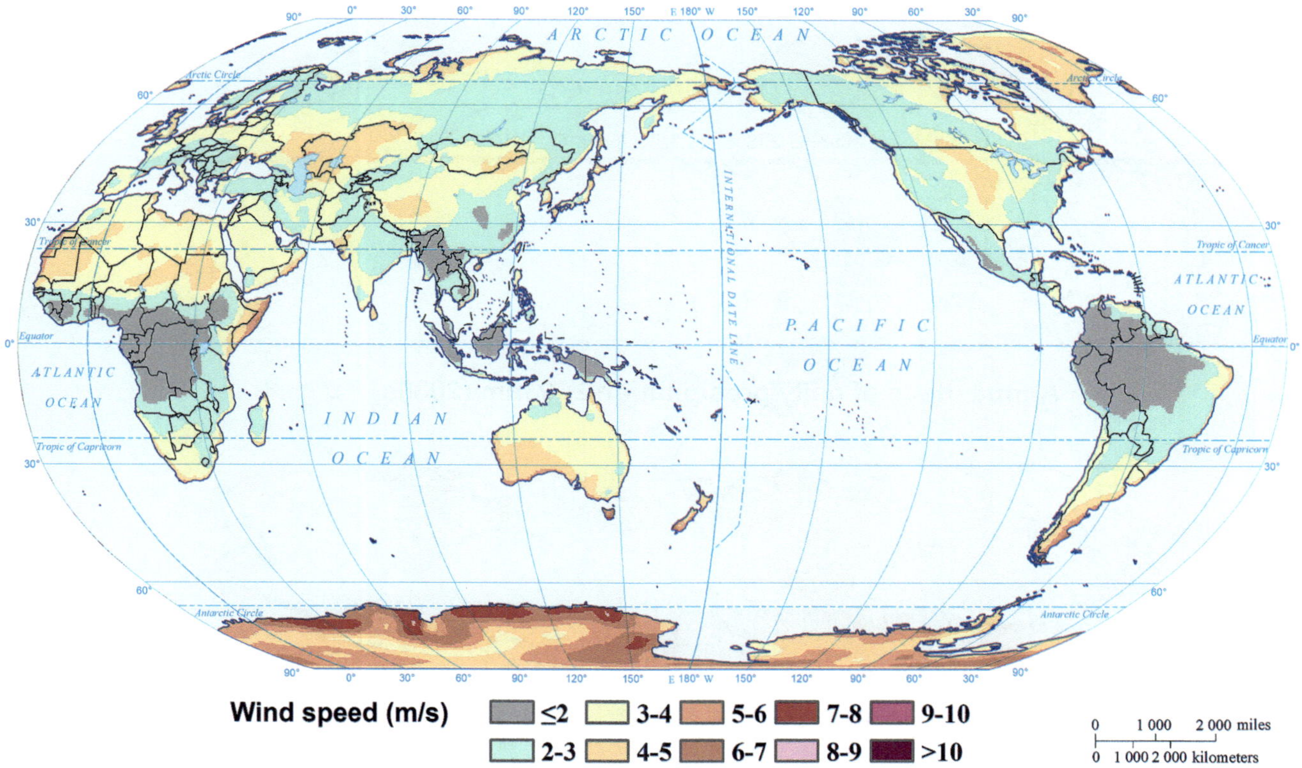

Annual mean wind speed (2050s, RCP2.6)

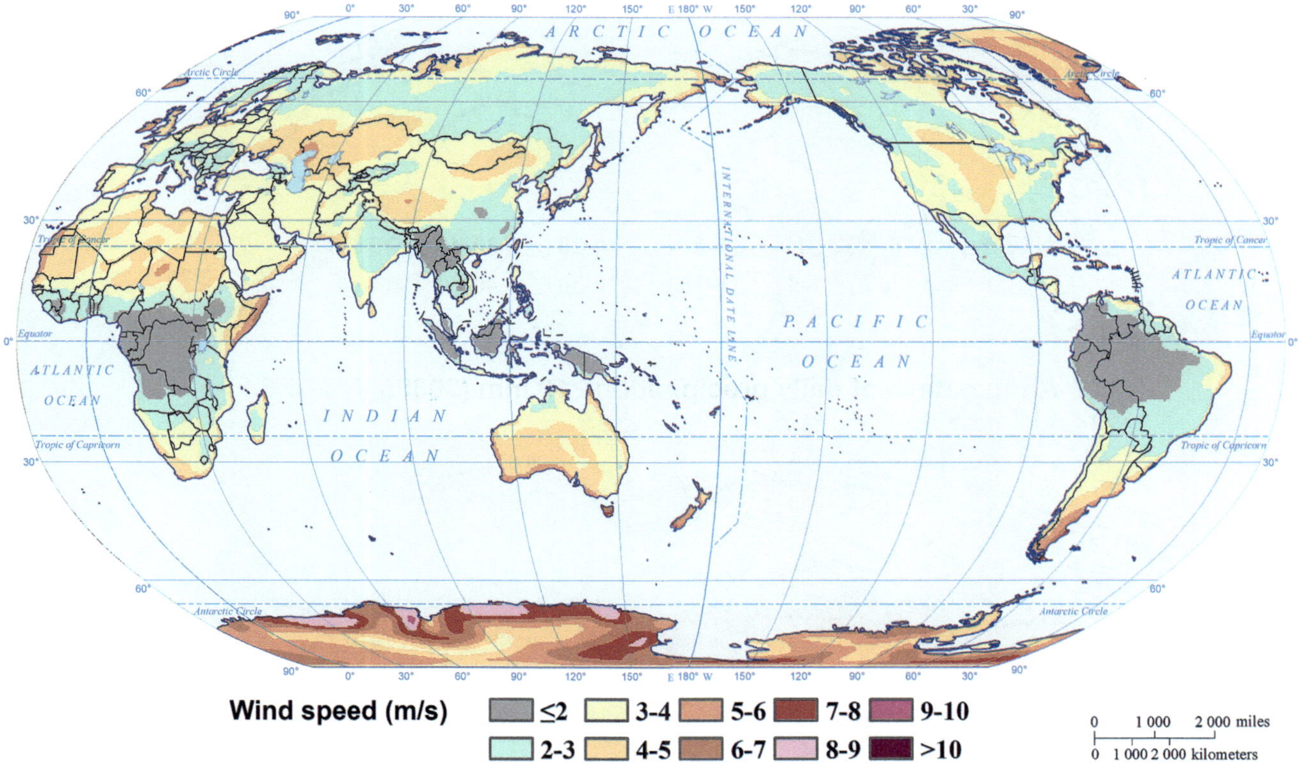

Annual mean wind speed (2050s, RCP4.5)

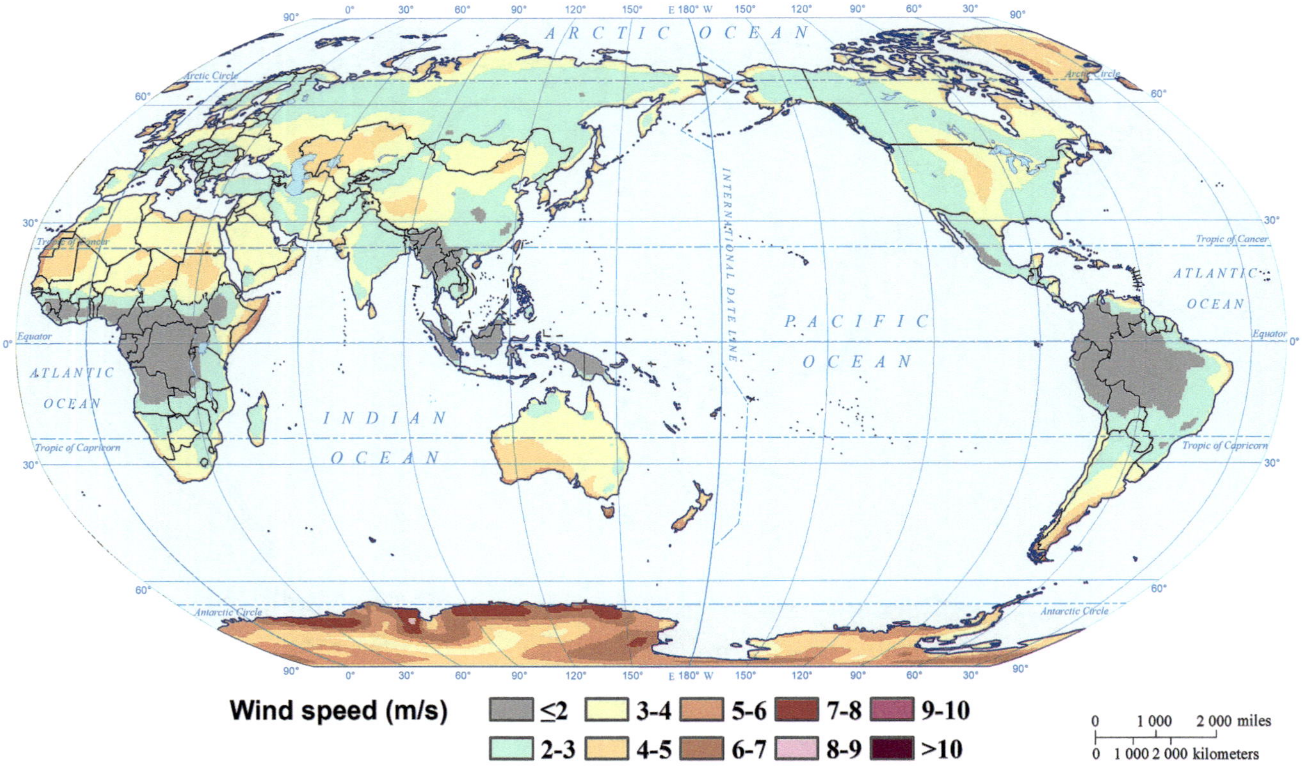

Annual mean wind speed (2050s, RCP8.5)

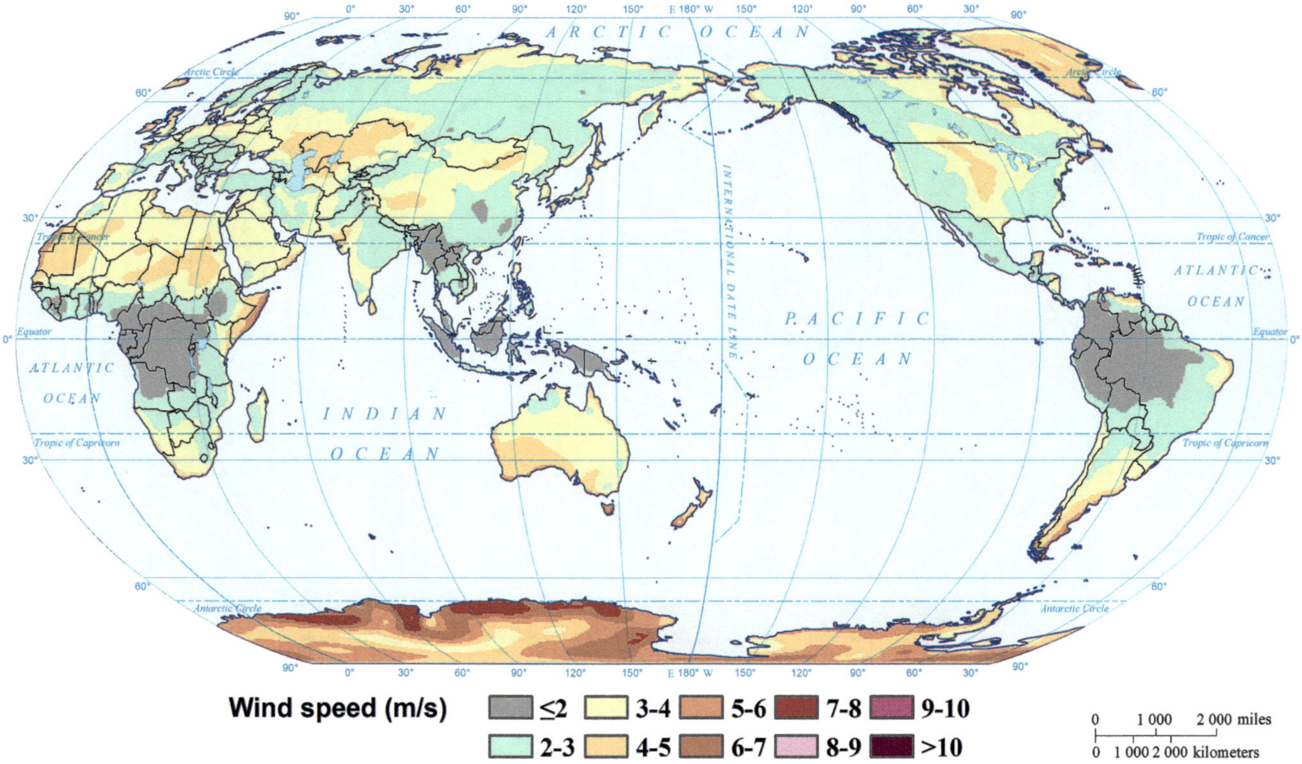

Variability of annual average near-surface wind speed (Baseline)

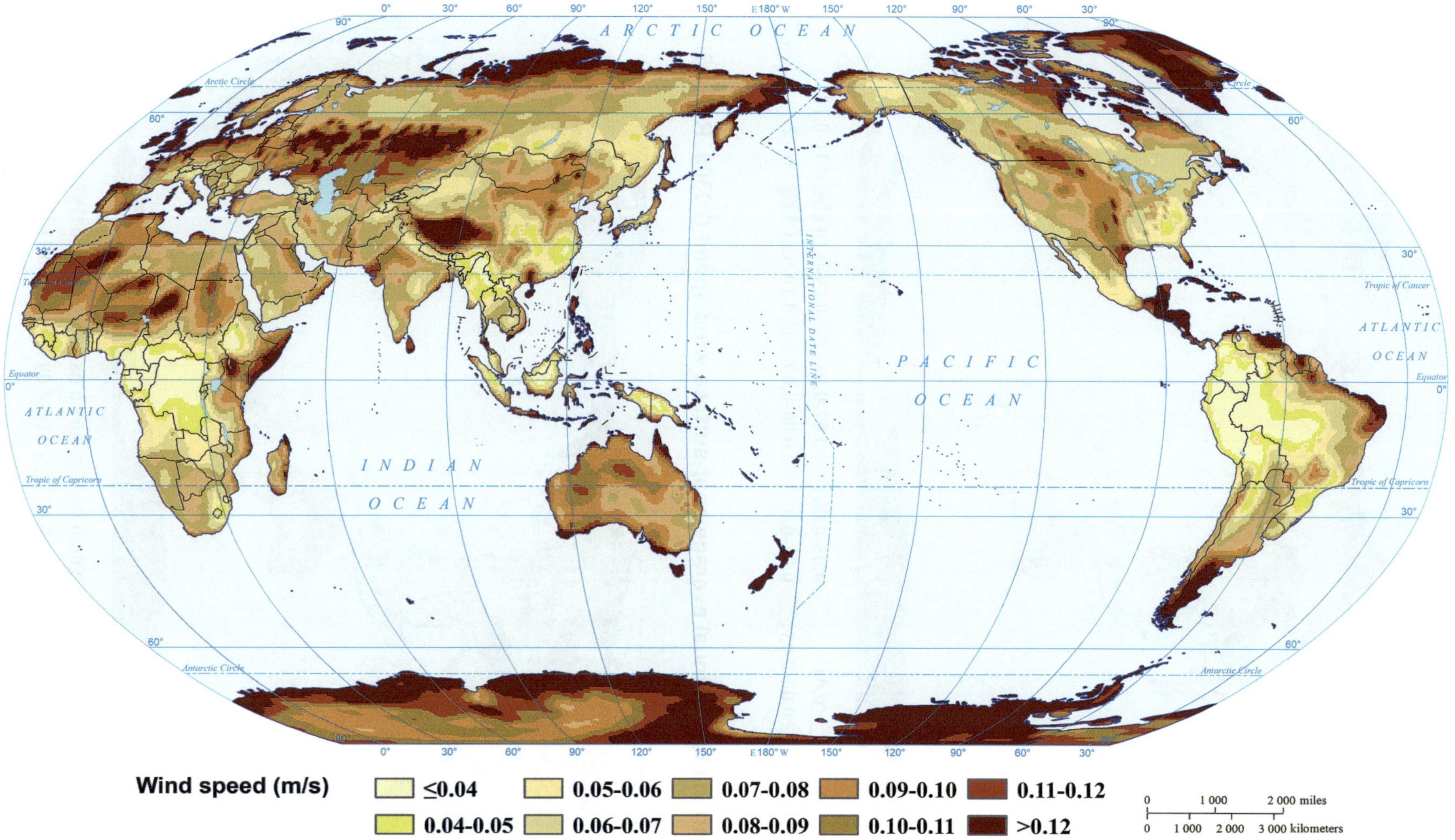

Variability of annual average near-surface wind speed (2030s, RCP2.6)

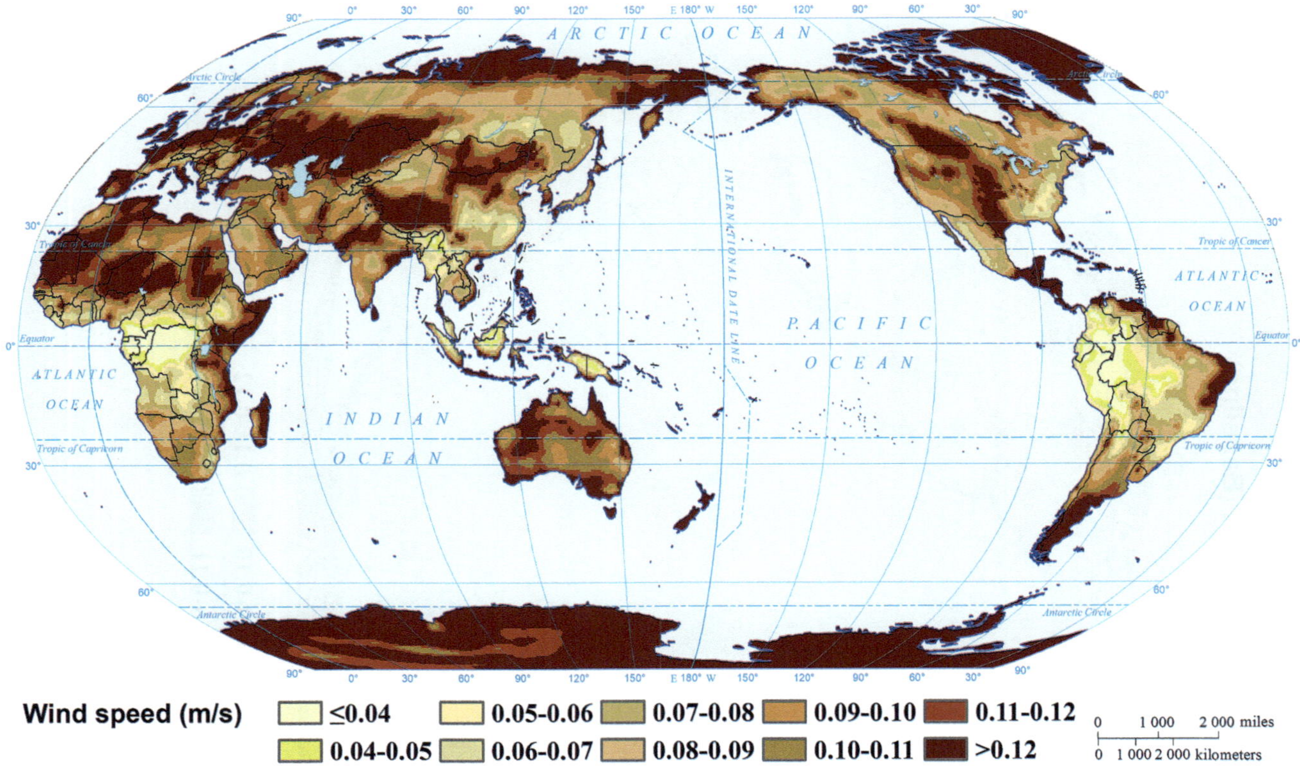

Variability of annual average near-surface wind speed (2030s, RCP4.5)

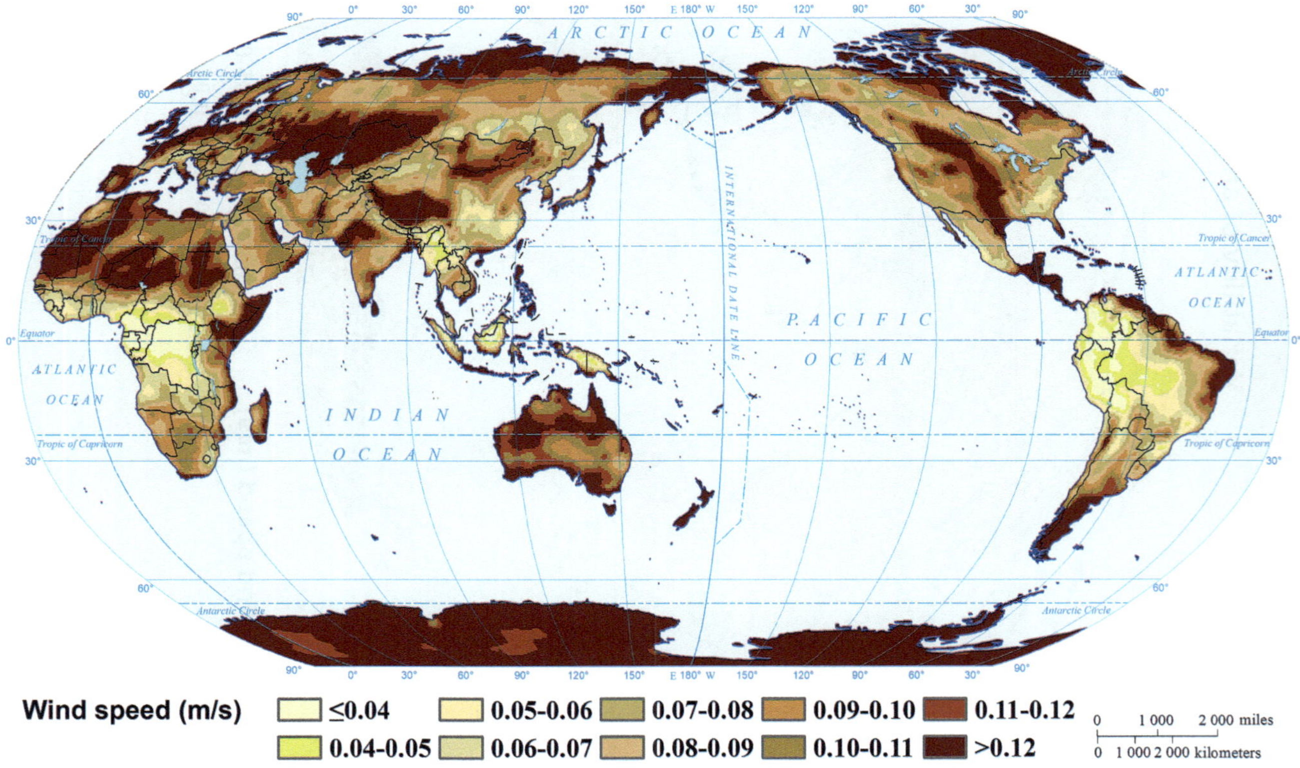

Variability of annual average near-surface wind speed (2030s, RCP8.5)

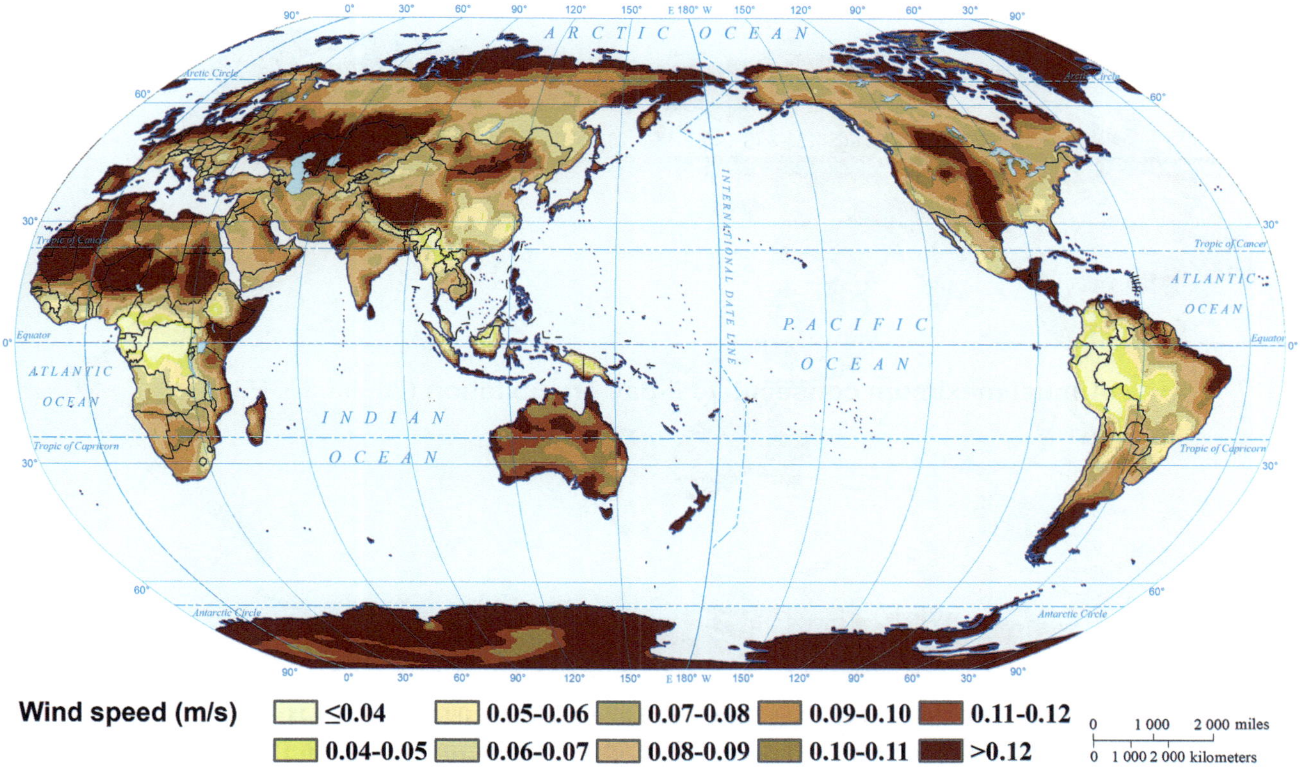

Variability of annual average near-surface wind speed (2050s, RCP2.6)

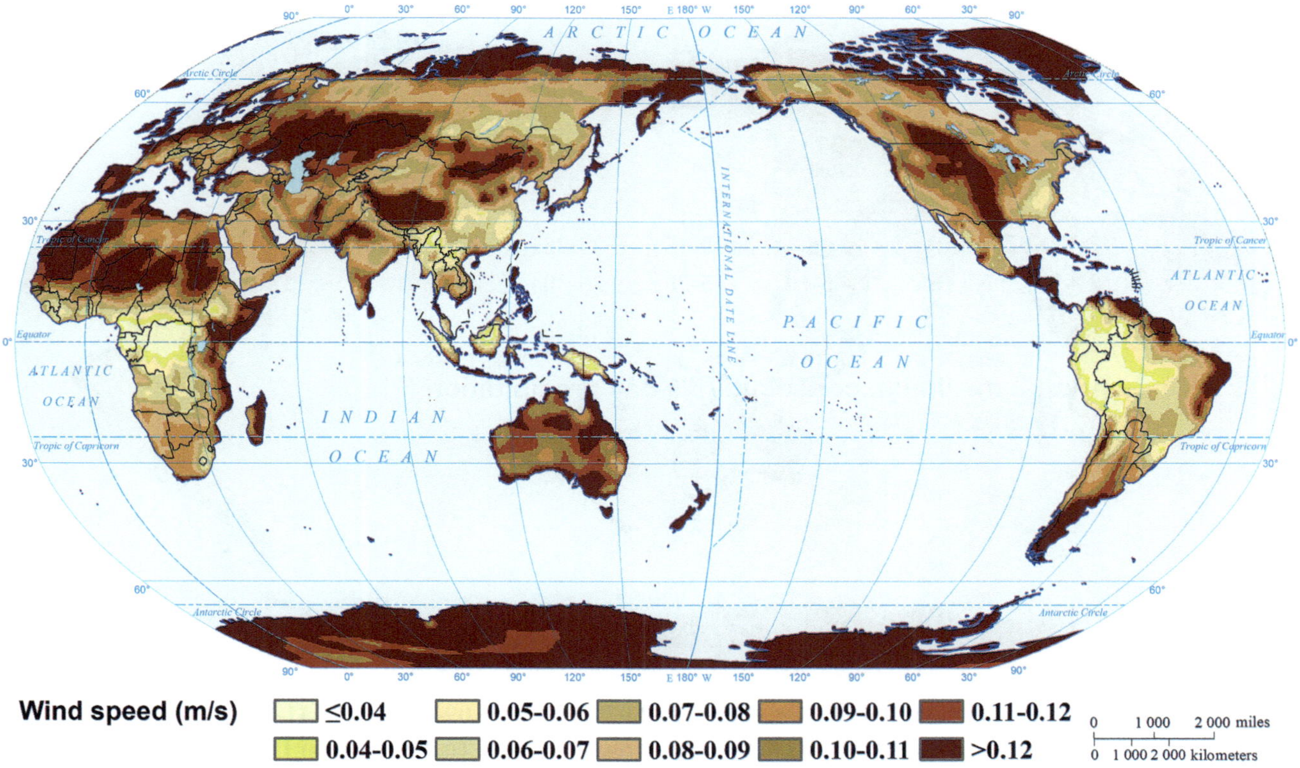

Variability of annual average near-surface wind speed (2050s, RCP4.5)

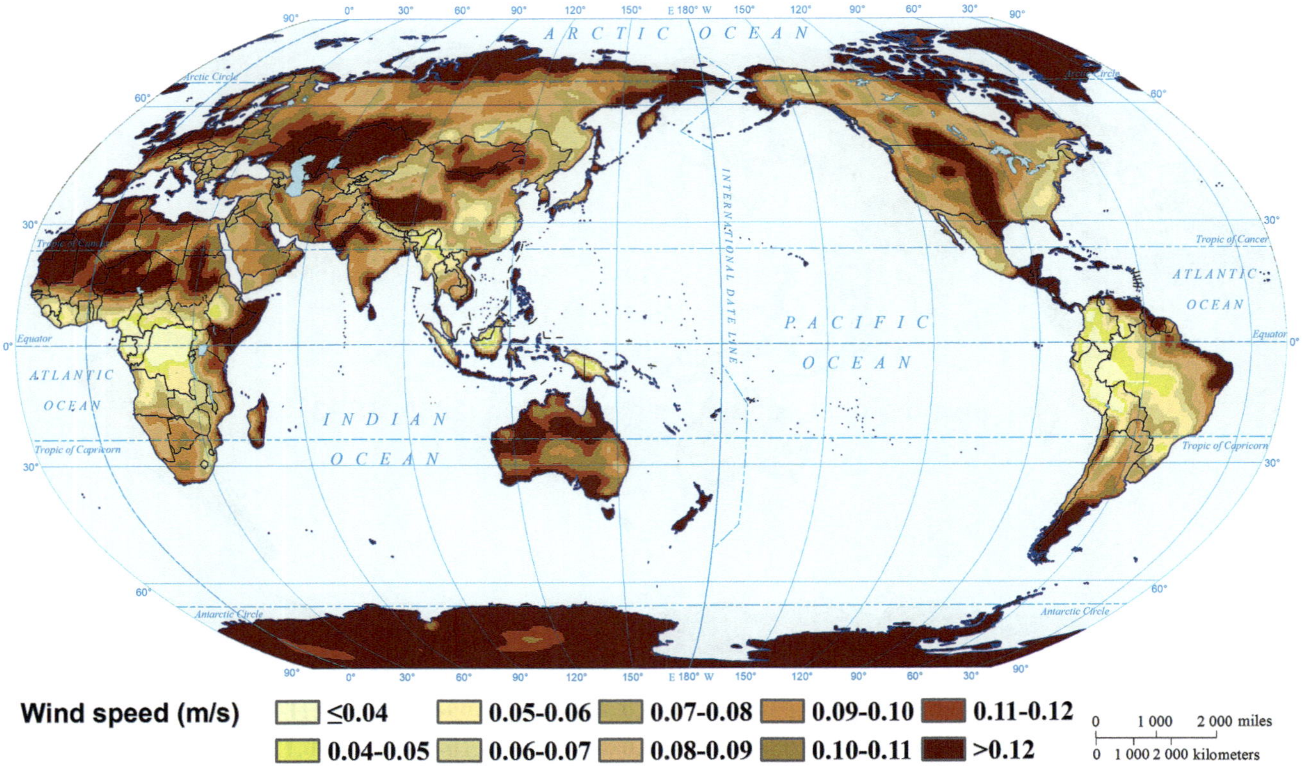

Variability of annual average near-surface wind speed (2050s, RCP8.5)

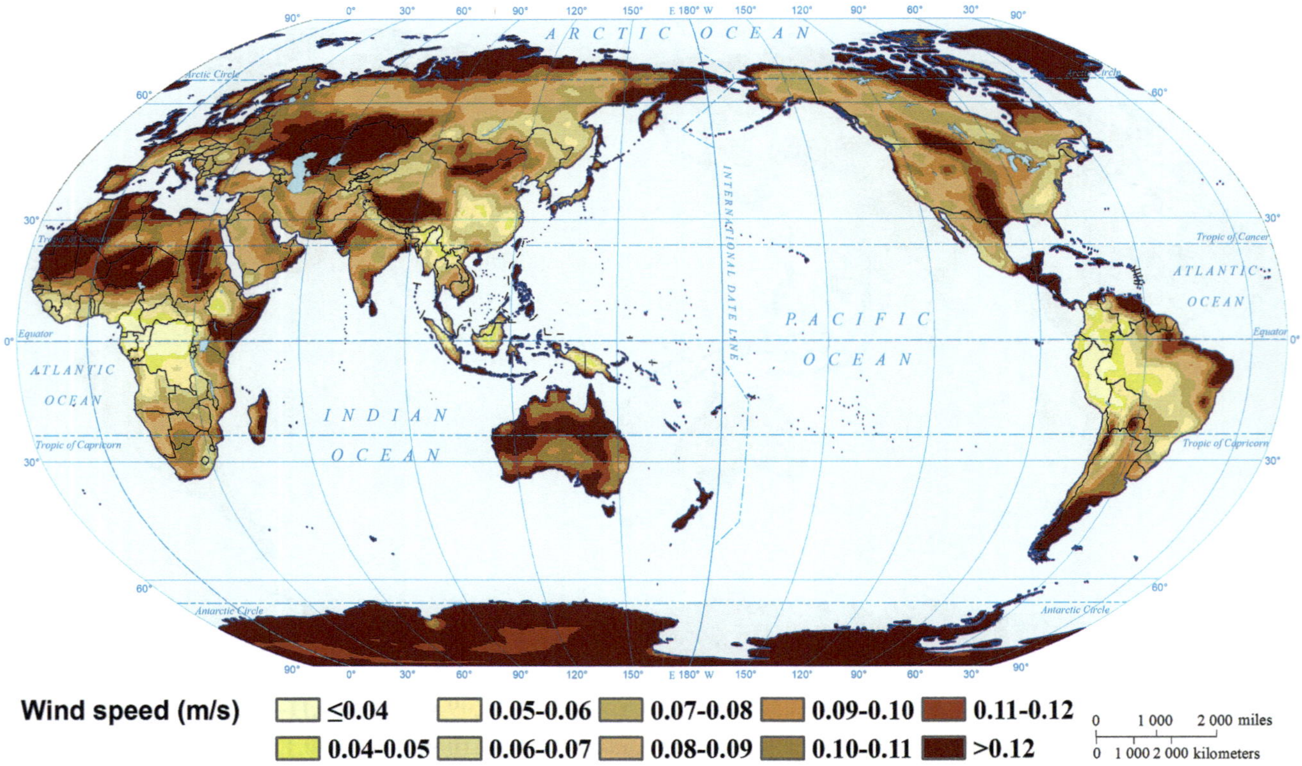

Annual maximum wind speed (Baseline)

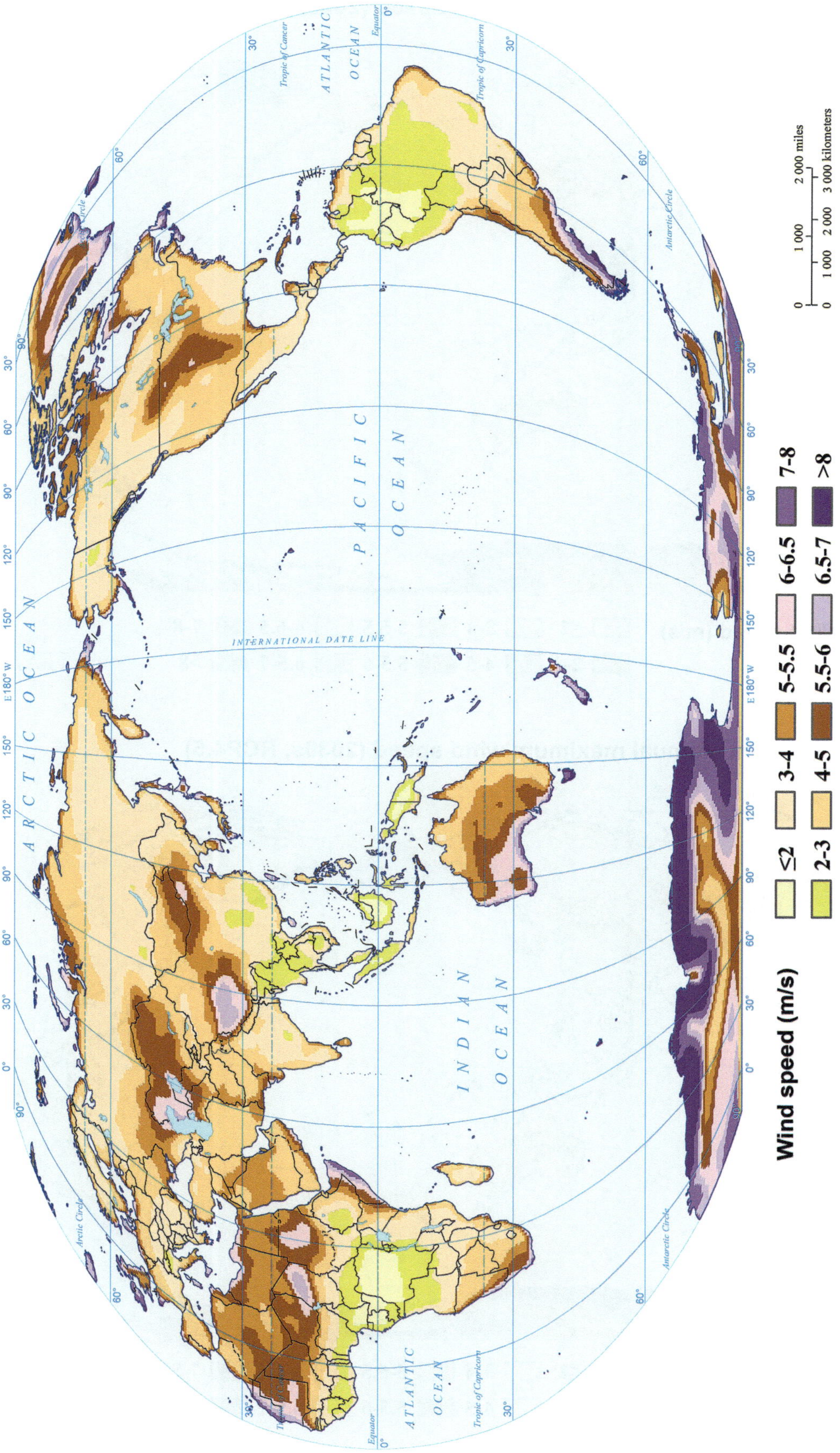

Annual maximum wind speed (2030s, RCP2.6)

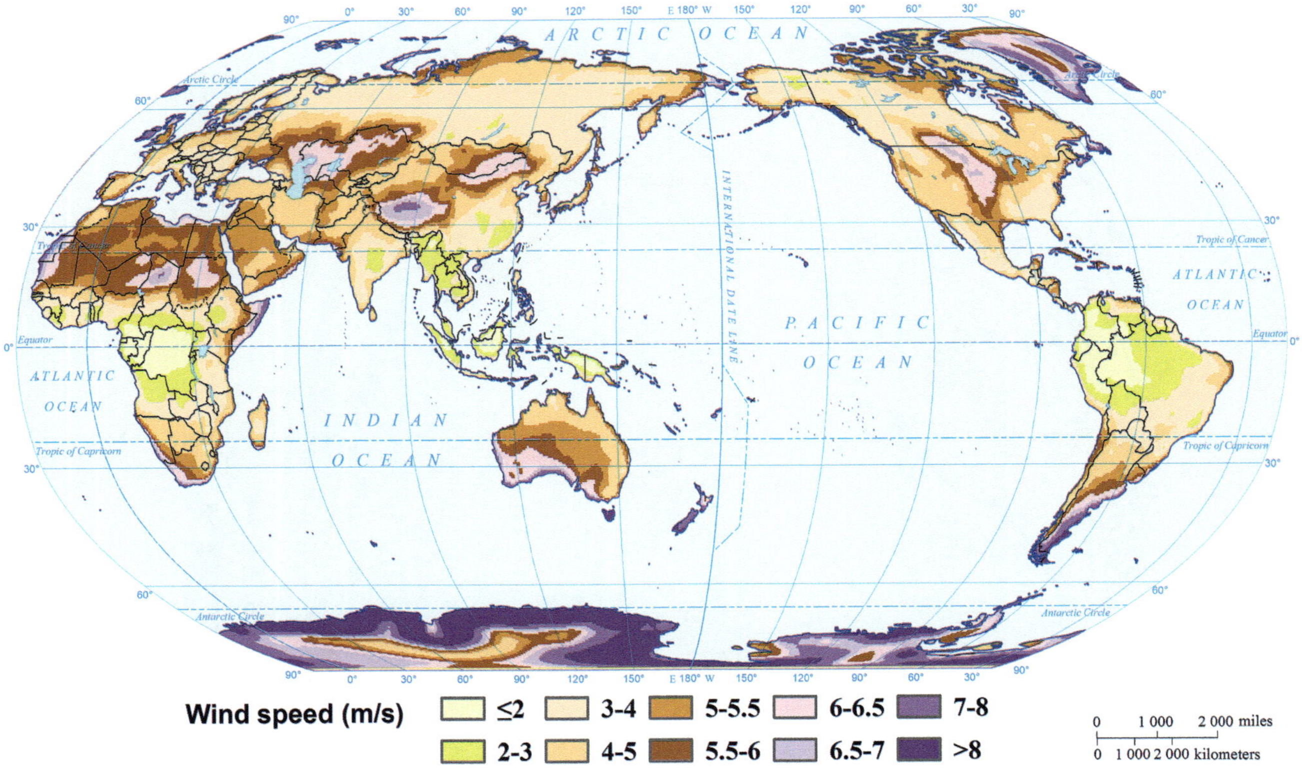

Annual maximum wind speed (2030s, RCP4.5)

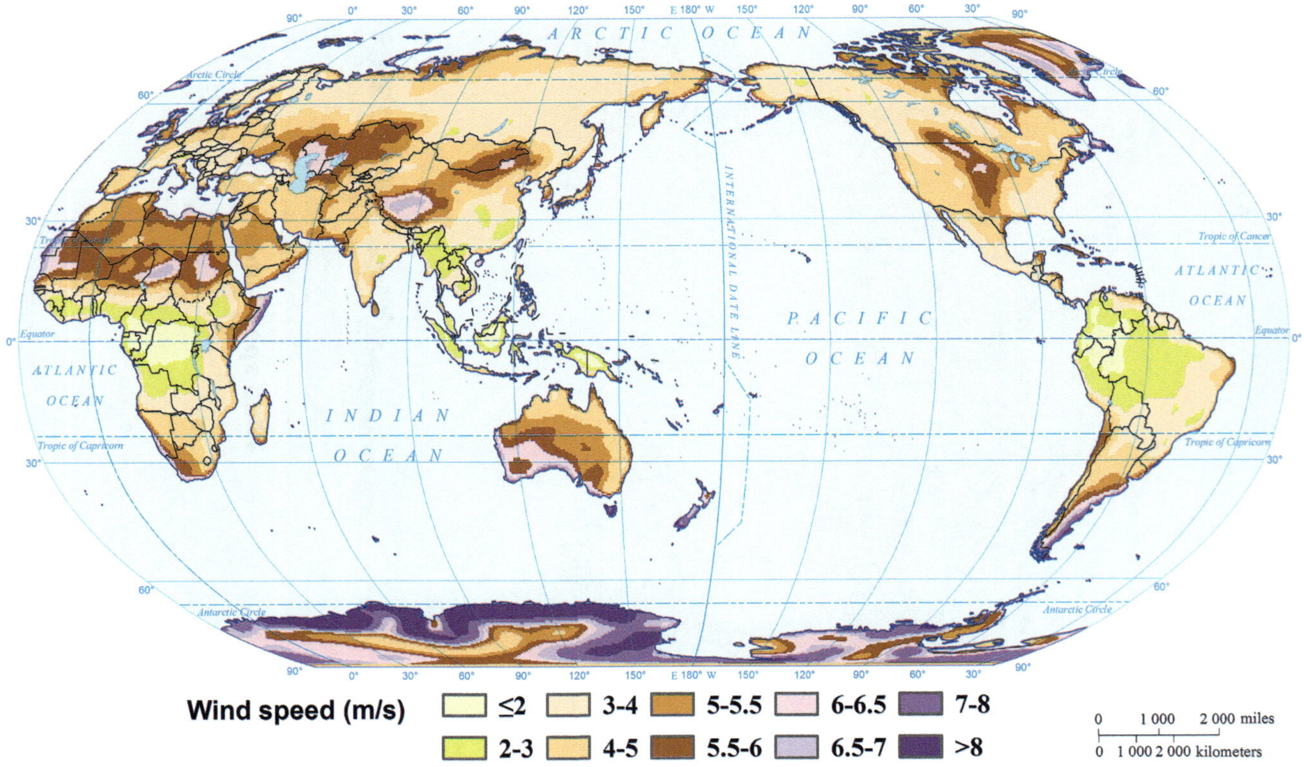

Annual maximum wind speed (2030s, RCP8.5)

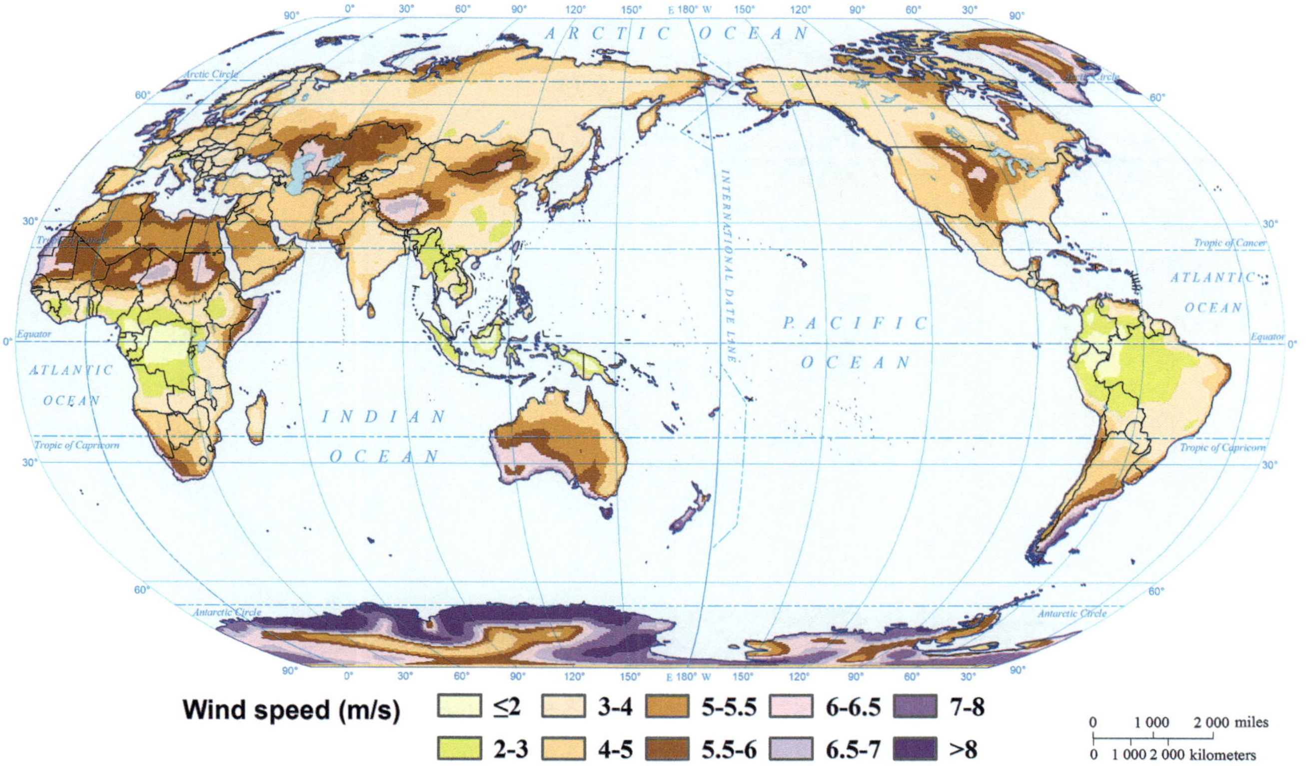

Annual maximum wind speed (2050s, RCP2.6)

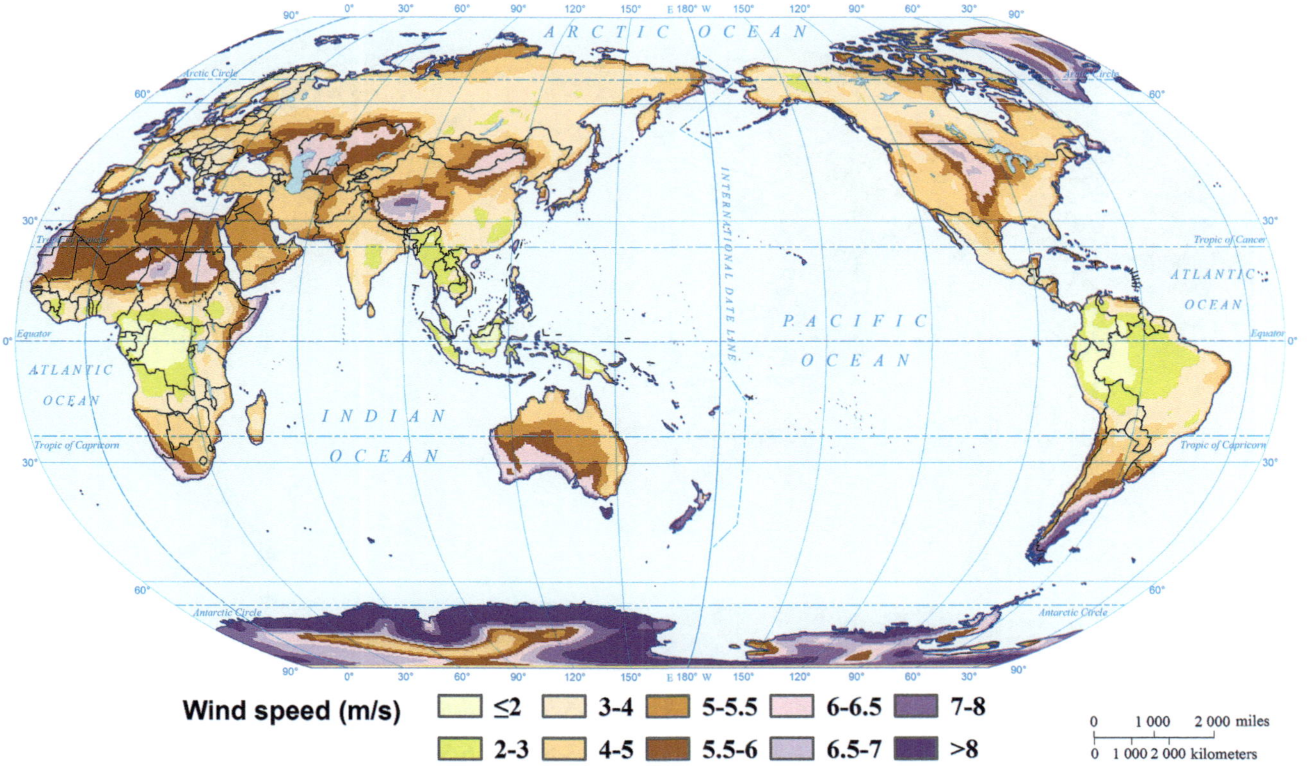

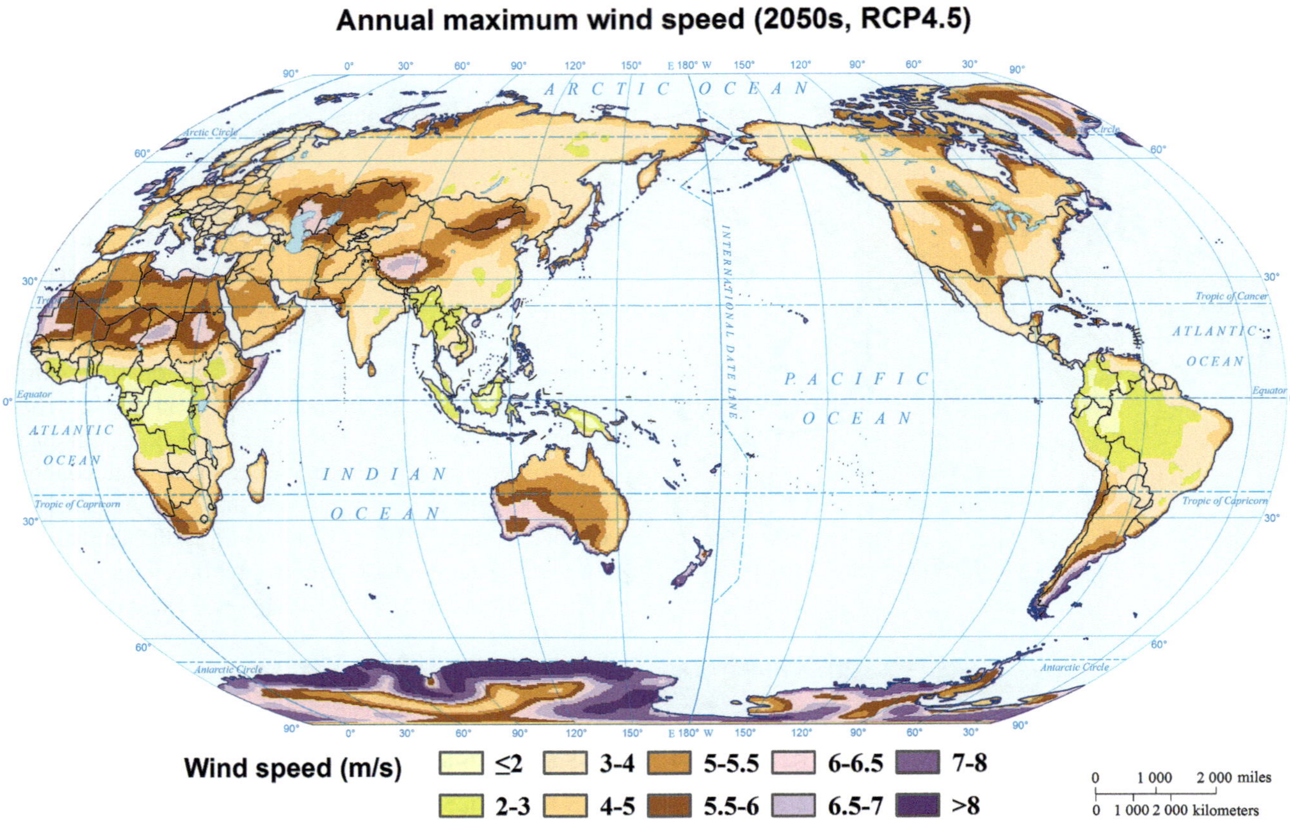
Annual maximum wind speed (2050s, RCP4.5)
ARCTIC OCEAN
PACIFIC OCEAN
INDIAN OCEAN
ATLANTIC OCEAN
INTERNATIONAL DATE LINE
Arctic Circle
Tropic of Cancer
Equator
Tropic of Capricorn
Antarctic Circle
Wind speed (m/s)
≤2
2-3
3-4
4-5
5-5.5
5.5-6
6-6.5
6.5-7
7-8
>8
0 1 000 2 000 miles
0 1 000 2 000 kilometers

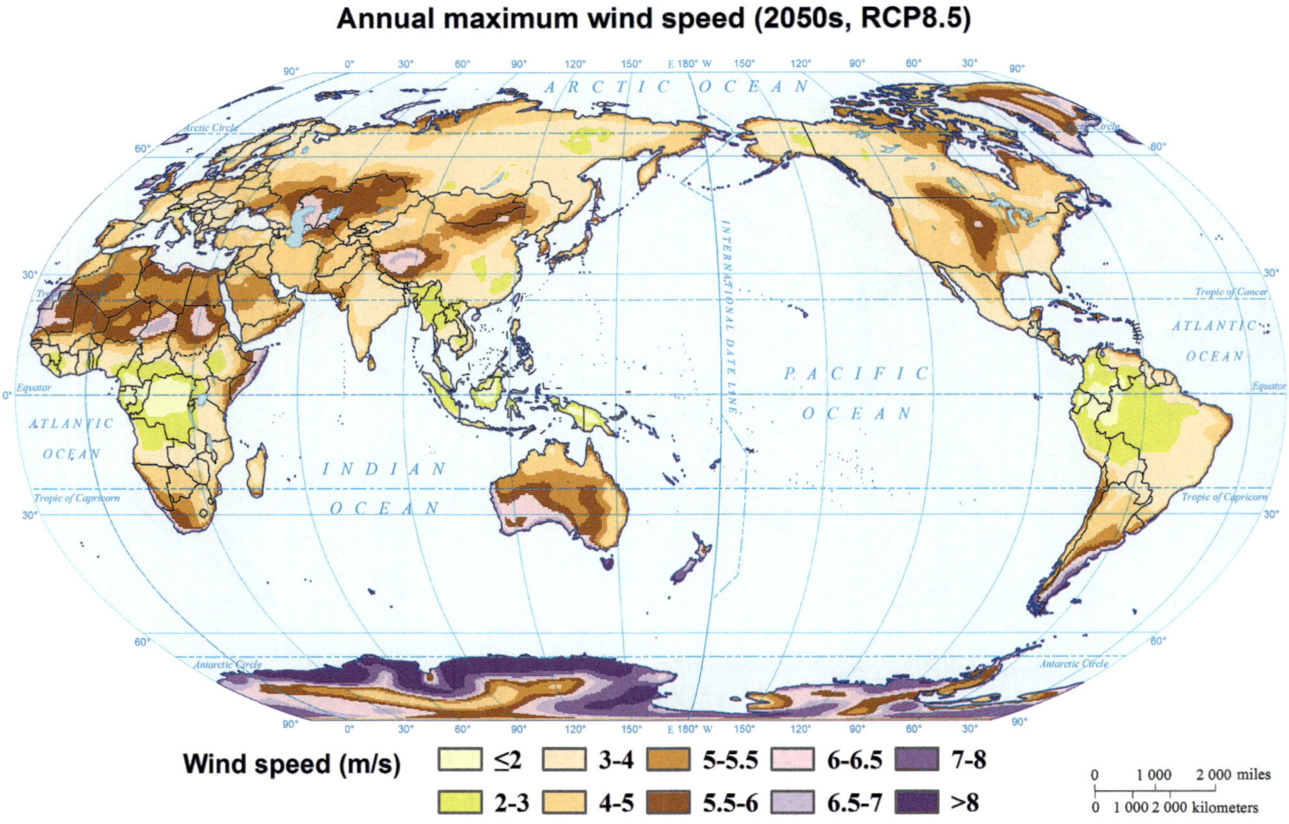
Annual maximum wind speed (2050s, RCP8.5)
ARCTIC OCEAN
PACIFIC OCEAN
INDIAN OCEAN
ATLANTIC OCEAN
INTERNATIONAL DATE LINE
Arctic Circle
Tropic of Cancer
Equator
Tropic of Capricorn
Antarctic Circle
Wind speed (m/s)
≤2
2-3
3-4
4-5
5-5.5
5.5-6
6-6.5
6.5-7
7-8
>8
0 1 000 2 000 miles
0 1 000 2 000 kilometers

References

Giorgi, F., and X. Bi. 2005. Regional changes in surface climate interannual variability for the 21st century from ensembles of global model simulations. Geophysical Research Letters 32: L13701.

Krishnan, A., and P.K. Bhaskaran. 2019. Performance of CMIP5 wind speed from global climate models for the Bay of Bengal region. *International Journal of Climatology*. https://doi.org/10.1002/joc.6404.

Kumar, D., V. Mishra, and A.R. Ganguly. 2015. Evaluating wind extremes in CMIP5 climate models. *Climate Dynamics* 45: 441–453.

Pryor, S.C., R.J. Barthelmie, N.E. Clausen, M. Drews, N. MacKellar, and E. Kjellstrom. 2012. Analyses of possible changes in intense and extreme wind speeds over northern Europe under climate change scenarios. *Climate Dynamics* 38: 189–208.

Tian, Q., G. Huang, K. Hu, and D. Niyogi. 2019. Observed and global climate model based changes in wind power potential over the Northern Hemisphere during 1979–2016. *Energy* 167: 1224–1235.

Vautard, R., J.L. Cattiaux, P. Yiou, J.N. Thepaut, and P. Ciais. 2010. Northern Hemisphere atmospheric stilling partly attributed to an increase in surface roughness. *Nature Geoscience* 3 (11): 756–761.

Wu, J., J. Zha, D. Zhao, and Q. Yang. 2018. Changes in terrestrial near-surface wind speed and their possible causes: An overview. *Climate Dynamics* 51: 2039–2078.

Yan, Z., S. Bate, R.E. Chandler, V. Isham, and H. Wheater. 2002. An analysis of daily maximum wind speed in Northwestern Europe using generalized linear models. *Journal of Climate* 15: 2073–2088.

Population and Economic System Changes

Mapping Global Population Changes

Yujie Liu and Jie Chen

1 Introduction

The increase in greenhouse gas emissions caused by human activities is considered as the main cause of global warming (Stocker et al. 2013). The rapid growth of population and economic activities in the twentieth century has brought unprecedented pressure on climate and the environment, and population has become an important topic in climate change research (Min et al. 2011; Diaz and Moore 2017; Forzieri et al. 2017). Accurate and robust predictions of population size and spatial distribution will help to assess the impact of climate change on socioeconomic development, human health, and resource demand and distribution, and provide a scientific basis for designing strategies to control greenhouse gas emissions and developing mitigation and adaptation policies (Lutz and Kc 2011; Field et al. 2014; Gerland et al. 2014).

Climate scenarios constitute the basis of climate change research, and the rational setting of socioeconomic development scenarios is the core of climate change impact assessment (Van Vuuren et al. 2012). Shared socioeconomic pathways (SSPs) are reference pathways that describe alternative trends that may emerge in social and economic system development in the twenty-first century in the absence of climate change or climate policies (O'Neill et al. 2014). Five such SSP schemes (SSP1–5) have been developed. In particular, SSP1 is a sustainable solution that advances in technology and reduces reliance on carbon energy; SSP2 describes a medium-sized solution that can maintain the current trend and gradually reduce the reliance on carbon energy; and SSP3 is a regionalization program that leads to reduced trade flows, adverse institutional development, and poor adaptation to climate change. Based on these SSPs, preliminary studies have been conducted on demographic and economic changes in more than 150 countries (O'Neill et al. 2014). It is expected that there will be more attention paid to the size, as well as the spatial distribution of future populations in the SSP scenarios, with the increasing demand for demographic analysis in small areas related to climate change (Chi 2009; Raymer et al. 2012).

This study aims to evaluate the changes in the global population under SSP1-2030, SSP1-2050, SSP2-2030, SSP2-2050, SSP3-2030, and SSP3-2050. We produced the data at the global and regional scales. The spatial resolution of global projections is $0.5° \times 0.5°$, and the spatial resolution of regional projections is $0.25° \times 0.25°$. The four hotspot regions are the Bohai Rim area in China, the Qinghai-Tibet Plateau in China, the northeastern United States, and western Europe. The present results may provide a scientific basis for climate risk assessment and effective adaptations.

2 Method

The methods for projecting population and its change for future SSP scenarios are as follows: (1) collecting the statistical data of national populations in 2005; (2) analyzing the projections of national populations under the three SSP

Authors: Yujie Liu, Jie Chen.
Map Designers: Yuanyuan Jing, Jing'ai Wang, Ying Wang.
Language Editor: Yujie Liu.

Y. Liu (✉) · J. Chen
Key Laboratory of Land Surface Pattern and Simulation, Institute of Geographic Sciences and Natural Resources Research, Chinese Academy of Sciences, Beijing, 100101, China
e-mail: liuyujie@igsnrr.ac.cn

J. Chen
University of Chinese Academy of Sciences, Beijing, 100049, China

P. Shi, *Atlas of Global Change Risk of Population and Economic Systems*, IHDP/Future Earth-Integrated Risk Governance Project Series, https://doi.org/10.1007/978-981-16-6691-9_5

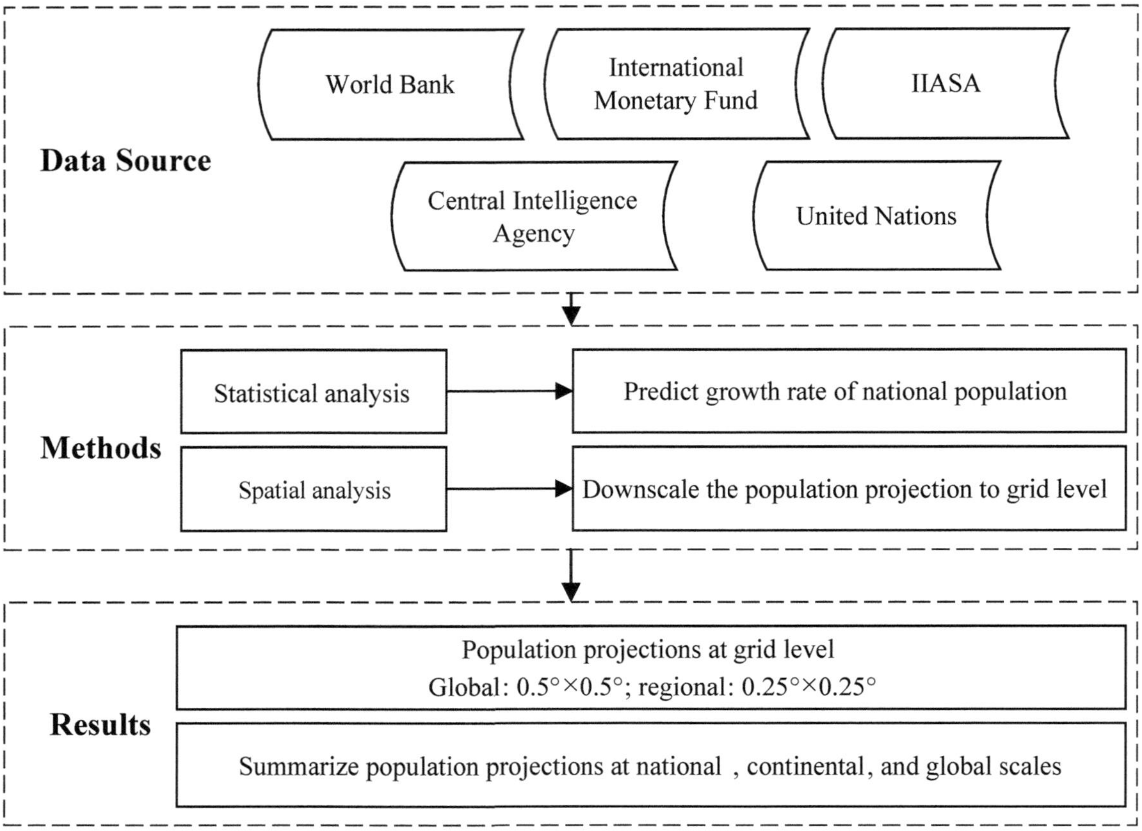

Fig. 1 Technical flowchart for the projection and mapping of the population of the world

scenarios in 2030 and 2050; (3) conducting the spatial downscaling of the populations at the country level to the grid level; and (4) summarizing the results of population projections. Figure 1 shows the technical flowchart for the projection and mapping of the populations of the world.

For population projections, the population at the country level in 2005 was obtained from the World Bank (https://data.worldbank.org.cn/), the International Monetary Fund (https://www.imf.org/en/Data), the Central Intelligence Agency (https://www.cia.gov/), and the United Nations (https://data.un.org/).

The global distribution of population in 2010 was assessed based on the population of the baseline period (2005) multiplied by the growth rate of population. The growth rate was calculated based on the population projections at the country level by the International Institute for Applied Systems Analysis (IIASA) (Samir and Lutz 2014). The statistical analysis was performed using ArcGIS. Then, the national population projections were spatially downscaled with reference to the spatial distribution of UN-WPP-Gridded Population of the World (GPW), v4 in 2005 (Doxsey-Whitfield et al. 2015). The spatial analysis was performed using ArcGIS. Then, the gridded population projections were summarized at the national, continental, and global scales, and the proportions of each continent to the world and the proportion of China were calculated for analysis.

3 Results

Global and continental populations under the SSP1–3 scenarios are shown in Fig. 2. The global population for the baseline period is 6575.50 $\times$ 10^6 persons, which increases to 9542.47 $\times$ 10^6 persons by 2050 under the SSP3 scenario. The increase in population is the largest for the SSP3 scenario and the smallest for SSP1. Among the continents, the population is the highest for Asia, followed by Africa, and the lowest in Oceania in both the baseline period (2010) and the SSP scenarios. Population in Asia and Africa in the baseline period is 61% and 15%, respectively. The percentage of population under the three SSP scenarios decreases for Asia and increases for Africa. The percentage of the population by 2050 under the SSP3 scenario decreases for Asia to 57% and increases for Africa to 24% of the global total. Population in China is projected to increase under all the SSP scenarios and time periods compared to the baseline period. However, the percentage of the Chinese population in Asia and the world is likely to decrease. In the baseline period, China accounted for 33% and 20% of the population in Asia and the world, while in 2050 under the SSP3 scenario, the percentage would decrease to 24% and 14%, respectively (Fig. 2).

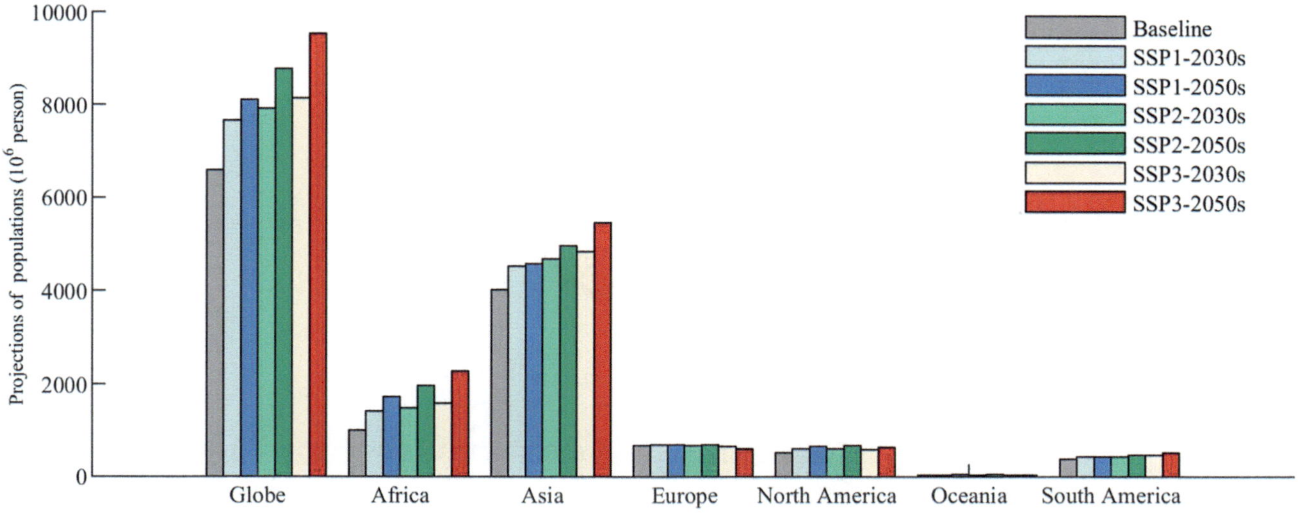

Fig. 2 Projections of global and continental populations under the Shared Socioeconomic Pathways

4 Maps

Global population distribution (Baseline)

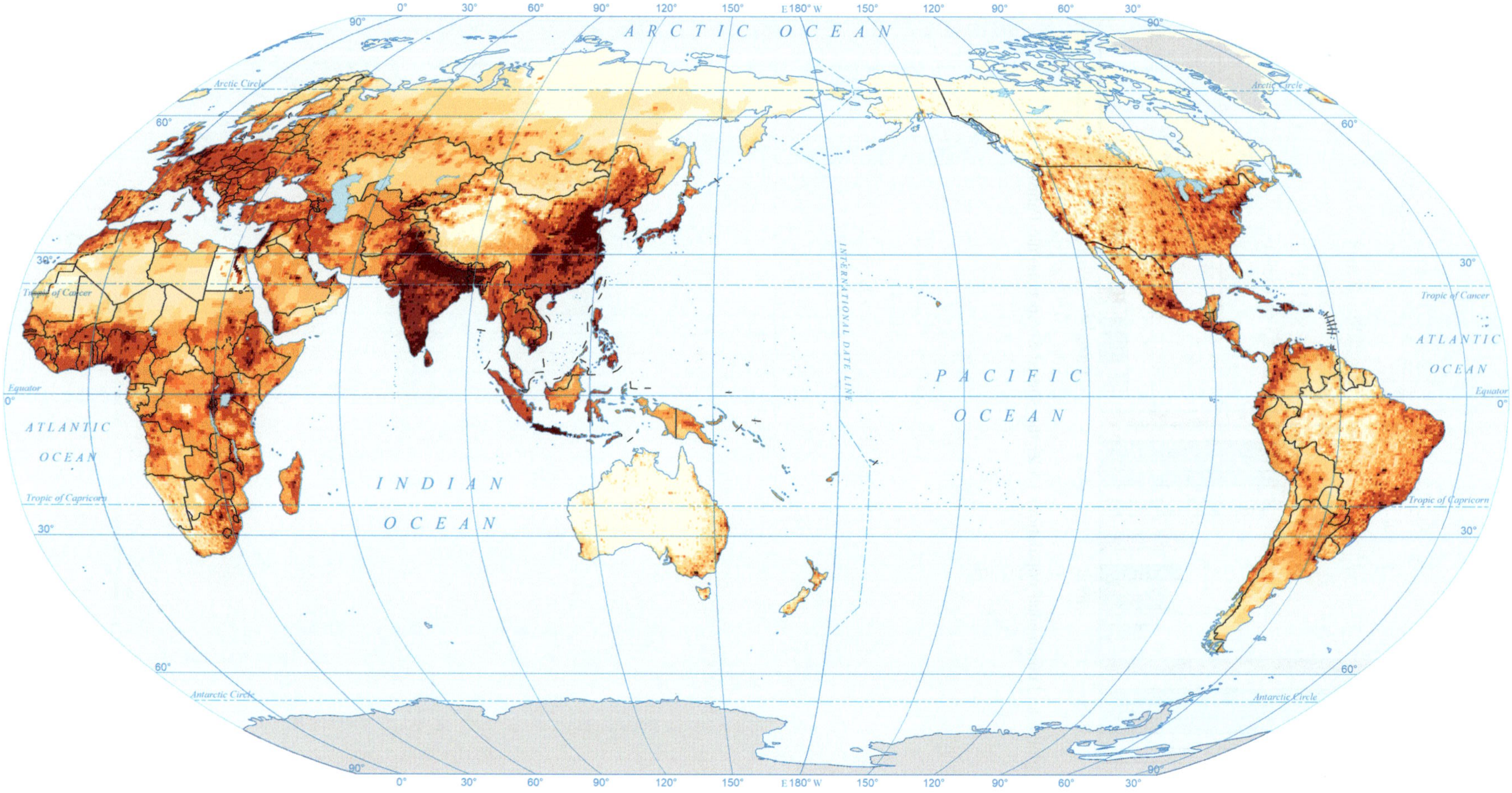

Pop. (10^3person) | ≤0.1 | 0.5-1 | 5-10 | 50-100 | 500-1000
No Data | 0.1-0.5 | 1-5 | 10-50 | 100-500 | >1000

0 1 000 2 000 miles
0 1 000 2 000 3 000 kilometers

Projected global population distribution (2030, SSP1)

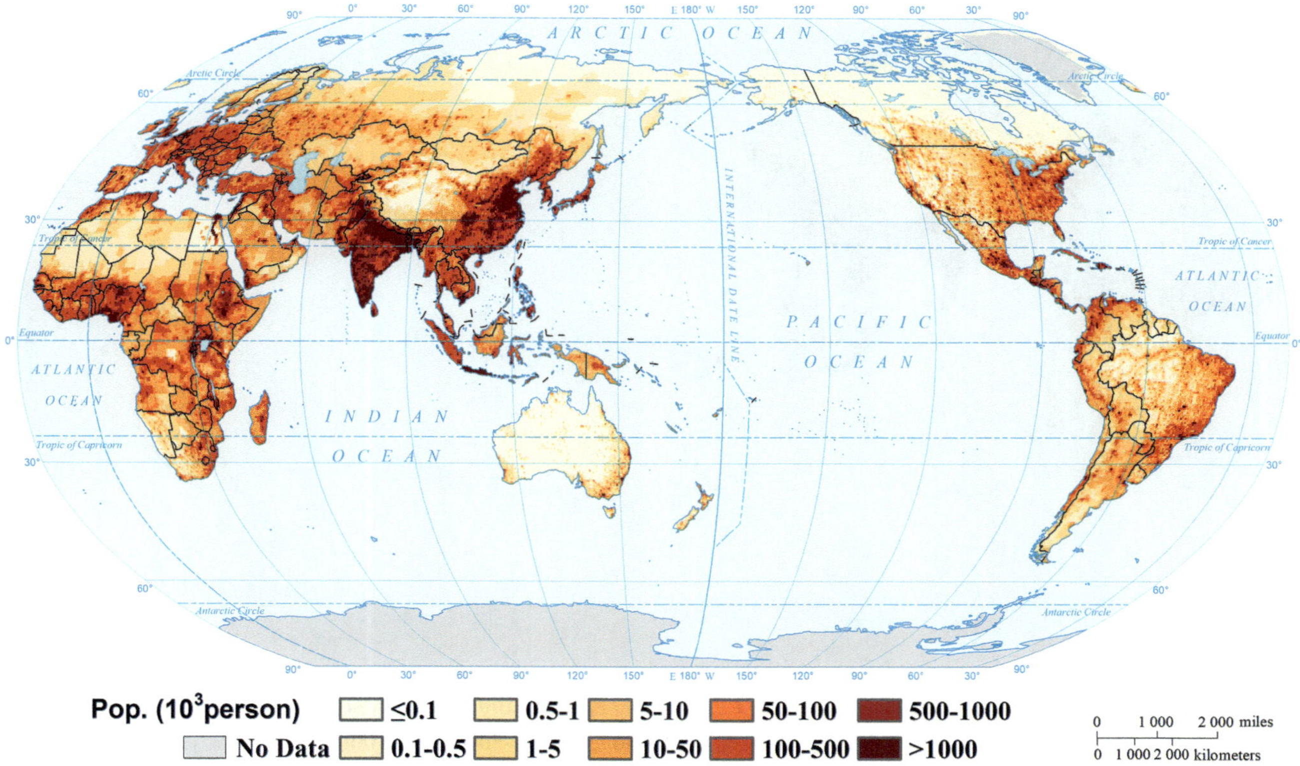

Projected global population distribution (2030, SSP2)

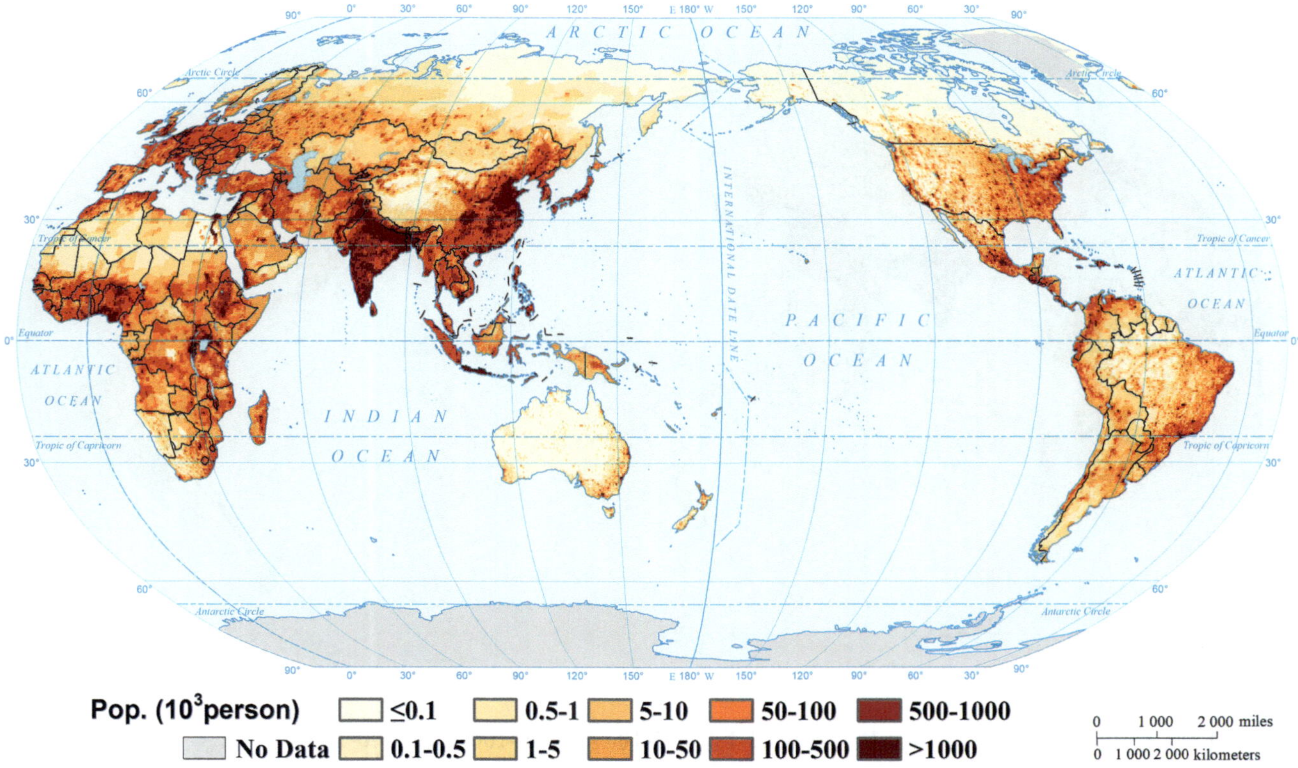

Projected global population distribution (2030, SSP3)

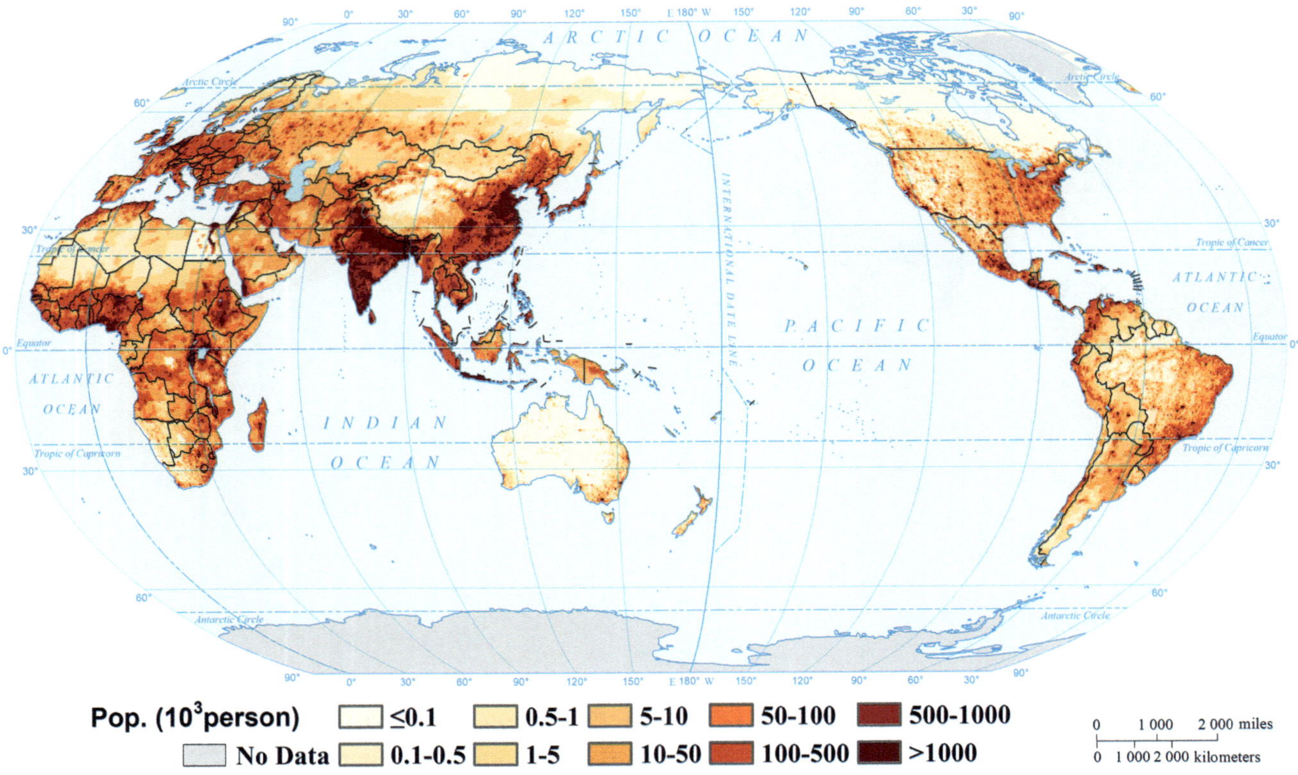

Projected global population distribution (2050, SSP1)

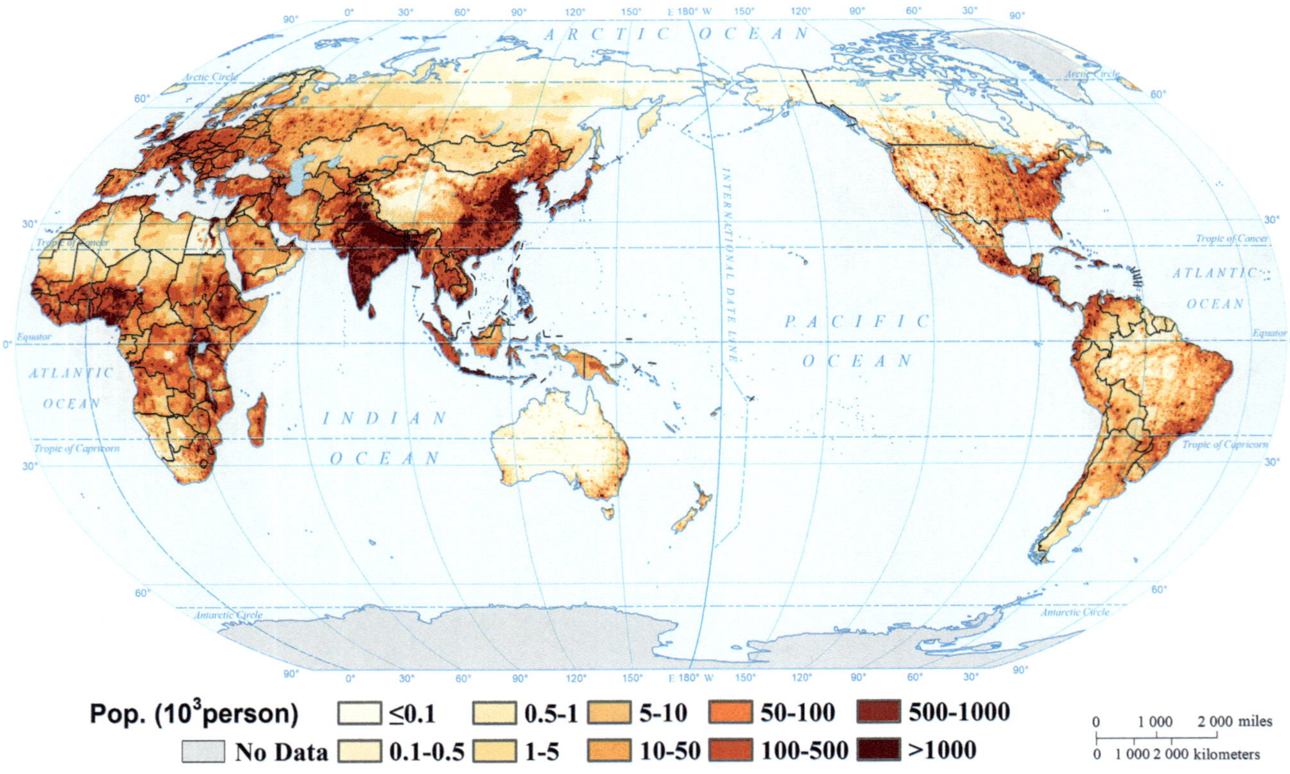

Projected global population distribution (2050, SSP2)

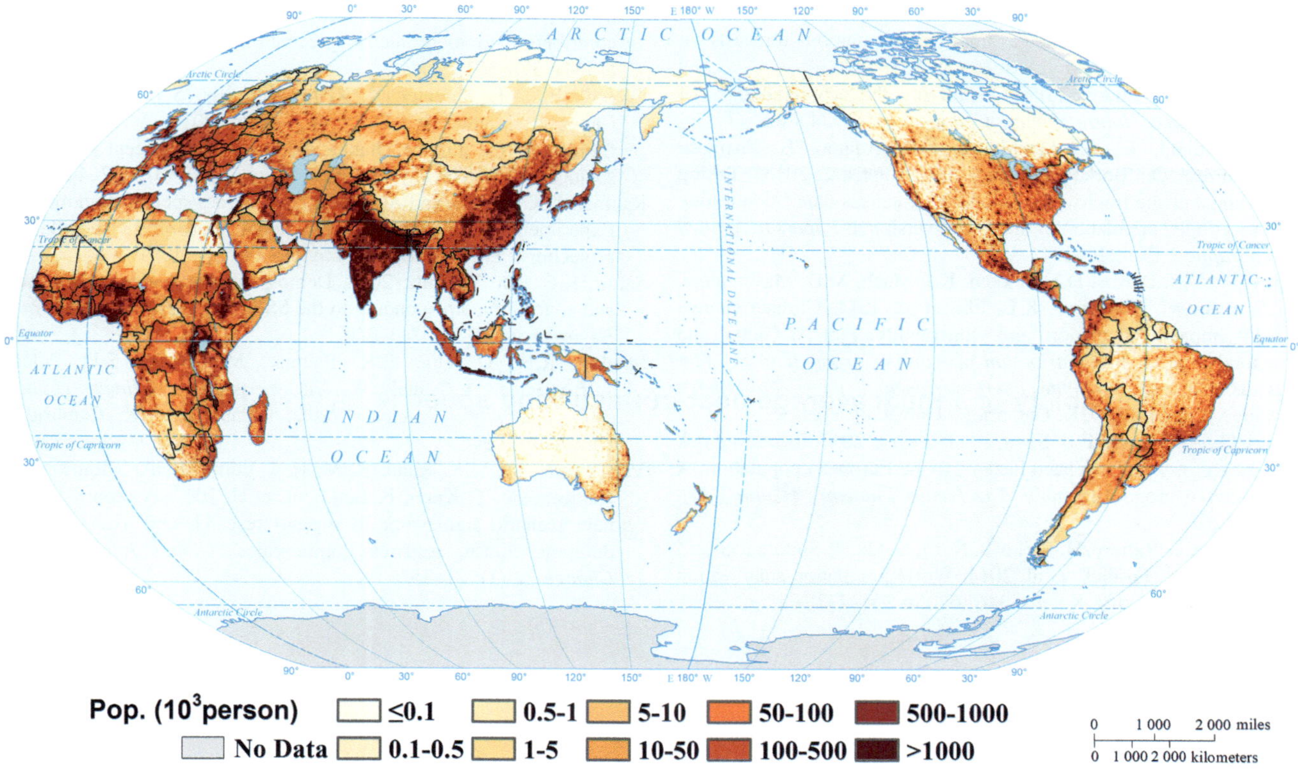

Projected global population distribution (2050, SSP3)

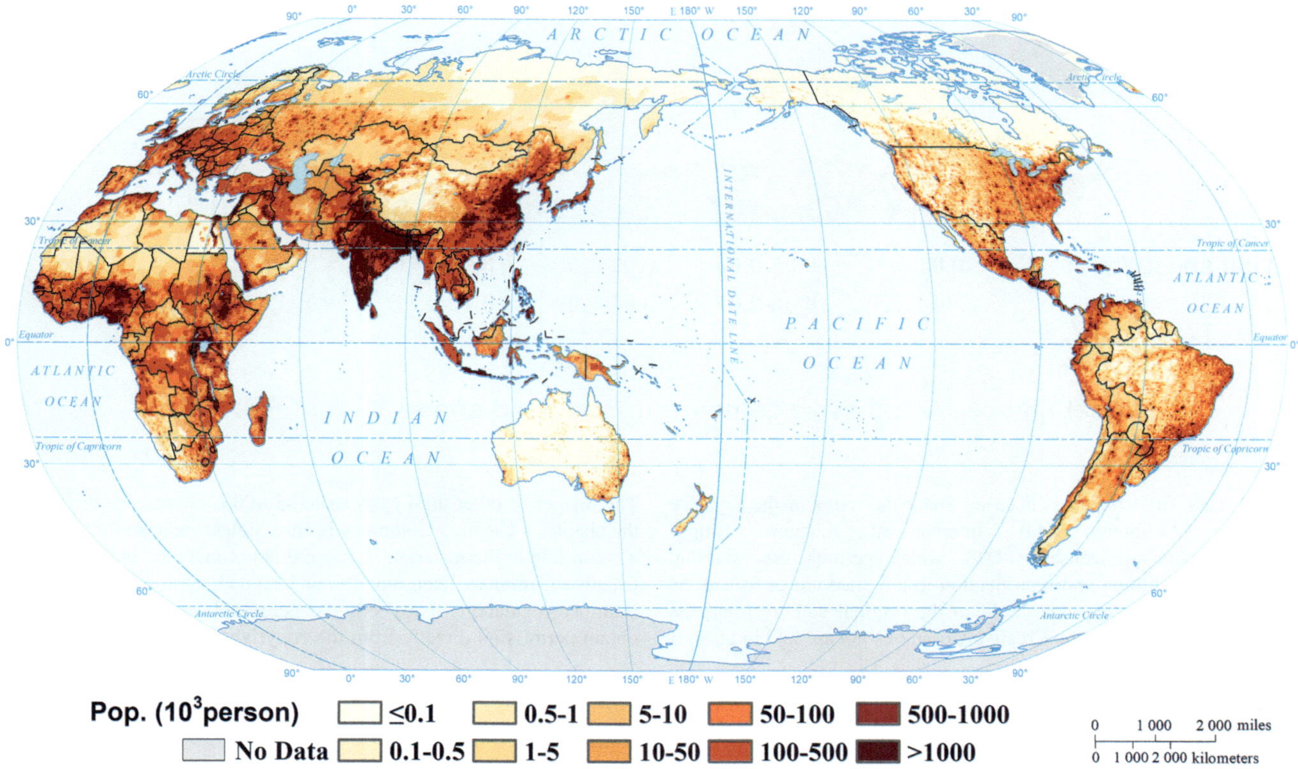

References

Chi, G. 2009. Can knowledge improve population forecasts at subcounty levels? *Demography* 46 (2): 405–427.

Diaz, D., and F. Moore. 2017. Quantifying the economic risks of climate change. *Nature Climate Change* 7 (11): 774–782.

Doxsey-Whitfield, E., K. MacManus, S.B. Adamo, L. Pistolesi, J. Squires, O. Borkovska, and S.R. Baptista. 2015. Taking advantage of the improved availability of census data: A first look at the gridded population of the world, version 4. *Papers in Applied Geography* 1 (3): 226–234.

Field, C.B., V.R. Barros, D.J. Dokken, K.J. Mach, M.D. Mastrandrea, T.E. Bilir, M. Chatterjee, K.L. Ebi, et al. 2014. Climate change 2014: Impacts, adaptation, and vulnerability. Part A: Global and sectoral aspects. In *Contribution of working group II to the fifth assessment report of the intergovernmental panel on climate change*. Cambridge, UK: Cambridge University Press.

Forzieri, G., A. Cescatti, F.B. Silva, and L. Feyen. 2017. Increasing risk over time of weather-related hazards to the European population: A data-driven prognostic study. *The Lancet Planetary Health* 1 (5): e200–e208.

Gerland, P., A.E. Raftery, H. Evikova, N. Li, D. Gu, T. Spoorenberg, L. Alkema, B.K. Fosdick, et al. 2014. World population stabilization unlikely this century. *Science* 346 (6206): 234–237.

Lutz, W., and S. Kc. 2011. Global human capital: Integrating education and population. *Science* 333 (6042): 587–592.

Min, S.K., X. Zhang, F.W. Zwiers, and G.C. Hegerl. 2011. Human contribution to more-intense precipitation extremes. *Nature* 470 (7334): 378–381.

O'Neill, B.C., E. Kriegler, K. Riahi, K.L. Ebi, S. Hallegatte, T.R. Carter, R. Mathur, and D.P. van Vuuren. 2014. A new scenario framework for climate change research: The concept of shared socioeconomic pathways. *Climatic Change* 122 (3): 387–400.

Raymer, J., G.J. Abel, and A. Rogers. 2012. Does specification matter? Experiments with simple multiregional probabilistic population projections. *Environment & Planning* 44 (11): 2664–2686.

Samir, K.C., and W. Lutz. 2014. Demographic scenarios by age, sex and education corresponding to the SSP narratives. *Population and Environment* 35 (3): 243–260.

Stocker, R.F., D. Qin, G.K. Plattner, M. Tignor, S.D. Allen, J. Boschung, A. Nauels, Y. Xia, et al. 2013. *Climate change 2013: The physical science basis*. Cambridge, UK: Cambridge University Press.

Van Vuuren, D.P., K. Riahi, R. Moss, J. Edmonds, A. Thomson, N. Nakicenovic, T. Kram, F. Berkhout, et al. 2012. A proposal for a new scenario framework to support research and assessment in different climate research communities. *Global Environmental Change* 22 (1): 21–35.

Mapping Global Population Exposure to Heatwaves

Qinmei Han, Wei Xu, and Peijun Shi

1 Background

Global warming has become a severe problem worldwide, where the average global temperature has steadily increased over recent decades, accompanied by the abnormally hot weather (IPCC 2013). Since the 1950s, heatwave events have increased in frequency, intensity, and duration and their impact on human health will also increase under enhanced global warming (Perkins-Kirkpatrick and Lewis 2020). Heatwaves have become one of the most serious climate events in the world. Thousands of people have died from exposing to heatwaves in recent years, for instance, the European heatwave of 2003 induced more than 70,000 additional deaths (Robine et al. 2008). Heat-related mortality and morbidity are not only attributed to natural hazards resulting from climate change (Seneviratne et al. 2012). Both climatic factors and socioeconomic factors such as population change and vulnerability of people exposed to heatwaves have impact on the number of deaths caused by heatwaves. Thus, a comprehensive and quantitative assessment of heatwave exposure is conducive to taking targeted measures to reduce the risk in hotspot regions of the world.

Compared to heat-related mortality risk assessment, assessment of exposure has received little attention in recent years. A few studies have examined global or regional population exposure to heatwaves (Jones et al. 2015; Liu et al. 2017; Chen et al. 2020), and all the existing results show that heatwave exposure will have a significant increase in the future compared to the historical periods.

We combined the future projections of temperature under the three Representative Concentration Pathways (RCP2.6, RCP4.5, and RCP8.5) and the future projections of population under the three Shared Socioeconomic Pathways (SSP1, SPP2, and SSP3) to evaluate future population exposure to heatwaves across the world at the grid level (0.25° × 0.25°) and the country level, respectively, in different time periods. The scenario combinations are RCP2.6-SSP1, RCP4.5-SSP2, and RCP8.5-SSP3. We also computed the decadal exposure change in the 2030s (2016–2035) and the 2050s (2046–2065) compared to the baseline period (1986–2005) using high spatial resolution climate and population data under different scenario combinations.

2 Method

This study used daily maximum temperature as metric to estimate heatwaves. The daily maximum temperature data were obtained from the NASA Earth Exchange Global Daily Downscaled Projections (NEX-GDDP), which was released in June 2015 (https://dataserver.nccs.nasa.gov/thredds/catalog/bypass/NEX-GDDP/catalog.html). The spatial resolution of the data is 0.25° × 0.25°. There are 21 general circulation models in this dataset, which contain two Representative Concentration Pathways (RCPs)—RCP4.5 and RCP8.5—for the period from 1950 to 2100. The temperature data of the lower emissions scenario RCP2.6 was computed by sub-project 1. The population projection data used in this study contained SSP1–3 computed by sub-project 2. We calculated heatwave intensity and heatwave exposure for two time periods (2016–2035 and 2046–2065) in the future

Authors: Qinmei Han, Wei Xu, Peijun Shi.
Map Designers: Yuanyuan Jing, Jing'ai Wang, Ying Wang.
Language Editor: Wei Xu.

Q. Han · W. Xu · P. Shi (✉)
Faculty of Geographical Science, Beijing Normal University, Beijing, 100875, China
e-mail: spj@bnu.edu.cn

W. Xu · P. Shi
State Key Laboratory of Earth Surface Processes and Resource Ecology, Beijing Normal University, Beijing, 100875, China

P. Shi, *Atlas of Global Change Risk of Population and Economic Systems*, IHDP/Future Earth-Integrated Risk Governance Project Series, https://doi.org/10.1007/978-981-16-6691-9_6

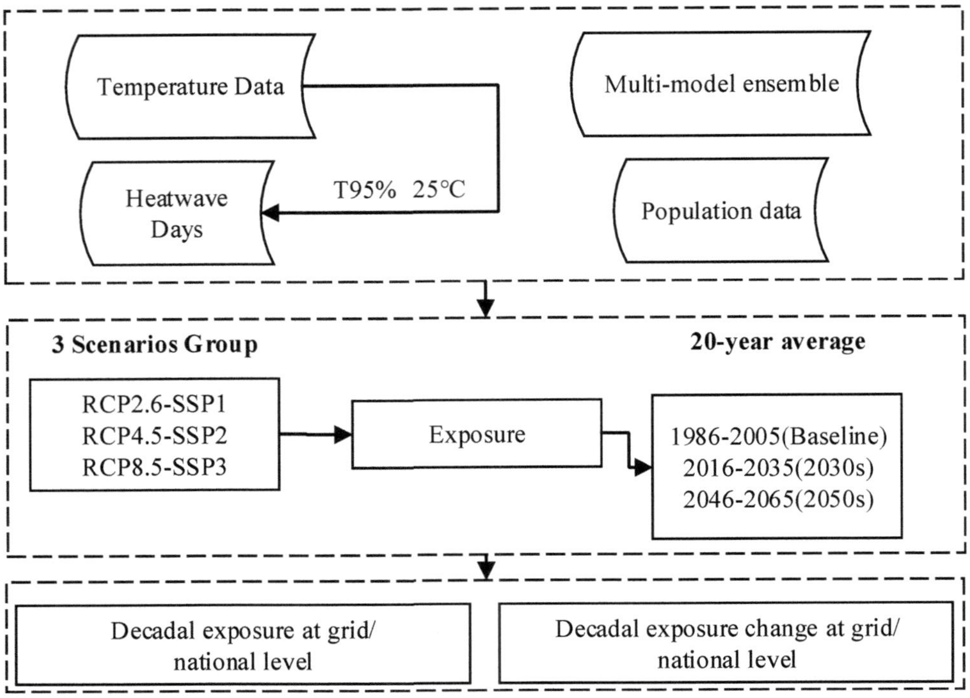

Fig. 1 Technical flowchart for mapping global heatwave exposure

compared to 1986–2005 under the three RCP-SSPs combinations—RCP2.6-SSP1, RCP4.5-SSP2, and RCP8.5-SSP3. Figure 1 shows the technical flowchart for mapping the global population exposure to heatwaves.

2.1 Heatwave Intensity Metrics

Heatwave was defined here as at least 3 consecutive days exceeding the given temperature threshold, which is the 95th percentile value of the daily maximum temperature series over the baseline period 1986–2005 at the grid level, and if the 95th percentile value was lower than 25 °C, we set 25 °C as the threshold in this grid. The annual heatwave intensity was quantified as the total heatwave days in a year.

2.2 Exposure

In this study, exposure was defined as the total population exposed to a heatwave, which was calculated by multiplying annual population and annual total number of heatwave days at the grid level. So the unit of exposure is person-day. For both the baseline period (1986–2005) and the future periods (2016–2035, 2046–2065), we calculated the 20-year average value of exposure to map the global exposure to heatwaves and compute the interdecadal change of exposure. In addition, we aggregated exposure of the grid level to the national level for further analysis.

3 Results

3.1 Heatwave Intensity

The spatial patterns of multi-model ensemble heatwave days for both the baseline period and the future periods are similar, whereas the regions with high heatwave days are mainly distributed at the equator and gradually decrease with increasing latitude. The number of heatwave days significantly increases over time. The global mean annual heatwave days in the baseline period are 18 days, which increase up to 50 days in the 2050s under high-emission scenarios. The highest heatwave days are 50.7 days in the baseline period, which is estimated to exceed 300 days in the 2050s under the RCP8.5 scenarios. The heatwave days are significantly higher in the 2050s (2046–2065) than the 2030s (2016–2035) under different scenario combinations.

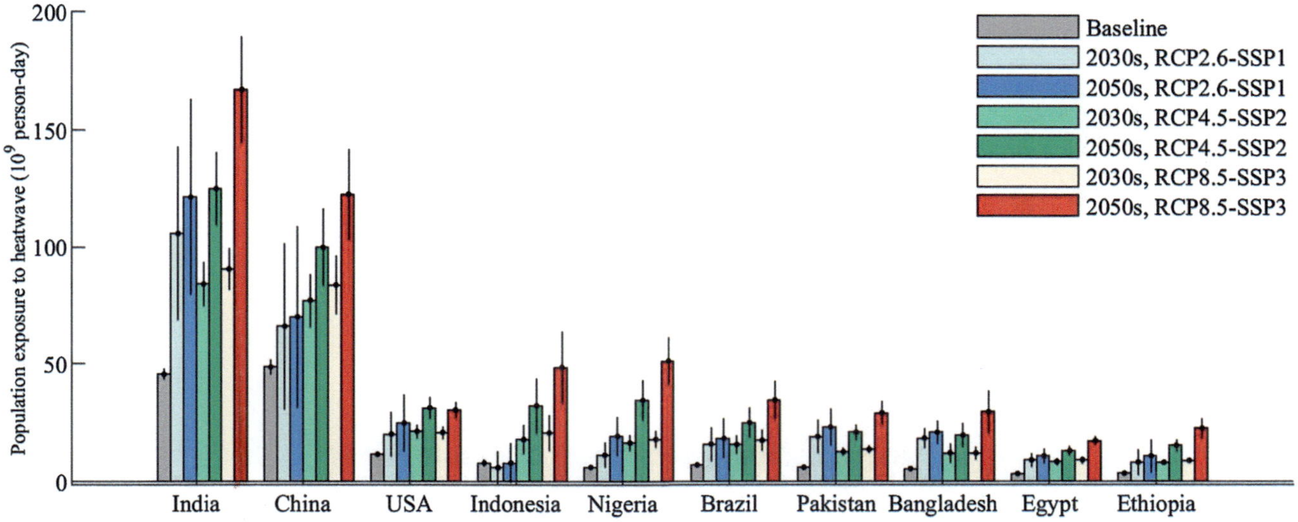

Fig. 2 Population exposure to heatwaves of the top 10 countries. The error bar represents the one standard deviation across the 21 (13 for Representative Concentration Pathway(RCP)2.6) general circulation models and 3 shared socioeconomic pathways (SSPs)

3.2 Heatwave Exposure

Regions with high heatwave exposure are primarily concentrated in densely populated areas, such as India, China, central Europe, and eastern United States for both the baseline and the future periods. It is estimated that in the 2050s (2046–2065), under the high-emission scenario (RCP8.5-SSP3), the global 20-year average exposure to heatwaves will increase by 3.4 times (~103.1 billion person-day) compared to the baseline period (~24 billion person-day), including 6.8 times in Africa, 4.8 times in South America, 2.7 times in Asia, 2.7 times in North America, 1.3 times in Europe, and 2.6 times in Oceania. Under the low-emission scenarios RCP4.5-SSP2 and RCP2.6-SSP1, the global average heatwave exposure increases by 2.2 times (~75.7 billion person-day) and 1.4 times (~56.0 billion person-day), respectively.

Exposure to heatwaves at the grid level was aggregated to the country scale for further analysis. The top 10 countries with high heatwave exposure are shown in Fig. 2. These countries are mainly located in Africa and Asia. Compared to the baseline period, the heatwave exposure of all countries increased significantly especially in the 2050s under different scenario combinations. For example, the annual total heatwave exposure of India is about 4.5 billon person-day in the historical period (1986–2005), which increases to 16.7 billion person-days for 2046–2065 under the RCP8.5-SSP3 scenario.

4 Maps

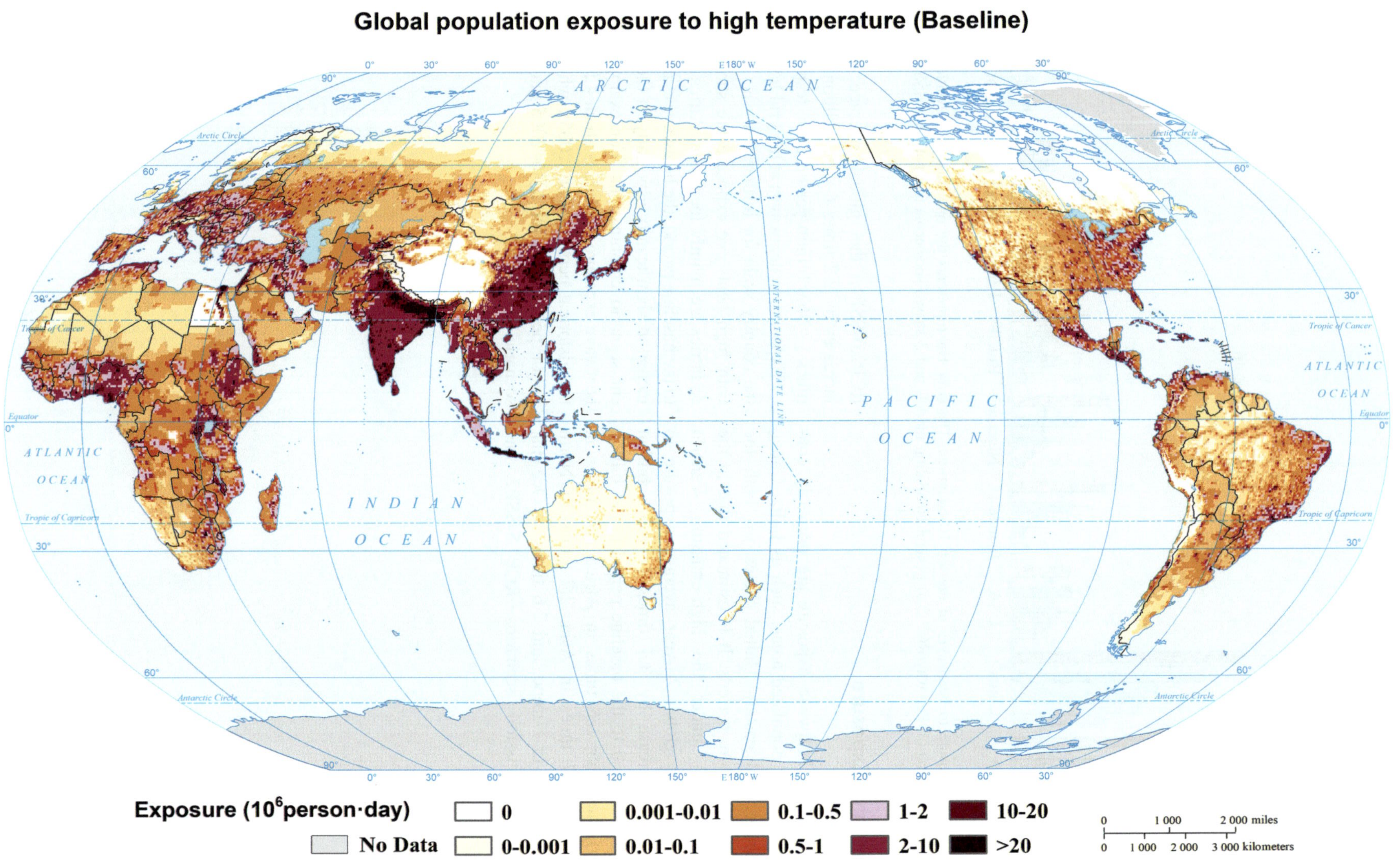
Global population exposure to high temperature (Baseline)
ARCTIC OCEAN
PACIFIC OCEAN
INDIAN OCEAN
ATLANTIC OCEAN
INTERNATIONAL DATE LINE
Arctic Circle
Tropic of Cancer
Equator
Tropic of Capricorn
Antarctic Circle
Exposure (10^6 person·day)
No Data
0
0-0.001
0.001-0.01
0.01-0.1
0.1-0.5
0.5-1
1-2
2-10
10-20
>20
0 1 000 2 000 miles
0 1 000 2 000 3 000 kilometers

Global population exposure to high temperature (2030s, RCP2.6-SSP1)

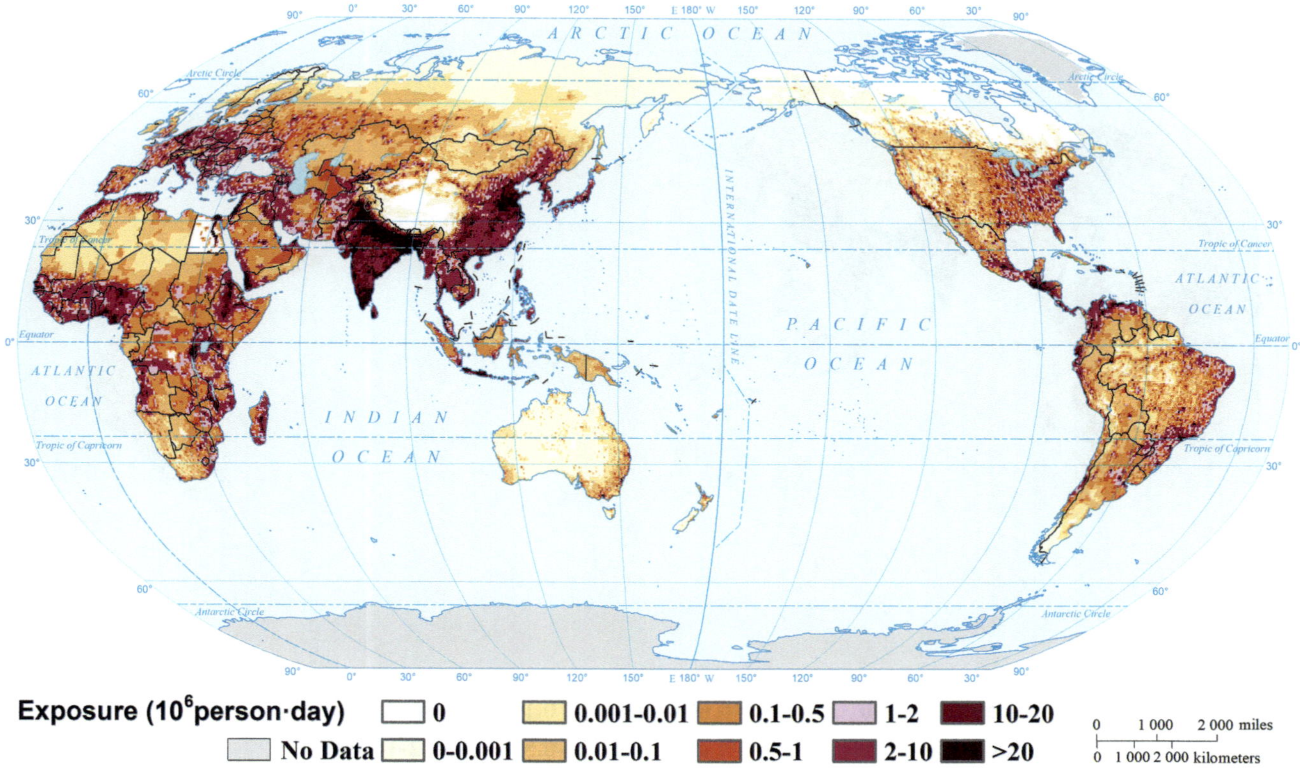

Global population exposure to high temperature (2030s, RCP4.5-SSP2)

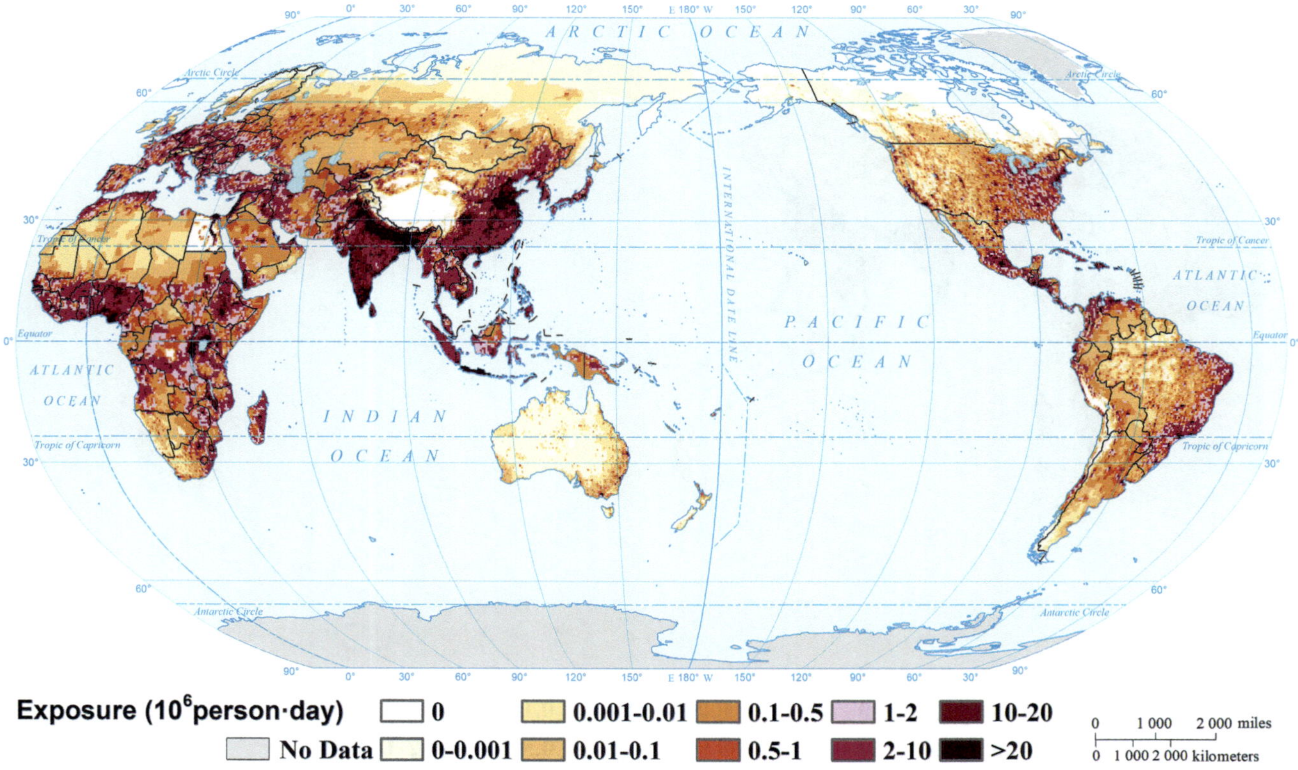

Global population exposure to high temperature (2030s, RCP8.5-SSP3)

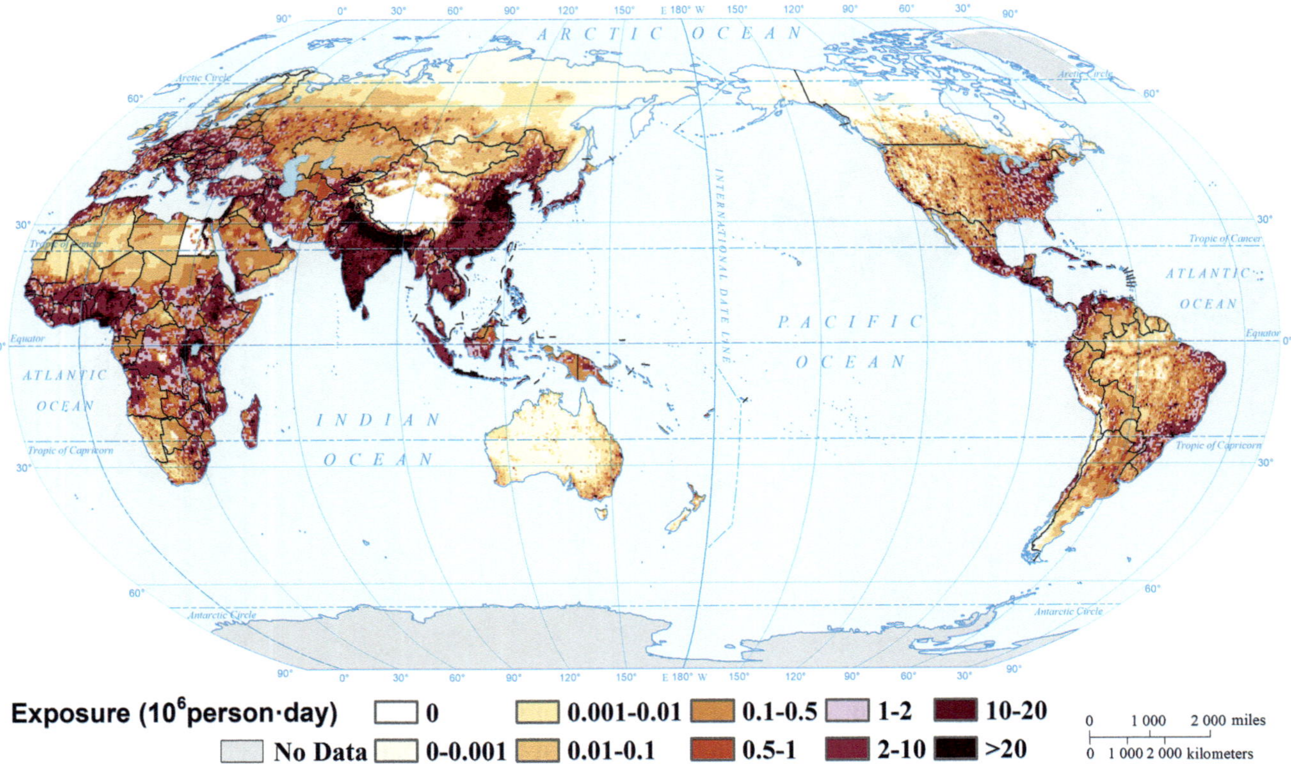

Global population exposure to high temperature (2050s, RCP2.6-SSP1)

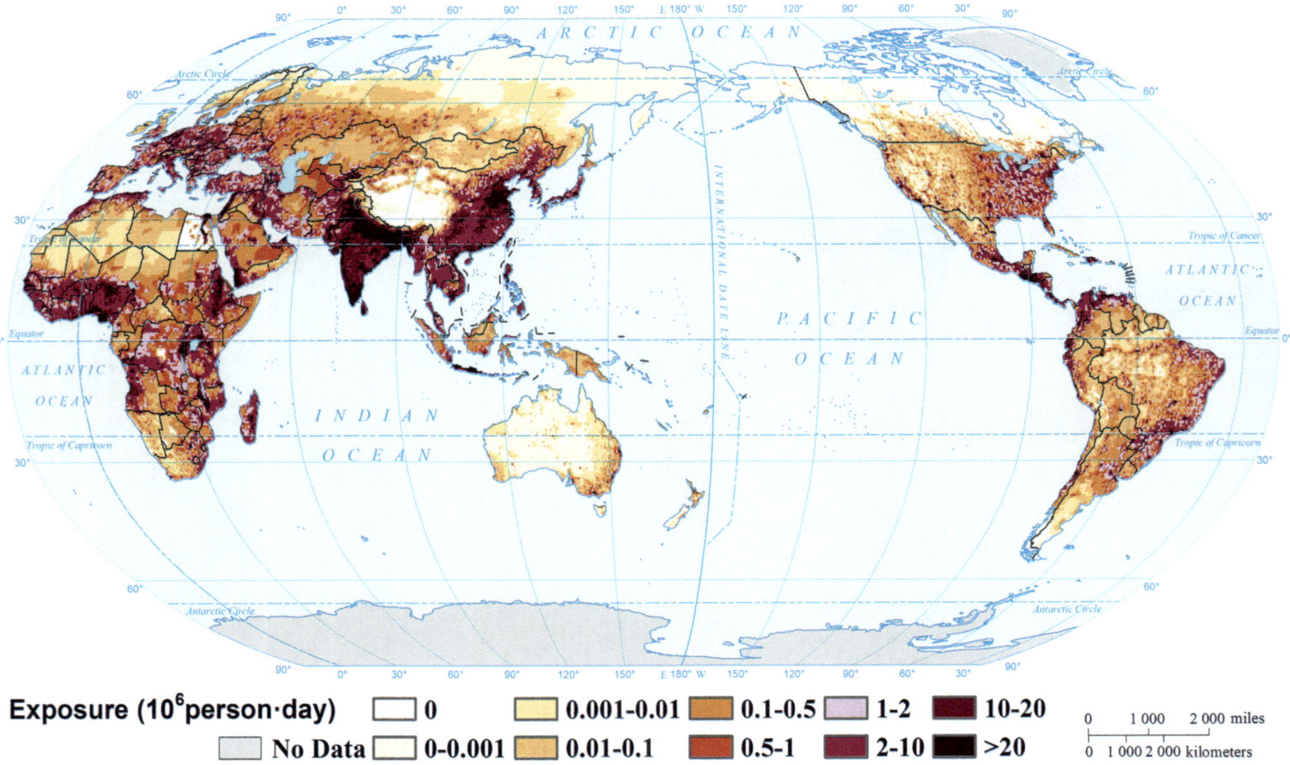

Global population exposure to high temperature (2050s, RCP4.5-SSP2)

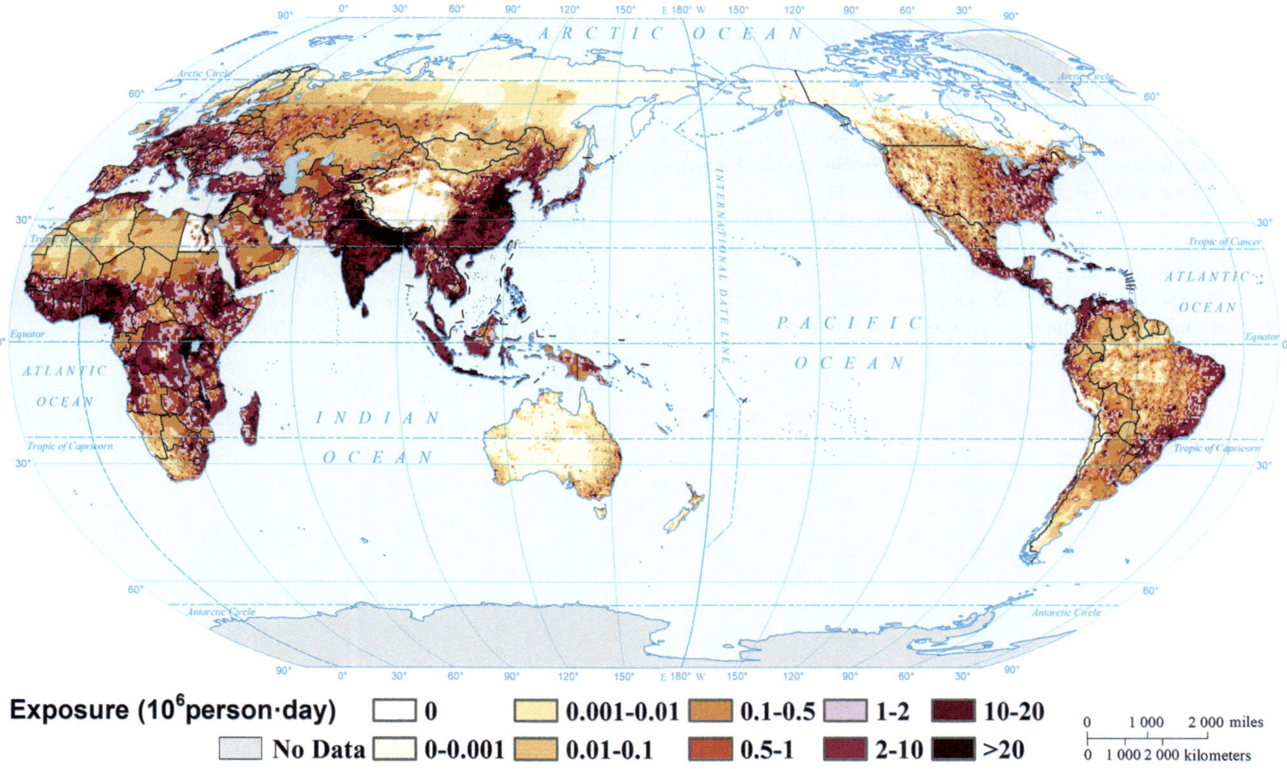

Global population exposure to high temperature (2050s, RCP8.5-SSP3)

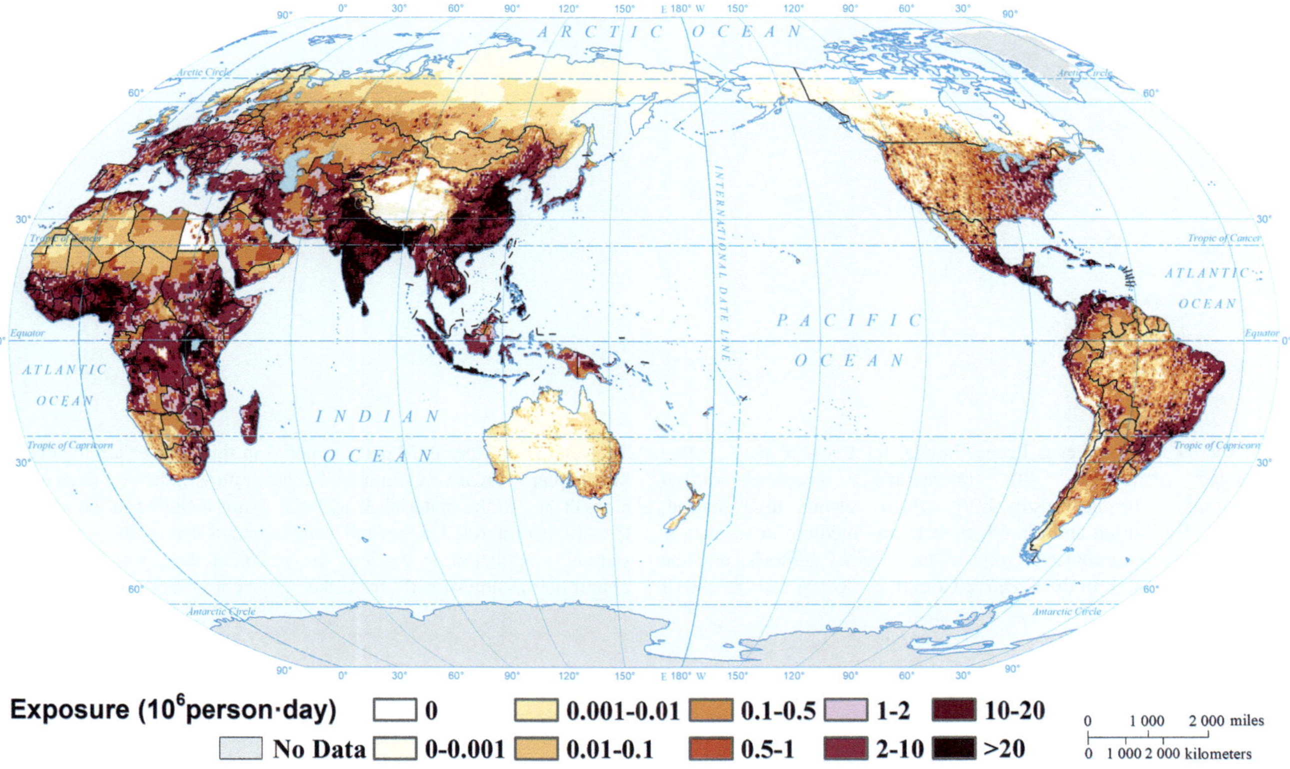

References

Chen, J., Y.J. Liu, T. Pan, P. Ciais, T. Ma, Y.H. Liu, D. Yamazaki, Q. Ge, et al. 2020. Global socioeconomic exposure of heat extremes under climate change.*Journal of Cleaner Production* 277. https://doi.org/10.1016/j.jclepro.2020.123275.

IPCC (Intergovernmental Panel on Climate Change). 2013. Summary for policymakers. In *Climate change 2013: The physical science basis*. Cambridge, UK: Cambridge University Press.

Jones, B., B.C. O'Neill, L. Mcdaniel, S. Mcginnis, L.O. Mearns, and C. Tebaldi. 2015. Future population exposure to US heat extremes. *Nature Climate Change* 5: 652–655.

Liu, Z., B. Anderson, K. Yan, W. Dong, H. Liao, and P. Shi. 2017. Global and regional changes in exposure to extreme heat and the relative contributions of climate and population change. *Scientific Reports* 7: 43909.

Perkins-Kirkpatrick, S.E., and S.C. Lewis. 2020. Increasing trends in regional heatwaves. *Nature Communications* 11: 1–8.

Robine, J.M., S.L.K. Cheung, S. Le Roy, H. Van Oyen, C. Griffiths, J. P. Michel, and F.R. Herrmann. 2008. Death toll exceeded 70,000 in Europe during the summer of 2003. *Comptes Rendus Biologies* 331 (2): 171–178.

Seneviratne, S.I., N. Nicholls, D. Easterling, C.M. Goodess, S. Kanae, J. Kossin, Y. Luo, J. Marengo, et al. 2012. Changes in climate extremes and their impacts on the natural physical environment. In *Managing the risks of extreme events and disasters to advance climate change adaptation. A special report of the intergovernmental panel on climate change*, 9781107025, 109–230. https://doi.org/10.1017/CBO9781139177245.006.

Mapping Global Population Exposure to Rainstorms

Xinli Liao, Junlin Zhang, Wei Xu, and Peijun Shi

1 Background

With global warming, the global hydrological cycle and the spatiotemporal pattern of extreme precipitation have changed greatly in recent years (IPCC 2013; Kong 2020). Evidence from both observations and climate model simulations suggests that the frequency and intensity of extreme precipitation showed an increasing trend and will further increase in the future with climate change. For example, Westra et al. (2013) examined data from 8326 high-quality land-based observing stations globally and found that about two-thirds of the stations showed a significant increase in extreme precipitation. Using the same dataset but for a longer time period, Sun et al. (2021) found that a larger percentage of stations showed statistically significant increasing trends. Climate projections from the Coupled Model Intercomparison Project Phase 5 show continued intensification of daily precipitation extremes from 1951 to 2099 (Donat et al. 2016). Given that extreme precipitation and its induced disasters (e.g., floods and landslides) are one of the most serious consequences of climate change and pose a great threat to life and property, it has aroused widespread attention. Understanding exposure is necessary for disaster risk reduction (IPCC 2012). Likewise, with the social and economic developments and the acceleration of urbanization, the population exposed to natural hazards is increasing and demographic changes, such as an increase in the elderly population, will also amplify exposure of vulnerable people (Qin et al. 2015; Liang et al. 2017). Different levels of global warming (e.g., 1.5 °C and 2 °C) would cause different population exposure to extreme precipitation, and an 0.5 °C less warming would reduce exposures remarkably (Zhang et al. 2018; Shi et al. 2021). From the baseline period to the end of the twenty-first century, under Representative Concentration Pathway (RCP) 4.5-shared socioeconomic pathway (SSP) 2 scenario, population exposure to rainstorms shows an increasing trend in most regions of the world, and the areas with high exposure are mainly distributed in Asia (Liao et al. 2019). By the end of the twenty-first century, although China's estimated population will drop, the population exposure to extreme precipitation will increase significantly under the RCP4.5-SSP2 scenario and the increase in the RCP8.5-SSP3 scenario is even larger (Chen and Sun 2020). While there exist many studies on population exposure to extreme precipitation, the resolution of climate models used in most studies is relatively coarse and there are few studies on RCP2.6 scenarios.

In this section, the population exposure to rainstorms of the world was calculated based on daily precipitation data (RCP2.6, RCP4.5 and RCP8.5) from the Institute of Atmospheric Physics, Chinese Academy of Sciences and population data (SSP1, SSP2, and SSP3) from the Institute of Geographic Sciences and Natural Resources Research, Chinese Academy of Sciences. The data cover three time periods and three scenario combinations, namely, the baseline (1986–2005), 2030s (2016–2035), and 2050s (2046–2065) and the RCP2.6-SSP1 scenario, RCP4.5-SSP2 scenario, and RCP8.5-SSP3 scenario, respectively. Based on these, the global population exposure to rainstorms was assessed and mapped at the grid level.

Authors: Xinli Liao, Junlin Zhang, Wei Xu, Peijun Shi.
Map Designers: Yuanyuan Jing, Jing'ai Wang, Ying Wang.
Language Editor: Wei Xu.

X. Liao · J. Zhang · W. Xu (✉) · P. Shi
Faculty of Geographical Science, Beijing Normal University, Beijing, 100875, China
e-mail: xuwei@bnu.edu.cn

W. Xu · P. Shi
State Key Laboratory of Earth Surface Processes and Resource Ecology, Beijing Norml University, Beijing, 100875, China

P. Shi, *Atlas of Global Change Risk of Population and Economic Systems*, IHDP/Future Earth-Integrated Risk Governance Project Series, https://doi.org/10.1007/978-981-16-6691-9_7

2 Method

In this section, rainstorm was defined by daily precipitation exceeding the 95th percentile value for each grid. The method improved by Bonsal et al. (2001) was used when calculating the 95th percentile, in which daily precipitation data for each year were first ranked in ascending order X_1, X_2, …, X_N, and the probability P_{ro} that a random value is less than or equal to the rank of that value X_m was estimated by Eq. (1).

$$P_{ro} = (m - 0.31) / (N + 0.38) \quad (1)$$

where m is the rank and N is the number of samples. For example, if there are 120 values, the 95th percentile value is linearly interpolated between the 115th-ranked value (corresponding to $P_{ro} = 95.27\%$) and the 114th-ranked value ($P_{ro} = 94.44\%$).

Figure 1 shows the technical flowchart for mapping the population exposure to rainstorms. We first calculated the threshold of each model for each year, then calculated the average threshold values of the 20 years in each period, which represents the threshold of each model for each period. Multi-model ensembles (MMEs) are widely used in research especially for global-scale climate change studies because it is generally found to have a better performance than single models. So, in this study, the average of the individual rainstorm thresholds of the 21 models (RCP2.6 has only 13 models) was calculated and used as the rainstorm threshold for each grid. Based on the threshold value, we calculated the annual rainfall from rainstorms (R95pTOT) by accumulating daily precipitation exceeding the threshold in a certain year for each model, then used the 20-year averages of the rainfall in each period to represent the rainstorm intensity, and rainfall from rainstorms was the average of the 21 models (RCP2.6 has only 13 models).

In this study, population exposure to rainstorms is defined as the population in rainstorm-prone areas. The exposure can be computed by multiplying the rainstorm intensity and population for each grid. Then by zonal statistics, we derived national population exposures to rainstorms. We obtained the change of exposures by subtracting exposures in the baseline period from exposures in the future scenarios and then calculated exposures' change of the MME by the same method as for exposures. We used standard deviation to represent the uncertainty between models.

3 Results

By zonal statistics of the population exposure to rainstorms, we obtained the national population exposures to rainstorms of each country/region and present the data of the top 10 countries/regions in Fig. 2. Figure 2 shows that most of these countries are in Asia. The exposures of India and China are more than 2.5×10^{11} person · mm, which is 2 to 18 times of the other eight countries.

The spatial distribution patterns of the population exposure to rainstorms are similar for each scenario and time period combination. The areas with high exposure are mainly distributed in East Asia (e.g., southeastern China, South Korea, and Japan), South Asia (e.g., India and Bangladesh), and Southeast Asia (e.g., the Philippines and Indonesia), and scattered in Africa (e.g., Uganda, Nigeria, and Ethiopia). The spatial patterns in Asia and Africa have changed the most.

Fig. 1 Technical flowchart for mapping global population exposure to rainstorms

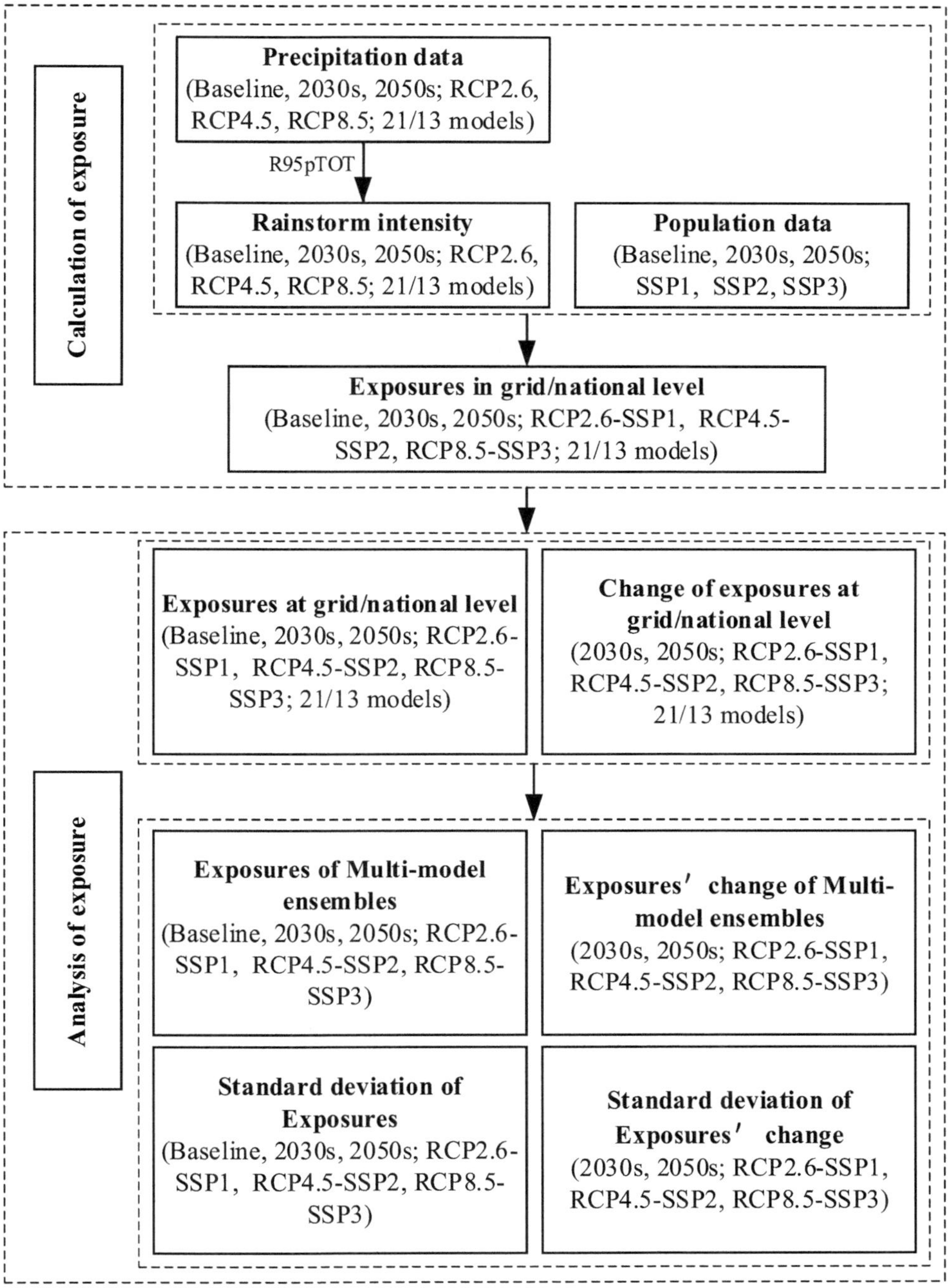

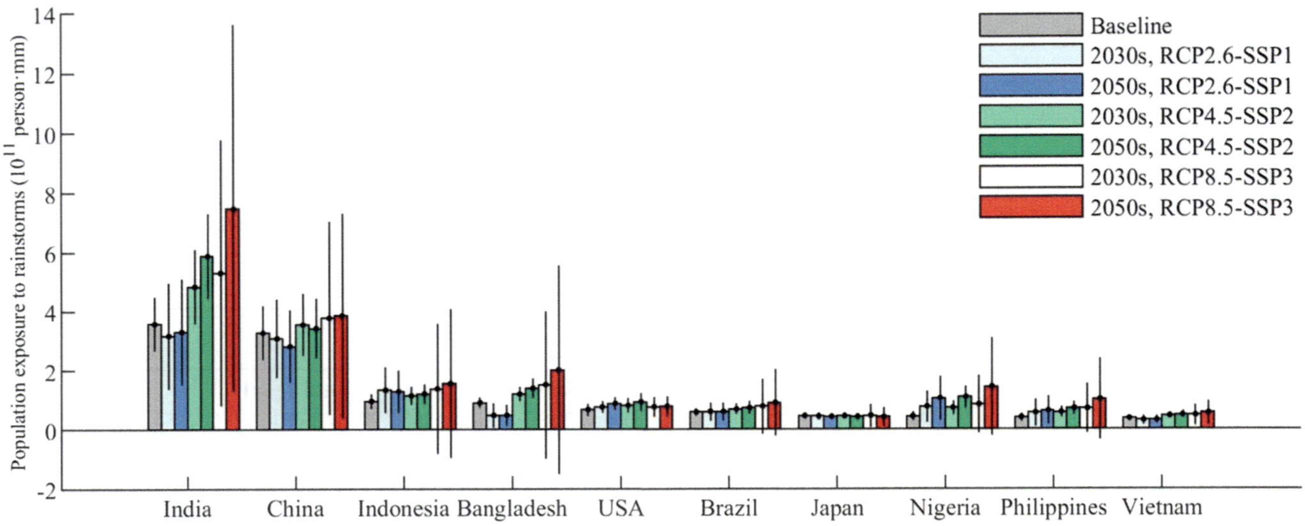

Fig. 2 Population exposure to rainstorms of the top 10 countries/regions (in descending order by total exposure). The error bar represents the one standard deviation across the combination of 21 (13 for Representative Concentration Pathway (RCP) 2.6) general circulation models (GCMs) and 3 shared socioeconomic pathways (SSPs)

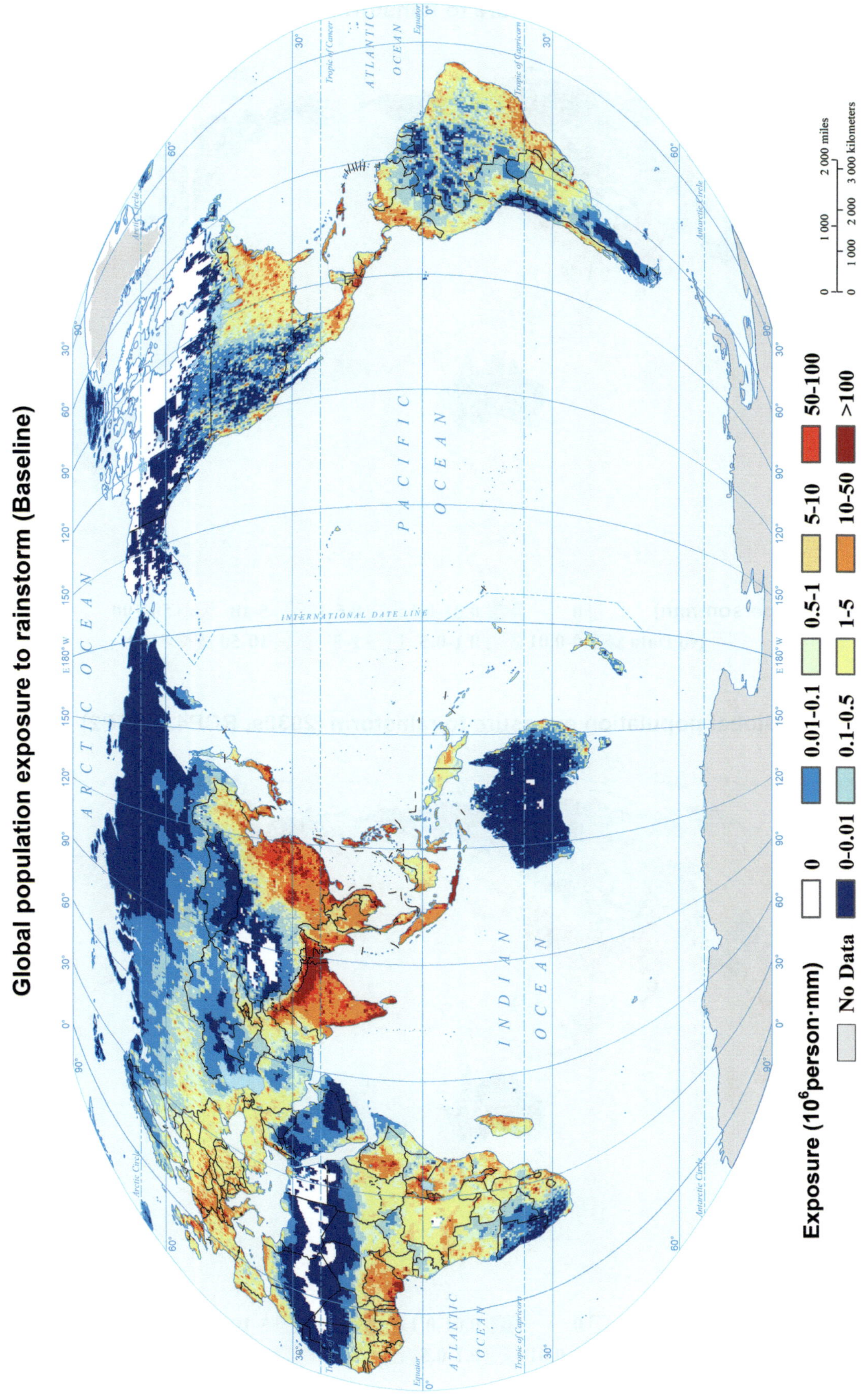
Global population exposure to rainstorm (Baseline)
ARCTIC OCEAN
PACIFIC OCEAN
ATLANTIC OCEAN
INDIAN OCEAN
INTERNATIONAL DATE LINE
Tropic of Cancer
Tropic of Capricorn
Equator
Arctic Circle
Antarctic Circle
Exposure (10^6 person·mm)
No Data
0
0-0.01
0.01-0.1
0.1-0.5
0.5-1
1-5
5-10
10-50
50-100
>100
0 1 000 2 000 miles
0 1 000 2 000 3 000 kilometers

Global population exposure to rainstorm (2030s, RCP2.6-SSP1)

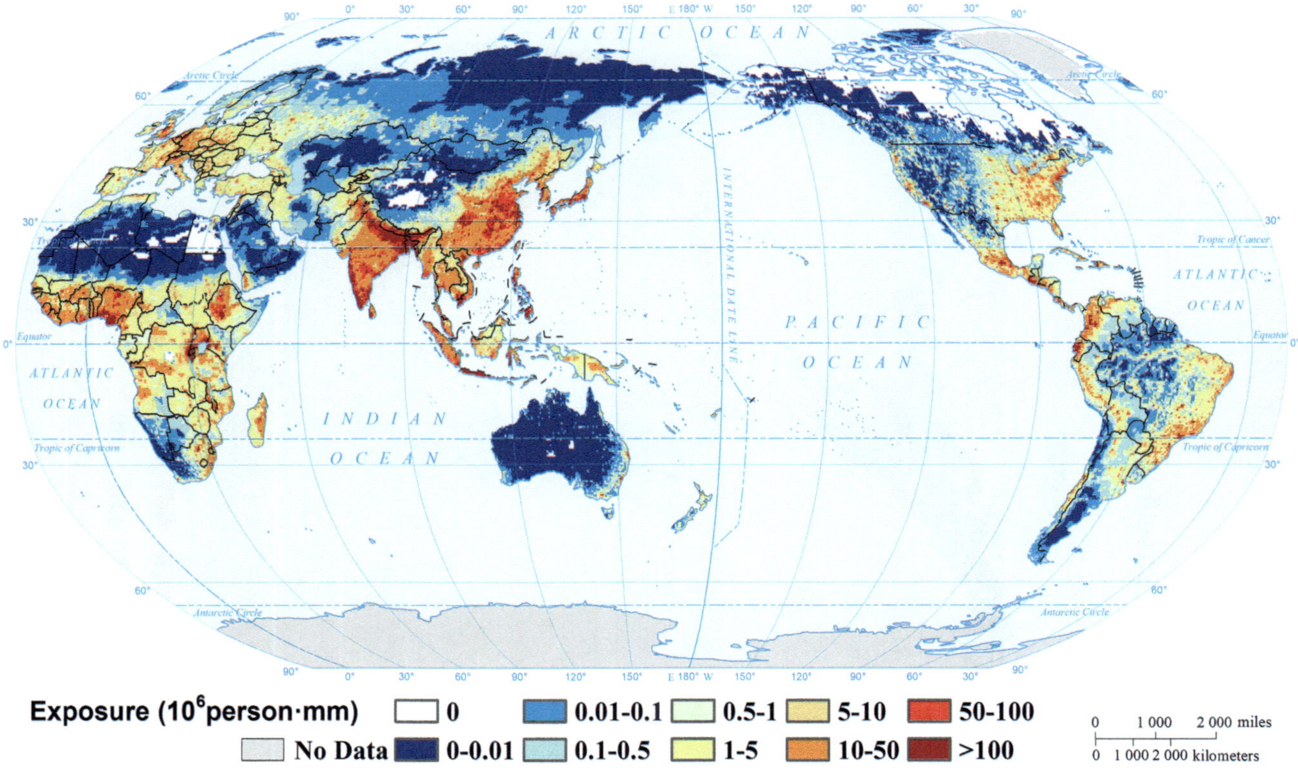

Global population exposure to rainstorm (2030s, RCP4.5-SSP2)

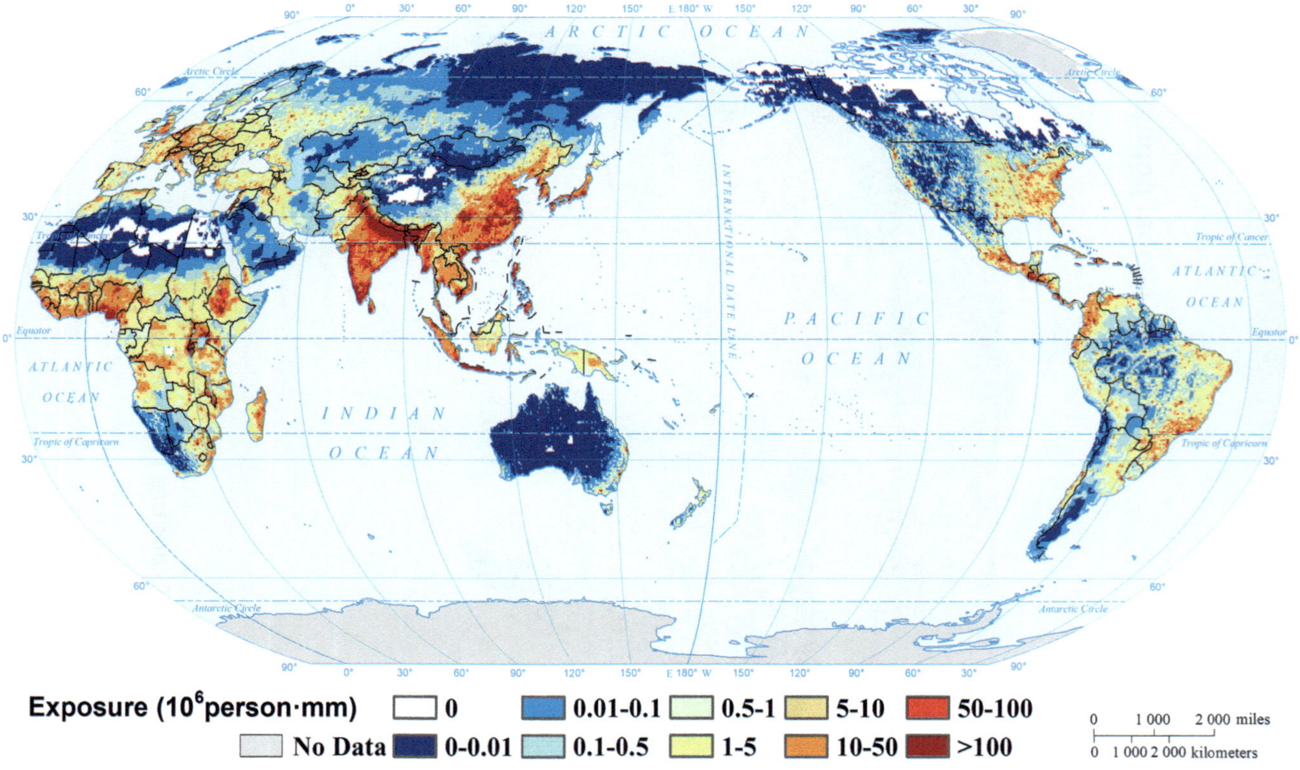

Global population exposure to rainstorm (2030s, RCP8.5-SSP3)

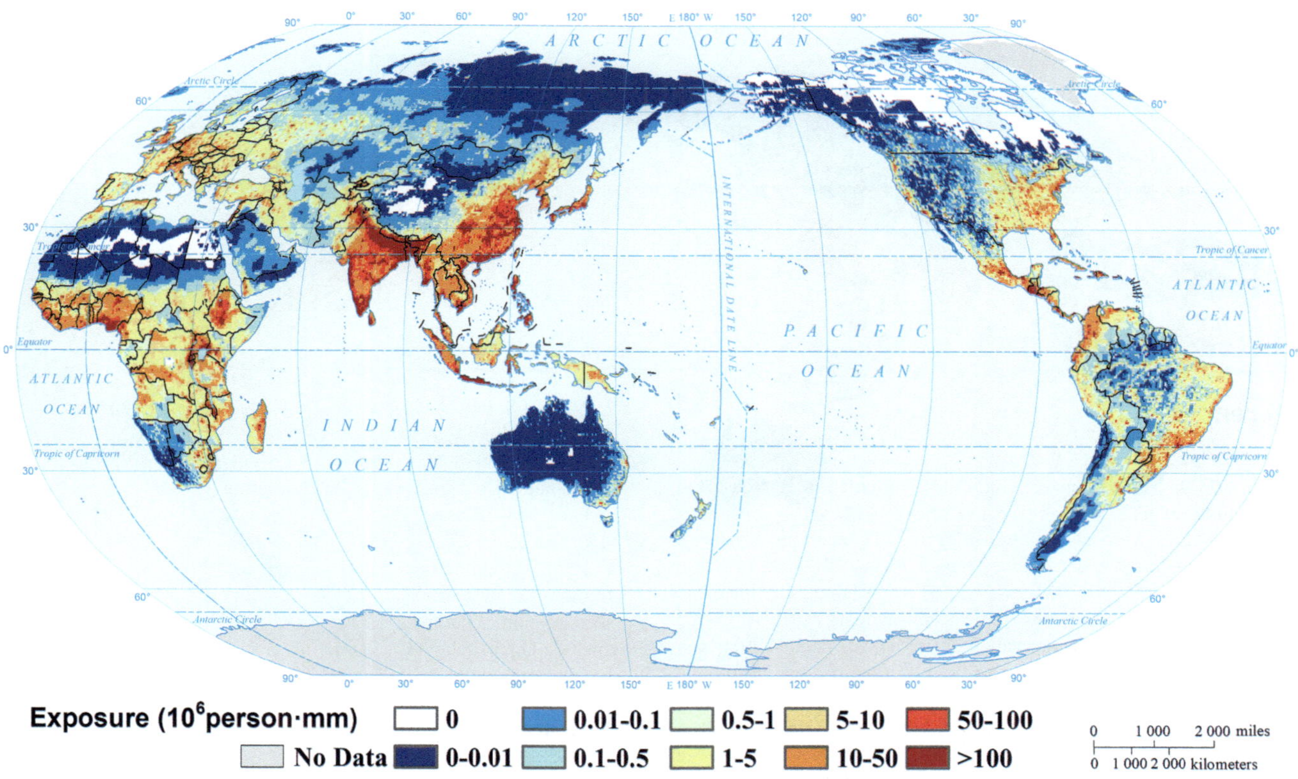

Global population exposure to rainstorm (2050s, RCP2.6-SSP1)

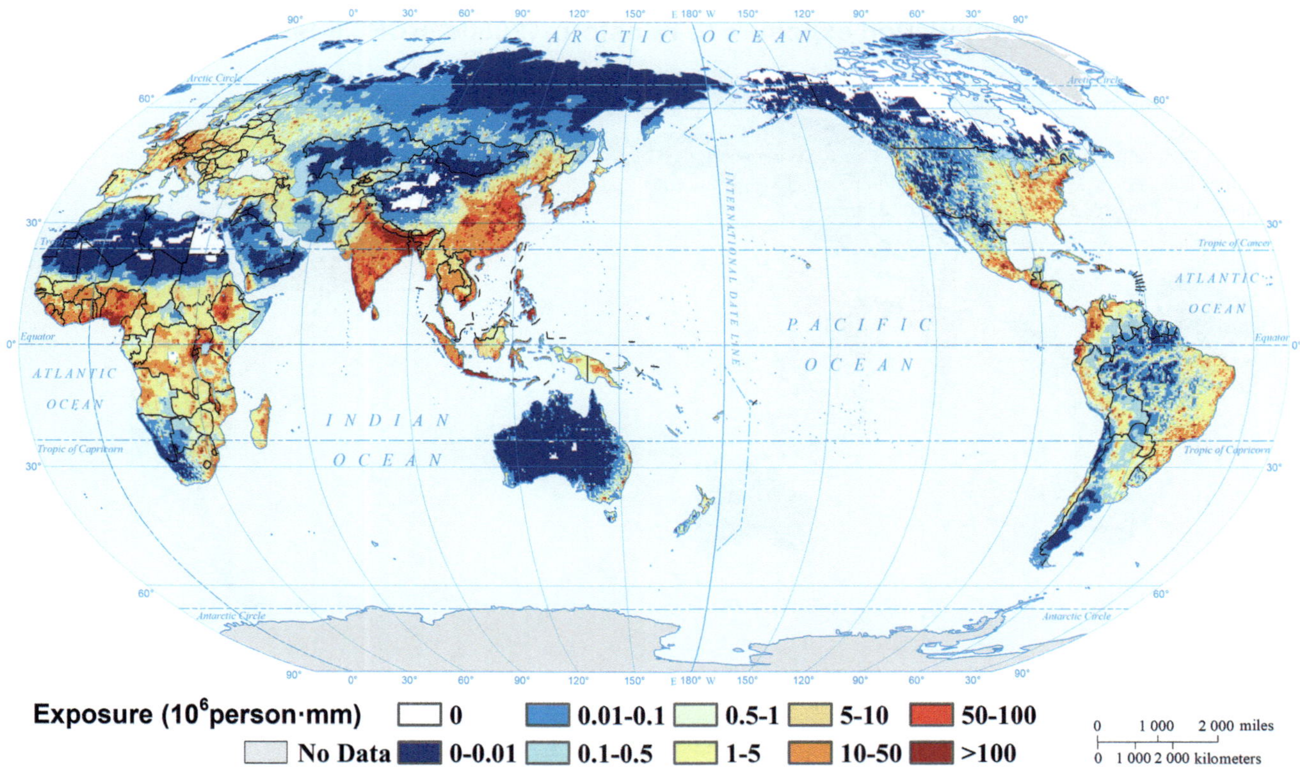

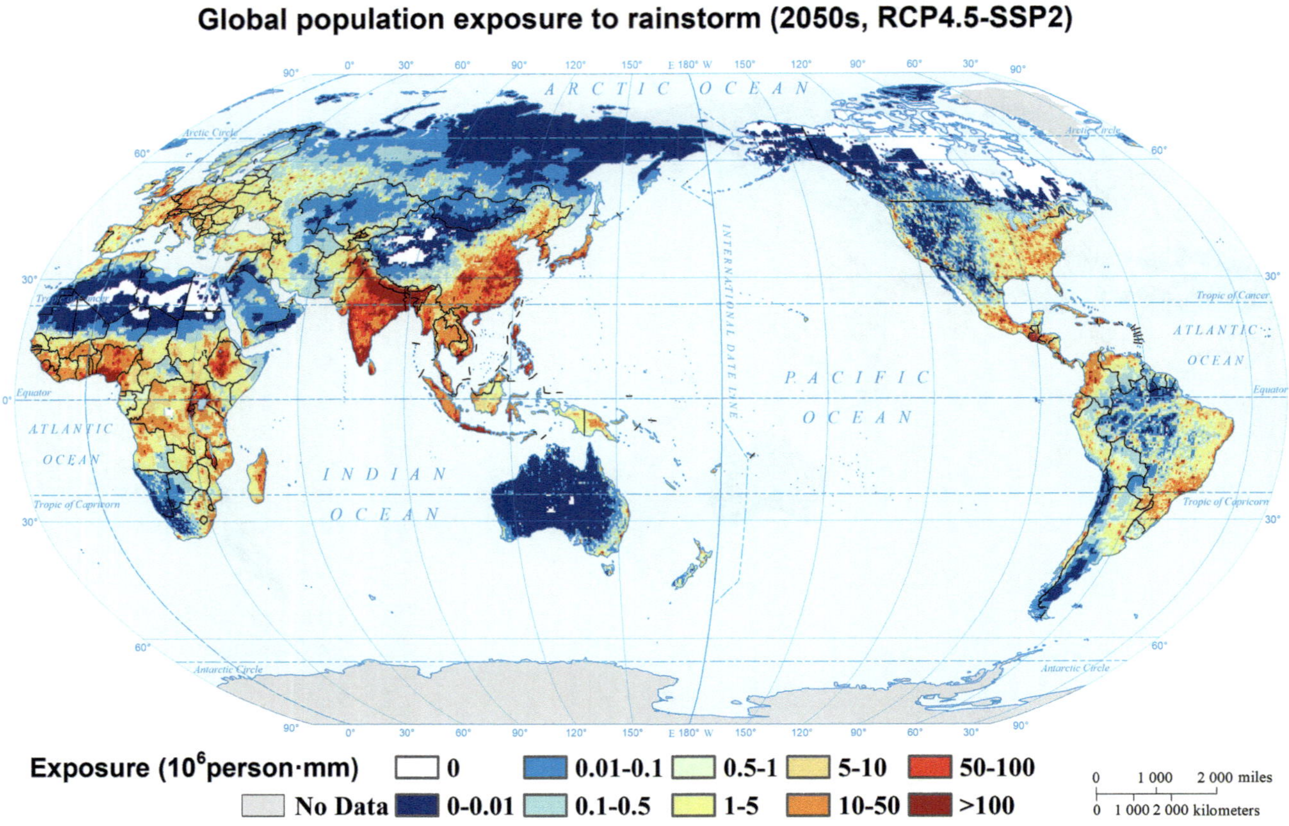
Global population exposure to rainstorm (2050s, RCP4.5-SSP2)
ARCTIC OCEAN
PACIFIC OCEAN
ATLANTIC OCEAN
INDIAN OCEAN
INTERNATIONAL DATE LINE
Arctic Circle
Tropic of Cancer
Equator
Tropic of Capricorn
Antarctic Circle
Exposure (10^6 person·mm)
0
No Data
0-0.01
0.01-0.1
0.1-0.5
0.5-1
1-5
5-10
10-50
50-100
>100
0 1 000 2 000 miles
0 1 000 2 000 kilometers

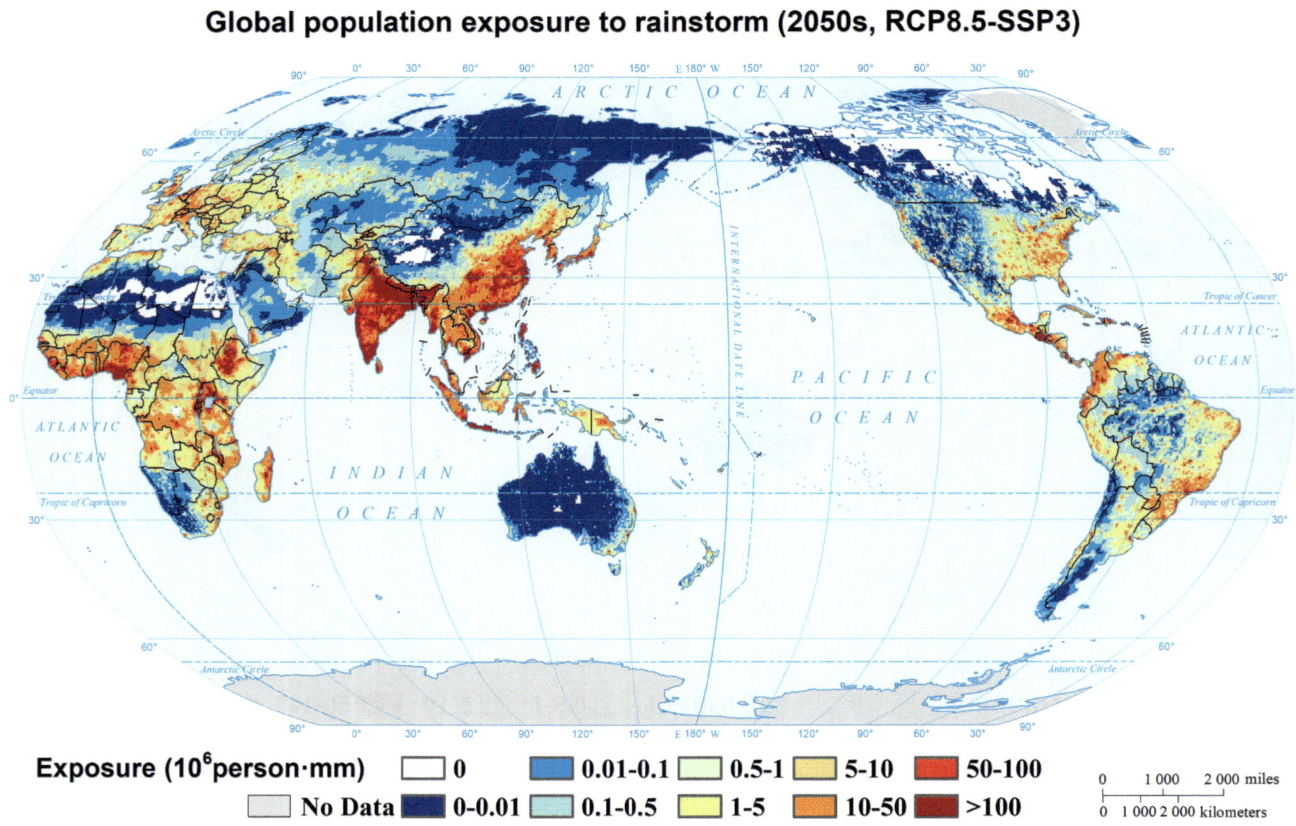
Global population exposure to rainstorm (2050s, RCP8.5-SSP3)
ARCTIC OCEAN
PACIFIC OCEAN
ATLANTIC OCEAN
INDIAN OCEAN
INTERNATIONAL DATE LINE
Arctic Circle
Tropic of Cancer
Equator
Tropic of Capricorn
Antarctic Circle
Exposure (10^6 person·mm)
0
No Data
0-0.01
0.01-0.1
0.1-0.5
0.5-1
1-5
5-10
10-50
50-100
>100
0 1 000 2 000 miles
0 1 000 2 000 kilometers

References

Bonsal, B.R., X. Zhang, L.A. Vincent, and W.D. Hogg. 2001. Characteristics of daily and extreme temperatures over Canada. *Journal of Climate* 14 (9): 1959–1976.

Chen, H.P., and J.Q. Sun. 2020. Increased population exposure to precipitation extremes in China under global warming scenarios. *Atmospheric and Oceanic Science Letter* 13 (1): 63–70.

Donat, M.G., A.L. Lowry, L.V. Alexander, P.A. O'Gorman, and N. Maher. 2016. More extreme precipitation in the world's dry and wet regions. *Nature Climate Change* 6 (5): 508–513.

IPCC (Intergovernmental Panel on Climate Change). 2012. Managing the Risks of Extreme Events and Disasters to Advance Climate Change Adaptation. A Special Report of Working Groups I and II of the Intergovernmental Panel on Climate Change. Cambridge, UK: Cambridge University Press.

IPCC (Intergovernmental Panel on Climate Change). 2013. Climate Change 2013: The Physical Science Basis. Contribution of Working Group I to the Fifth Assessment Report of the Intergovernmental Panel on Climate Change. Cambridge, UK: Cambridge University Press.

Kong, F. 2020. SSPs scenarios-based evolution comparison and mutation characteristics pre-estimate of global sea-land rainstorm time series. *Water Resources and Hydropower Engineering* 51 (10): 1–9.

Liang, P.J., W. Xu, Y.J. Ma, X.J. Zhao, and L.J. Qin. 2017. Increase of elderly population in the rainstorm hazard areas of China. *International Journal of Environmental Research and Public Health* 14 (9): 963.

Liao, X.L., W. Xu, J.L. Zhang, Y. Li, and Y.G. Tian. 2019. Global exposure to rainstorms and the contribution rates of climate change and population change. *Science of the Total Environment* 663: 644–653.

Qin, D.H., J.Y. Zhang, C.C. Shan, and L.C. Song. 2015. *China national assessment report on risk management and adaptation of climate extremes and disasters*. Beijing: Science Press.

Shi, X.Y., J. Chen, L. Gu, C.Y. Xu, H. Chen and L.P. Zhang. 2021. Impacts and socioeconomic exposures of global extreme precipitation events in 1.5 and 2.0°C warmer climates. *Science of The Total Environment* 766: 142665.

Sun, Q.H., X.B. Zhang, F. Zwiers, S. Westra, and L.V. Alexander. 2021. A global, continental, and regional analysis of changes in extreme precipitation. *Journal of Climate* 34 (1): 243–258.

Westra, S., L.V. Alexander, and F.W. Zwiers. 2013. Global increasing trends in annual maximum daily precipitation. *Journal of Climate* 26 (11): 3904–3918.

Zhang, W.X., T.J. Zhou, L.W. Zou, L.X. Zhang and X.L. Chen. 2018. Reduced exposure to extreme precipitation from 0.5°C less warming in global land monsoon regions. *Nature Communications* 9(1): 3153.

Mapping Global GDP Distribution

Fubao Sun, Tingting Wang, and Hong Wang

1 Introduction

Socioeconomic projections are crucial in climate change impact, mitigation, and adaptation research and risk assessments for future scenarios (O'Neill et al. 2014). The climate projections in the Scenario Model Intercomparison Project (ScenarioMIP) are formed based on different Shared Socioeconomic Pathways (SSPs) and specific Representative Concentration Pathways (RCPs) of Phase 6 of the Coupled Model Intercomparison Project (CMIP6) (O'Neill et al. 2016). The development of research on future socioeconomic impact and on reduction in exposure and vulnerability and increase in resilience to climate extremes (Wilbanks and Ebi 2014; Chen et al. 2020) can benefit from more spatially explicit socioeconomic data of higher spatial resolution and precision under five SSPs (O'Neill et al. 2016).

The GDP is a standard socioeconomic indicator in economic development assessment within and across countries (Nordhaus 2011; Kummu et al. 2018; Tobias 2018), and is usually provided at the national scale from several global institutions such as the World Bank, the Organization for Economic Co-operation and Development (OECD), and so on. But regional GDP data at a finer spatial resolution (e.g., at the state, city, or county level) are often unavailable, especially in many developing countries (Nordhaus 2011; Kummu et al. 2018; Huang et al. 2018). The recent development of satellite-derived nighttime light (NTL) images and gridded population data helps disaggregate global and regional GDP into gridded datasets at various spatial scales (Doll et al. 2006; Ghosh et al. 2010; Nordhaus 2011; Zhao et al. 2017) to support current research on climate change and risk assessments. These disaggregated GDP data are, however, often biased due to saturation problem of NTL images or the assumption of even GDP per capita within a given administrative boundary when using population datasets. Besides, the global gridded GDP projections of future SSPs are rather limited (Jones and O'Neill, 2016; Jiang et al. 2017,2018; Kummu et al. 2018).

The objective of this research is to present a set of spatially explicit global gridded GDP data that are comparable and represent substantial long-term changes of GDP for both the historical period and for future projections under five SSPs (Wang and Sun 2020).

Authors: Fubao Sun, Tingting Wang, Hong Wang.
Map Designers: Qingyuan Ma, Jing'ai Wang, Ying Wang.
Language Editor: Tingting Wang.

F. Sun (✉) · T. Wang · H. Wang
Key Laboratory of Water Cycle and Related Land Surface Processes, Institute of Geographic Sciences and Natural Resources Research, Chinese Academy of Sciences, Beijing, 100101, China
e-mail: sunfb@igsnrr.ac.cn

F. Sun
State Key Laboratory of Desert and Oasis Ecology, Xinjiang Institute of Ecology and Geography, Chinese Academy of Sciences, Urumqi, 830011, China

Akesu National Station of Observation and Research for Oasis Agro-ecosystem, Akesu, 843000, China

College of Resources and Environment, University of Chinese Academy of Sciences, Beijing, 100049, China

P. Shi, *Atlas of Global Change Risk of Population and Economic Systems*, IHDP/Future Earth-Integrated Risk Governance Project Series, https://doi.org/10.1007/978-981-16-6691-9_8

2 Method

2.1 Global GDP Distribution

National GDP purchasing power parity (PPP) of 195 countries/regions for 2005 was obtained from Geiger (2018), which was mainly from the Penn World Tables, and data of missing countries were taken from another version and the World Bank after rescaling from 2011 to 2005 PPP in U.S. dollars.

2.2 Population-Based GDP Disaggregation

Broad literature has emphasized the role of human capital as a key driver of economic growth, and population can well capture the link between the human and economic systems in various models. Hence, gridded population dataset has been widely applied in spatial allocation of global and regional GDP based on several well-known population datasets, e.g., the LandScan Global Population database, the Gridded Population of the World dataset, Version 4 (GPWv4), the Worldpop, and so on. The GDP disaggregation based on population dataset (denoted as GDP_{Pop}) can be expressed as.

$$\mathrm{GDP_{pop}} = \mathrm{Pop_{pixel}} \times \mathrm{P_{cap}} = \mathrm{Pop_{pixel}} \times \frac{\mathrm{GDP_i}}{\mathrm{Pop_i}} \quad (1)$$

where GDP_{pop} and Pop_{pixel} are the GDP and population in each pixel in administrative unit i, and $Pcap$ is the GDP per capita, which is the ratio of GDP total (GDP_i) to population total (Pop_i) in a given administrative area i.

2.3 NTL-Based GDP Disaggregation

Numeric modelings have shown that the satellite-derived NTL data are well correlated with GDP at all examined scales, and such data has been widely used in the spatial allocation of GDP across the globe. The Defense Meteorological Satellite Program's Operational Linescan System (DMSP-OLS) NTL images in 2005 (average visible, stable lights, and cloud-free coverages, satellites F14 and F15 simultaneously collected global NTL images, and data from F15 were chosen as newer sensor would have less degradation of data quality) were used to disaggregate global GDP to a spatial resolution of 30 arc seconds. Based on the results of relevant studies (Ghosh et al. 2010; Nordhaus 2011; Zhao et al. 2017; Eberenz et al. 2020), the GDP totals were directly distributed to each pixel in proportion to the DN values of pixels in a given administrative area, and the NTL-based GDP disaggregation (denoted as GDP_{Lit}) can be described as

$$\mathrm{GDP_{Lit}} = \mathrm{GDP_{per\_light}} \times \mathrm{DN_{pixel}} = \frac{\mathrm{GDP_i}}{\mathrm{SL_i}} \times \mathrm{DN_{pixel}} \quad (2)$$

where GDP_i is the total GDP, SL_i is the sum of DN values, GDP_{per_light} is the constant in administrative unit i, and DN_{pixel} and GDP_{pixel} are the DN value and corresponding GDP in each pixel in administrative unit i.

The saturation problem in the DMSP-OLS NTL images, however, has resulted in biased GDP disaggregation (Zhao et al. 2017). Zhao et al. (2017) improved its accuracy by incorporating the gridded population data into NTL-based GDP disaggregation in each pixel, and Eberenz et al. (2020) further improved this method. By multiplying the NTL image with the LandScan population data, a LitPop image was produced based on Eq. (3), which was then used to spatialize GDP at the global scale (denoted as $\mathrm{GDP_{Lit\text{-}Pop}}$) using Eq. (4):

$$\mathrm{LitPop_{pix}} = \begin{cases} \mathrm{Pop} & \mathrm{Lit} = 0 \\ \mathrm{Lit} \cdot \mathrm{pop} & \mathrm{Lit} > 0\ \&\mathrm{Pop} > 0 \\ \mathrm{Lit} & \mathrm{Pop} = 0 \end{cases} \quad (3)$$

$$\mathrm{GDP_{LitPop}} = \frac{\mathrm{GDP_i}}{\mathrm{SLP_i}} \times \mathrm{LitPop_{pix}} \quad (4)$$

where $LitPop_{pix}$ is the GDP value of each pixel of the *LitPop* data and SLP_i is the sum of values of the LitPop image in administrative unit i.

The technical flow of mapping global GDP is shown in Fig. 1.

3 Results

To examine the performance of the GDP_{Pop}, GDP_{lit}, and GDP_{LitPop}, comparisons were made between GDP in 205 countries using the World Bank data (Fig. 2a); in 476 states (provinces) of 36 OECD countries, the United States, and China (Fig. 2b); and in 5231 counties in the United States and China (Fig. 2c) in 2005 that are spatially joined to the corresponding GIS-based administrative boundaries, respectively. The comparisons show that the accuracy of the three disaggregated GDP datasets decreases with the change

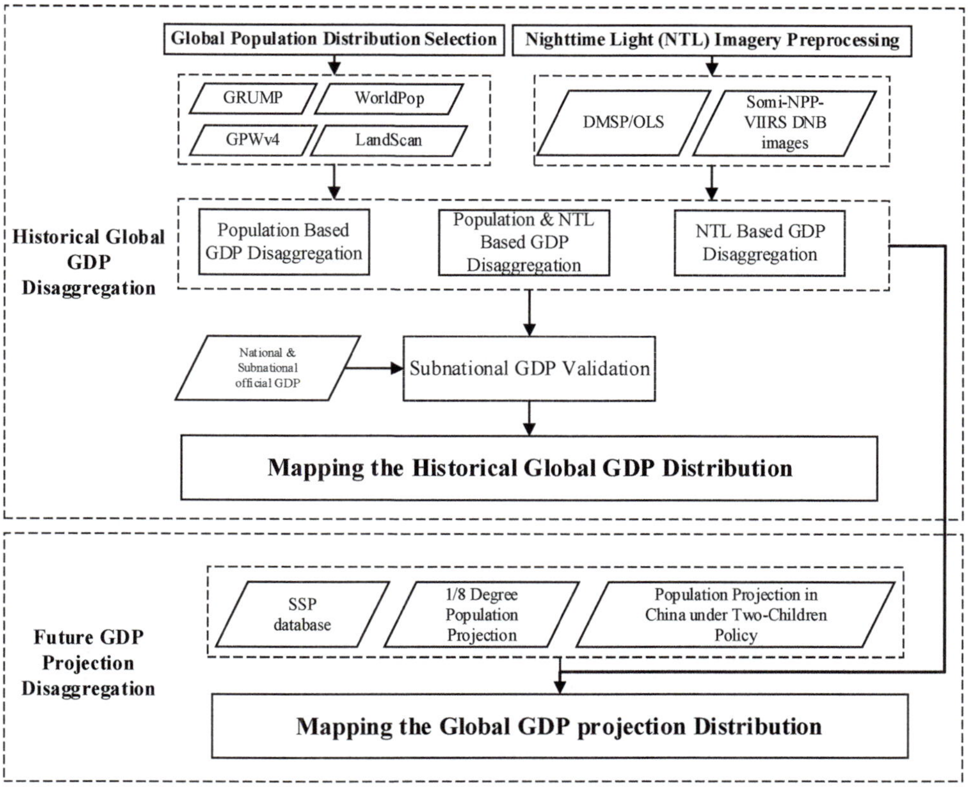

Fig. 1 Technical flowchart of mapping global GDP

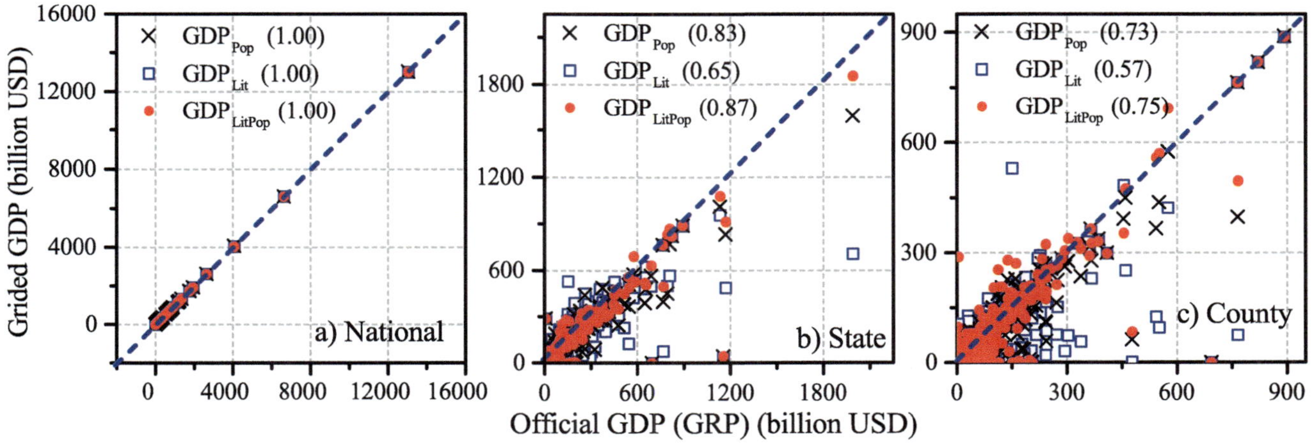

Fig. 2 Comparison between official GDP and disaggregated results at the national **a**, state **b**, and county **c** scales in 2005, using population-based disaggregation method, NTL-based approach, and LitPop approach, respectively, and the value in each brackets is their corresponding R^2

Table 1 GDP statistics (trillion USD in 2005 PPP)

		High income	Upper middle income	Middle income	Low and middle income	Lower middle income	Low income
2005		37.01	10.73	0.27	1.14	0.55	0.09
2030	SSP1	70.28	83.63	0.95	5.93	2.67	0.58
	SSP2	62.28	66.36	0.84	5.06	2.29	0.47
	SSP3	70.28	83.63	0.95	5.93	2.67	0.58
2050	SSP1	95.07	136.08	1.95	14.90	8.69	2.72
	SSP2	85.06	101.78	1.64	11.27	6.33	1.71
	SSP3	68.63	76.92	1.46	7.96	4.51	1.11

(decrease) of their spatial scales, and GDP_{LitPop} is superior to GDP_{Pop} and GDP_{Lit} at the national, state (provincial), and county levels with clear advantages evaluated by their R^2 and RMSE (Fig. 2).

Gross National Income (GNI) includes the nation's GDP plus the income from overseas sources, and has been widely used to measure and track a nation's wealth from year to year. Based on GNI per capita (current international dollars in PPP) in 2019 level from the World Bank, the global and regional GDP growth depicts major differences under the three SSP scenarios in 2030 and 2050 by income level of countries (Table 1). For countries with lower middle income (≤ 6761 GNI per capita) and low income (≤ 2458 GNI per capita), in 2030 the GDP is about 6 times that of 2005 for SSP1-3, and reaches as high as 30, 19, and 12 times in 2050, respectively. In developed countries with high income ($\geq 52{,}412$ GNI per capita), in 2030 the GDP will only be about 1.7–.9 times that of 2005, and increase to about 2.6, 2.3, and 1.8 times in 2050 for SSP1–3. Meanwhile, for countries with middle income ($\leq 11{,}934$ GNI per capita) and low and middle income ($\leq 10{,}937$ GNI per capita), in 2030 the GDP will be 3.5 and 6.2 times that of 2005, showing an unequal growth rate among regions.

Global gross domestic production (Baseline)

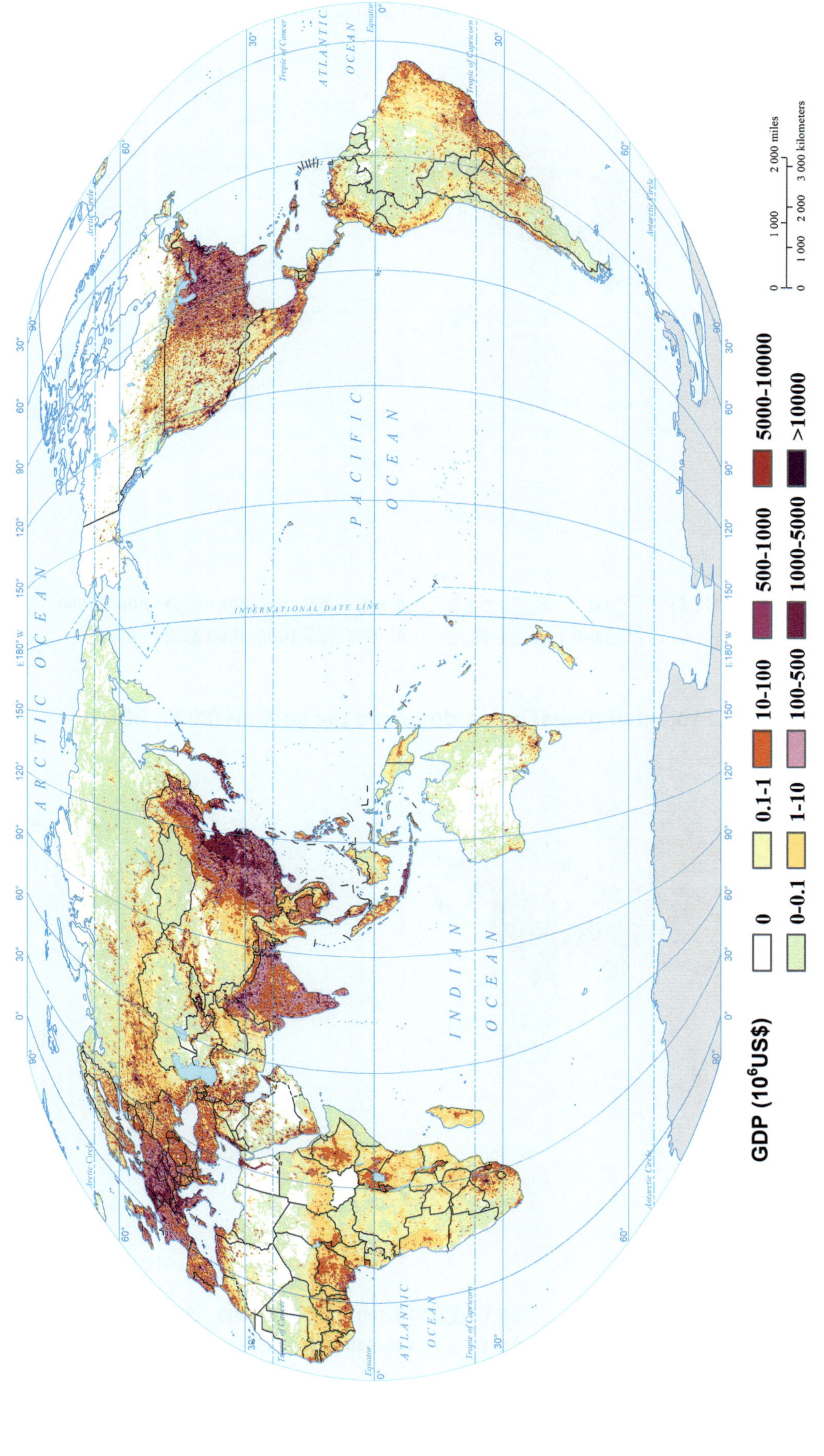

Projected global gross domestic production (2030, SSP1)

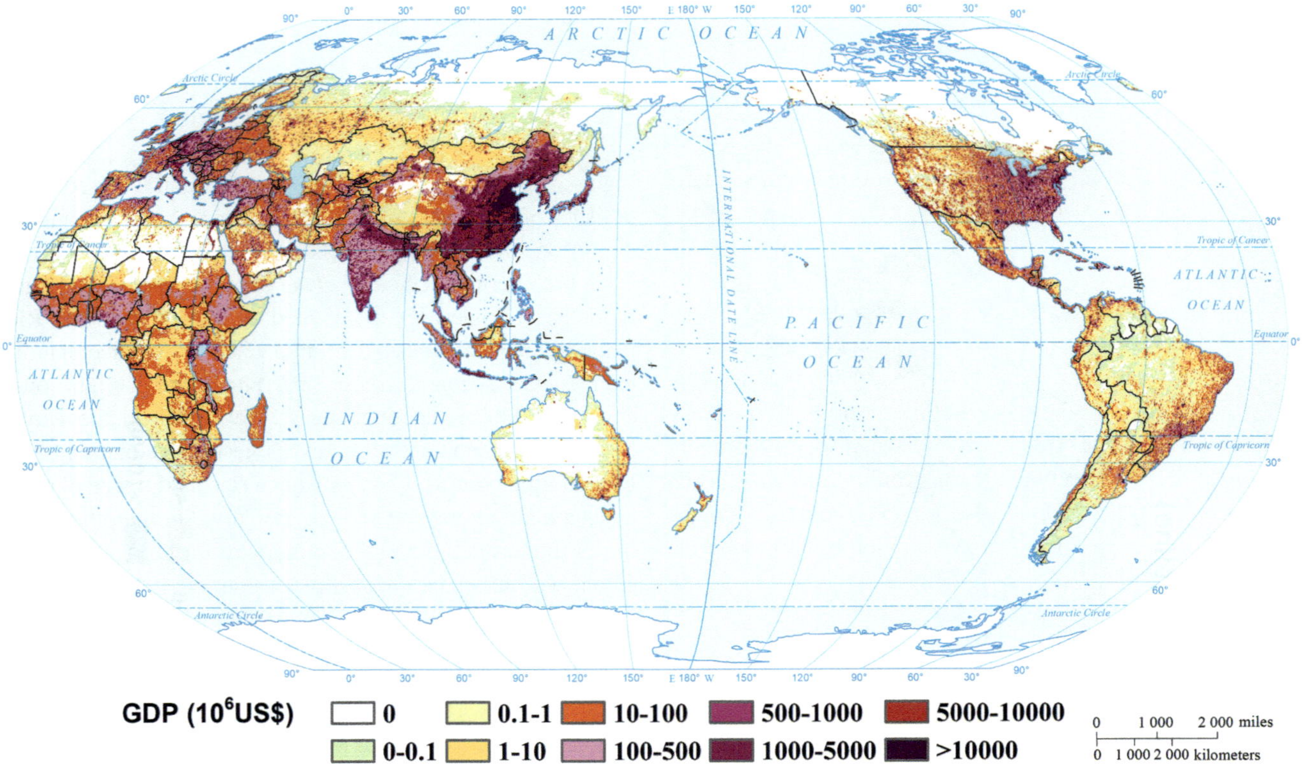

Projected global gross domestic production (2030, SSP2)

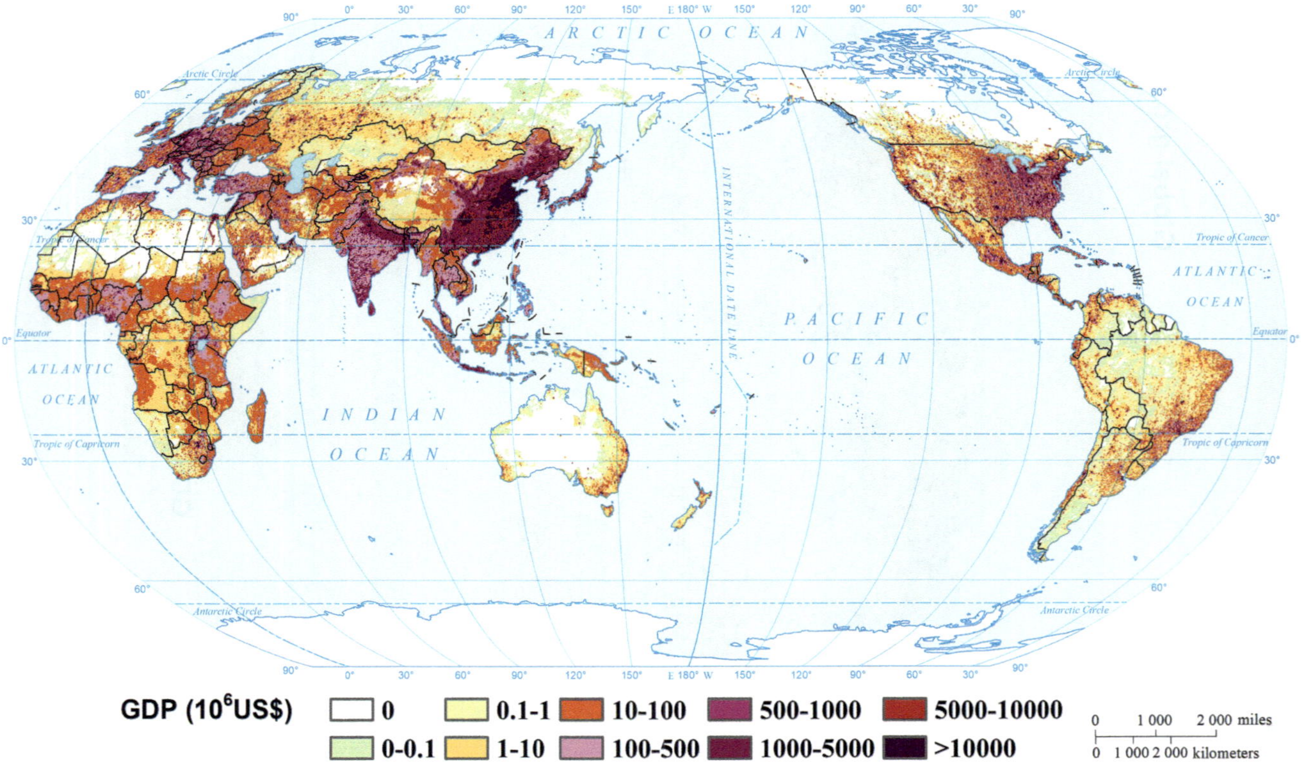

Projected global gross domestic production (2030, SSP3)

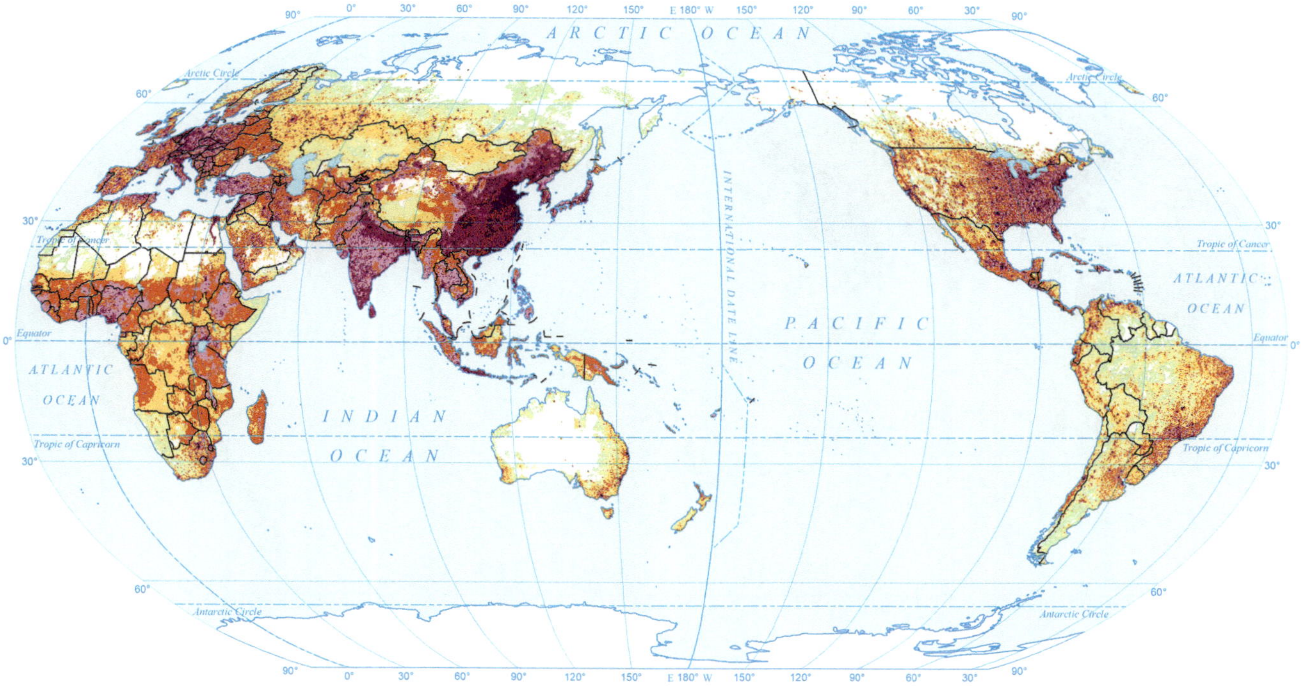

GDP (10^6US$) 0; 0-0.1; 0.1-1; 1-10; 10-100; 100-500; 500-1000; 1000-5000; 5000-10000; >10000

0 1 000 2 000 miles
0 1 000 2 000 kilometers

Projected global gross domestic production (2050, SSP1)

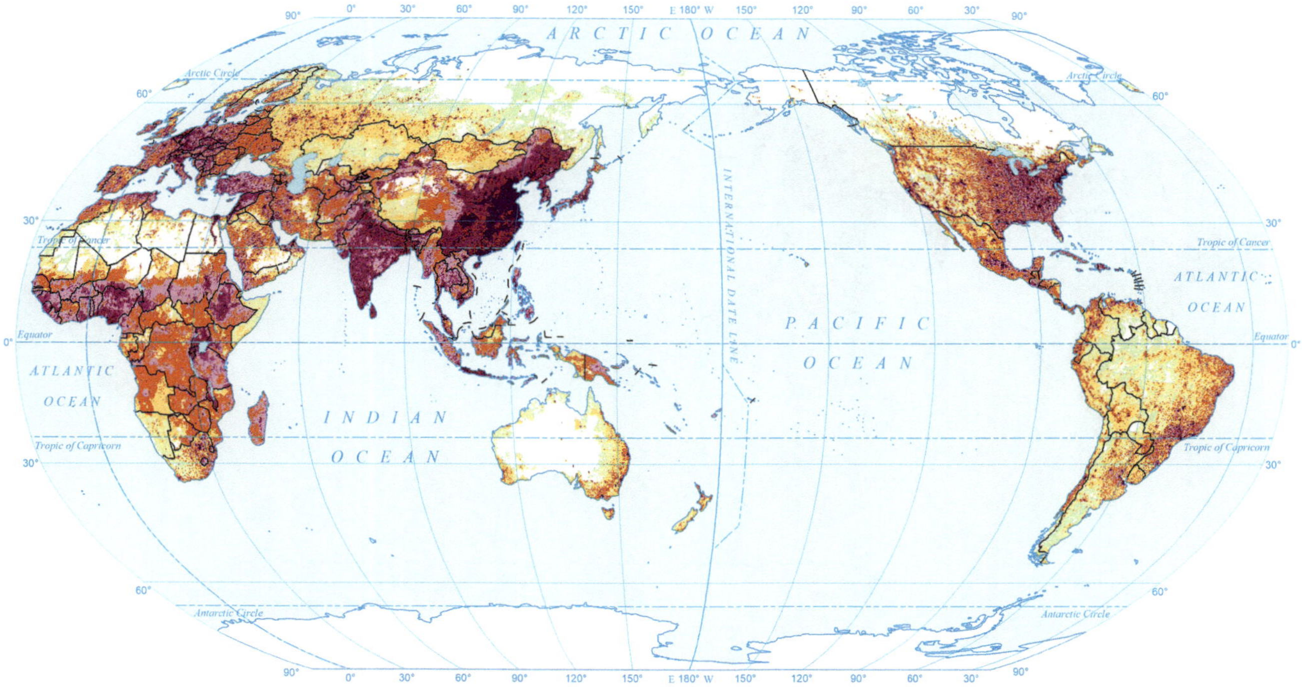

GDP (10^6US$) 0; 0-0.1; 0.1-1; 1-10; 10-100; 100-500; 500-1000; 1000-5000; 5000-10000; >10000

0 1 000 2 000 miles
0 1 000 2 000 kilometers

Projected global gross domestic production (2050, SSP2)

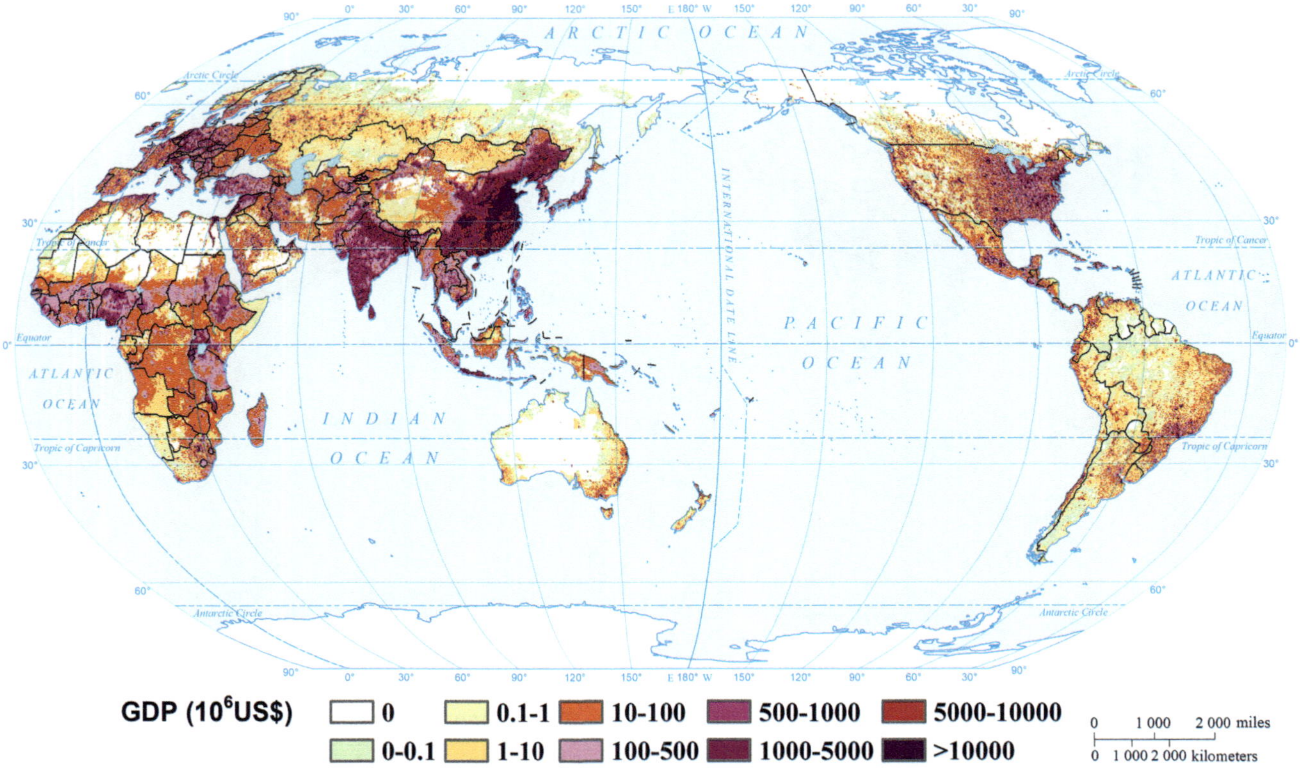

Projected global gross domestic production (2050, SSP3)

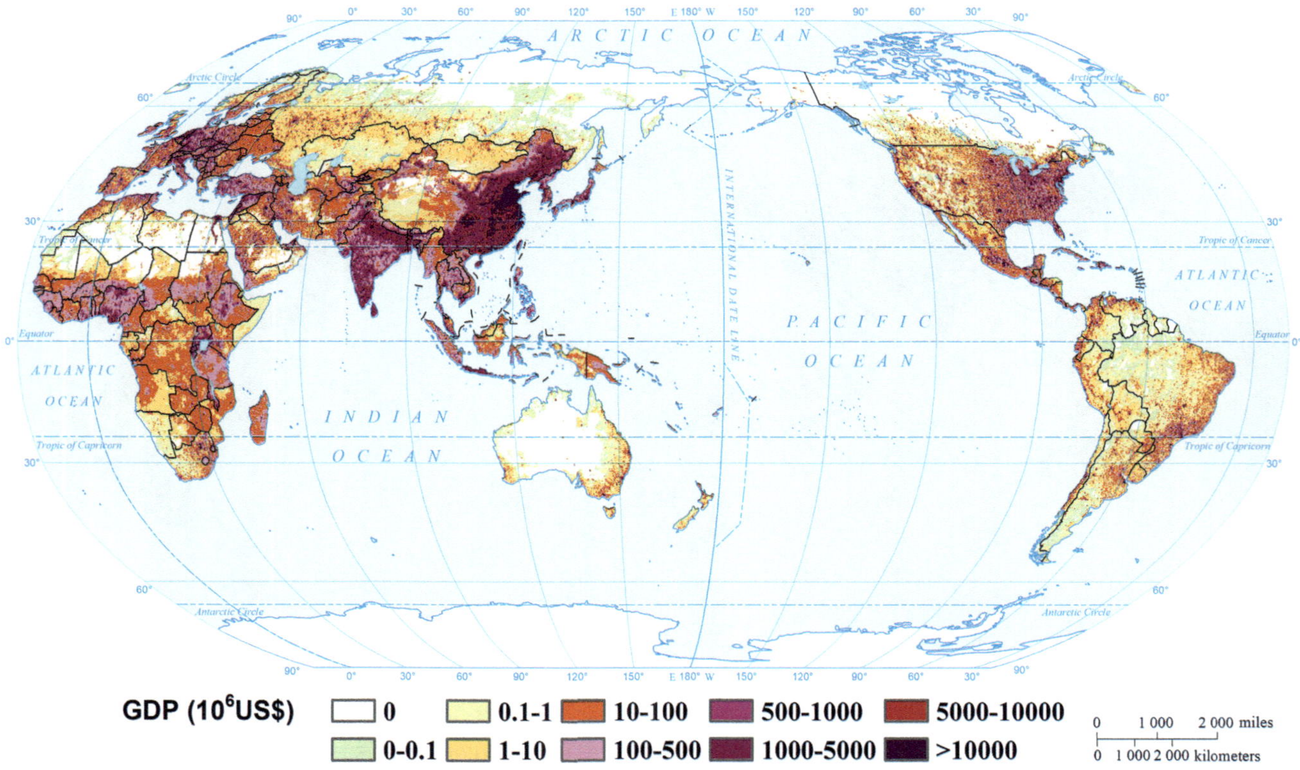

References

Chen, Q., T. Ye, N. Zhao, M. Ding, and X. Yang. 2020. Mapping China's regional economic activity by integrating points-of-interest and remote sensing data with random forest. *Environment and Planning B Urban Analytics and City Science* 3: 1–19.

Doll, C.N.H., J.P. Muller, and J.G. Morley. 2006. Mapping regional economic activity from night-time light satellite imagery. *Ecological Economics* 57: 75–92.

Eberenz, S., D. Stocker, T. Rsli, et al. 2020. Asset exposure data for global physical risk assessment. *Earth System Science Data* 12: 817–833.

Geiger, T. 2018. Continuous national gross domestic product (gdp) time series for 195 countries: past observations (1850–2005) harmonized with future projections according to the shared socio-economic pathways (2006–2100). *Earth System Science Data* 10: 847–856.

Ghosh, T., R.L. Powell, C.D. Elvidge, K.E. Baugh, P.C. Sutton, and S. Anderson. 2010. Shedding light on the global distribution of economic activity. *The Open Geography Journal* 3: 147–160.

Huang, J., D. Qin, T. Jiang, Y. Wang, Z. Feng, J. Zhai, L. Cao, Q. Chao, X. Xu, and G. Wang. 2019. Effect of fertility policy changes on the population structure and economy of China: From the perspective of the shared socioeconomic pathways. *Earth's Future* 7: 250–265.

Jiang, T., J. Zhao, L. Cao, Y. Wang, B. Su, C. Jing, R. Wang, and C. Gao. 2018. Projection of national and provincial economy under the shared socioeconomic pathways in China. *Climate Change Research* 14: 50–58.

Jiang, T., J. Zhao, C. Jing, L. Cao, Y. Wang, H. Sun, A. Wang, J. Huang, et al. 2017. National and provincial population projected to 2100 under the shared socioeconomic pathways in China. *Climate Change Research* 13: 128–137.

Jones, B., and B.C. O'Neill. 2016. Spatially explicit global population scenarios consistent with the shared socioeconomic pathways. *Environmental Research Letters* 11: 084003.

Kummu, M., M. Taka, and J.H. Guillaume. 2018. Gridded global datasets for gross domestic product and human development index over 1990–2015. *Scientific Data* 5: 180004.

Nordhaus, C.W.D. 2011. Using luminosity data as a proxy for economic statistics. *Proceedings of the National Academy of Sciences of the United States of America* 108: 8589–8594.

O'Neill, B.C., E. Kriegler, K. Riahi, K.L. Ebi, S. Hallegatte, T.R. Carter, R. Mathur, and D.P. Van Vuuren. 2014. A new scenario framework for climate change research: The concept of shared socioeconomic pathways. *Climatic Change* 122: 387–400.

O'Neill, B.C., C. Tebaldi, D.P. Van Vuuren, V. Eyring, P. Friedlingstein, G. Hurtt, R. Knutti, E. Kriegler, et al. 2016. The scenario model intercomparison project (ScenarioMIP) for CMIP6. *Geoscientific Model Development* 9: 3461–3482.

Tobias, G. 2018. Continuous national gross domestic product (GDP) time series for 195 countries: Past observations (1850–2005) harmonized with future projections according to the shared socio-economic pathways (2006–2100). *Earth System Science Data* 10: 847–856.

Wang, T., and F. Sun. 2020. Spatially explicit global gross cell product (GCP) data set consistent with the shared socioeconomic pathways. *Zenodo, Dataset,*. https://doi.org/10.5281/zenodo.5004249.

Wilbanks, T.J., and K.L. Ebi. 2014. SSPs from an impact and adaptation perspective. *Climatic Change* 122: 473–479.

Zhao, N., Y. Liu, G. Cao, E.L. Samson, and J. Zhang. 2017. Forecasting China's GDP at the pixel level using nighttime lights time series and population images. *Mapping Ences & Remote Sensing* 54: 407–425.

Mapping Global GDP Exposure to Drought

Fubao Sun, Tingting Wang, and Hong Wang

1 Introduction

Accumulative evidences have shown that anthropogenic climatic changes are already influencing the frequency, magnitude, and duration of droughts (Mann and Gleick 2015). Severe drought events during the past decades, e.g., the East Africa drought, the California drought, and a series of severe drought events recently in southern China have profound impacts on global and regional water resources, agriculture activities, and the ecosystem (Wada et al. 2013), and have resulted in huge losses to the society. The climate warming has intensified the magnitude and severity of drought conditions, posing considerable economic, societal, and environmental challenges globally (Carrão et al. 2016; Su et al. 2018; Ahmadalipour et al. 2019; Gu et al. 2020; Cook et al. 2020). Drought losses have significantly increased in recent years, for a range of reasons, including nonclimatic factors around the world. Enhanced drying has been observed and projected over many land areas under a warming climate, due to increasing atmospheric concentrations of greenhouse gases. In the context of climate change, drought exposure is likely to increase in many historically drought-prone regions (Dai 2011; Liu et al. 2020; Su et al. 2018). A better understanding of changes in global drought characteristics and their socioeconomic impacts in the twenty-first century should feed into long-term climate adaptation and mitigation plans.

With expected increases in widespread severe drought events and rapid socioeconomic development, more GDP will be exposed to droughts, resulting in higher drought risks and more potential GDP losses in the future. Exposure of socioeconomic activities is one of the most important aspects of drought risk assessment (Su et al. 2018). The changing vulnerability, exposure of socioeconomic activities to climate extremes are driving a need to move beyond administrative unit-based analyses to enable flexible integration with spatially explicit datasets of population and economic systems of long-term SSPs (Jones et al. 2015; Chen et al. 2017; Su et al. 2018; Liu et al. 2018; Liu et al. 2020). Yet, the existing analyses are very limited and mainly focused on the population exposure to droughts, and the magnitude of drought impacts on GDP in a warming climate is poorly addressed across the globe.

Hence, the GDP exposure to droughts, especially to severe and extreme droughts, across the globe from baseline period (1986-2005) and future scenarios (2016–2035 and 2046–2065) was estimated and quantified based on the general circulation models (GCMs) from the Institute of Atmospheric Physics, Chinese Academy of Sciences, and the GDP data from the Institute of Geographic Sciences and Natural

Authors: Fubao Sun, Tingting Wang, Hong Wang.
Map Designers: Qingyuan Ma, Jing'ai Wang, Ying Wang.
Language Editor: Tingting Wang.

F. Sun (✉) · T. Wang · H. Wang
Key Laboratory of Water Cycle and Related Land Surface Processes, Institute of Geographic Sciences and Natural Resources Research, Chinese Academy of Sciences, Beijing, 100101, China
e-mail: sunfb@igsnrr.ac.cn

F. Sun
State Key Laboratory of Desert and Oasis Ecology, Xinjiang Institute of Ecology and Geography, Chinese Academy of Sciences, Urumqi, 830011, China

Akesu National Station of Observation and Research for Oasis Agro-ecosystem, Akesu, 843000, China

College of Resources and Environment, University of Chinese Academy of Sciences, Beijing, 100049, China

P. Shi, *Atlas of Global Change Risk of Population and Economic Systems*, IHDP/Future Earth-Integrated Risk Governance Project Series, https://doi.org/10.1007/978-981-16-6691-9_9

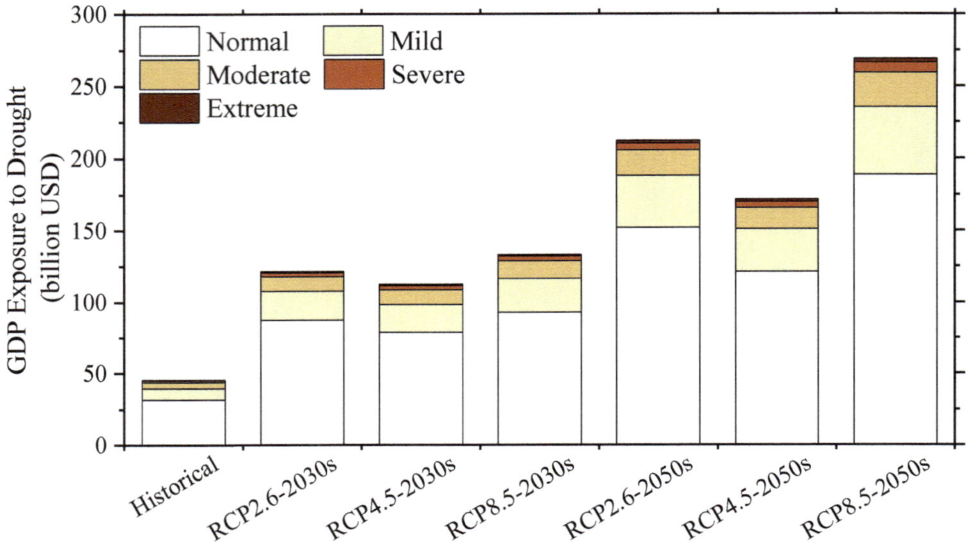

Fig. 1 Global GDP exposure to droughts for the historical period and the 2030s and 2050s under SSP1 -3, and RCP2.6, RCP4.5, and RCP8.5 scenarios

Resources Research of the Chinese Academy of Sciences. The GDP exposure to droughts is mapped at the grid level.

2 Method

2.1 Drought Estimation

The Palmer Drought Severity Index (PDSI) is a simple water balance model originally designed by Palmer under the framework of water balance between water supply and atmospheric evaporative demand (Palmer 1965), and is estimated by the difference between monthly precipitation, potential evaporation, and some other parameters. There were 13 out of 21 GCMs in CMIP5 (Coupled Model Intercomparison Project phase 5): CanESM2, CNRM-CM5, CSIRO-Mk3-6–0, GFDL-CM3, GFDL-ESM2M, IPSL-CM5A-MR, MIROC5, MIROC-ESM, MIROC-ESM-CHEM, MPI-ESM-LR, and MPI-ESM-MR under each of three Representative Concentration Pathway (RCP) scenarios, i.e., RCP2.6, RCP4.5 and RCP8.5, collected and used here.

The precipitation, maximum temperature, and minimum temperature from above GCMs for both historical period (1986–2005) and future of the 2030s (2016 –2035) and the 2050s (2046–2065) under the RCP 2.6, RCP4.5, and RCP8.5 scenarios were obtained for PDSI calculation. The degrees of droughts are identified by PDSI values below a threshold value (PDSI < -1): PDSI ≥ -1 represents normal conditions, PDSI in the range of $[-2, -1)$ as mild drought, $[-3, -2)$ as moderate drought, $[-4, -3)$ as severe drought, and PDSI < -4 as extreme drought. Drought frequency is the times of each classification per grid cell in the selected 20-year period for all GCMs.

2.2 GDP Exposure to Droughts

The GDP exposure to varying degrees of droughts is defined as the GDP that is exposed to normal conditions, mild, moderate, severe, and extreme droughts, i.e., the frequency of each class multiplied by the GDP values exposed per grid cell for both the historical period and the future periods under the three RCP and SSP scenarios. For example, the global GDP in 2030 of SSP1 was multiplied by normal condition, mild, moderate, severe, and extreme drought frequencies per grid cell, respectively, under the RCP2.6 scenario to estimate the corresponding GDP exposure to each degree of drought in the 2030s at a spatial resolution of 0.25 degrees. The ensemble mean of GDP exposure of all GCMs was adopted in our analysis.

The unit of GDP exposure to droughts is purchasing power parity (PPP) in 2005 USD, which ensures spatiotemporal comparisons of substantial long-term changes of climate and economic activities for both the historical period and the future projections under different climate change and socioeconomic scenarios.

3 Results

The global GDP exposure to normal (no drought), mild, moderate, severe, and extreme droughts is shown in Fig. 1, and the proportions of GDP exposure to severe and extreme droughts by gross national income (GNI) per capita at 2019 level provided from the World Bank high-income countries ($\geq$ 52,412 GNI per capita), upper middle income countries ($\leq$ 17,196 GNI per capita), middle income countries ($\leq$ 11,934 GNI per capita), low and middle income countries ($\leq$10,937 GNI per capita), lower middle income countries ($\leq$6761 GNI per capita), and low income countries ($\leq$2458 GNI per capita) are shown in Fig. 2.

There is about 70% of global GDP exposed to normal conditions and less than 5% exposed to severe and extreme droughts for the historical period and the 2030s and the 2050s under SSP1-3 and RCP2.6, RCP4.5, and RCP8.5 scenarios. The global GDP exposure to severe droughts will be as high as about 2.8 –3.6 billion USD, and the GDP exposure to extreme droughts will be about 0.5–1.1 billion in the 2030s for SSP1–3, respectively. Furthermore, the GDP exposure to severe and extreme droughts increases in the 2050s, with exposure of 1.5 to 2.1 times for severe droughts and 1.7 to 2.3 times for extreme droughts for SSP1 –3 compared to their 2030s values (Fig. 1).

The GDP exposure to severe and extreme droughts depicts major differences around the globe (Fig. 2). The statistics show that the GDP exposure to droughts is the highest in upper middle income countries but the lowest in low income, lower middle income, and low and middle income countries.

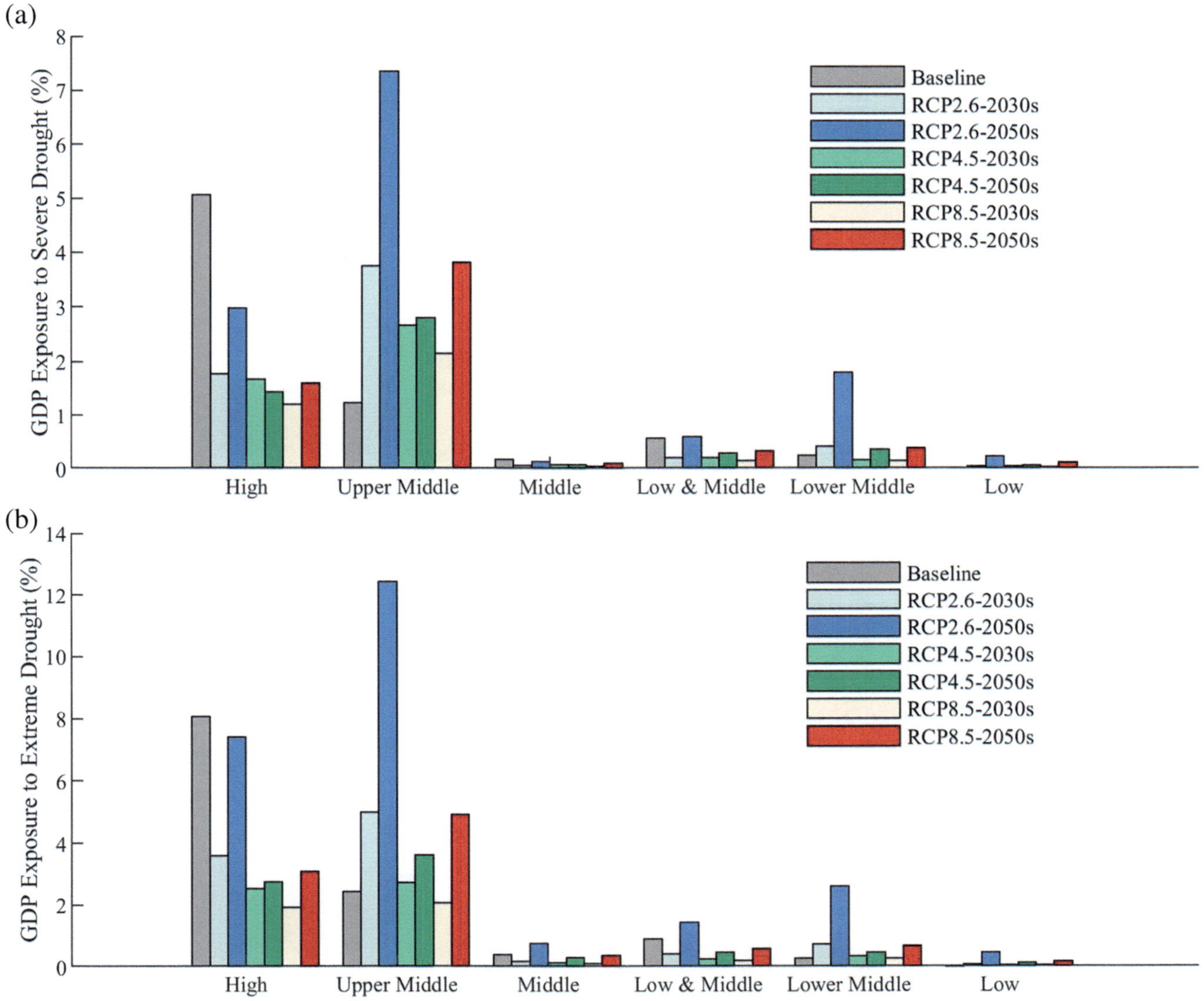

Fig. 2 Global GDP exposure (%) to severe droughts (**a**) and extreme droughts (**b**) by gross national income (GNI) for the historical period and the 2030s and 2050s under SSP1–3 and RCP2.6, RCP4.5, and RCP8.5 scenarios

4 Maps

Global GDP exposure to extreme drought (Baseline)

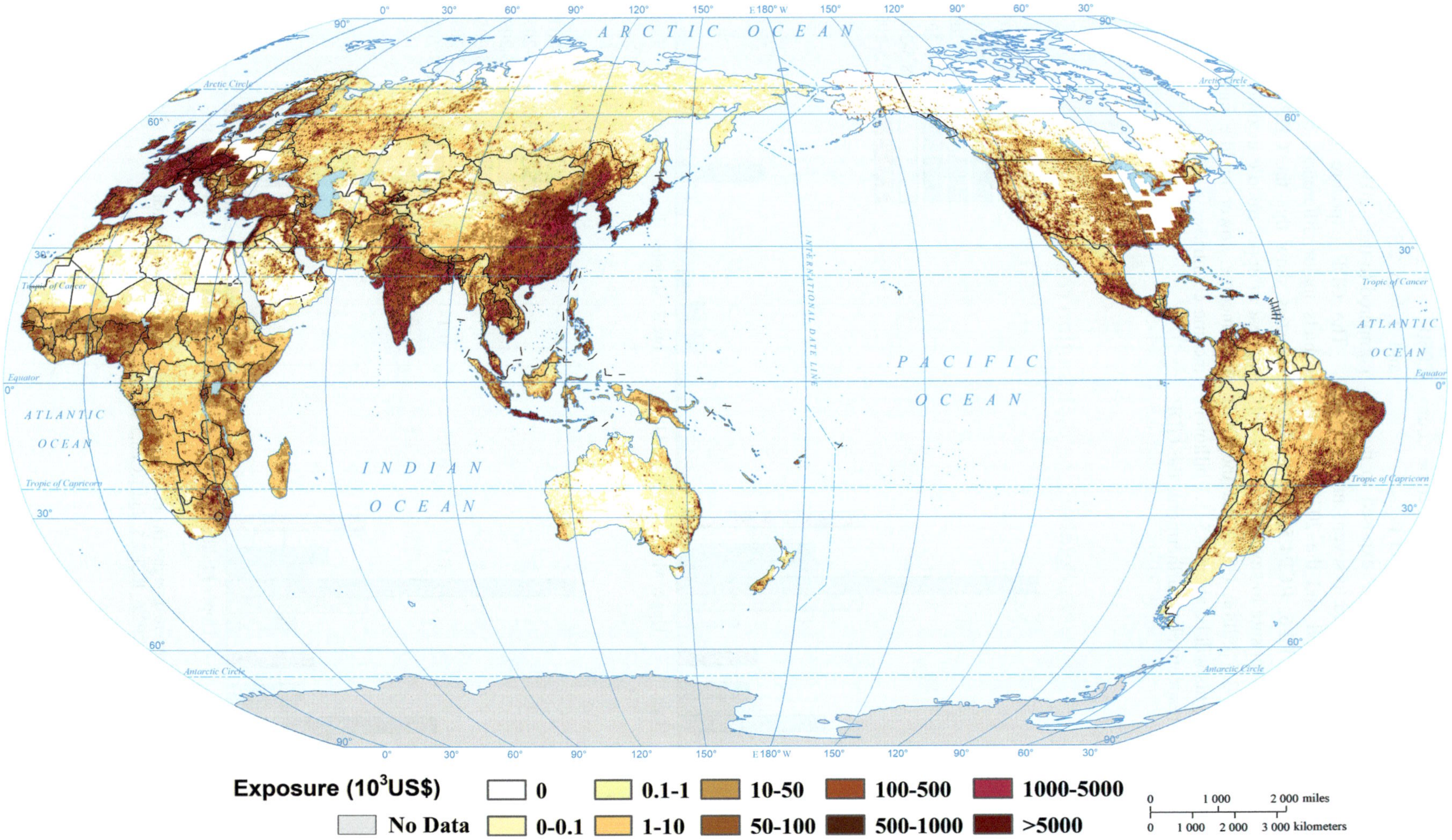

Global GDP exposure to extreme drought (2030s, RCP2.6-SSP1)

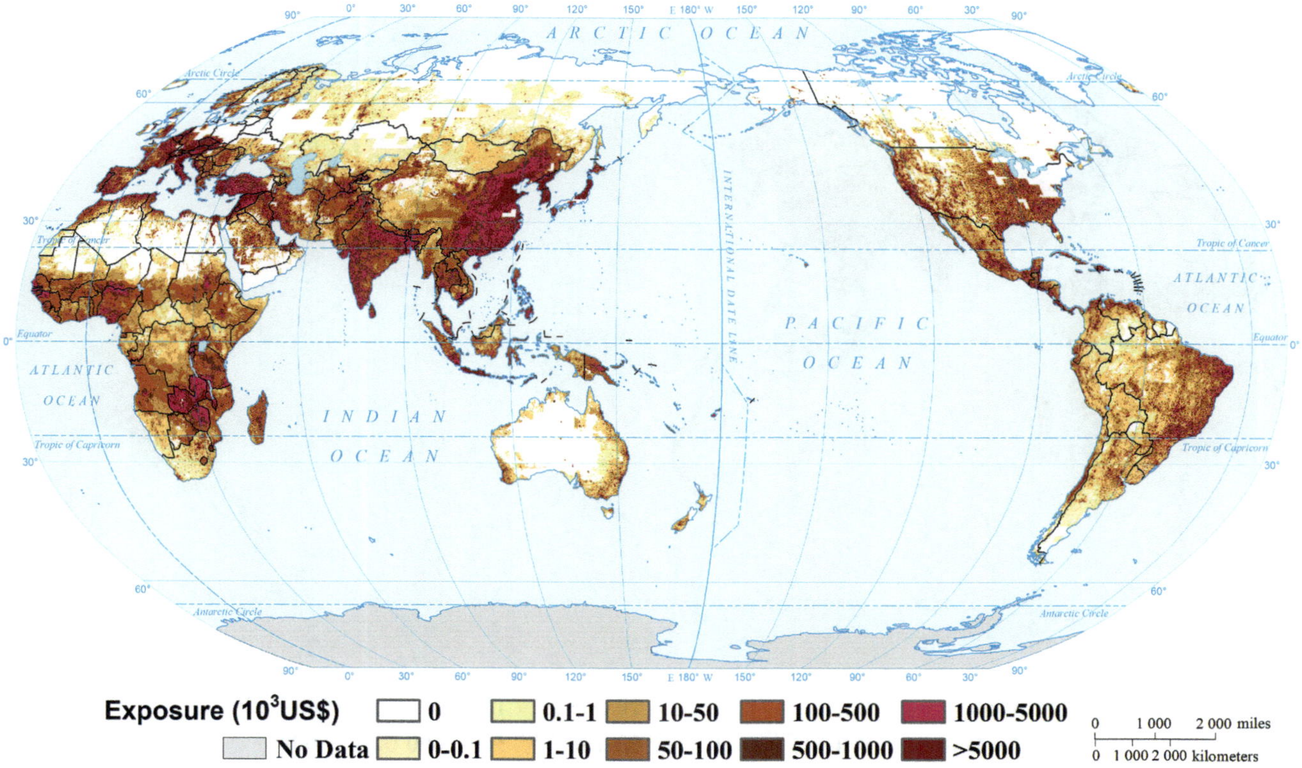

Global GDP exposure to extreme drought (2030s, RCP4.5-SSP2)

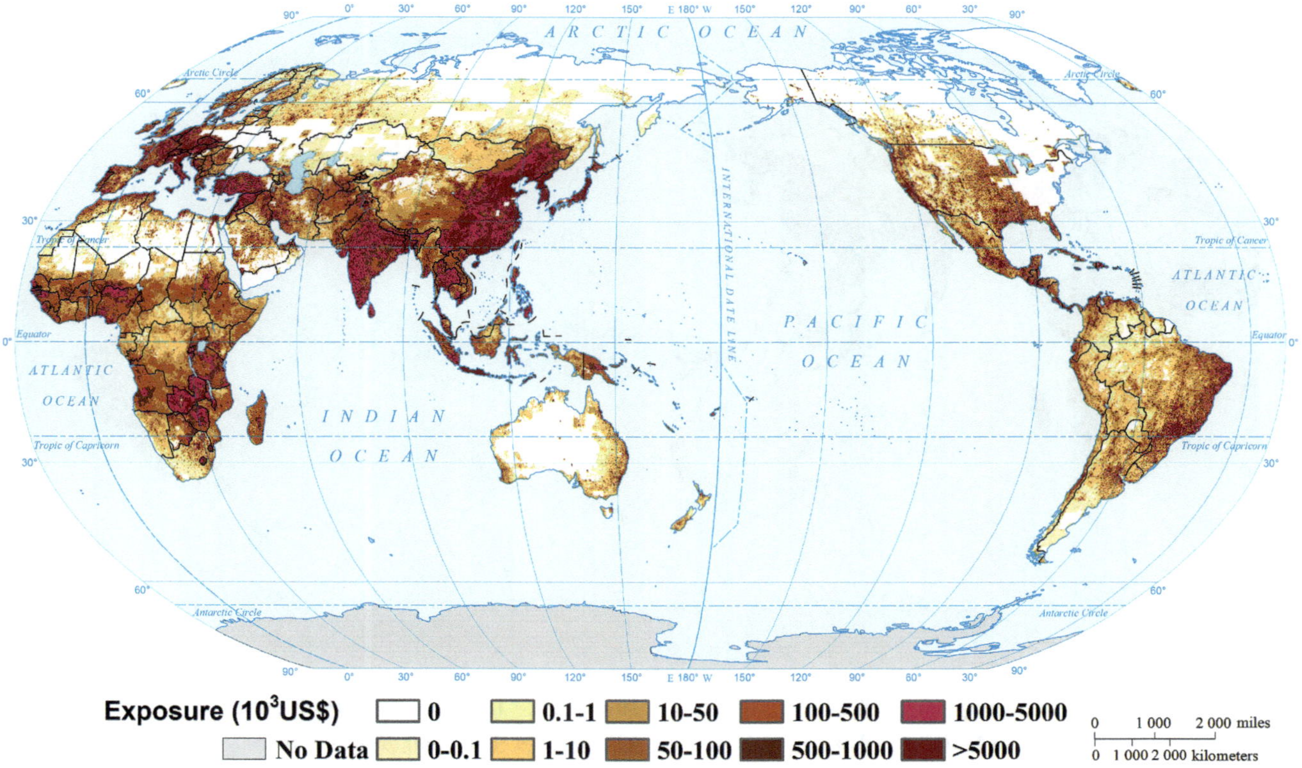

Global GDP exposure to extreme drought (2030s, RCP8.5-SSP3)

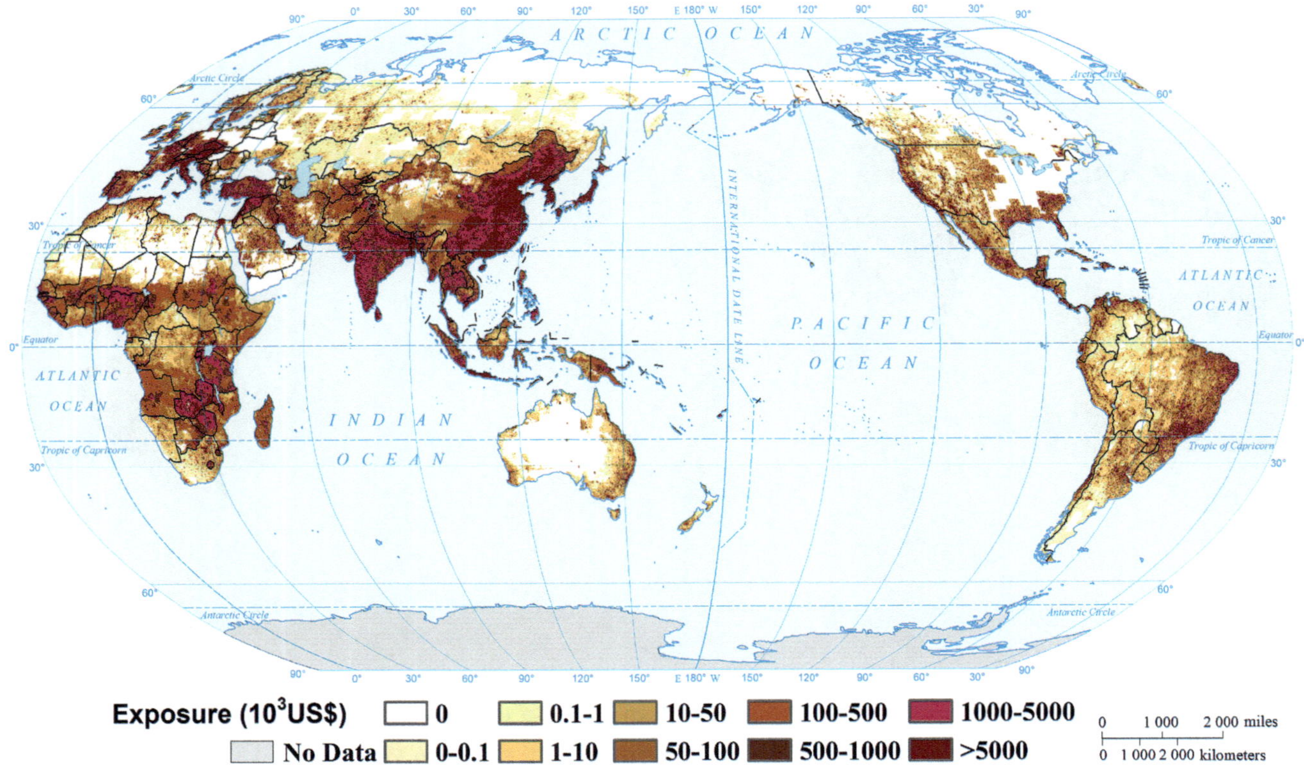

Global GDP exposure to extreme drought (2050s, RCP2.6-SSP1)

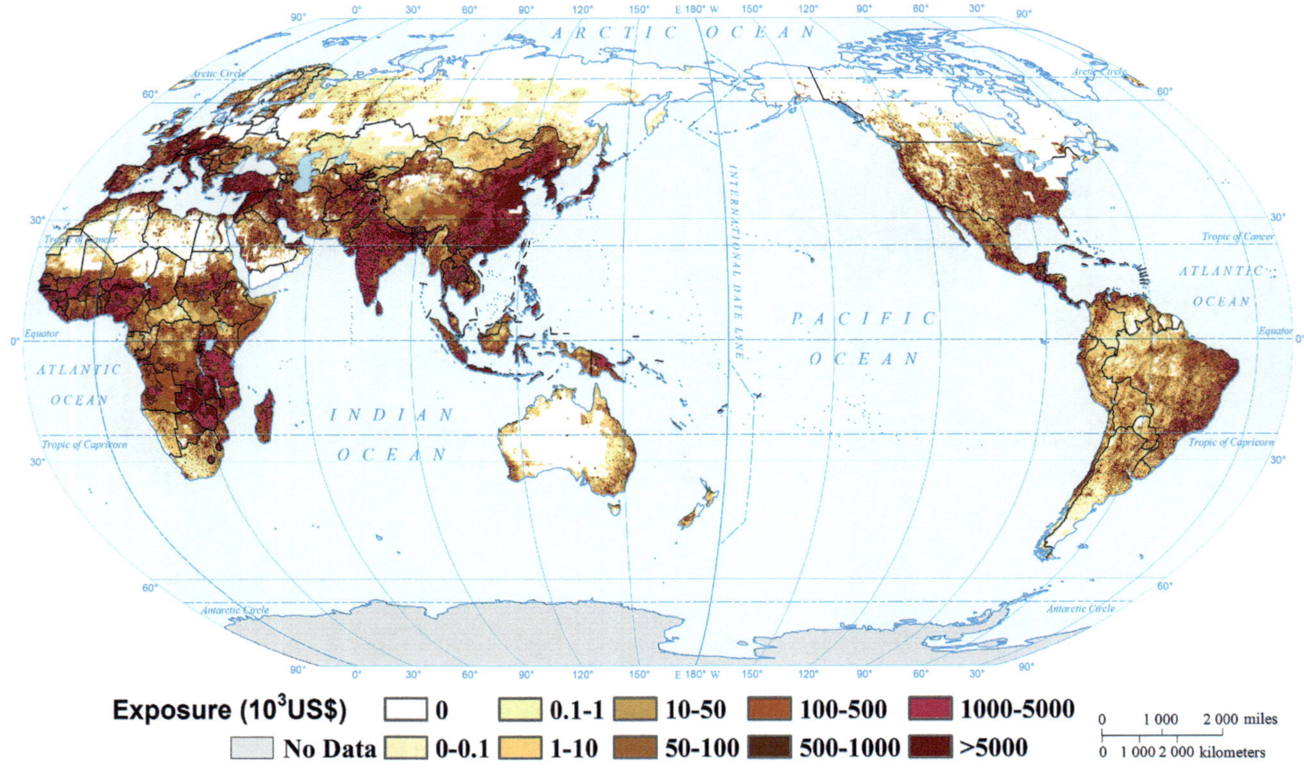

Global GDP exposure to extreme drought (2050s, RCP4.5-SSP2)

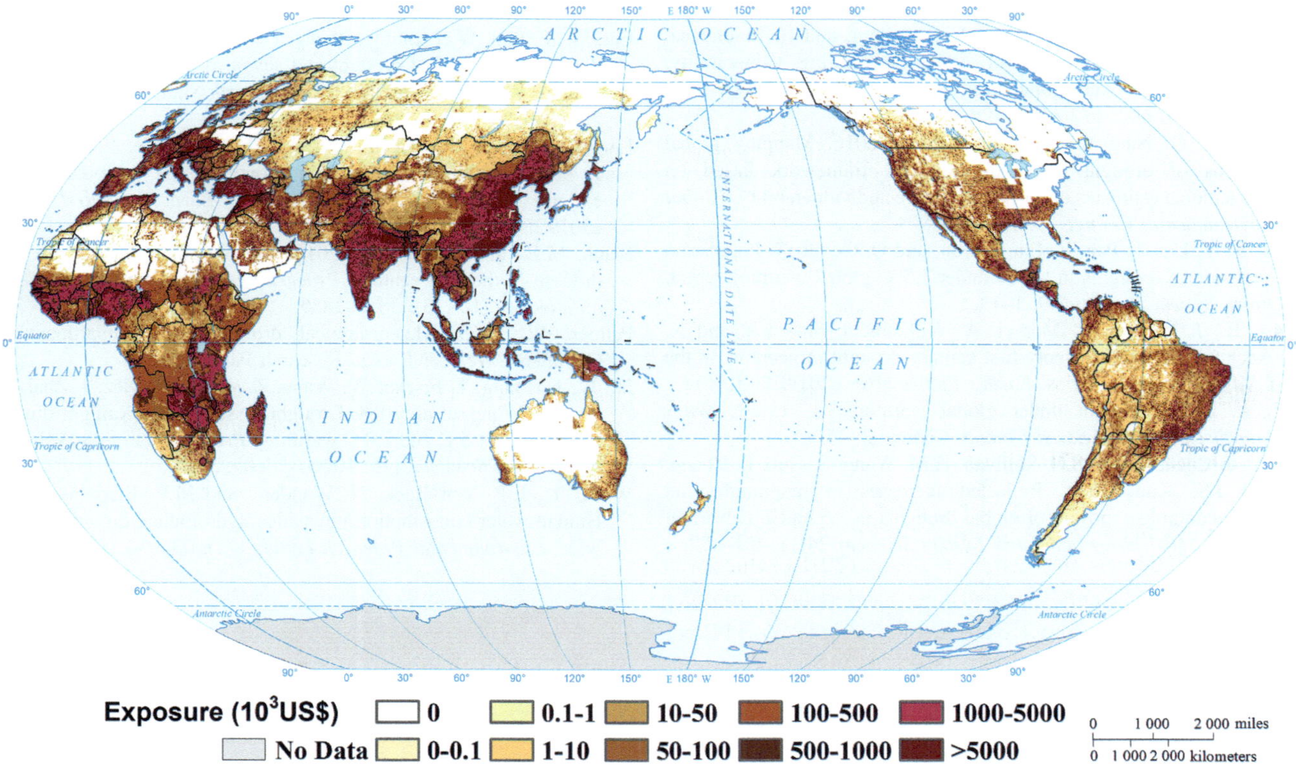

Global GDP exposure to extreme drought (2050s, RCP8.5-SSP3)

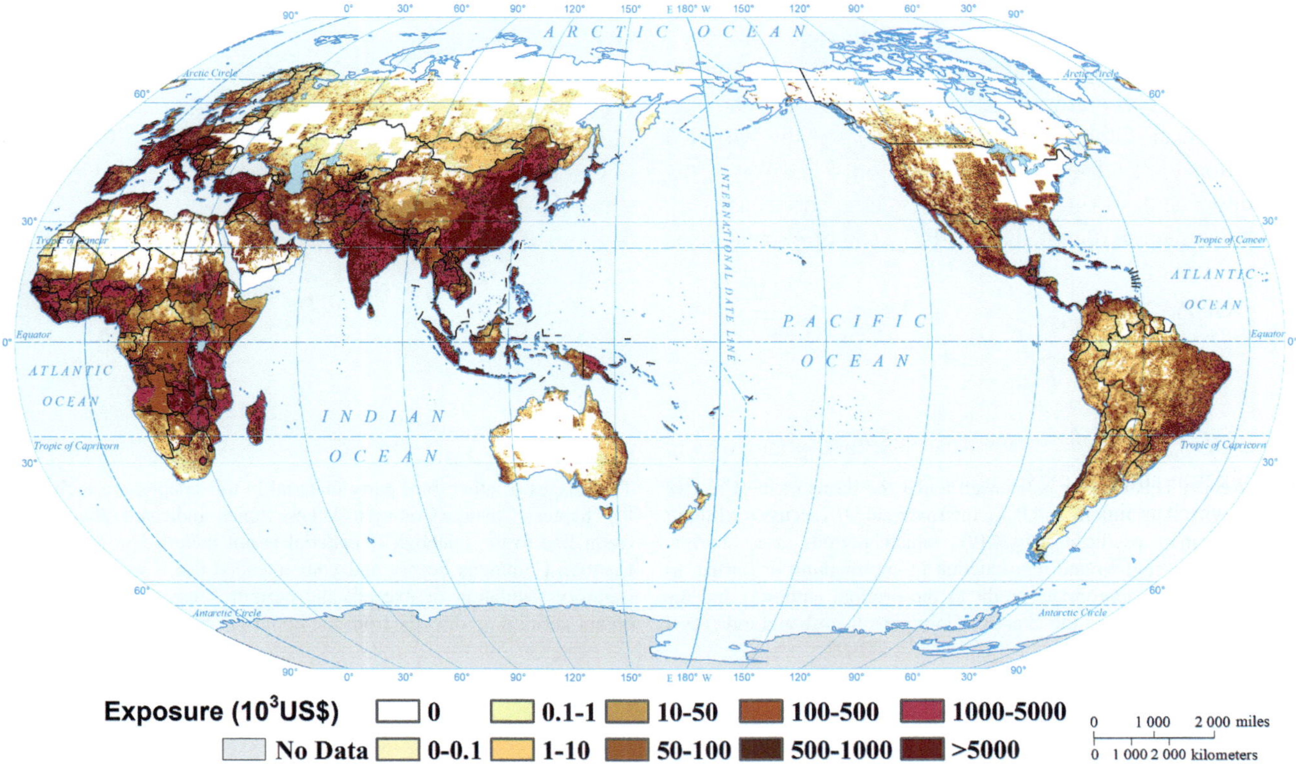

References

Ahmadalipour, A., H. Moradkhani, A. Castelletti, and N. Magliocca. 2019. Future drought risk in Africa: Integrating vulnerability, climate change, and population growth. *Science of the Total Environment* 662: 672–686.

Carrão, H., G. Naumann, and P. Barbosa. 2016. Mapping global patterns of drought risk: An empirical framework based on sub-national estimates of hazard, exposure and vulnerability. *Global Environmental Change* 39: 108–124.

Chen, J., Y. Liu, T. Pan, Y. Liu, F. Sun, and Q. Ge. 2017. Population exposure to droughts in China under 1.5°C global warming target. *Earth System Dynamics* 9: 1–13.

Cook, B., J. Mankin, K. Marvel, A. Williams, J. Smerdon, and K. Anchukaitis, 2020. Twenty-first century drought projections in the CMIP6 forcing scenarios. *Earth's Future* 8(6): e2019EF001461.

Dai, A. 2011. Drought under global warming: A review. *Wiley Interdisciplinary Reviews: Climate Change* 2: 45–65.

Gu, L., J. Chen, J. Yin, S.C. Sullivan, H.M. Wang, S. Guo, L. Zhang, and J.S. Kim. 2020. Projected increases in magnitude and socioeconomic exposure of global droughts in 1.5 and 2°C warmer climates. *Hydrology and Earth System Sciences* 24(1): 451–472.

Jones, B., B.C. O'Neill, L. Mcdaniel, S. Mcginnis, L.O. Mearns, and C. Tebaldi. 2015. Future population exposure to US heat extremes. *Nature Climate Change* 5: 592–597.

Liu, W., F. Sun, W.H. Lim, J. Zhang, H. Wang, H. Shiogama, and Y. Zhang. 2018. Global drought and severe drought-affected populations in 1.5 and 2°C warmer worlds. *Earth System Dynamics* 9(1): 267–283.

Liu, Y., J. Chen, T. Pan, Y. Liu, Y. Zhang, Q. Ge, P. Ciais, and J. Penuelas. 2020. Global socioeconomic risk of precipitation extremes under climate change. *Earth's future* 8(9): e2019EF001331.

Mann, M.E., and P.H. Gleick. 2015. Climate change and California drought in the 21st century. *Proceedings of the National Academy of Sciences* 112 (13): 3858–3859.

Palmer, W. 1965. Meteorological droughts. US Department of Commerce Weather Bureau Research Paper 45, 58.

Su, B., J. Huang, T. Fischer, Y. Wang, Z.W. Kundzewicz, J. Zhai, H. Sun, A. Wang, et al. 2018. Drought losses in China might double between the 1.5 C and 2.0 C warming. *Proceedings of the National Academy of Sciences* 115: 10600–10605.

Wada, Y., L.P. Van Beek, N. Wanders, and M.F. Bierkens. 2013. Human water consumption intensifies hydrological drought worldwide. *Environmental Research Letters* 8(3): 034036.

Mapping Global Crop Distribution

Yaojie Yue, Peng Su, Yuan Gao, Puying Zhang, Ran Wang, Anyu Zhang, Qinghua Jiang, Weidong Ma, Yuantao Zhou, and Jing'ai Wang

1 Introduction

The latest special report from the Intergovernmental Panel on Climate Change (IPCC), published in 2018 (Anandhi et al. 2016), estimates a 1.5 °C increase in global temperature in 2040 at the current rate of global warming. Such a rise has serious implications for major cereal crop cultivation: unless crop varieties adapted to higher temperatures become available, the areas suitable for cropping are bound to shift in the future. Therefore, to safeguard food security, we need to predict such changes in spatial and temporal terms, which can intuitively reflect the potential distribution of crops under different climate change scenarios and for different time periods, helping to reduce losses according to local conditions (Deng et al. 2009).

Species distribution models (SDMs) have been widely applied in evaluation of land suitability for crop cultivation and the potential distributions of some cereal crops under climate change scenarios (Stansbury and Pretorius 2001; Subbarao et al. 2001; Sun et al. 2012; He and Zhou 2016; Shabani and Kotey 2016), but socioeconomic scenarios have rarely been considered. As an important part of the atlas, the maps of global crop distribution were produced to show the potential distribution of major cereal crops, including wheat, maize, and rice, under the influence of global climate change. These maps cover global distribution of wheat, rice, and maize in the 2000s (baseline period), 2030s (near term), and 2050s (medium term) under the RCP2.6, RCP4.5, and RCP8.5 scenarios.

2 Method

2.1 Wheat Distribution Prediction Method

Environmental variables: This study considered that the cultivation of wheat is affected by climate and soil (Motuma et al. 2016). Hence, on the basis of previous research, we established an environmental variables system, including the following variables: ≥ 0 °C cumulative temperature, annual precipitation, annual average temperature, average temperature of the coldest month (Sun et al. 2012; Wang et al. 2016), pH, drainage, conductivity, exchangeable sodium percentage, soil property, and soil depth (Wandahwa and Van Ranst 1996; Mendas and Delali 2012; Wang et al. 2016). Considering that terrain has a great influence on the convenience of wheat cultivation, this study also chose slope as an environmental variable.

Samples: According to the *Harvested Area and Yield for 175 Crops of Year 2000* (Monfreda et al. 2008), 15,500 samples, which accounted for approximately 5% of all grids, were selected. A total of 75% of these samples were used for

Authors: Yaojie Yue, Peng Su, Yuan Gao, Puying Zhang, Ran Wang, Anyu Zhang, Qinghua Jiang, Weidong Ma, Yuantao Zhou, Jing'ai Wang.**Map Designers**: Yichen Li, Qingyuan Ma, Jing'ai Wang, Ying Wang.
Language Editor: Tao Ye, Peng Su.

Y. Yue · Y. Gao · P. Zhang · R. Wang · A. Zhang · Q. Jiang·
J. Wang (✉)
Key Laboratory of Environmental Change and Natural Disaster of Ministry of Education, Faculty of Geographical Science, Beijing Normal University, Beijing, 100875, China
e-mail: jwang@bnu.edu.cn

P. Su · W. Ma · Y. Zhou
School of Geographic Science, Qinghai Normal University, Xining, 810008, China

J. Wang
State Key Laboratory of Earth Surface Processes and Resource Ecology, Beijing Normal University, Beijing, 100875, China

P. Shi, *Atlas of Global Change Risk of Population and Economic Systems*, IHDP/Future Earth-Integrated Risk Governance Project Series, https://doi.org/10.1007/978-981-16-6691-9_10

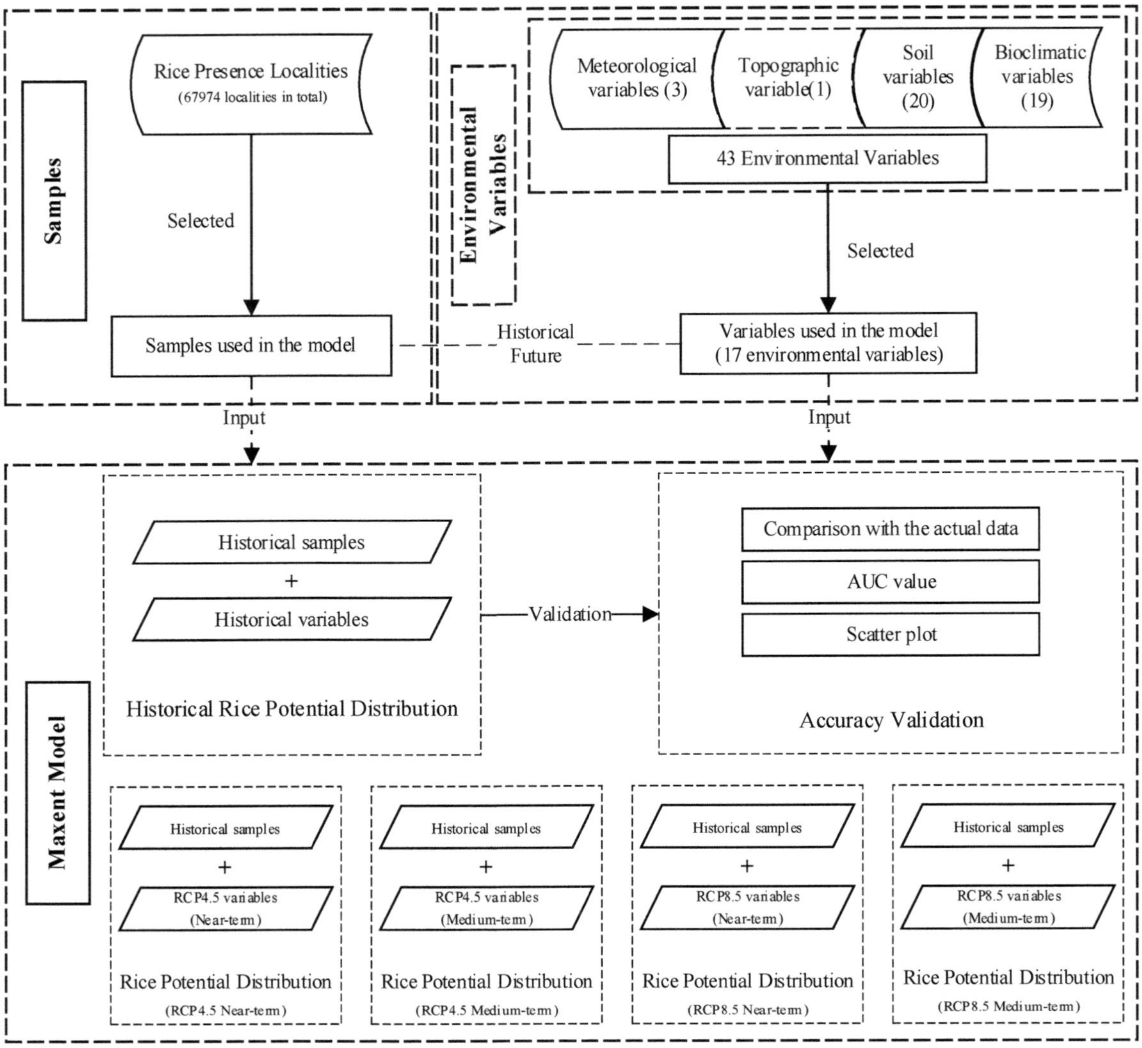

Fig. 1 Research framework for predicting rice suitability. RCP = Representative concentration pathways; ROC = Area under the curve

model training and the remaining 25% were used for validation.

The Maxent model was trained to predict future land suitability for wheat cultivation for the 2000s, 2030s, and 2050s under different future climate scenarios. The specific process of modeling can be found in Yue et al. (2019).

2.2 Rice Distribution Prediction Method

The Maxent model combines the data on the presence of a given species within a grid with environmental variables representing different environmental gradients within that grid to determine whether the area is suitable for a particular species. The overall research framework for the study is shown in Fig. 1.

Environmental variables and Samples: After the selection of sample points and environment variables, a total of 2228 samples and 17 environmental variables were used in predicting future suitable areas for rice cultivation.

The Maxent model was trained to predict future land suitability for rice cultivation for the 2000s, 2030s, and 2050s under different future climate scenarios. In order to reduce errors, the simulation was run for 30 times, and the average result was used as the final result. The specific process of modeling can be found in Su et al. (2021).

To more realistically represent future global rice cultivation distribution, this study also considered the balance of the supply and demand of rice among countries. On the supply side, we used suitability changes to predict the initial rice harvest area as the starting value of the extended simulation. The rice harvest area was adjusted by comparing the countries' rice supply and demand. If the supply exceeded demand, we reduced the rice harvest area, and vice versa. Besides, we calculated irrigated and non-irrigated areas separately. This model can be expressed as a multi-objective optimization model (Eqs. 1–4).

$$D = S + I \quad (1)$$

where D, S, and I represent the countries' rice demand, supply, and net import, respectively,

$$A_{i,\text{rice}} \leq A_{i,c3} \quad (2)$$

where $A_{i,\text{rice}}$, $A_{i,c3}$, respectively, represent the rice harvest area and C3 crop harvest area of the ith grid.

$$\sum \left(A_{i,n+1} \cdot Su_i\right) \geqslant \sum \left(A_{i,n} \cdot Su_i\right) \quad (n = 1, 2, 3 \ldots) \quad (3)$$

where n is the number of iterations, $n = 1$ is the initial situation inferred from the change in the suitable area, and Su is the suitability.

$$\text{Ir}_{\text{FAO}} = \frac{\sum A_{i,\text{Ir}}}{\sum A_{i,\text{rice}}} \quad (4)$$

where Ir_{FAO} is the national rice irrigation rate predicted by FAO and $A_{i,\text{Ir}}$, $A_{i,\text{rice}}$ are the rice irrigation area and rice harvest area in grid i. Finally, socioeconomic changes in crop planting are mapped in the atlas to reflect the adjusted global distribution of rice cultivation.

2.3 Maize Distribution Prediction Method

Applying the same prediction method using the Maxent model and following the procedure described in Fig. 1, we selected 5548 maize sample points and 16 environment variables to predict the potential distribution of maize. Then we redistributed maize cultivation among countries on the basis of natural maize potential distribution by following the redistribution method of rice cultivation.

3 Results

3.1 Latitudinal Distribution of Crops

The latitudinal distributions of the cultivated areas of the three crops in the historical period (red bar) and the RCP8.5 scenario (blue bar) are shown in Fig. 2.

During the two study periods, the three crops are cultivated mainly in the northern hemisphere. The most important areas for wheat, rice, and maize planting are between 20°N–60°N, 20°N – 30°N, and 30°N–60°N, respectively. The comparison of the planting areas in the two periods shows that under the RCP8.5 scenario, the cultivated areas of the three crops in the northern hemisphere will move to higher latitudes and the cultivated areas of rice and maize will have a significant increase.

3.2 Country Distribution of Crops

Figure 3 shows the ranking of the top 20 countries based on the statistics of the cultivated areas of the crops in each country in the historical period and under the RCP8.5 scenario.

In both periods, China's wheat planting area is the largest globally, followed by the United States and Russia. Rice planting area is the largest in India, and China's rice planting area is the second largest and much larger than other countries. Maize planting area is the largest in the United States, and China has the second largest maize planting area. In both time periods, China's wheat, rice, and maize planting areas are all among the highest globally. On the other hand, in the future once the high temperature for crops intensifies, China will also face greater losses than other countries.

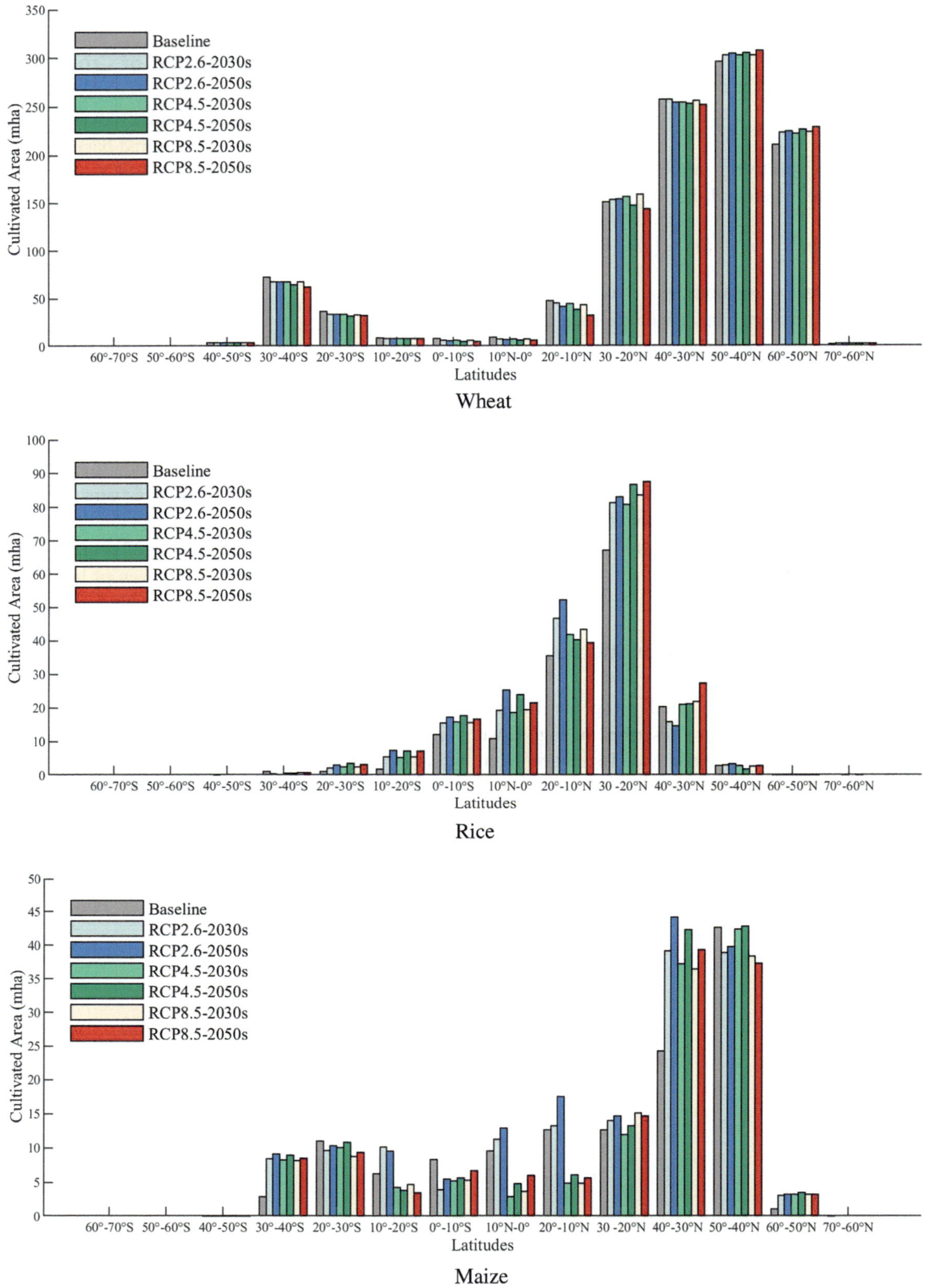

Fig. 2 Latitudinal distribution of the cultivated areas of wheat, rice, and maize (unit: million ha)

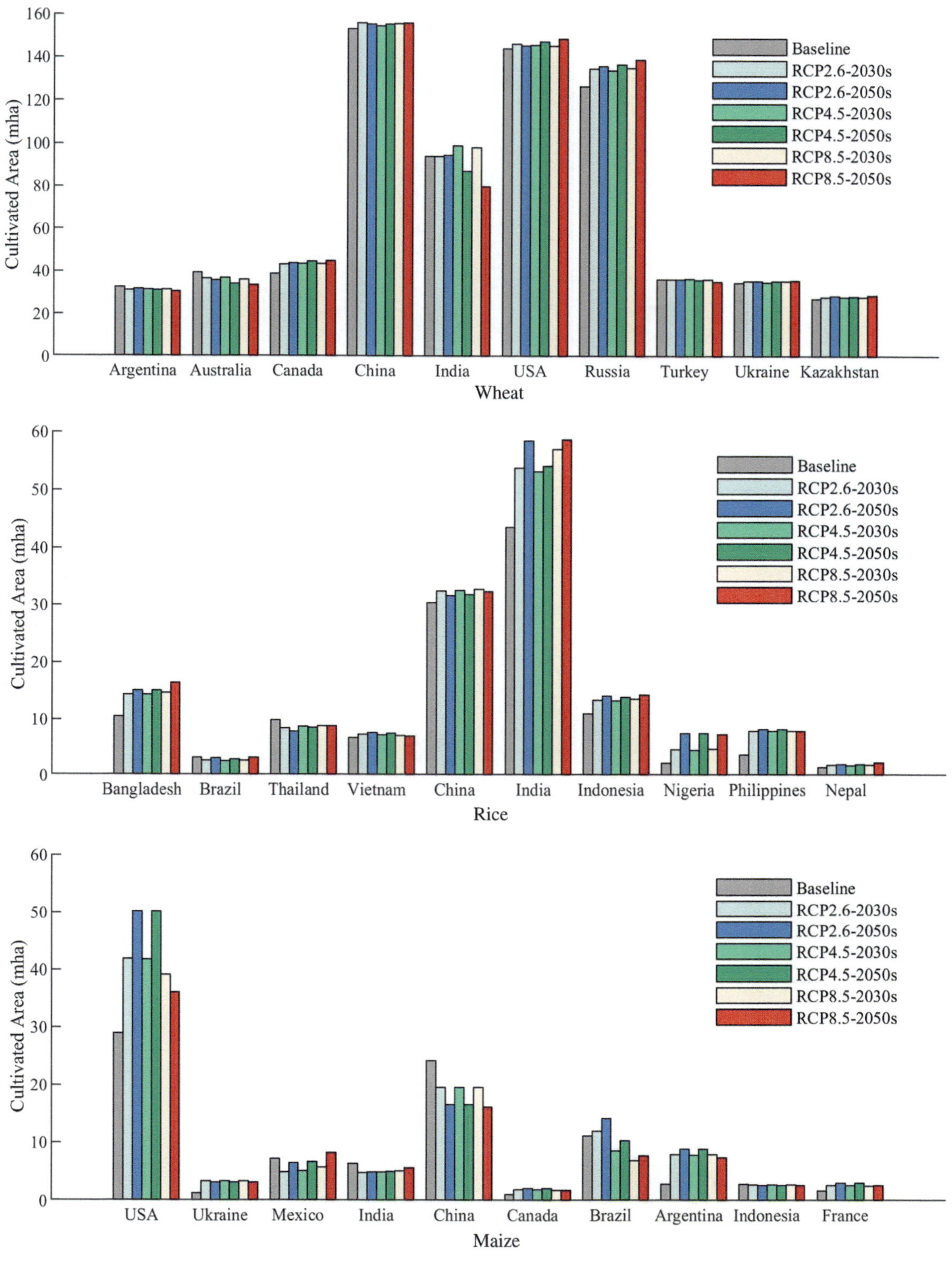

Fig. 3 Ranking of the top 20 countries by cultivated area of wheat, rice, and maize (unit: million ha)

4 Maps

Global wheat distribution (Baseline)

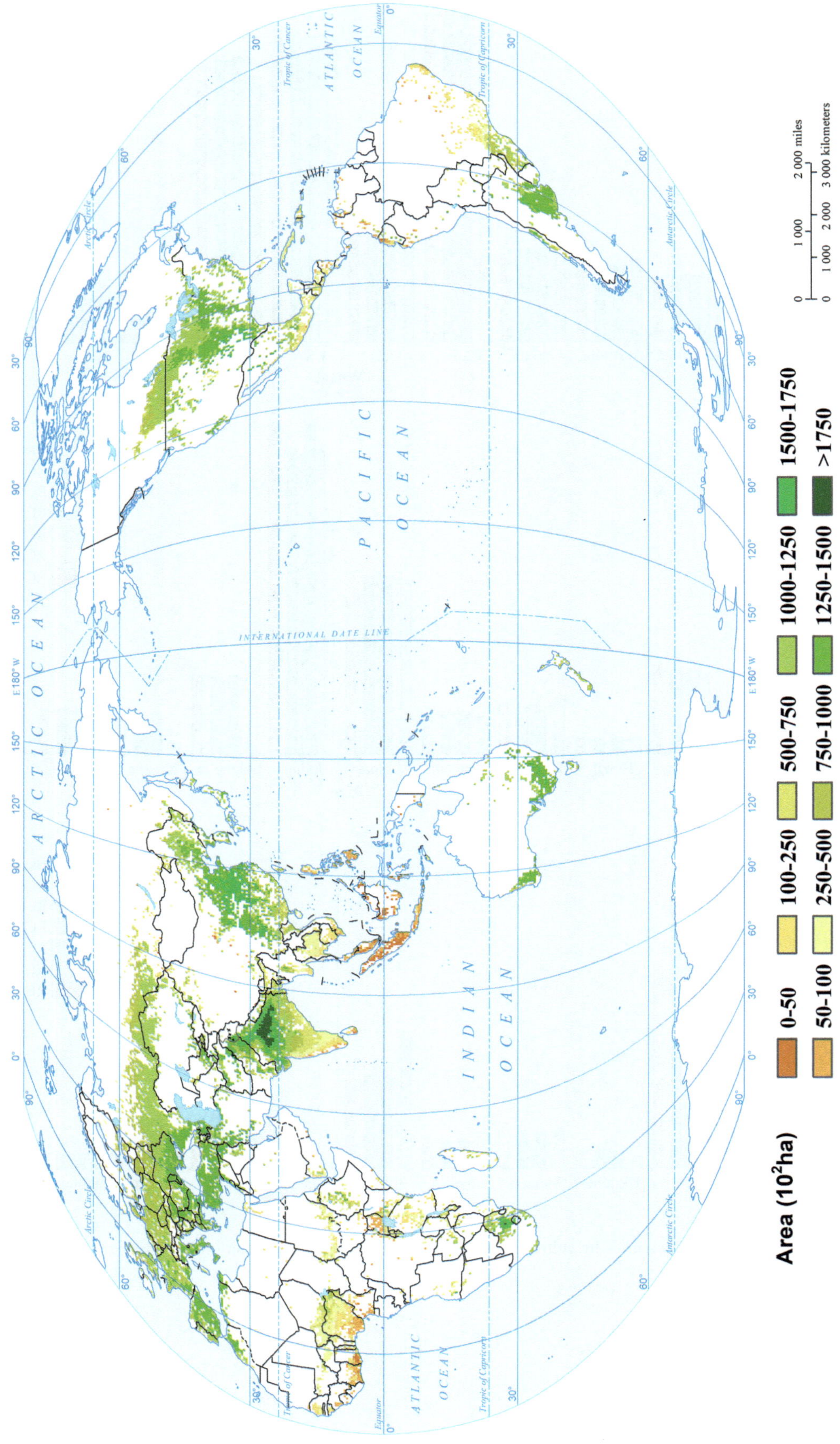

Projected global wheat distribution (2030s, RCP2.6)

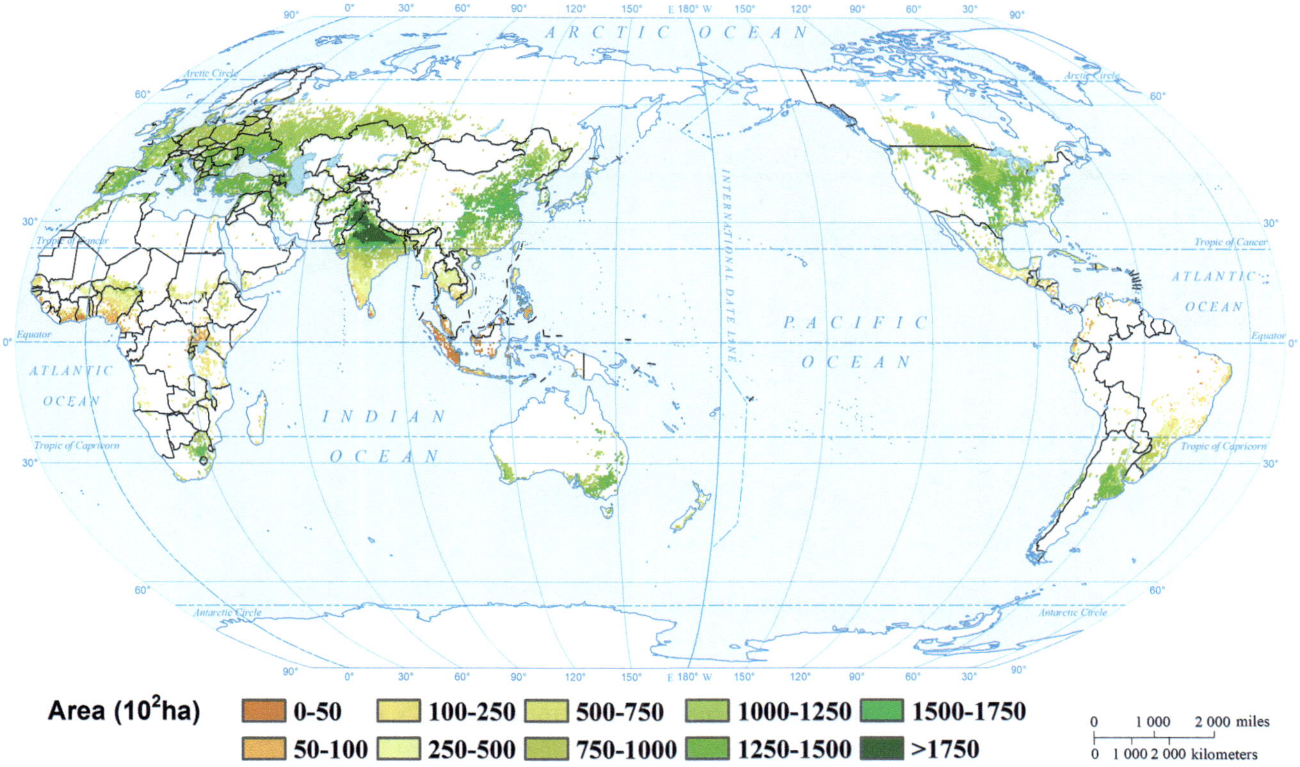

Projected global wheat distribution (2030s, RCP4.5)

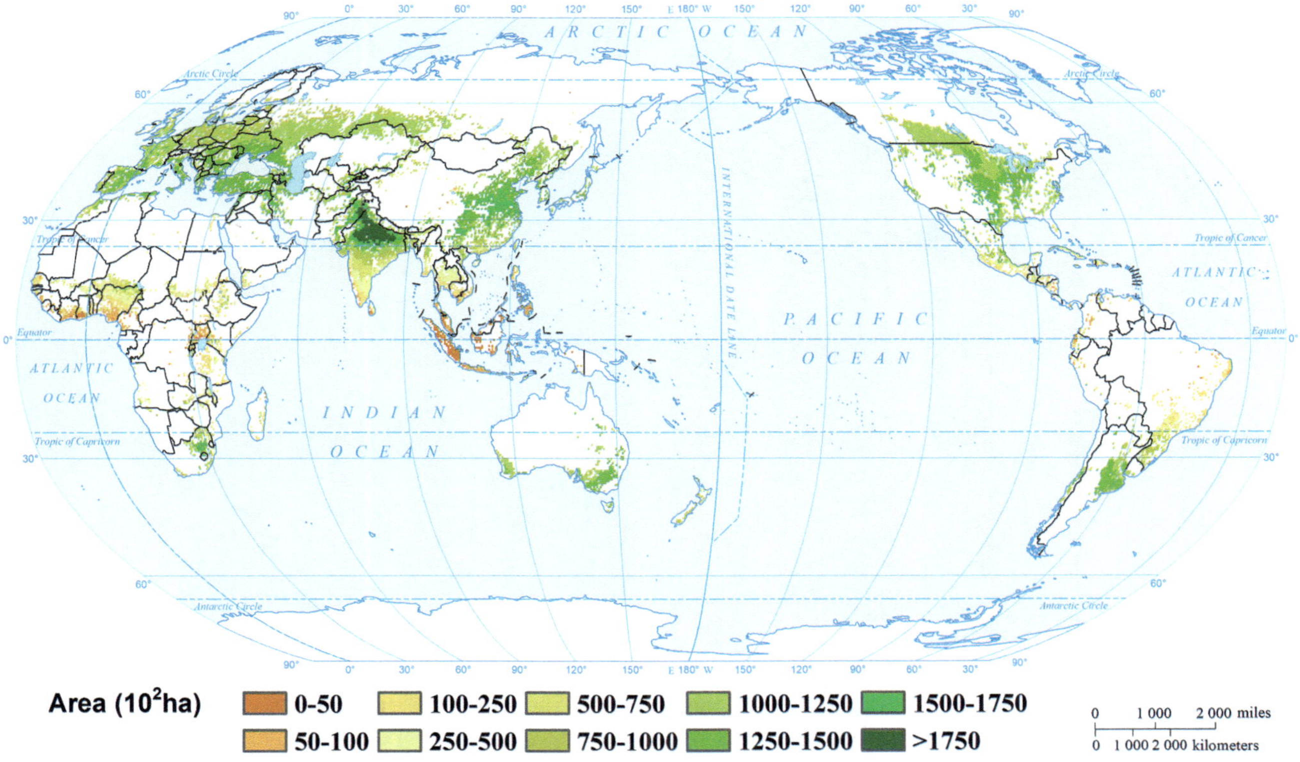

Projected global wheat distribution (2030s, RCP8.5)

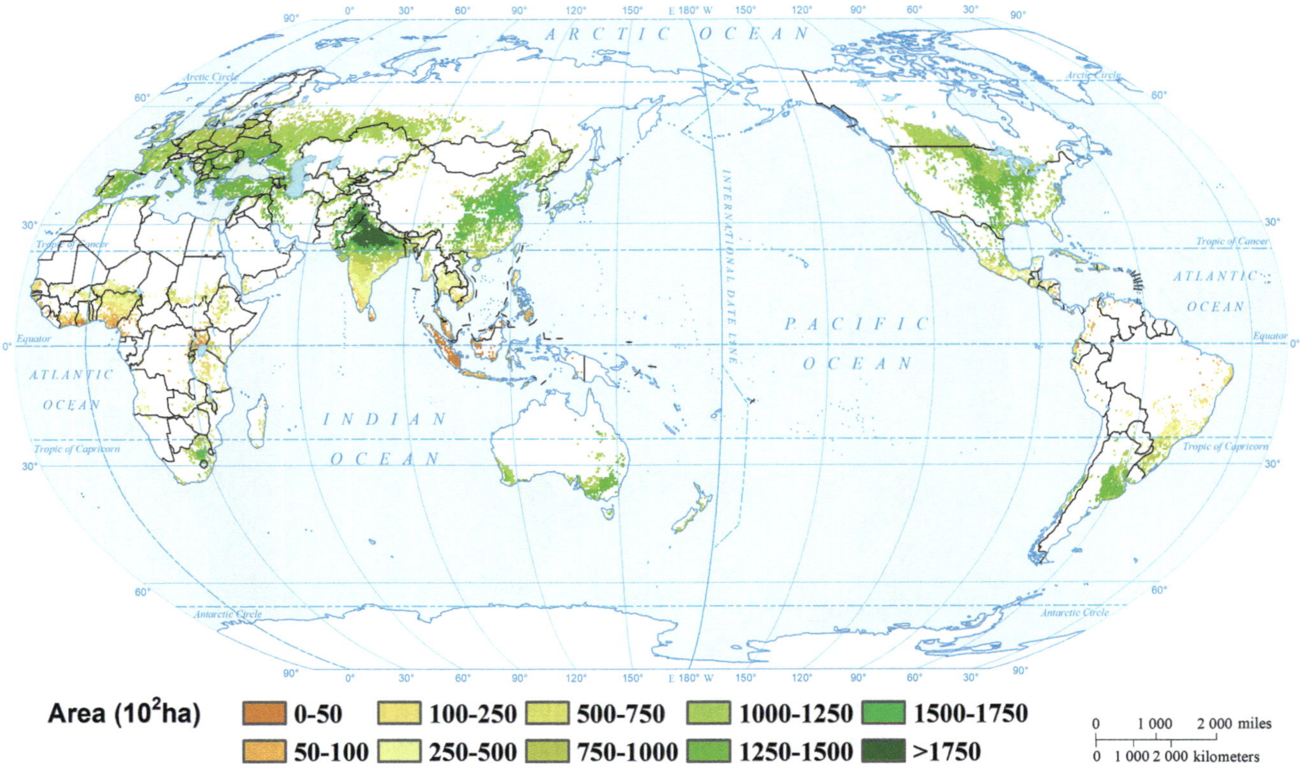

Projected global wheat distribution (2050s, RCP2.6)

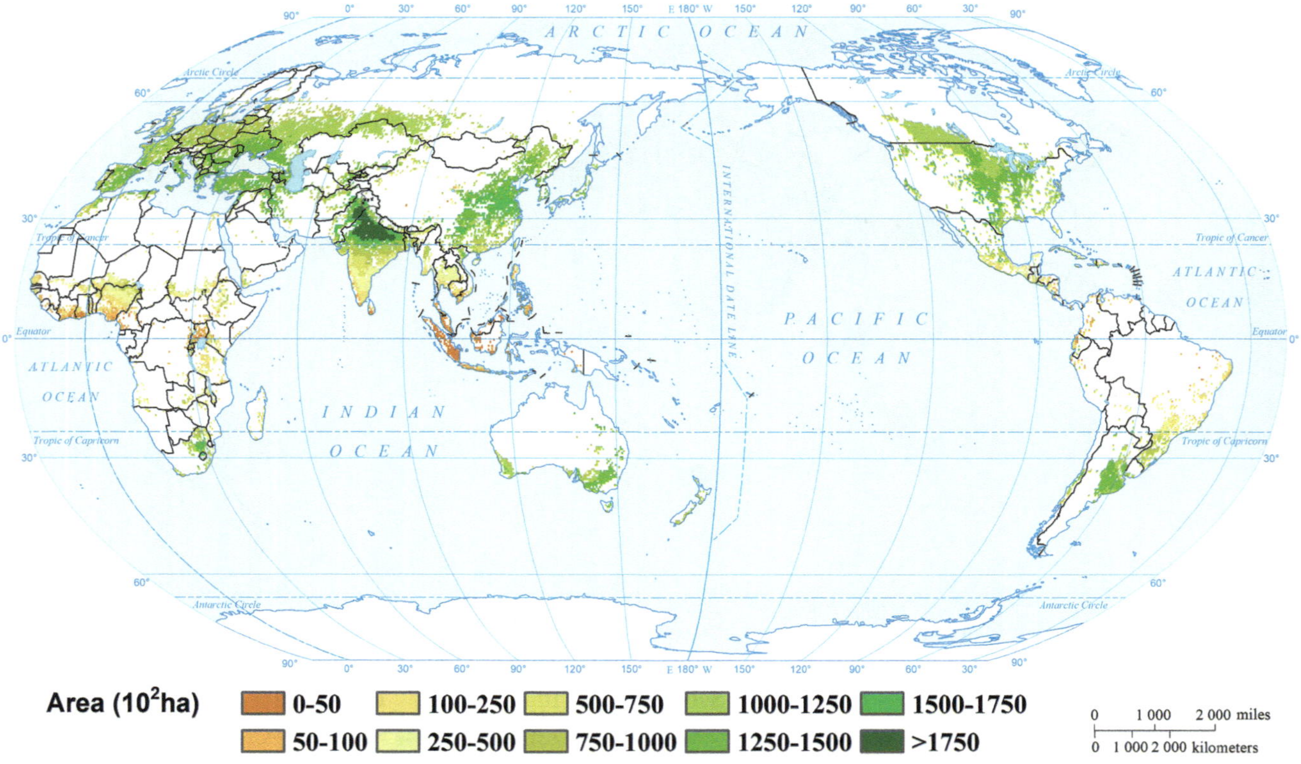

Projected global wheat distribution (2050s, RCP4.5)

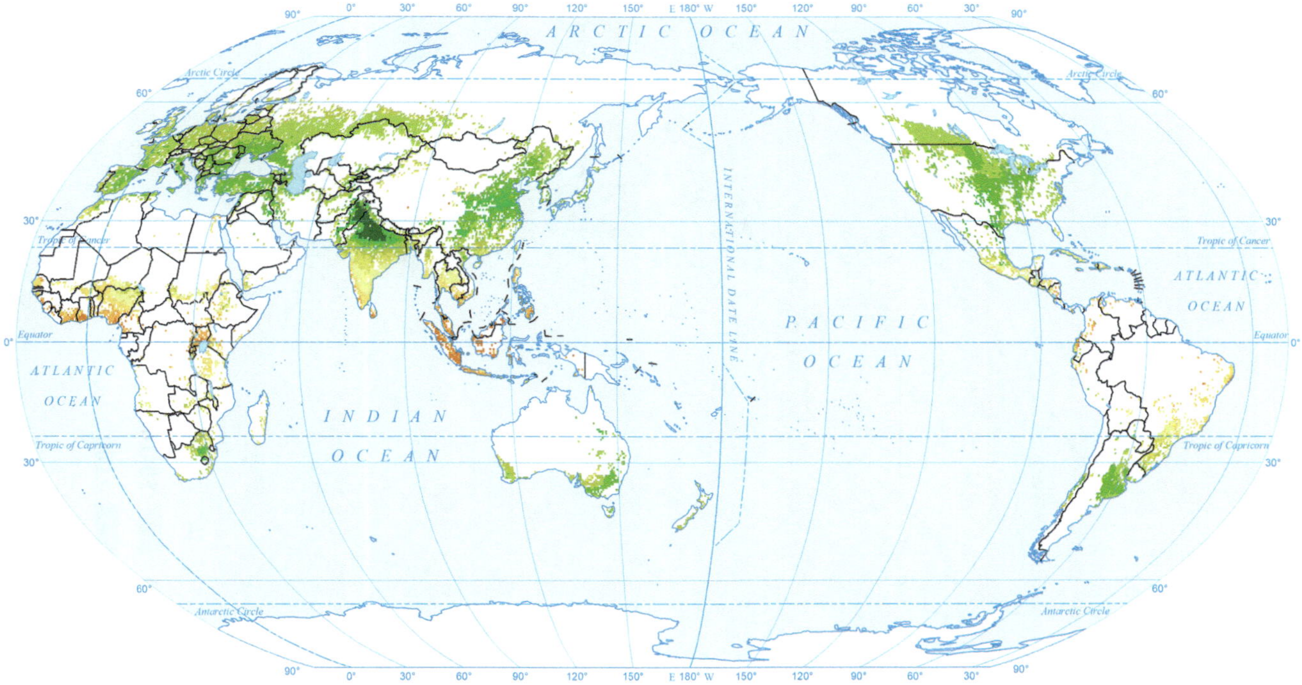

Area (10^2ha) 0-50 | 50-100 | 100-250 | 250-500 | 500-750 | 750-1000 | 1000-1250 | 1250-1500 | 1500-1750 | >1750

0 1 000 2 000 miles
0 1 000 2 000 kilometers

Projected global wheat distribution (2050s, RCP8.5)

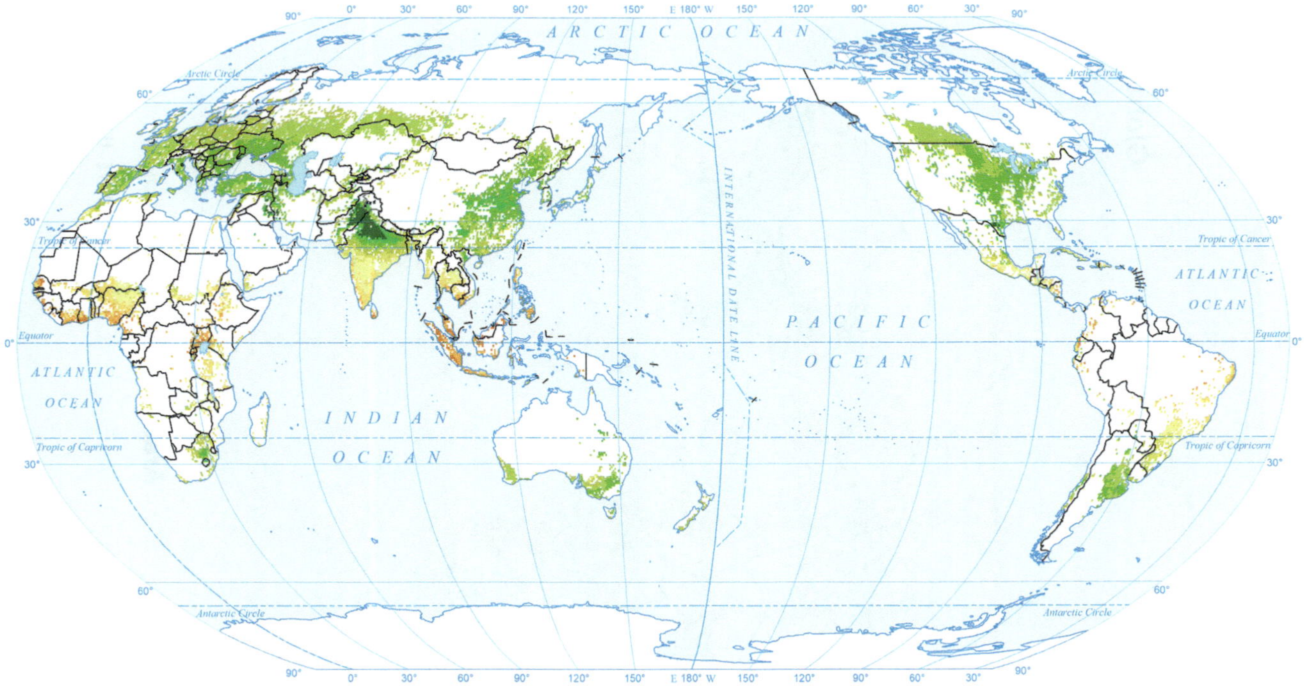

Area (10^2ha) 0-50 | 50-100 | 100-250 | 250-500 | 500-750 | 750-1000 | 1000-1250 | 1250-1500 | 1500-1750 | >1750

0 1 000 2 000 miles
0 1 000 2 000 kilometers

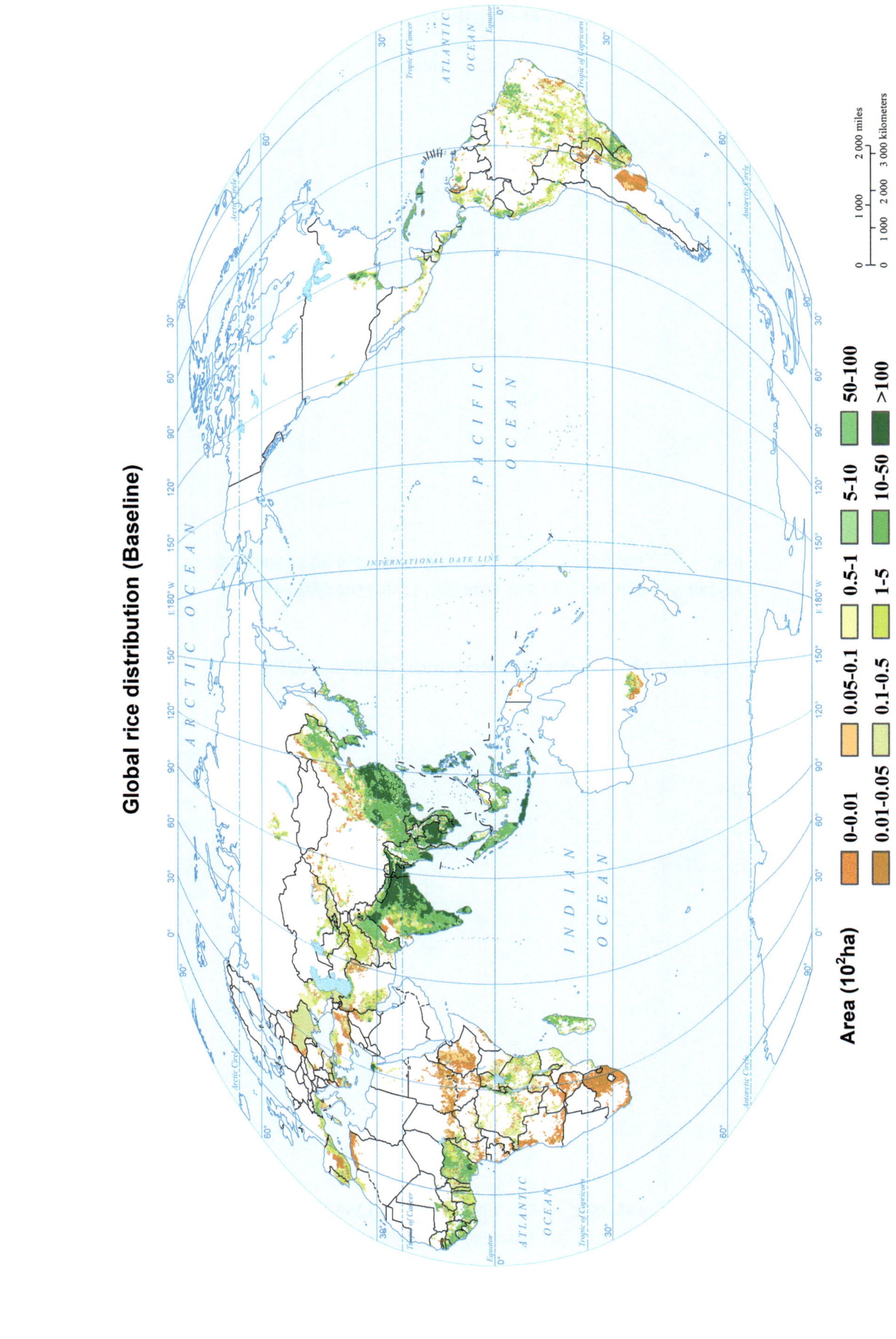

Global rice distribution (Baseline)
Area (10²ha)
0-0.01
0.01-0.05
0.05-0.1
0.1-0.5
0.5-1
1-5
5-10
10-50
50-100
>100
ARCTIC OCEAN
PACIFIC OCEAN
INDIAN OCEAN
ATLANTIC OCEAN
INTERNATIONAL DATE LINE
Tropic of Cancer
Tropic of Capricorn
Equator
Arctic Circle
Antarctic Circle
2 000 miles
3 000 kilometers

Projected global rice distribution (2030s, RCP2.6)

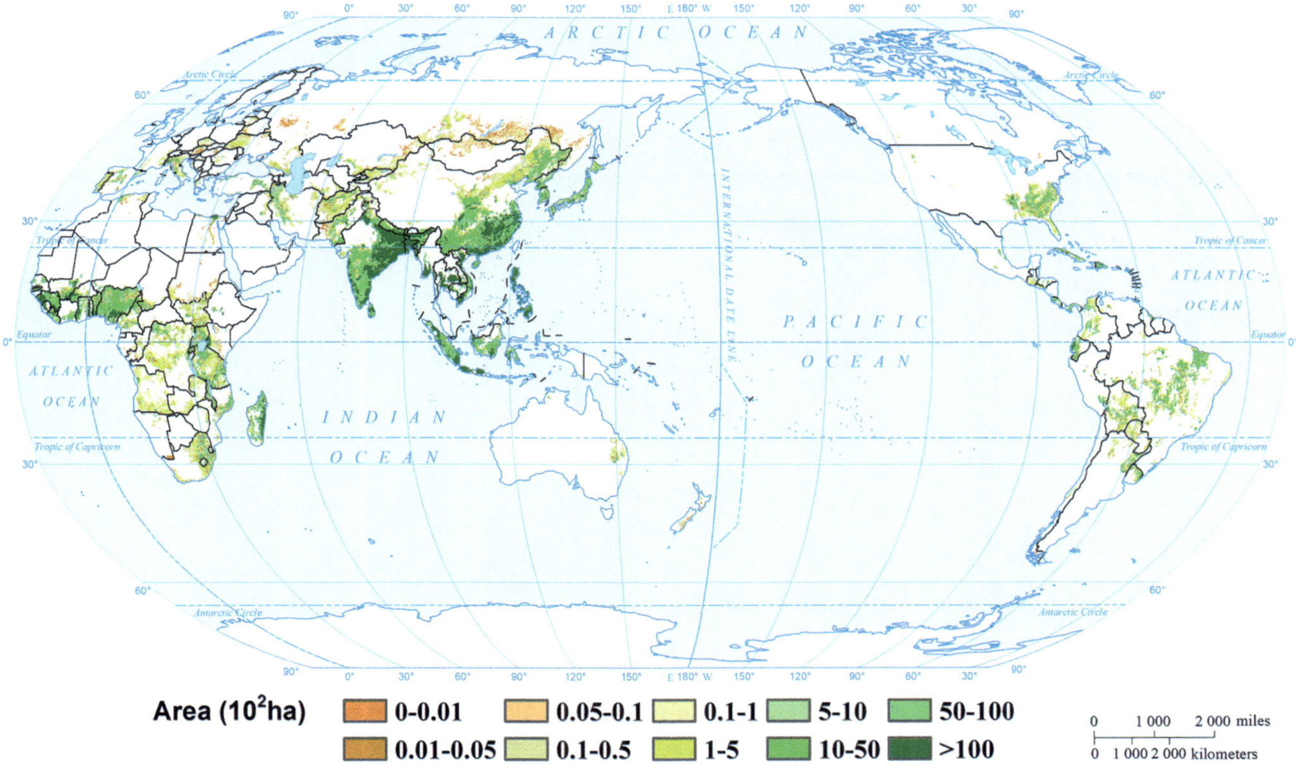

Projected global rice distribution (2030s, RCP4.5)

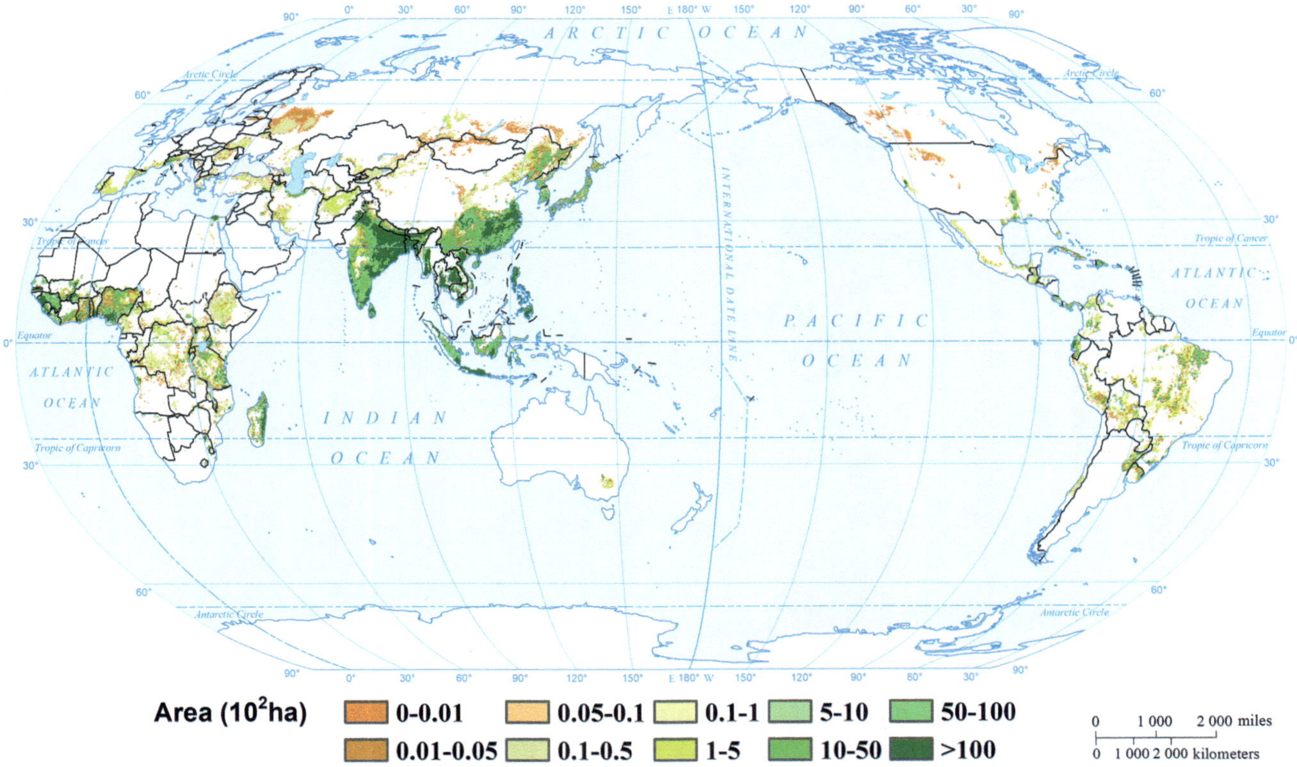

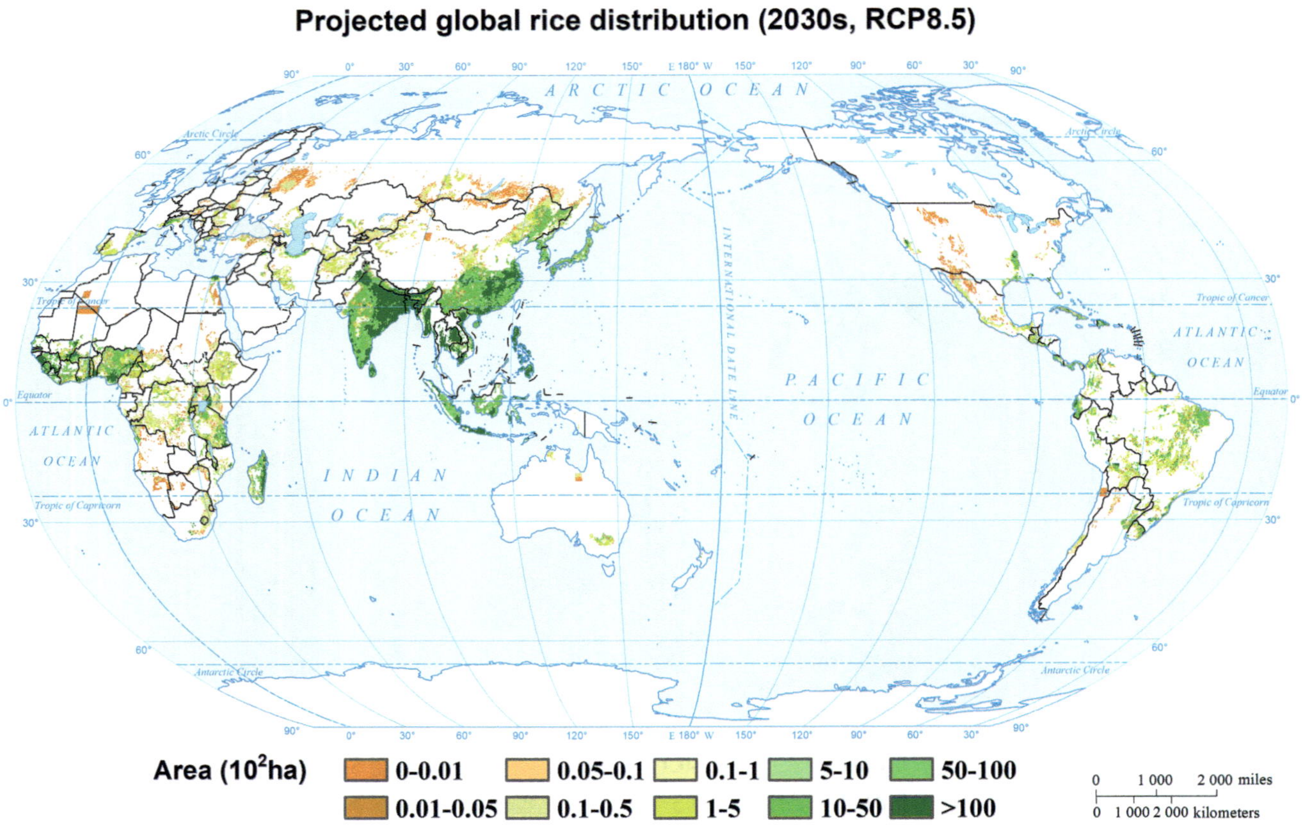
Projected global rice distribution (2030s, RCP8.5)
ARCTIC OCEAN
PACIFIC OCEAN
INDIAN OCEAN
ATLANTIC OCEAN
INTERNATIONAL DATE LINE
Arctic Circle
Tropic of Cancer
Equator
Tropic of Capricorn
Antarctic Circle
Area (10²ha)
0-0.01
0.01-0.05
0.05-0.1
0.1-0.5
0.1-1
1-5
5-10
10-50
50-100
>100
0 1 000 2 000 miles
0 1 000 2 000 kilometers

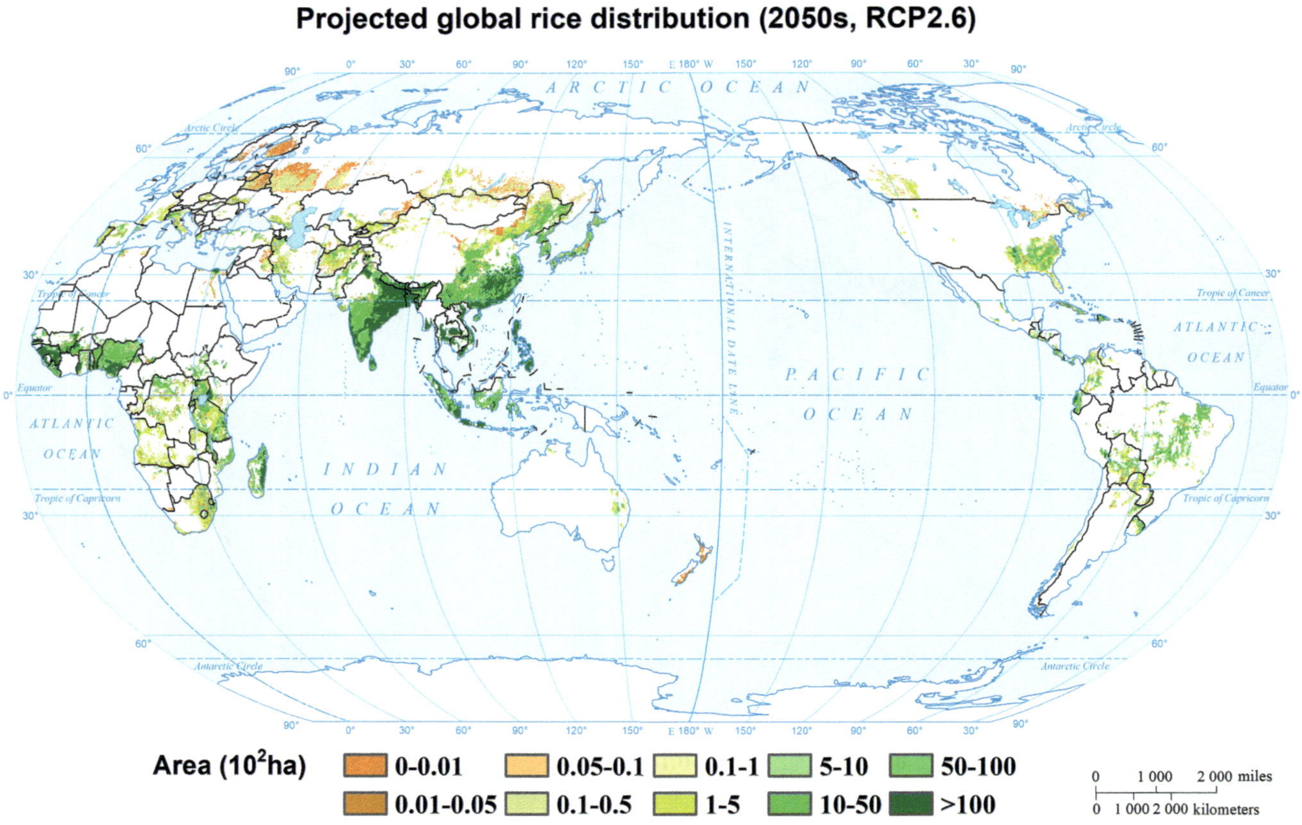
Projected global rice distribution (2050s, RCP2.6)
ARCTIC OCEAN
PACIFIC OCEAN
INDIAN OCEAN
ATLANTIC OCEAN
INTERNATIONAL DATE LINE
Arctic Circle
Tropic of Cancer
Equator
Tropic of Capricorn
Antarctic Circle
Area (10²ha)
0-0.01
0.01-0.05
0.05-0.1
0.1-0.5
0.1-1
1-5
5-10
10-50
50-100
>100
0 1 000 2 000 miles
0 1 000 2 000 kilometers

Projected global rice distribution (2050s, RCP4.5)

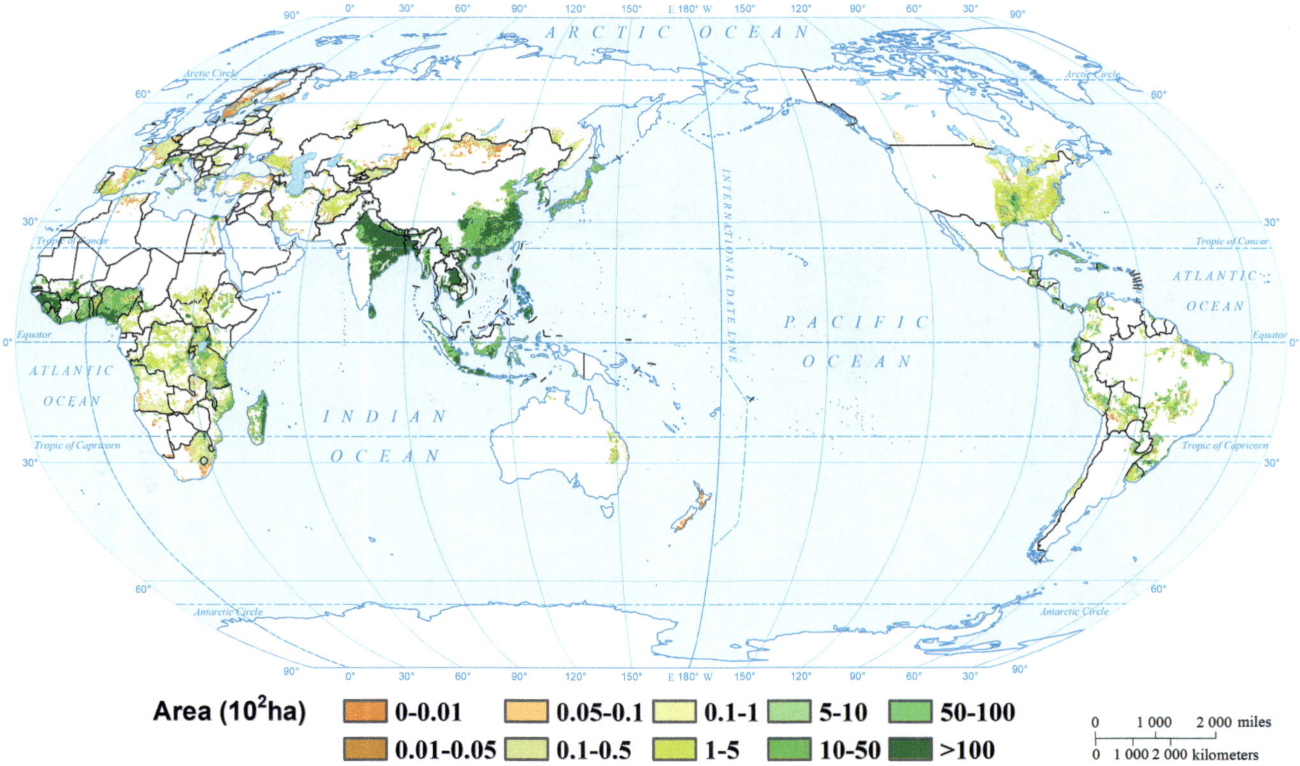

Projected global rice distribution (2050s, RCP8.5)

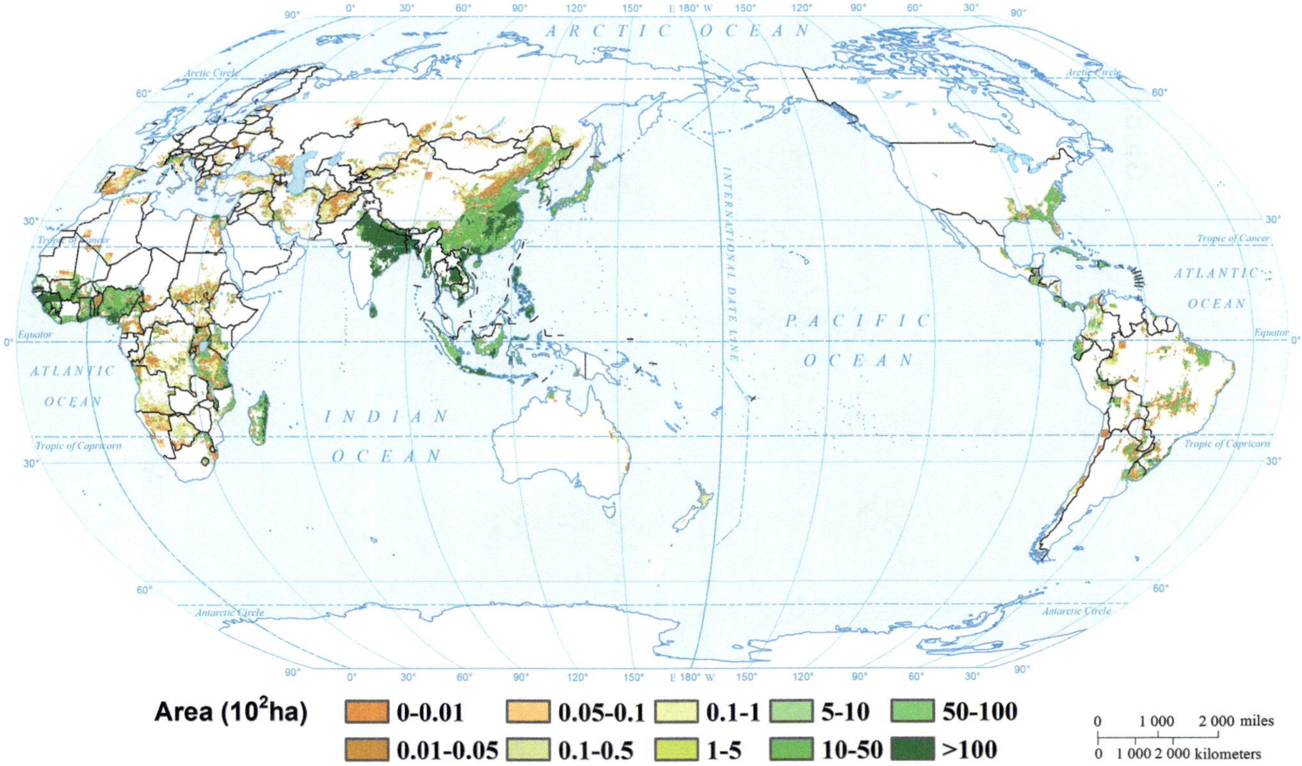

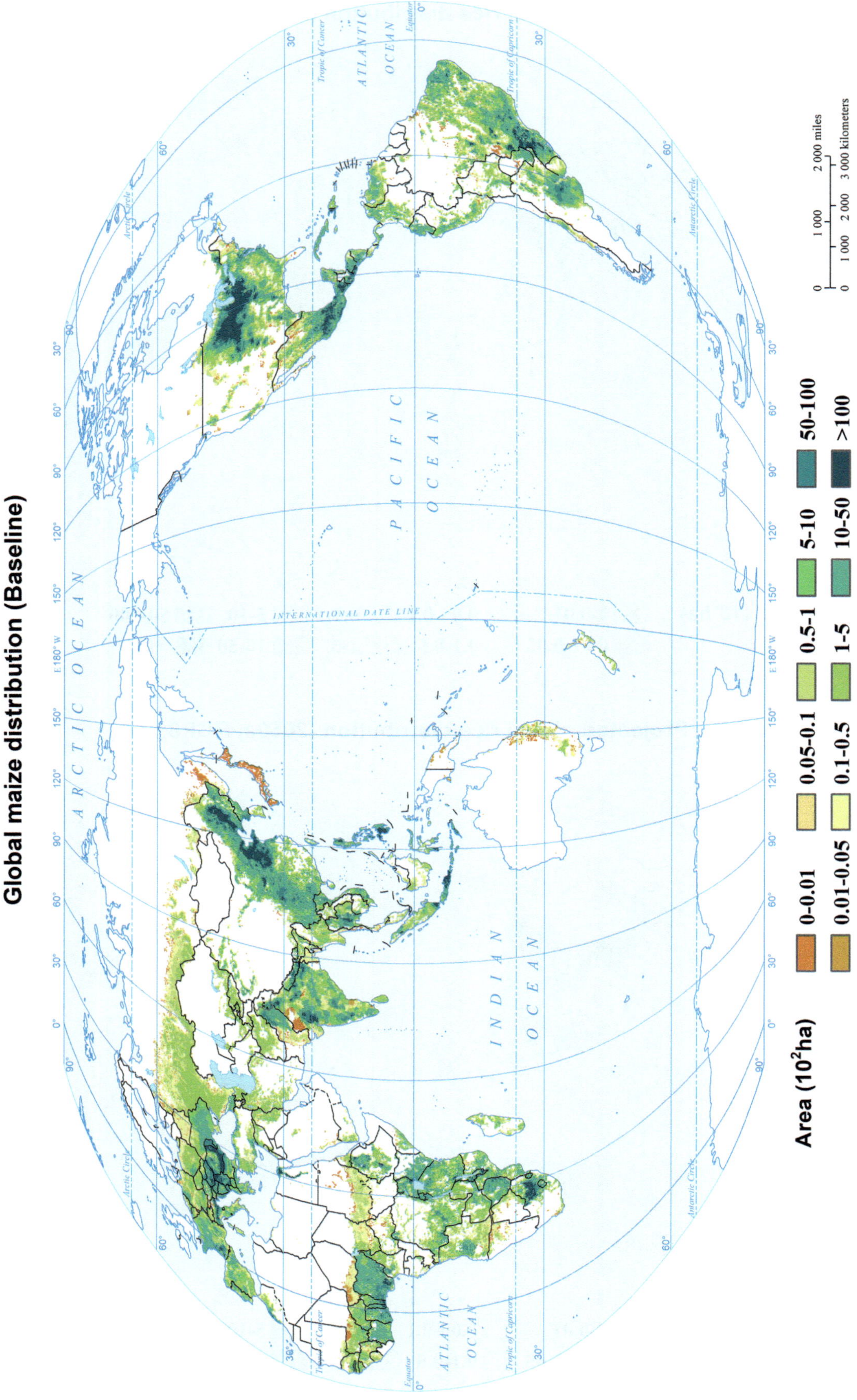
Global maize distribution (Baseline)
Area (10^{2}ha)
0-0.01
0.01-0.05
0.05-0.1
0.1-0.5
0.5-1
1-5
5-10
10-50
50-100
>100
0 1000 2000 miles
0 1000 2000 3000 kilometers
ARCTIC OCEAN
PACIFIC OCEAN
INDIAN OCEAN
ATLANTIC OCEAN
INTERNATIONAL DATE LINE
Arctic Circle
Tropic of Cancer
Equator
Tropic of Capricorn
Antarctic Circle

Projected global maize distribution (2030s, RCP2.6)

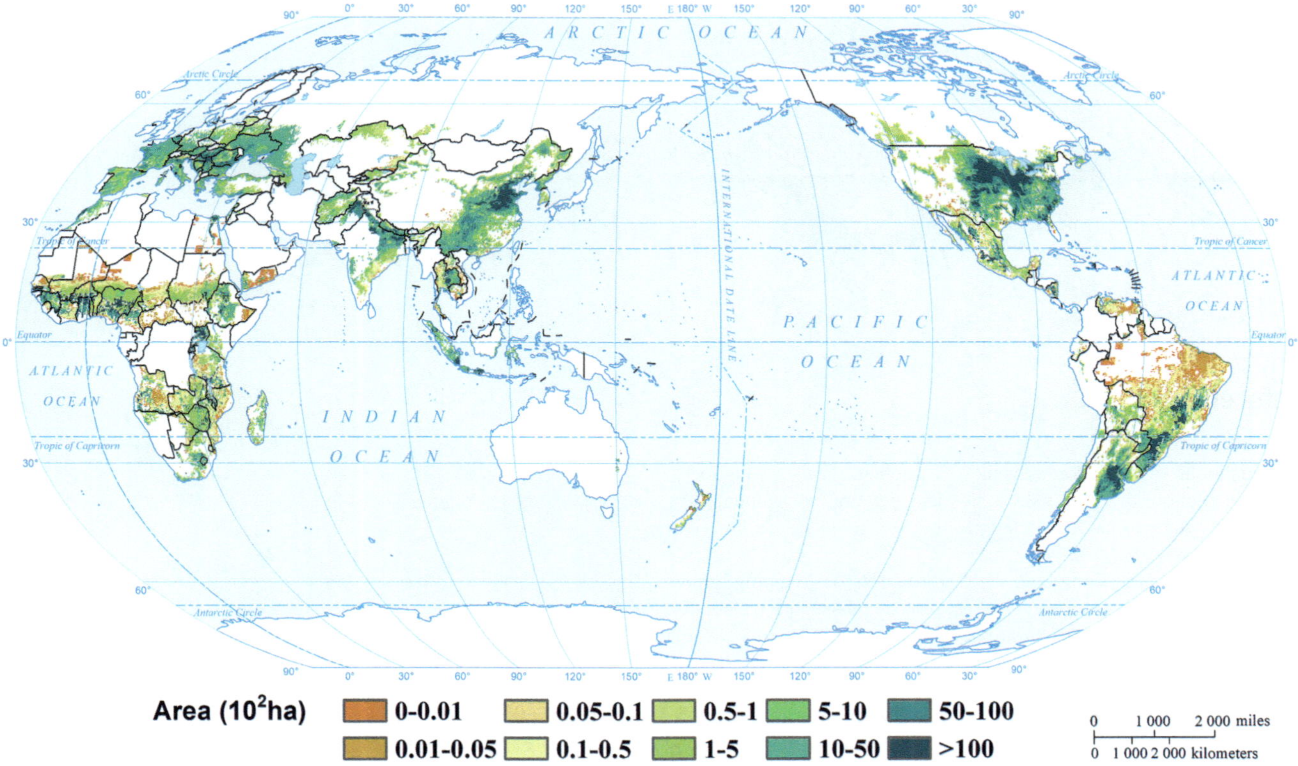

Projected global maize distribution (2030s, RCP4.5)

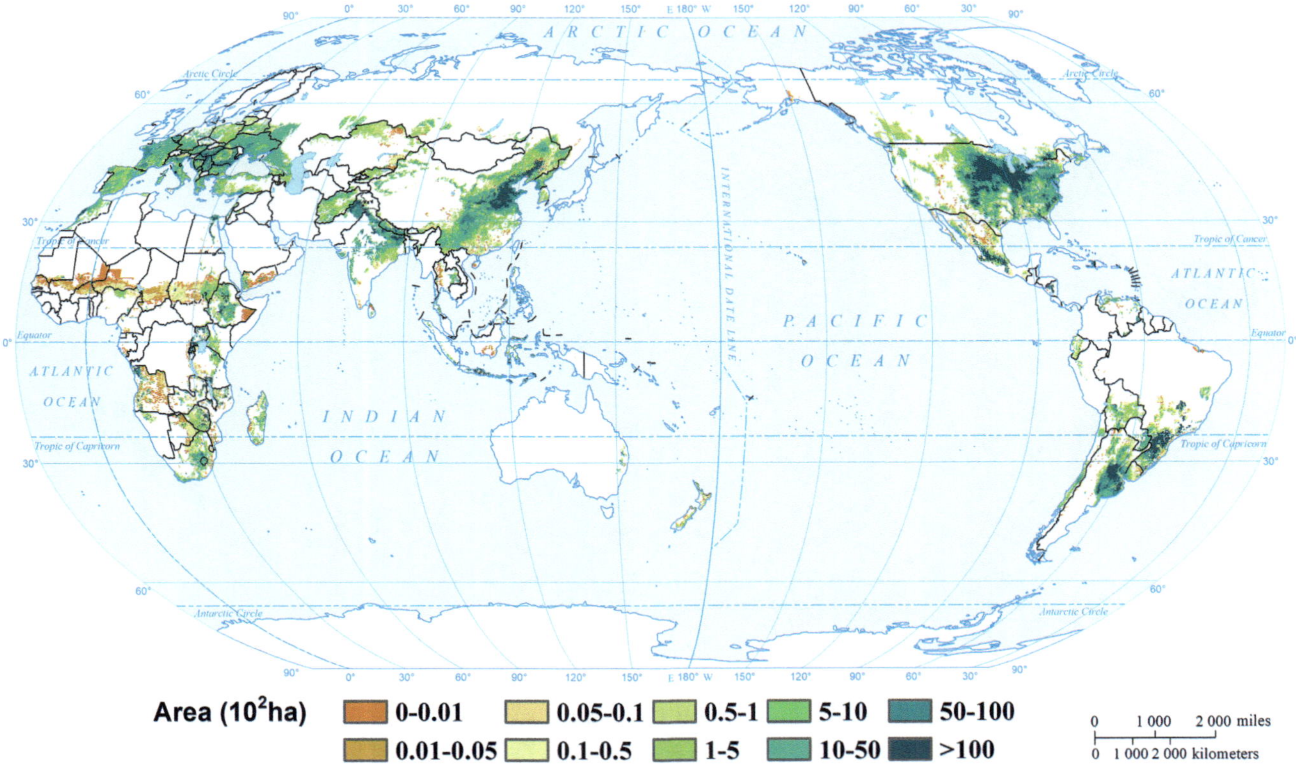

Projected global maize distribution (2030s, RCP8.5)

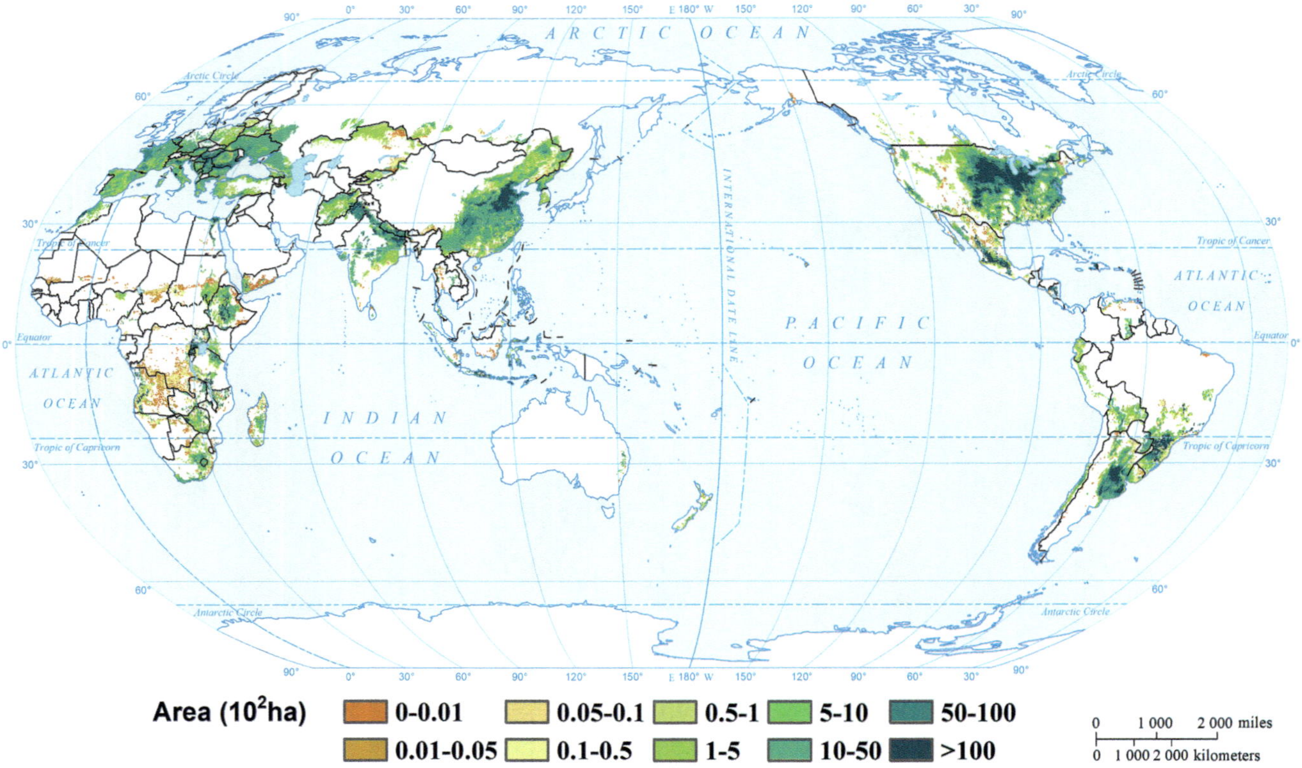

Projected global maize distribution (2050s, RCP2.6)

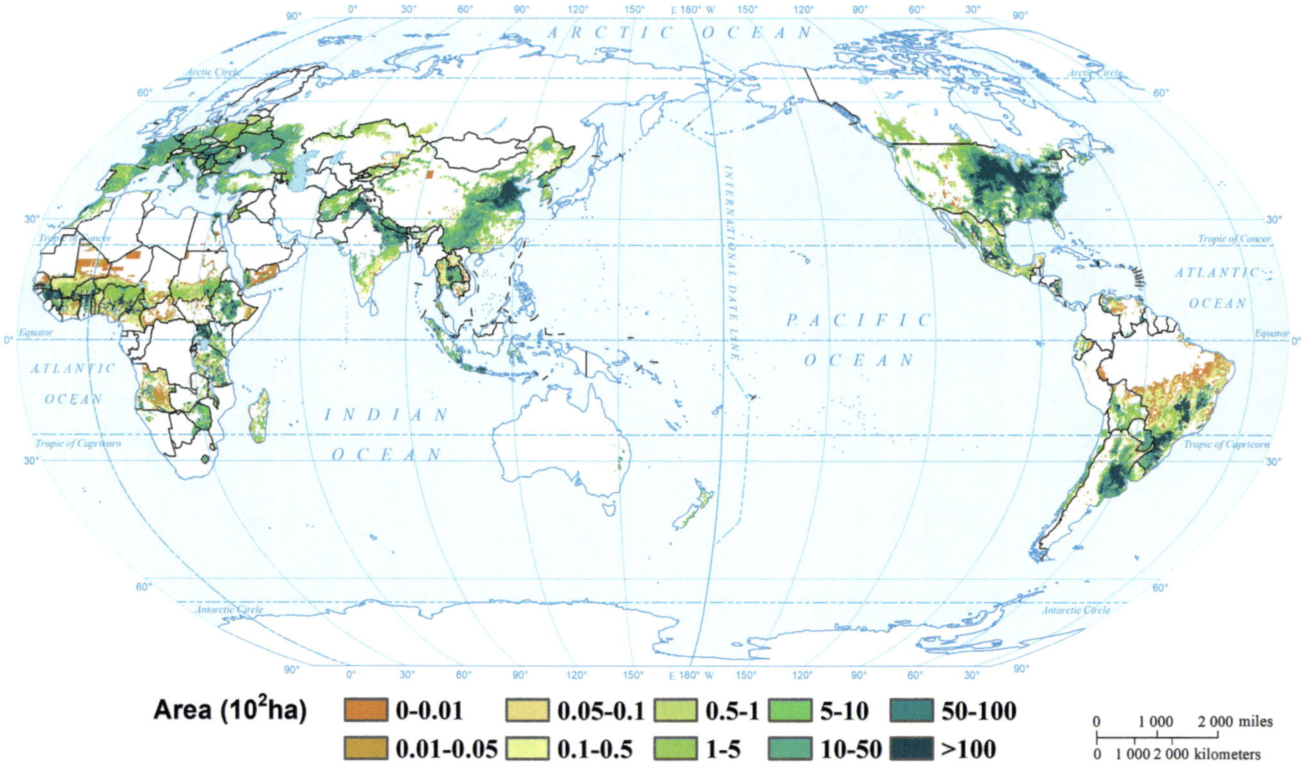

Projected global maize distribution (2050s, RCP4.5)

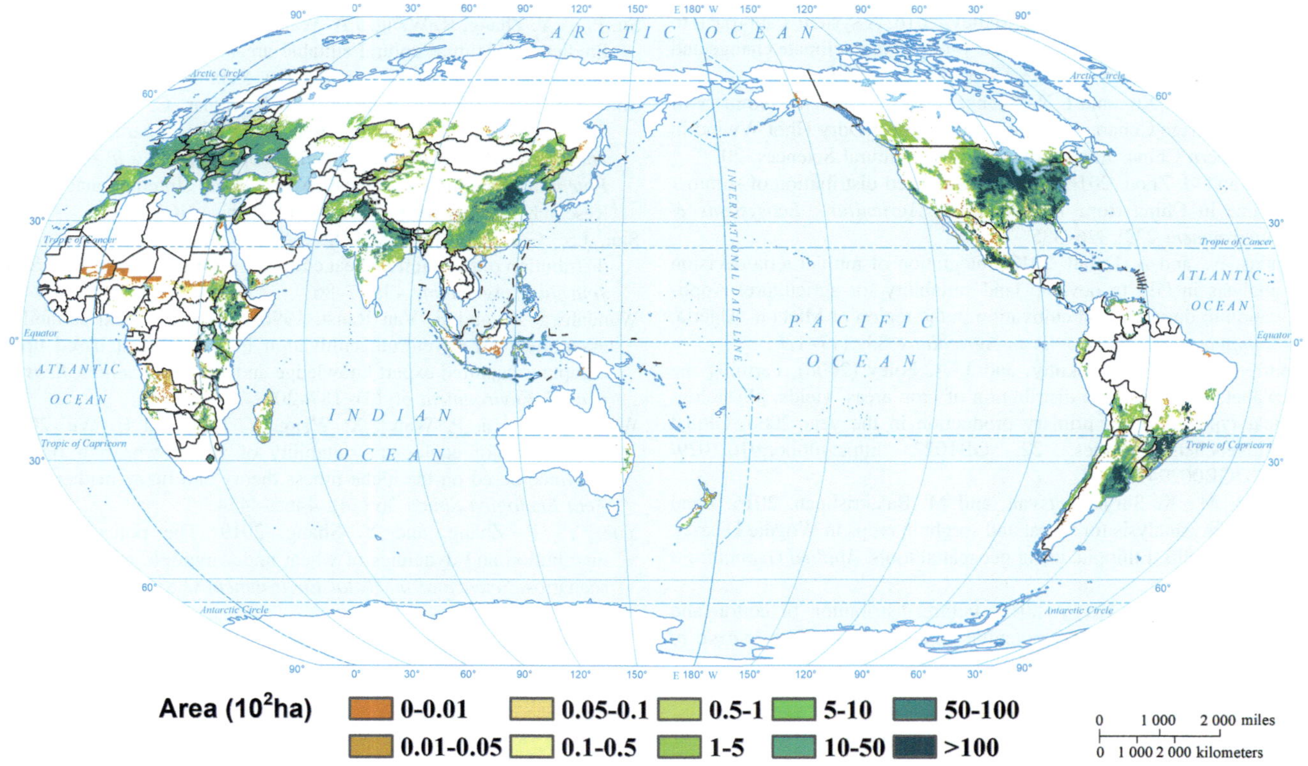

Projected global maize distribution (2050s, RCP8.5)

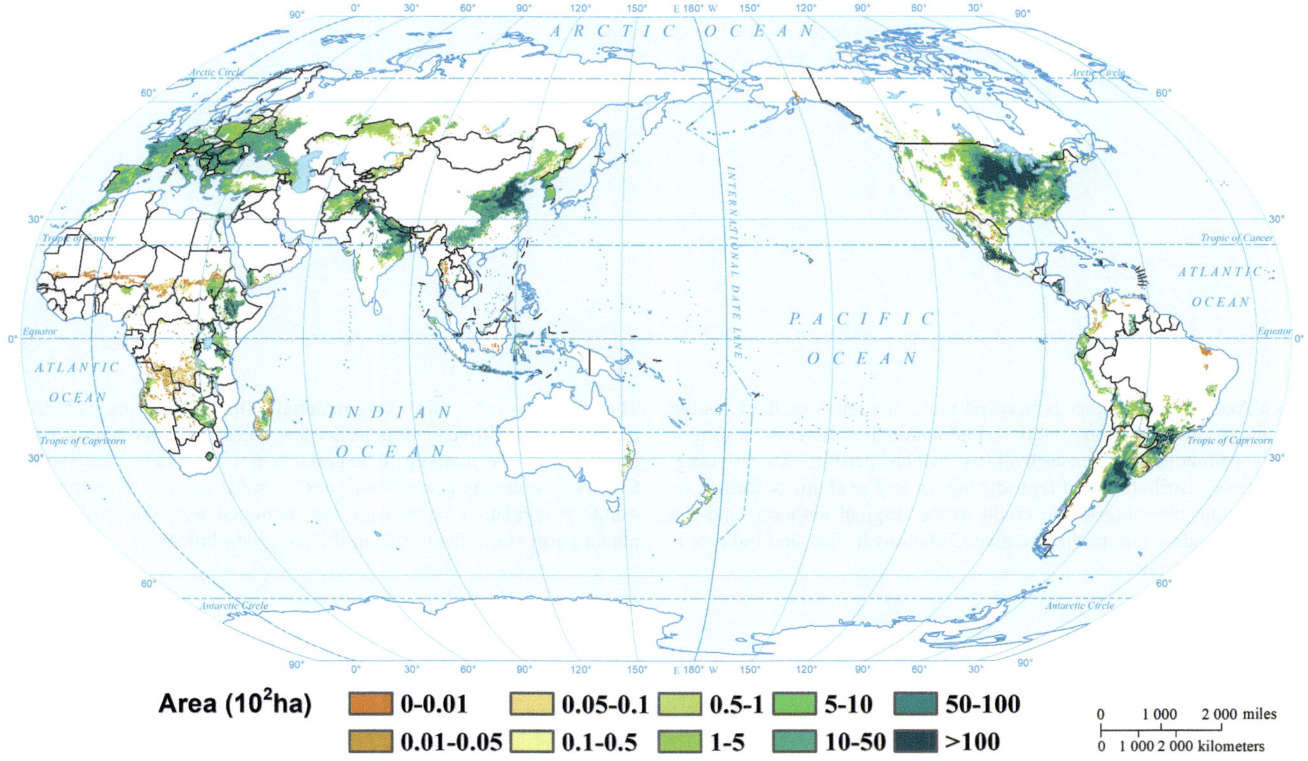

References

Anandhi, A., J.L. Steiner, and N. Bailey. 2016. A system's approach to assess the exposure of agricultural production to climate change and variability. *Climatic Change* 136 (3–4): 647–659.

Deng, Z.Y., J.F, Xu., and L.N, Huang. (2009). Research Ssummary of the Ddamage Ccharacteristics of the Wwheat Ddry Hhot Wwind in Nnorthern China. Journal of Anhui Agricultural Sciences, 20.

He, Q., and G. Zhou. 2016. Climate-associated distribution of summer maize in China from 1961 to 2010. *Agriculture, Ecosystems & Environment* 232: 326–335.

Mendas, A., and A. Delali. 2012. Integration of multicriteria decision analysis in GIS to develop land suitability for agriculture: Application to durum wheat cultivation in the region of Mleta in Algeria. *Computers and Electronics in Agriculture* 83: 117–126.

Monfreda, C., N. Ramankutty, and J. A. Foley (2008), Farming the planet: 2. Geographic distribution of crop areas, yields, physiological types, and net primary production in the year 2000, Global Biogeochem. Cycles, 22, GB1022. https://doi.org/10.1029/2007GB002947.

Motuma, M., K. Suryabhagavan, and M. Balakrishnan. 2016. Land suitability analysis for wheat and sorghum crops in Wogdie District, South Wollo, Ethiopia, using geospatial tools. *Applied Geomatics* 8 (1): 57–66.

Shabani, F., and B. Kotey. 2016. Future distribution of cotton and wheat in Australia under potential climate change. *The Journal of Agricultural Science* 154 (2): 175–185.

Stansbury, C., and Z. Pretorius. 2001. Modelling the potential distribution of Karnal bunt of wheat in South Africa. *South African Journal of Plant and Soil* 18 (4): 159–168.

Su, P., A.Y. Zhang, R. Wang, J.A. Wang, Y. Gao, and F.G. Liu. 2021. Prediction of future natural suitable areas for rice under representative concentration pathways (RCPs). *Sustainability* 13: 1580.

Subbarao, G.V., J.V.D.K. Kumar Rao, J. Kumar, C. Johansen, U.K. Deb, I. Ahmed, M.V. Krishna Rao, L. Venkataratnam, et al. 2001. *Spatial distribution and quantification of rice-fallows in south Asia: Potential for legumes*. Patancheru, India: International Crops Research Institute for the Semi-Arid Tropics (ICRISAT).

Sun, J.S., G.S. Zhou, and X.H. Sui. 2012. Climatic suitability of the distribution of the winter wheat cultivation zone in China. *European Journal of Agronomy* 43: 77–86.

Wandahwa, P., and E. Van Ranst. 1996. Qualitative land suitability assessment for pyrethrum cultivation in west Kenya based upon computer-captured expert knowledge and GIS. *Agriculture, Ecosystems & Environment* 56 (3): 187–202.

Wang, L., Y. Li, P. Wang, X. Wang, Y. Luo, and H. Wu. 2016. Assessment of ecological suitability of winter wheat in Jiangsu Province based on the niche-fitness theory and fuzzy mathematics. *Acta Ecologica Sinica* 36 (14): 4465–4474.

Yue, Y., P. Zhang, and Y. Shang. 2019. The potential global distribution and dynamics of wheat under multiple climate change scenarios. *Science of the Total Environment* 688: 1308–1318.

Mapping Global Crop Exposure to Extremely High Temperature

Yaojie Yue, Peng Su, Yuan Gao, Puying Zhang, Ran Wang, Anyu Zhang, Qinghua Jiang, Weidong Ma, Yuantao Zhou, and Jing'ai Wang

1 Introduction

The increasing temperature with global warming will have great impacts on major cereal crop cultivation (Peng et al. 2004; Fahad et al. 2019). Among various impacts, the crop exposure to extremely high temperature, which is based on the land suitability for crop cultivation, may ultimately pose a great threat to food security.

The crop disaster exposure is one of the primary drivers of food system instability (Change 2018), and its response to climate change is critical to understanding the impact of climate change on food security. Recent studies have provided evidence for changes in the yield of major cereal crops due to disaster exposure and identified significant impacts of climate change globally, either at the country level or at the 0.5° grid level (Osborne and Wheeler 2013; Iizumi and Ramankutty 2016). However, from the perspective of disaster exposure, research on crop exposure to certain hazards (e.g., extreme heat) with the changes of climate and social conditions is still lacking.

As an important part of the atlas, the maps of global crop exposure to extremely high temperature are intended to evaluate the changes in the global staple crop yield risk from three aspects—mean yield, interannual yield variability, and lower extreme yield—under the RCP2.6-2030s, RCP2.6-2050s, RCP4.5-2030s, RCP4.5-2050s, RCP8.5-2030s, and RCP8.5-2050s scenarios. The crop yield risk was measured by multi-model ensemble (MME) simulation using global high spatial resolution (0.25°) climate forcing data. To enable such MME simulation, the development of emulators of global gridded crop models (GGCM) is required (Lobell and Burke 2010; Holzkämper et al. 2012; Oyebamiji et al. 2015; Raimondo et al. 2020). The present results may provide crucial information for climate risk assessment and effective adaptations.

2 Method

2.1 Wheat Exposure Calculation

The quantification process of wheat exposure to extreme high-temperature hazard is as follows (Jiang et al. 2019): A high-temperature day is recorded when the maximum temperature of the current day reaches or exceeds 30°C, and if this lasts for 3 days or more, it is recorded as a heatwave event (Deng et al. 2009; Chen et al. 2016).

The high-temperature days (HD) refer to the accumulation of total high-temperature days of heatwave events in each grid. The calculation formula is as follows:

$$\mathrm{HD} = \sum_{i=1}^{n} D_i \tag{1}$$

Authors: Yaojie Yue, Peng Su, Yuan Gao, Puying Zhang, Ran Wang, Anyu Zhang, Qinghua Jiang, Weidong Ma, Yuantao Zhou, Jing'ai Wang
Map Designers: Yichen Li, Qingyuan Ma, Yuanyuan Jing, Jing'ai Wang, Ying Wang. **Language Editor**: Tao Ye, Peng Su.

Y. Yue · Y. Gao · P. Zhang · R. Wang · A. Zhang · Q. Jiang · J. Wang (✉)
Key Laboratory of Environmental Change and Natural Disaster of Ministry of Education, Faculty of Geographical Science, Beijing Normal University, Beijing, 100875, China
e-mail: jwang@bnu.edu.cn

P. Su · W. Ma · Y. Zhou
School of Geographic Science, Qinghai Normal University, Xining, 810008, China

J. Wang
State Key Laboratory of Earth Surface Processes and Resource Ecology, Beijing Normal University, Beijing, 100875, China

P. Shi, *Atlas of Global Change Risk of Population and Economic Systems*, IHDP/Future Earth-Integrated Risk Governance Project Series, https://doi.org/10.1007/978-981-16-6691-9_11

where n refers to the frequency of heatwaves and Di refers to the days corresponding to the ith heatwave event.

Estimating the intensity of the heatwave hazard impacts on wheat involves first calculating the intensity (I), which represents the daily maximum temperature exceeding the temperature threshold (30°C) of each high-temperature day within heatwave events and for each grid; then, we sum up all Is. The details of this calculation are as follows:

$$\mathrm{HI} = \sum\nolimits_{d_h}^{d_m} I \quad (2)$$

$$I = \begin{cases} 0 & T_{max} < T_h \\ T_{max} - T_h & T_{max} \geq T_h \end{cases} \quad (3)$$

where d_h and d_m represent the heading period and mature period of wheat, respectively; T_{max} represents the daily maximum temperature; I represents the daily temperature (° C) exceeding the temperature threshold during heatwave events; and T_h represents the high-temperature threshold.

The extreme high-temperature exposure of wheat is calculated as follows: The exposure range of wheat is the spatial superposition of the distribution range of wheat and the hazard range, and the areal value in the range represents the exposure value, which is the annual wheat harvest area.

2.2 Rice Exposure Calculation

The quantification process of rice exposure to extreme high-temperature hazard is as follows: A high-temperature event is recorded when the maximum temperature of the current day reaches or exceeds 30 °C (Melillo et al. 1995; Janetos 1997). The accumulative high-temperature stress of high-temperature events during the growth period (GHTS) was used as the high-temperature hazard intensity. The calculation process can be divided into two steps: (1) according to the abovementioned high-temperature standard, determine the high-temperature event and calculate the high-temperature stress (HTS) of the day, and then estimate the GHTS by calculating the total HTS during the whole growth period (Eqs. 4 and 5).

$$\mathrm{HTS}_i = \begin{cases} 0 & T_i \geq 30°\mathrm{C} \\ 1 - \sin\left[\frac{\pi}{2} \times \left(\frac{T_i - T_b}{T_o - T_b}\right)\right] & T_i \geq 30°\mathrm{C} \end{cases} \quad (4)$$

$$\mathrm{GHTS} = \sum_{i=1}^{n} \mathrm{HTS}_i \quad (5)$$

where Ti is the daily average temperature of the ith high-temperature event during the growth period; Tb is the base temperature during the growth period; To is the optimum temperature during the growth period; n is the frequency of high-temperature events during the growth period; and HTSi is the daily high-temperature stress value of the ith high-temperature event.

The extreme high-temperature exposure of rice is calculated as follows: The exposure range of rice is the spatial superposition of the distribution range of rice and the hazard range, and the areal value in the range represents the exposure value, which is the annual rice harvest area.

2.3 Maize Exposure Calculation

Maize exposure calculation applies the same method as rice exposure calculation. A high-temperature event is recorded when the maximum temperature of the current day reaches or exceeds 37 °C (Melillo et al. 1995; Janetos 1997). The accumulative high-temperature stress of high-temperature events during the growth period (GHTS) was used as the high-temperature hazard intensity.

The extreme high-temperature exposure of maize is calculated as follows: The exposure range of maize is the spatial superposition of the distribution range of maize and the hazard range, and the areal value in the range represents the exposure value, which is the annual maize harvest area.

3 Results

3.1 Latitudinal Distribution of Crop Exposure to High Temperature

The latitudinal distributions of the exposure of the three crops to high temperature in the historical period (red bar) and the RCP8.5-SSP3 scenario (medium term) (blue bar) are shown in Fig. 1.

During the two study periods, the exposures of the three crops are mainly concentrated in the northern hemisphere. The most important areas for wheat, rice, and maize exposure to extreme high temperature are between 40°N–50°N, 20°N–30°N, and 40°N–50°N, respectively. The comparison of the exposure areas in the two periods shows that under the RCP8.5-SSP3 scenario, the exposure areas of the three crops in the northern hemisphere will move to higher latitudes and the exposure areas of rice and maize will have a significant increase.

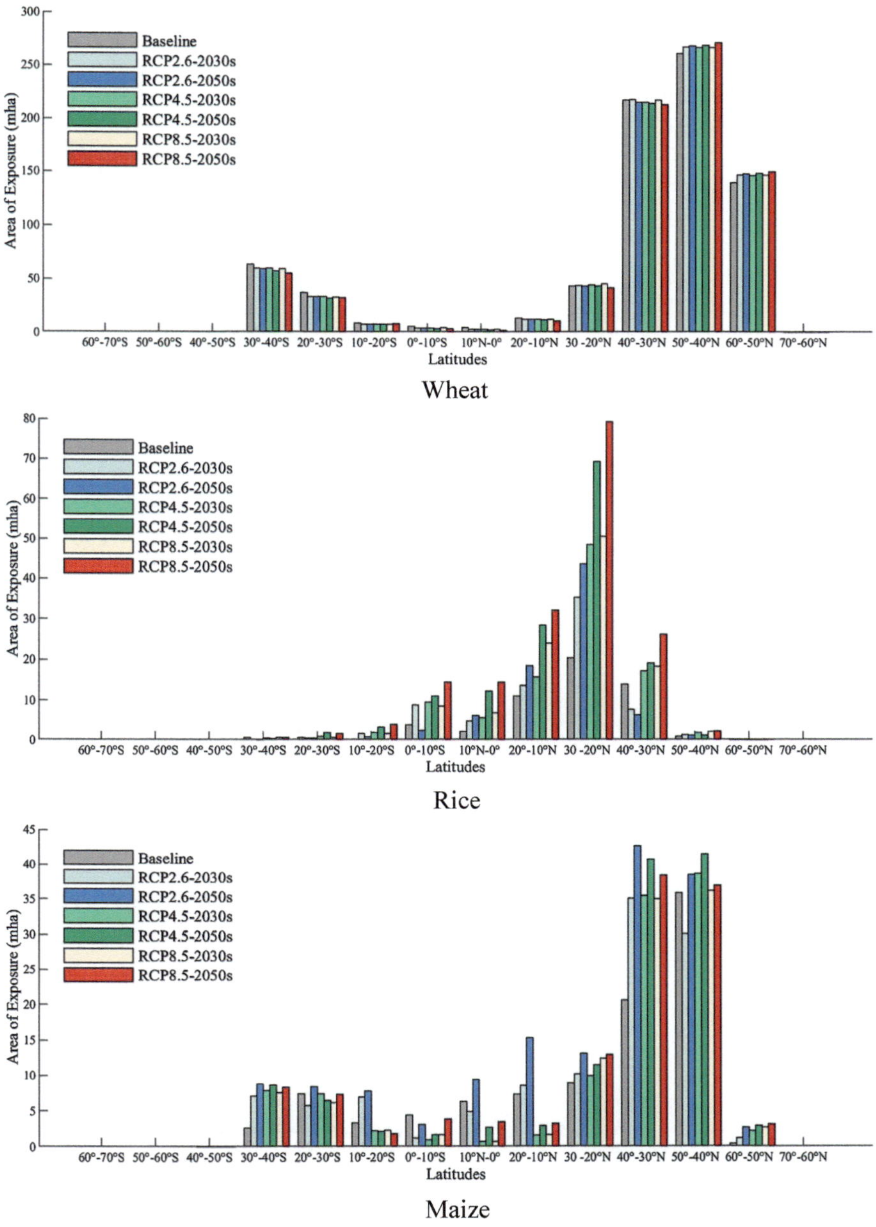

Fig. 1 Latitudinal distribution of the exposure of wheat, rice, and maize to extreme high temperature (unit: million ha)

3.2 Countries Distribution of Crop Exposure to High Temperature

Figure 2 shows the ranking of the top 20 countries based on the statistics of the area of crop exposure to high temperature in each country in the historical period and under the RCP8.5-SSP3 scenario.

In both periods, the United States' wheat exposure area ranks the highest, followed by China. The rice exposure area of India ranks the highest, and China's rice exposure area is the second largest. The maize exposure area of the United States ranks the highest, and China's maize exposure area is the second largest. Under the RCP8.5-SSP3 scenario, the exposure area of all crops in China will increase, and the exposure area of rice is about three times that of the historical period. It indicates that China will be seriously exposed to extreme high temperature in the future, and the country should be prepared to prevent risks and reduce losses.

Fig. 2 Ranking of the top 20 countries by the area of crop exposure to extreme high temperature (unit: million ha)

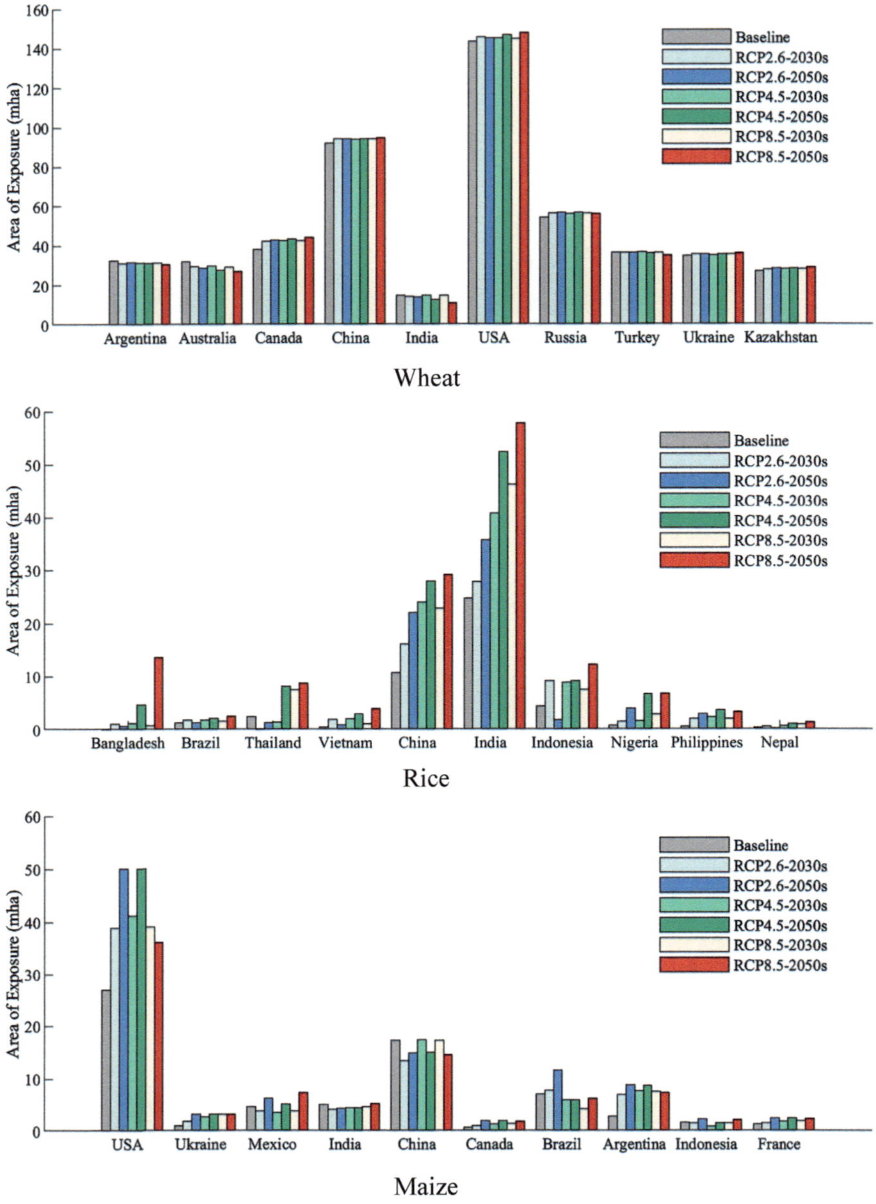

Global wheat exposure to extremely high temperature (Baseline)

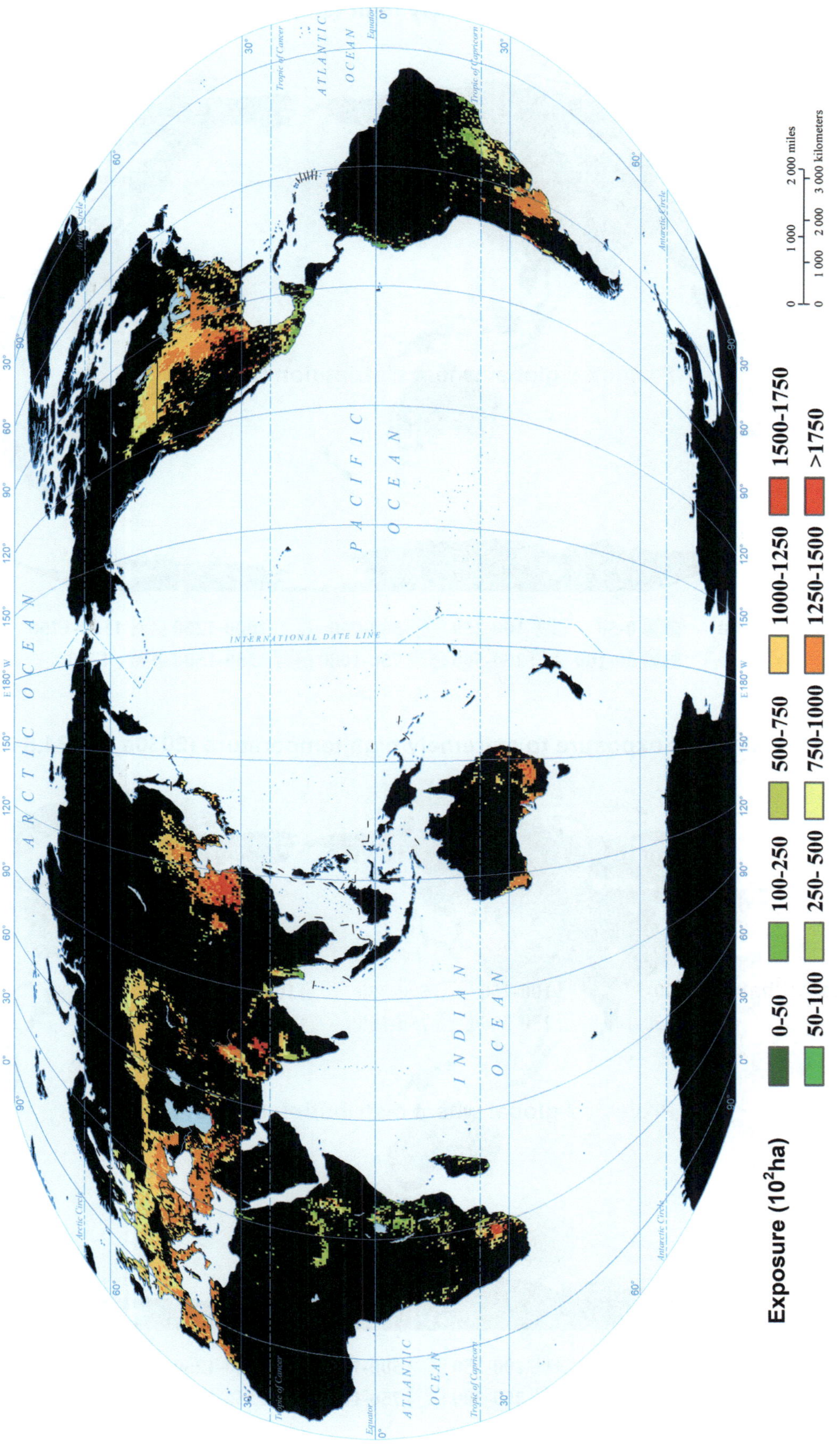

Global wheat exposure to extremely high temperature (2030s, RCP2.6-SSP1)

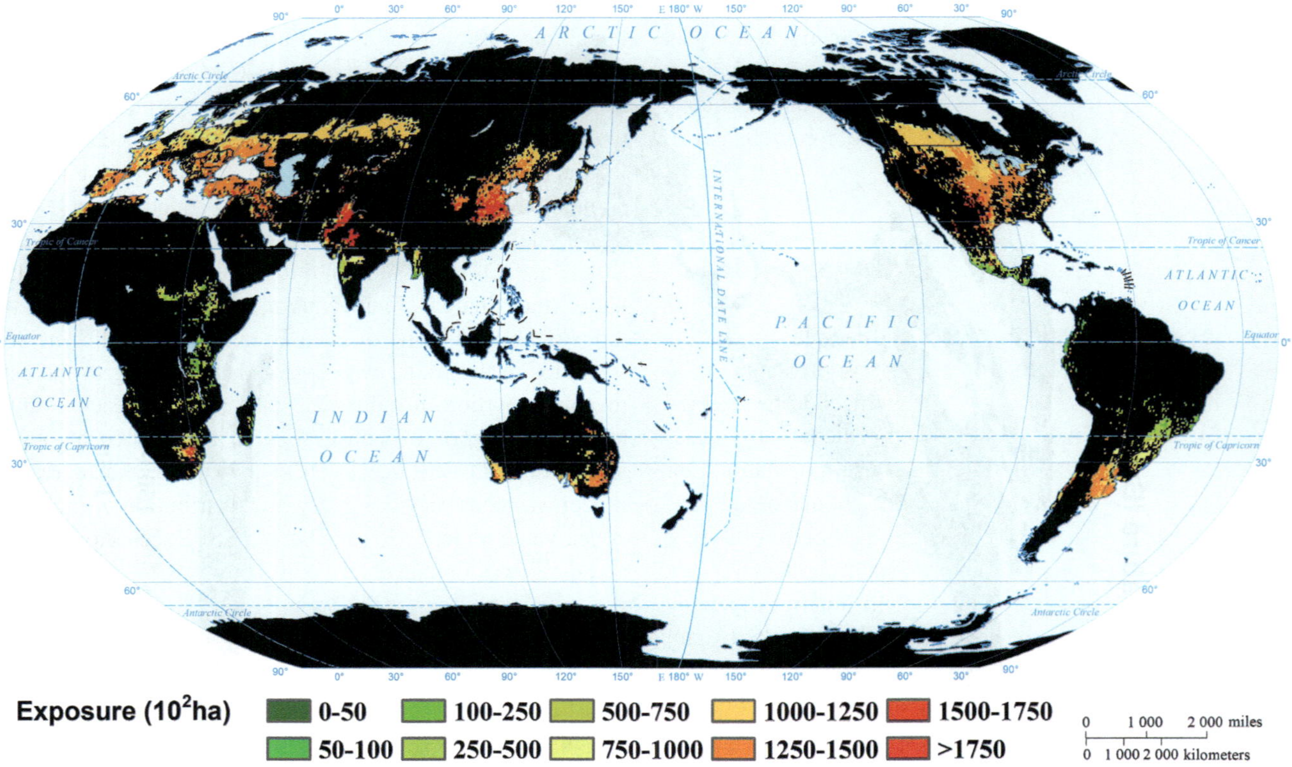

Global wheat exposure to extremely high temperature (2030s, RCP4.5-SSP2)

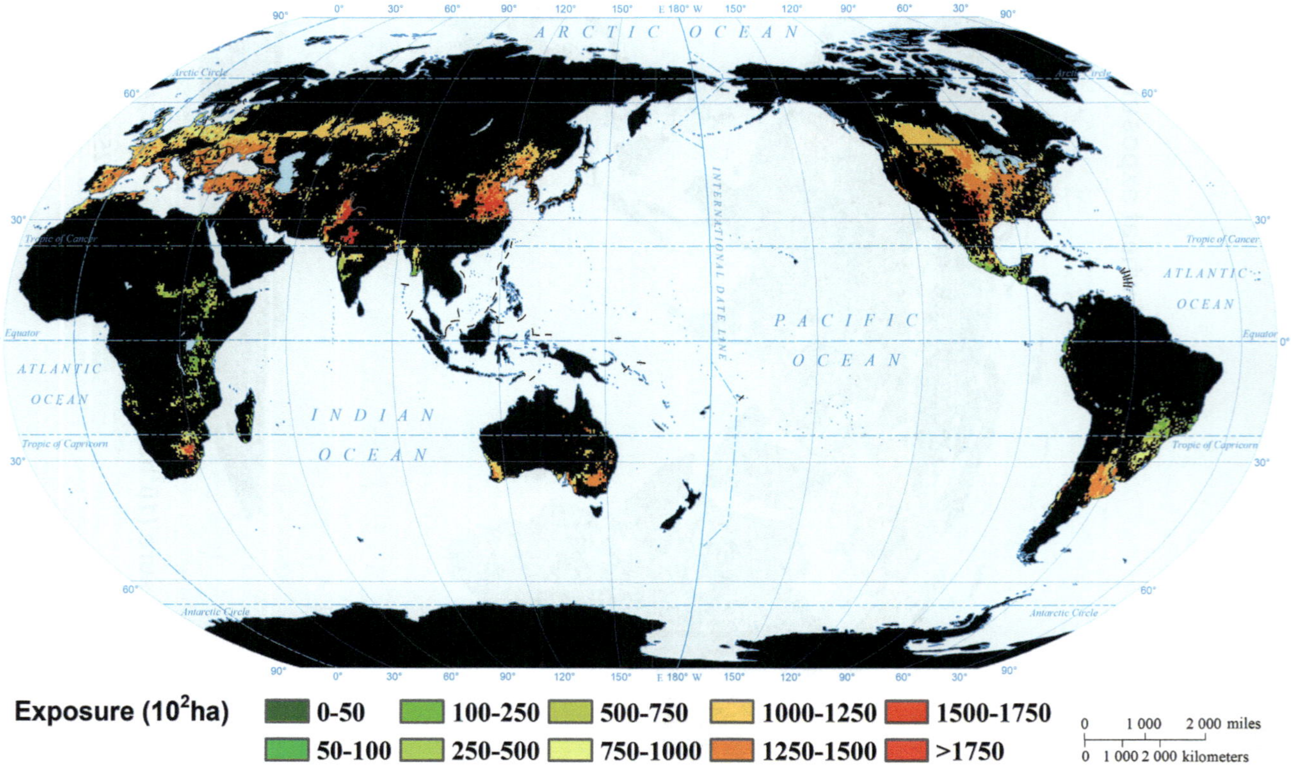

Global wheat exposure to extremely high temperature (2030s, RCP8.5-SSP3)

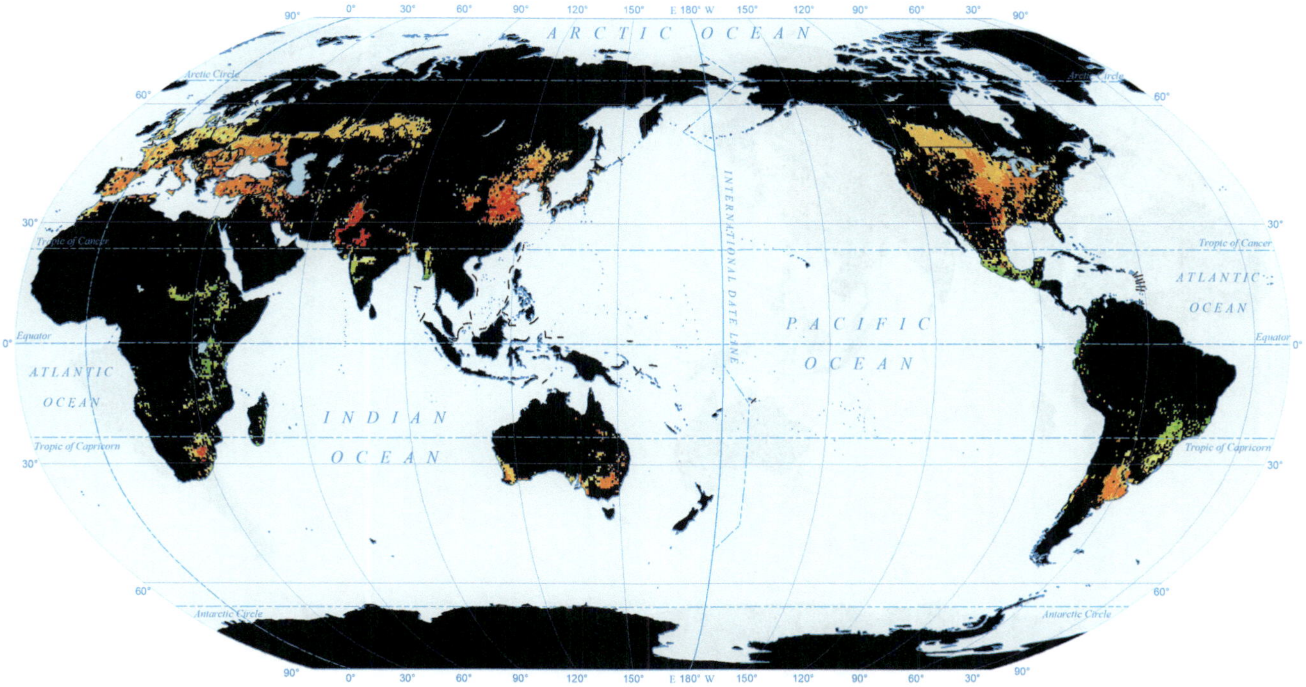

Exposure (10^2ha) 0-50 | 50-100 | 100-250 | 250-500 | 500-750 | 750-1000 | 1000-1250 | 1250-1500 | 1500-1750 | >1750

0 1 000 2 000 miles
0 1 000 2 000 kilometers

Global wheat exposure to extremely high temperature (2050s, RCP2.6-SSP1)

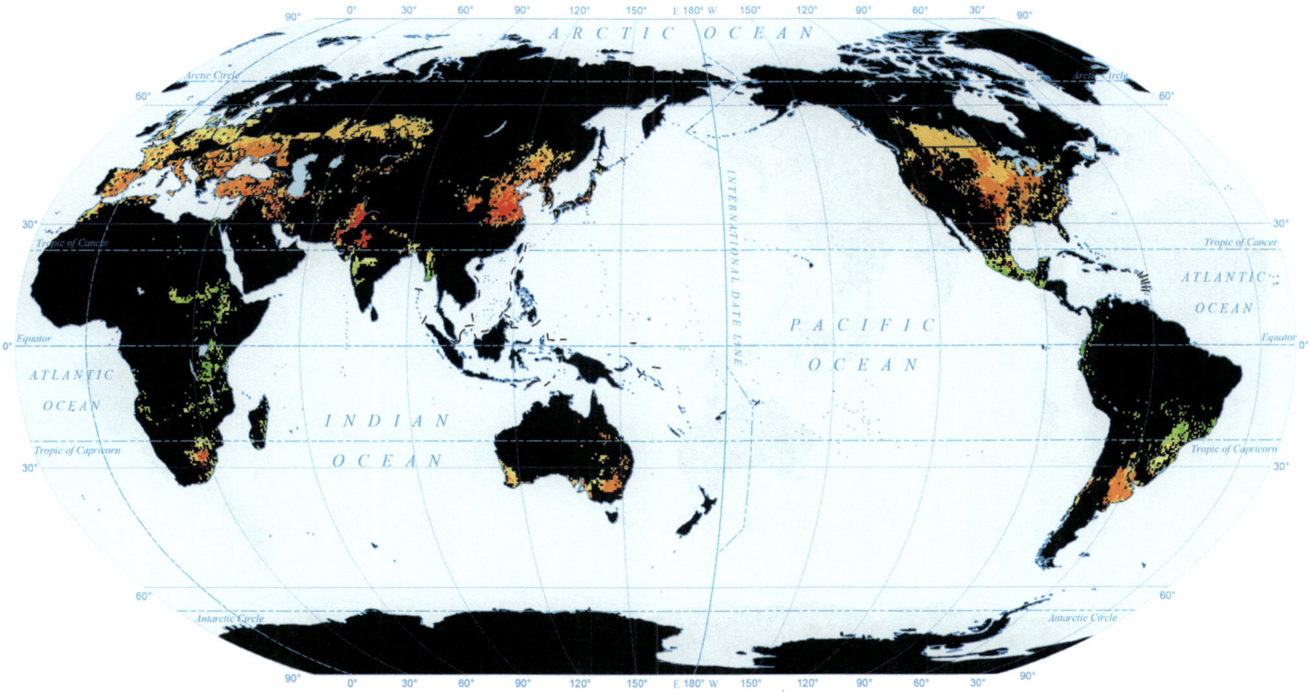

Exposure (10^2ha) 0-50 | 50-100 | 100-250 | 250-500 | 500-750 | 750-1000 | 1000-1250 | 1250-1500 | 1500-1750 | >1750

0 1 000 2 000 miles
0 1 000 2 000 kilometers

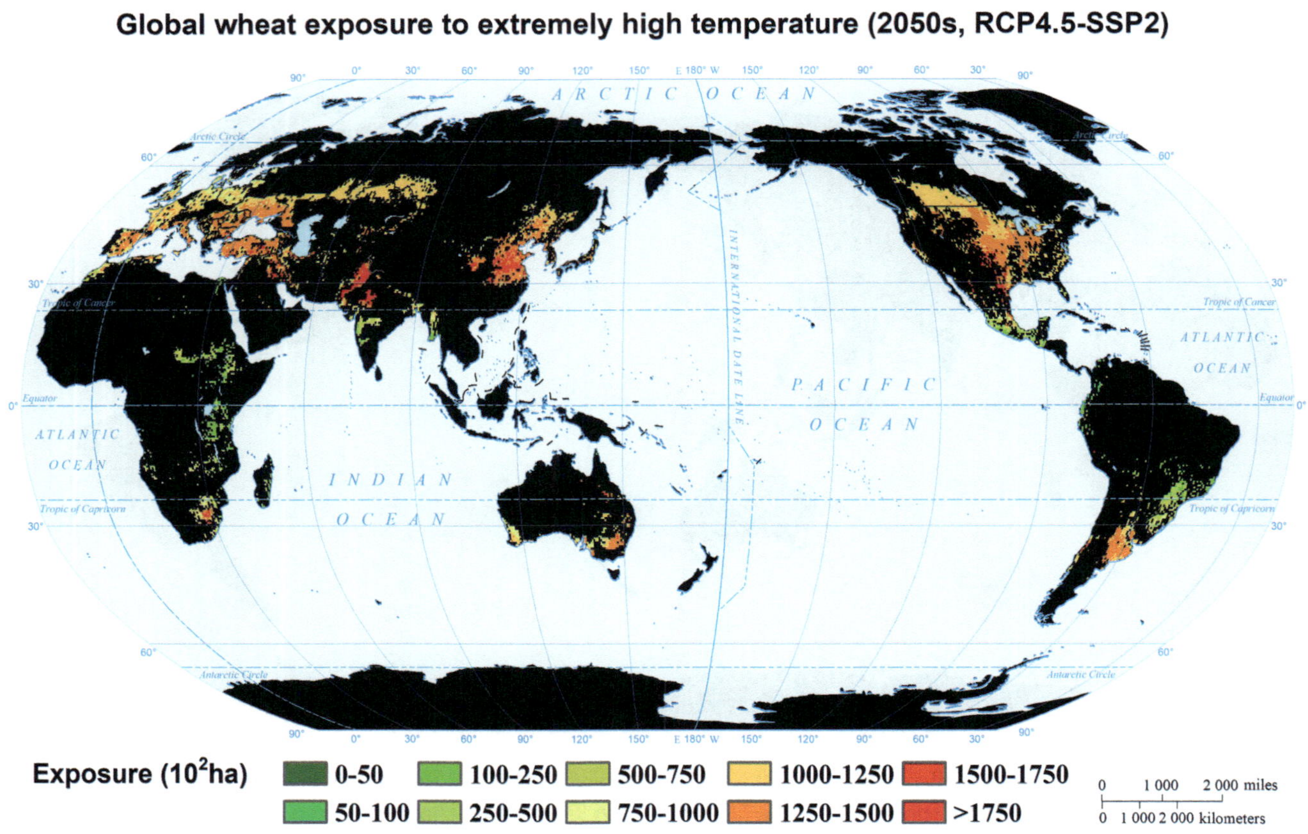
Global wheat exposure to extremely high temperature (2050s, RCP4.5-SSP2)
ARCTIC OCEAN
PACIFIC OCEAN
INDIAN OCEAN
ATLANTIC OCEAN
ATLANTIC OCEAN
INTERNATIONAL DATE LINE
Arctic Circle
Tropic of Cancer
Equator
Tropic of Capricorn
Antarctic Circle
Exposure (10²ha)
0-50
50-100
100-250
250-500
500-750
750-1000
1000-1250
1250-1500
1500-1750
>1750
0 1 000 2 000 miles
0 1 000 2 000 kilometers

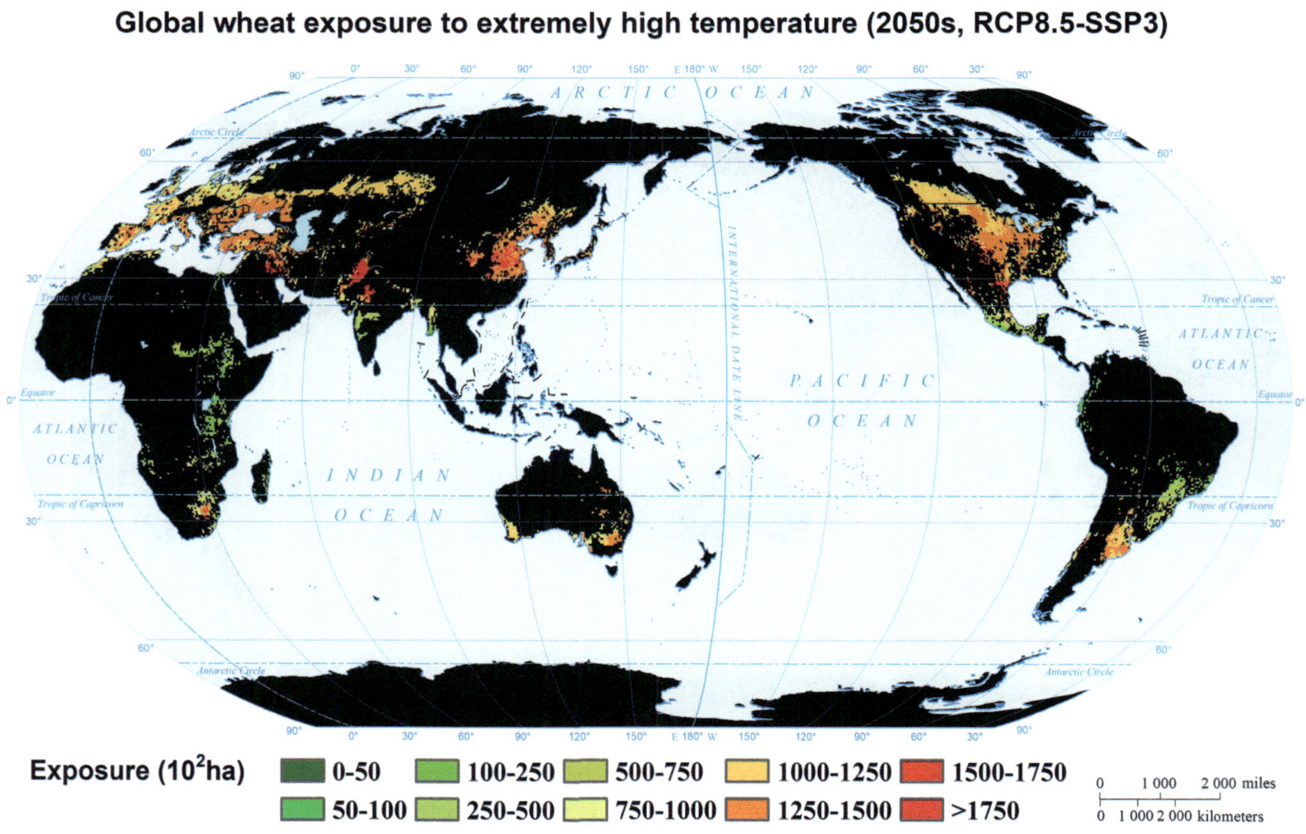
Global wheat exposure to extremely high temperature (2050s, RCP8.5-SSP3)
ARCTIC OCEAN
PACIFIC OCEAN
INDIAN OCEAN
ATLANTIC OCEAN
ATLANTIC OCEAN
INTERNATIONAL DATE LINE
Arctic Circle
Tropic of Cancer
Equator
Tropic of Capricorn
Antarctic Circle
Exposure (10²ha)
0-50
50-100
100-250
250-500
500-750
750-1000
1000-1250
1250-1500
1500-1750
>1750
0 1 000 2 000 miles
0 1 000 2 000 kilometers

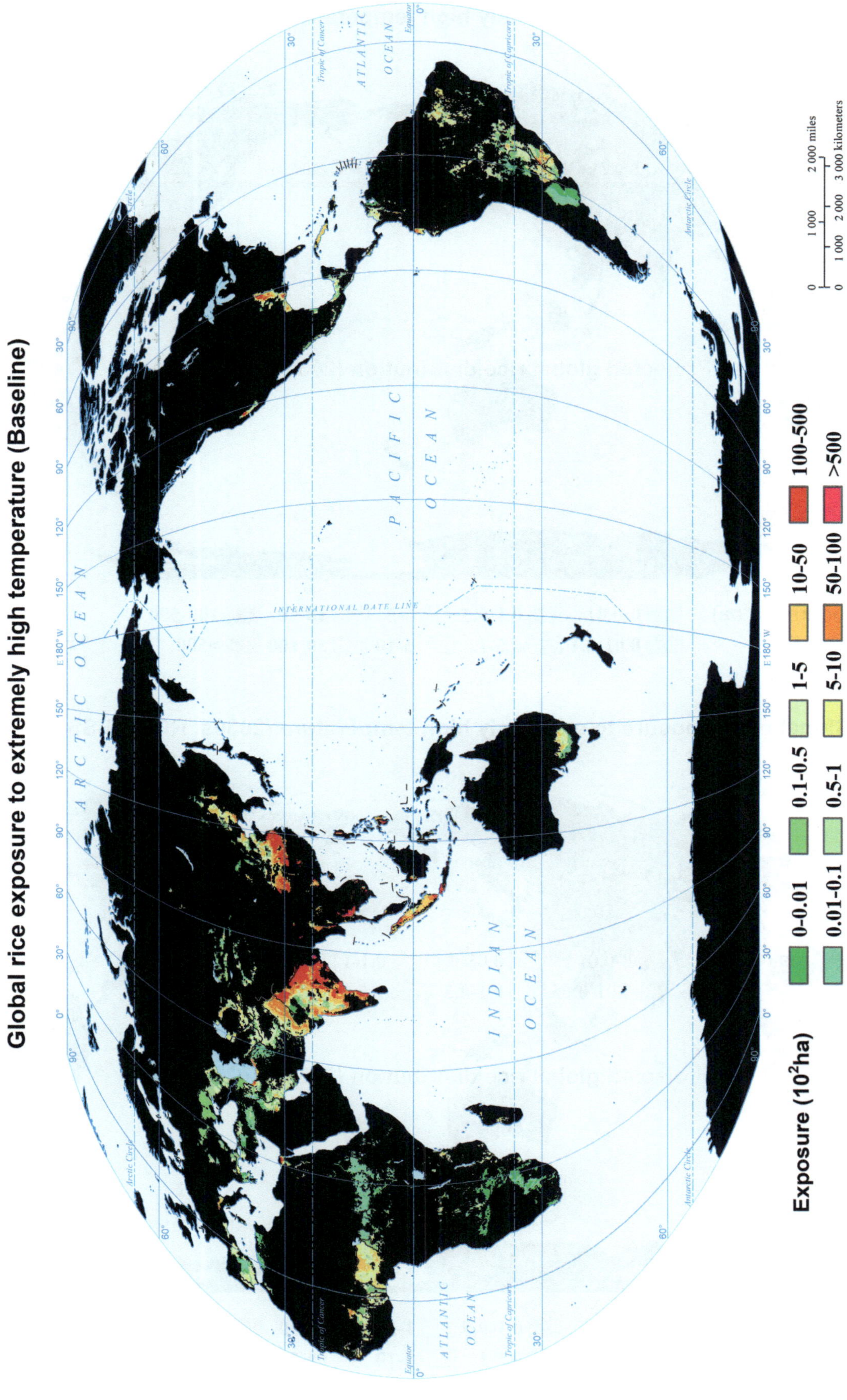
Global rice exposure to extremely high temperature (Baseline)
Exposure (10^2ha)
0-0.01
0.01-0.1
0.1-0.5
0.5-1
1-5
5-10
10-50
50-100
100-500
>500
ARCTIC OCEAN
PACIFIC OCEAN
INDIAN OCEAN
ATLANTIC OCEAN
INTERNATIONAL DATE LINE
Arctic Circle
Tropic of Cancer
Equator
Tropic of Capricorn
Antarctic Circle
0 1 000 2 000 miles
0 1 000 2 000 3 000 kilometers

Global rice exposure to extremely high temperature (2030s, RCP2.6-SSP1)

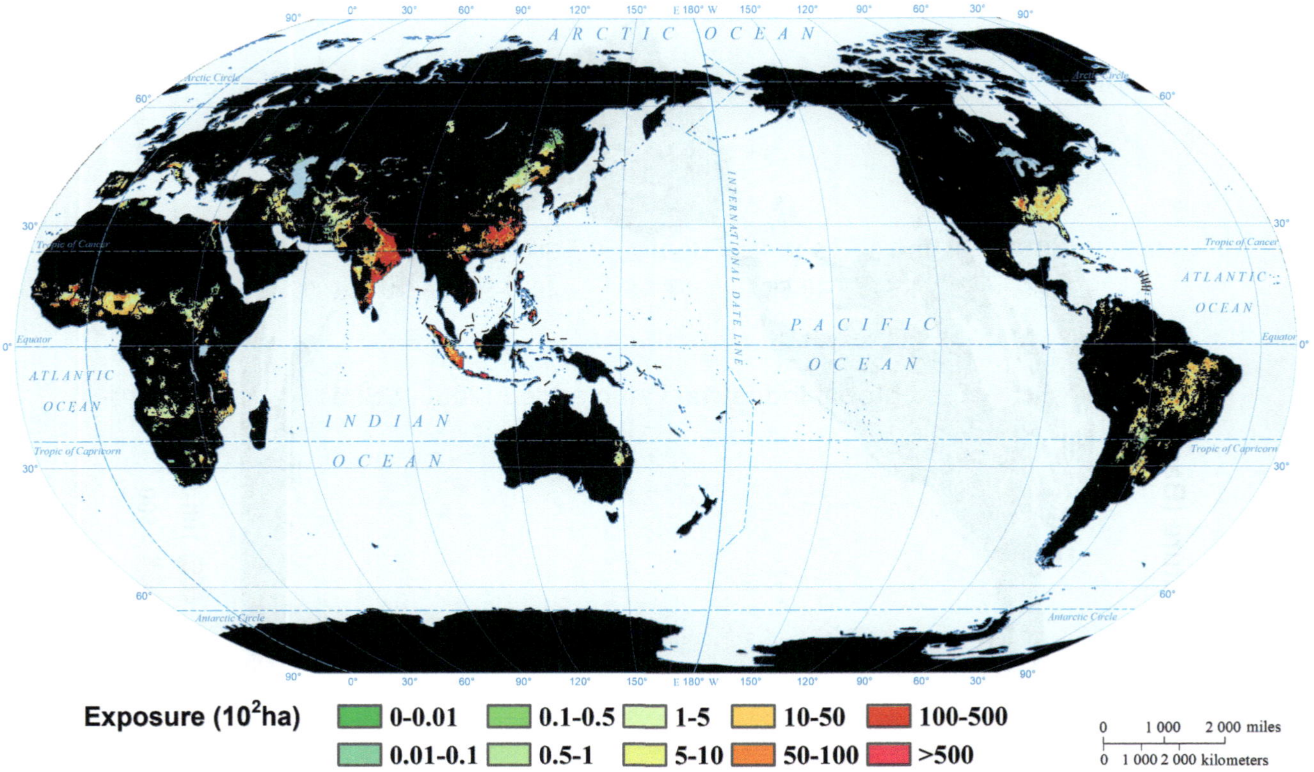

Global rice exposure to extremely high temperature (2030s, RCP4.5-SSP2)

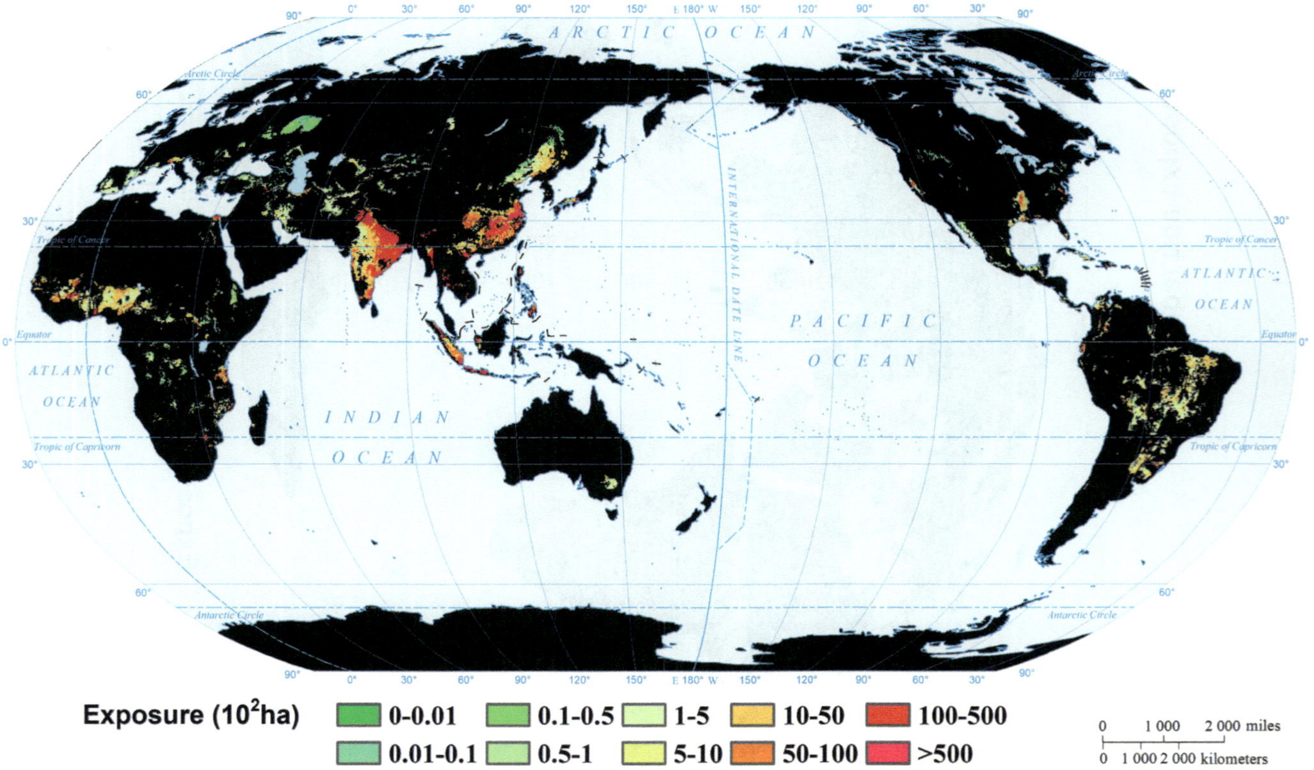

Global rice exposure to extremely high temperature (2030s, RCP8.5-SSP3)

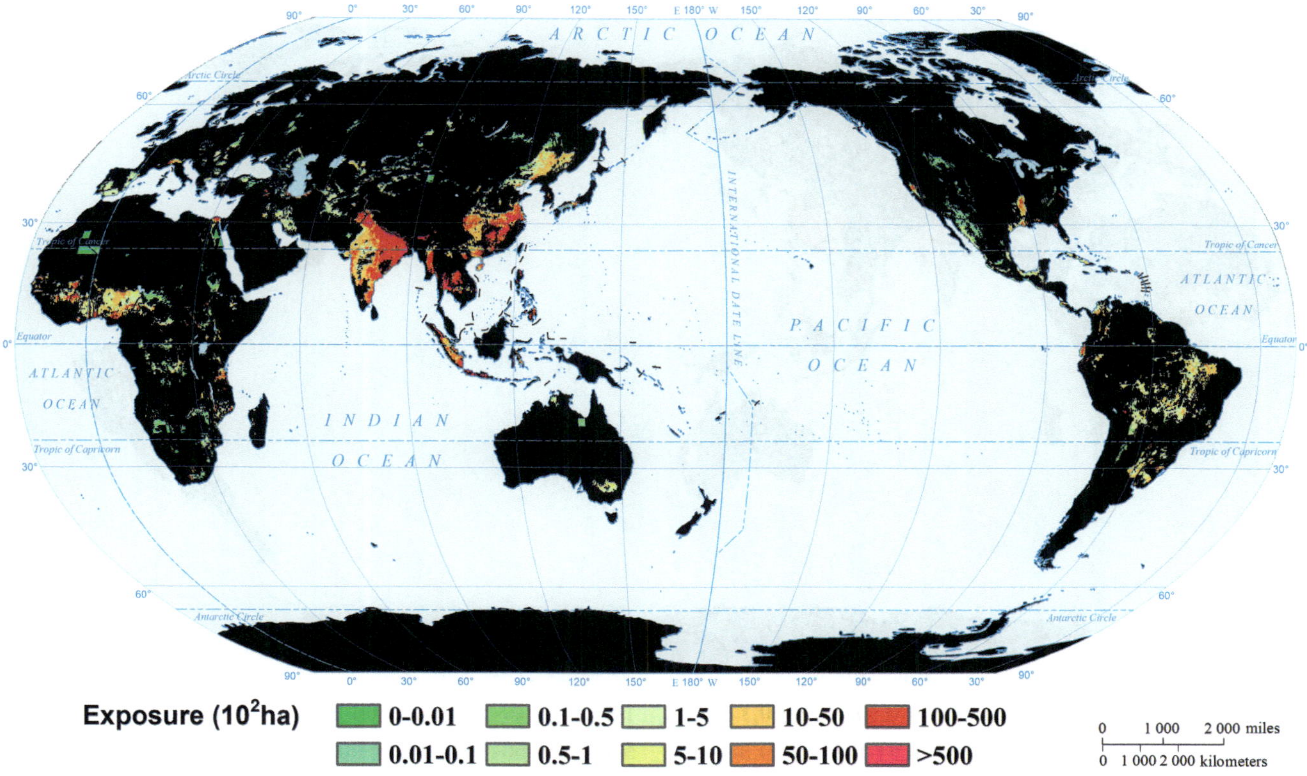

Global rice exposure to extremely high temperature (2050s, RCP2.6-SSP1)

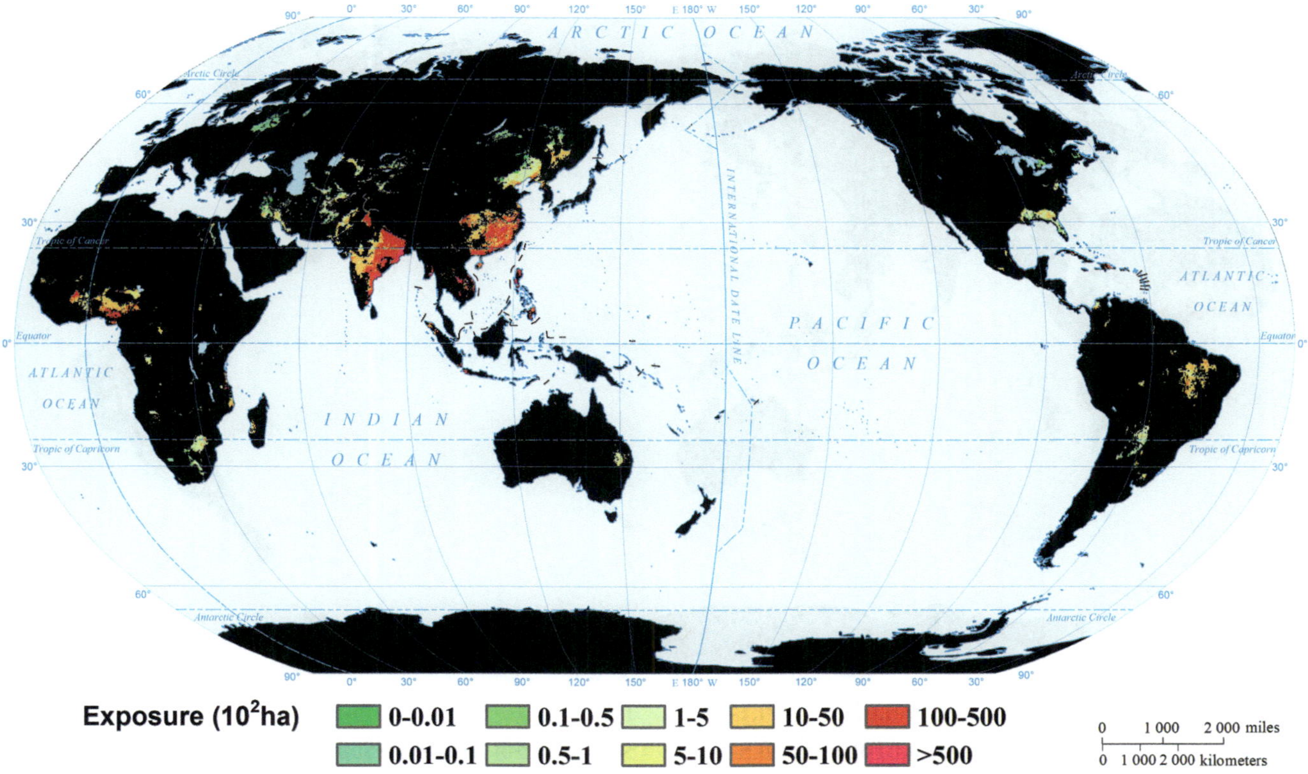

Global rice exposure to extremely high temperature (2050s, RCP4.5-SSP2)

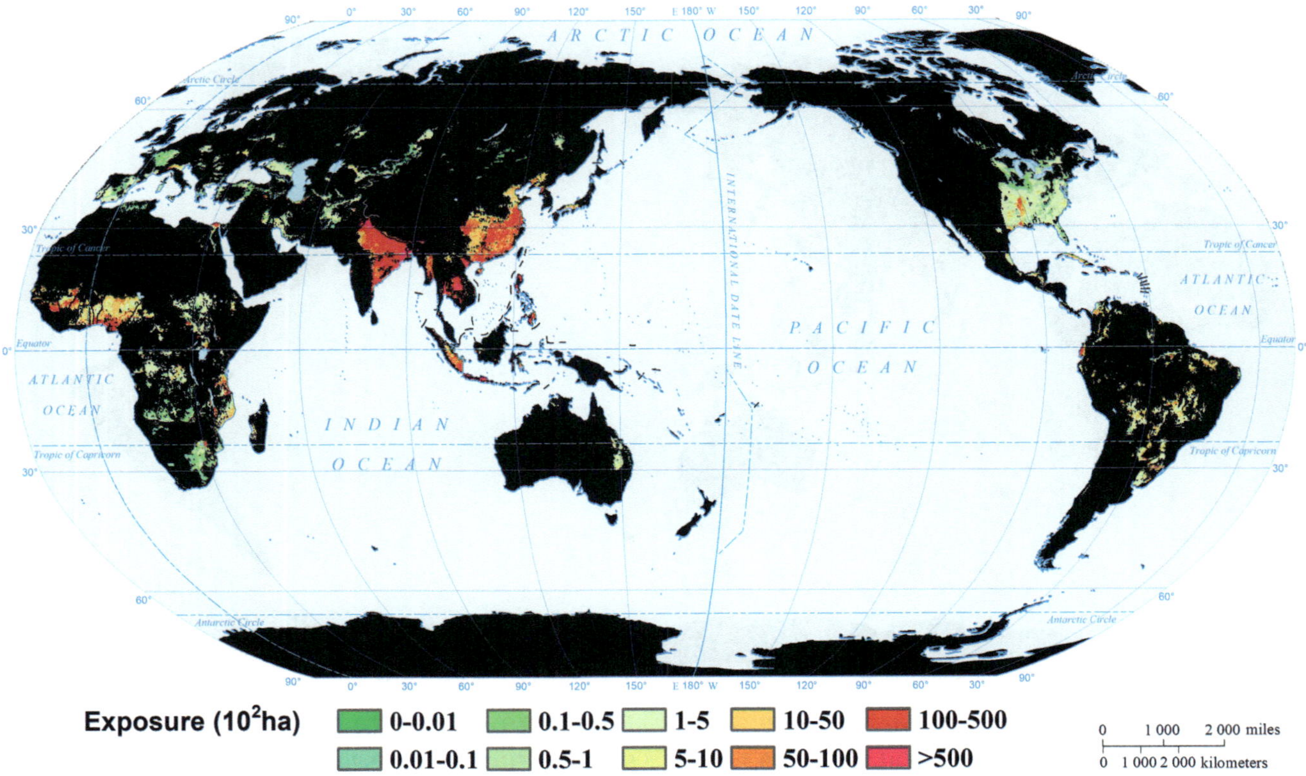

Global rice exposure to extremely high temperature (2050s, RCP8.5-SSP3)

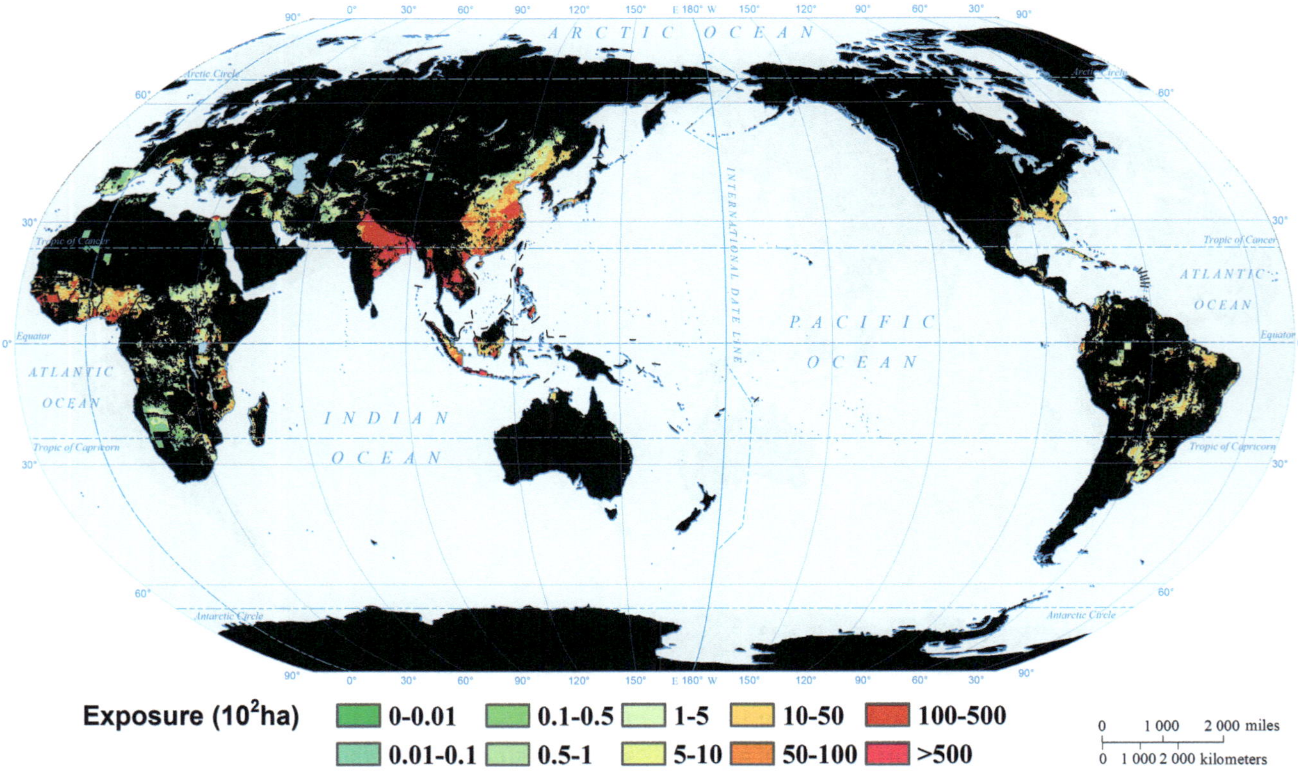

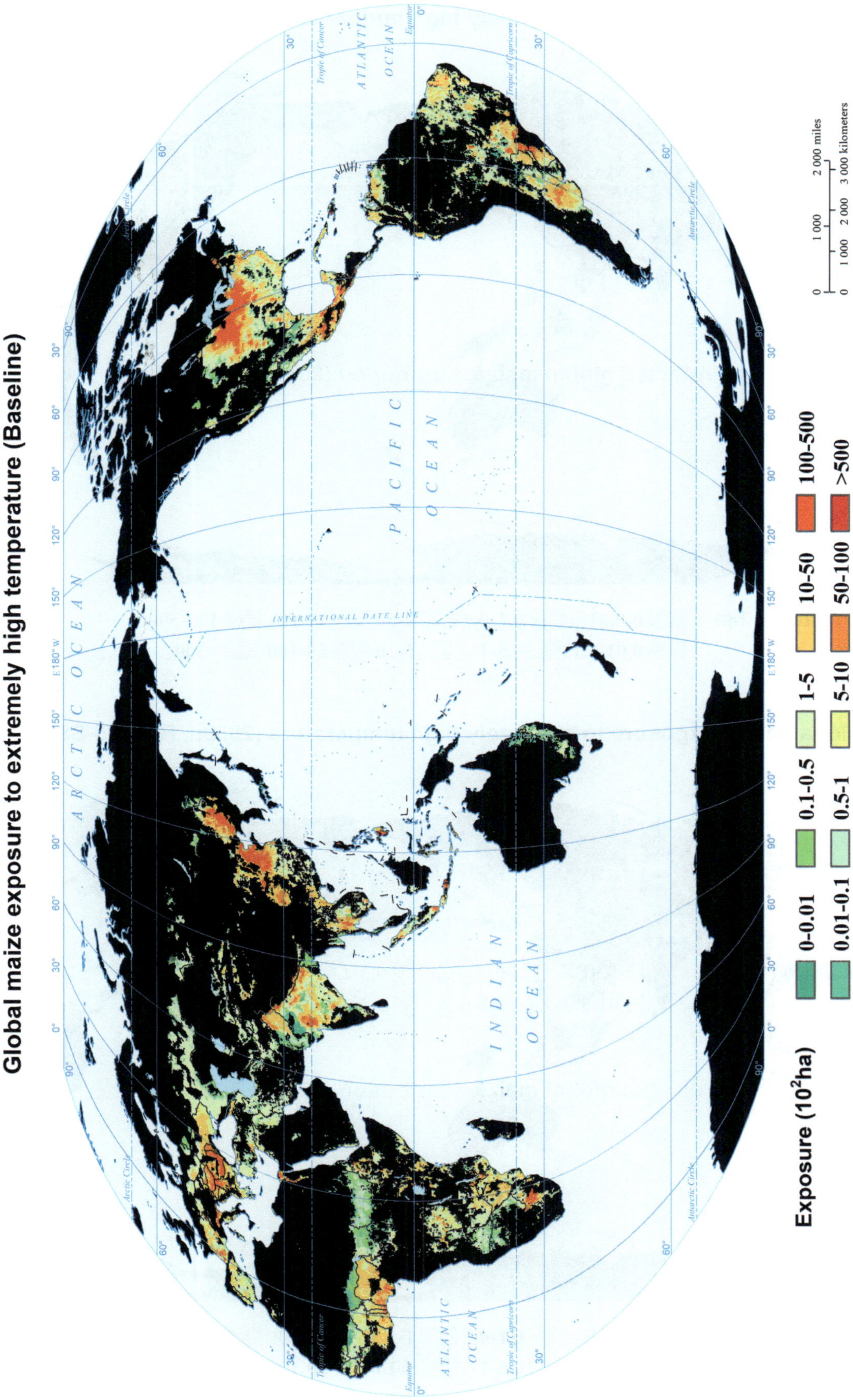
Global maize exposure to extremely high temperature (Baseline)
Exposure (10^{2}ha)
0-0.01
0.01-0.1
0.1-0.5
0.5-1
1-5
5-10
10-50
50-100
100-500
>500
0 1 000 2 000 miles
0 1 000 2 000 3 000 kilometers
ARCTIC OCEAN
PACIFIC OCEAN
INDIAN OCEAN
ATLANTIC OCEAN
INTERNATIONAL DATE LINE
Tropic of Cancer
Tropic of Capricorn
Equator
Arctic Circle
Antarctic Circle

Global maize exposure to extremely high temperature (2030s, RCP2.6-SSP1)

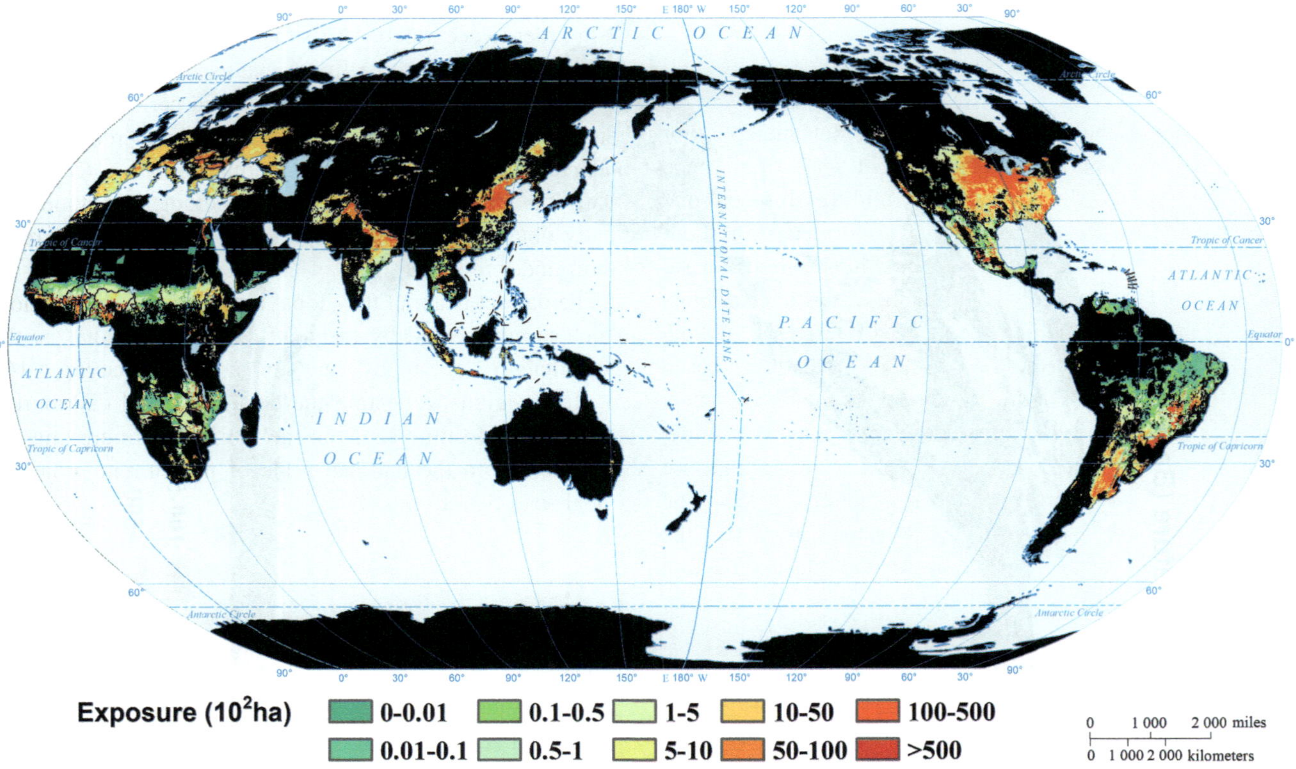

Global maize exposure to extremely high temperature (2030s, RCP4.5-SSP2)

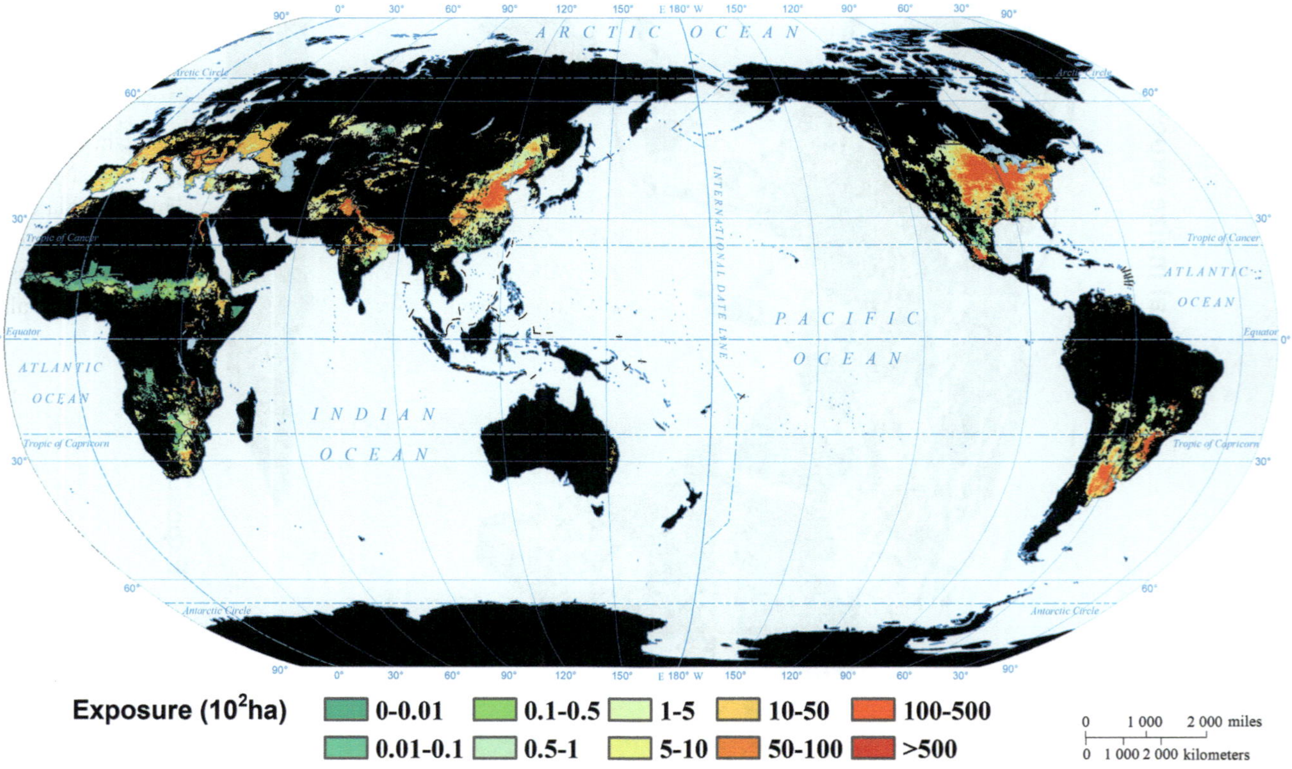

Global maize exposure to extremely high temperature (2030s, RCP8.5-SSP3)

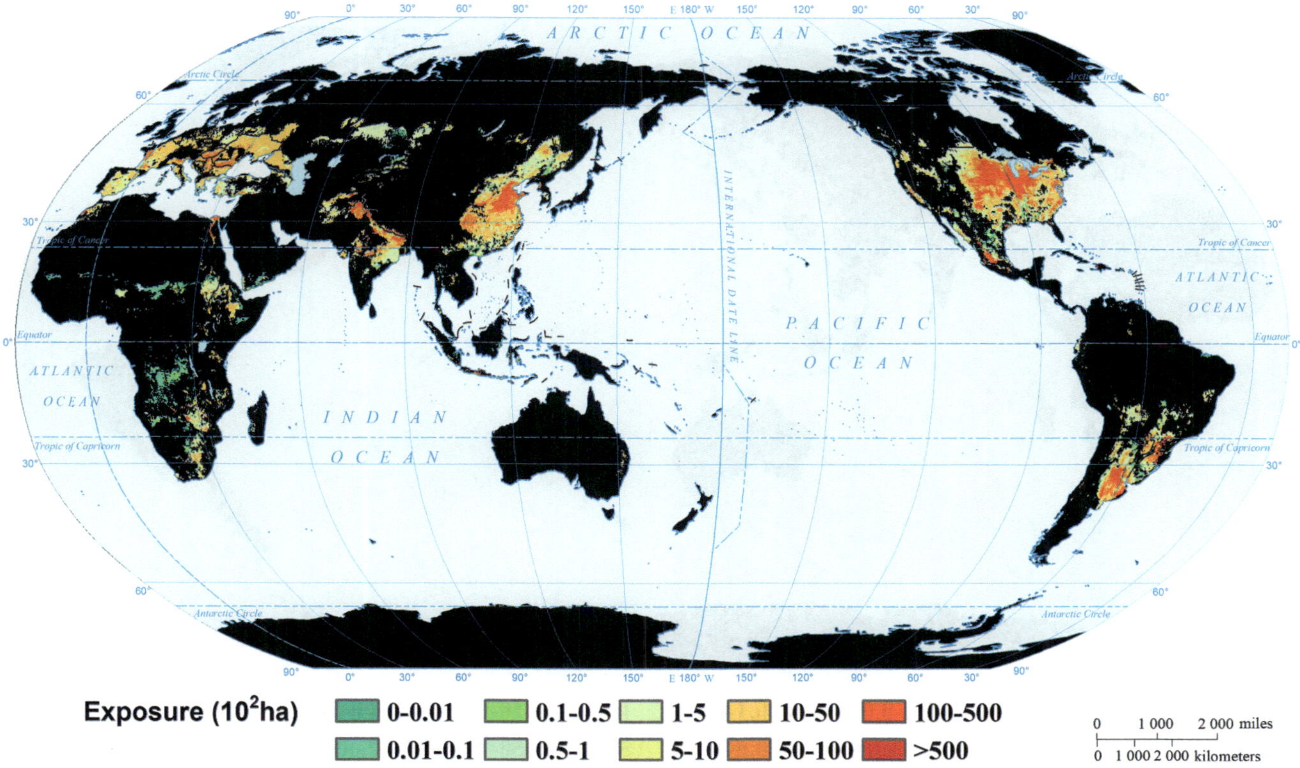

Global maize exposure to extremely high temperature (2050s, RCP2.6-SSP1)

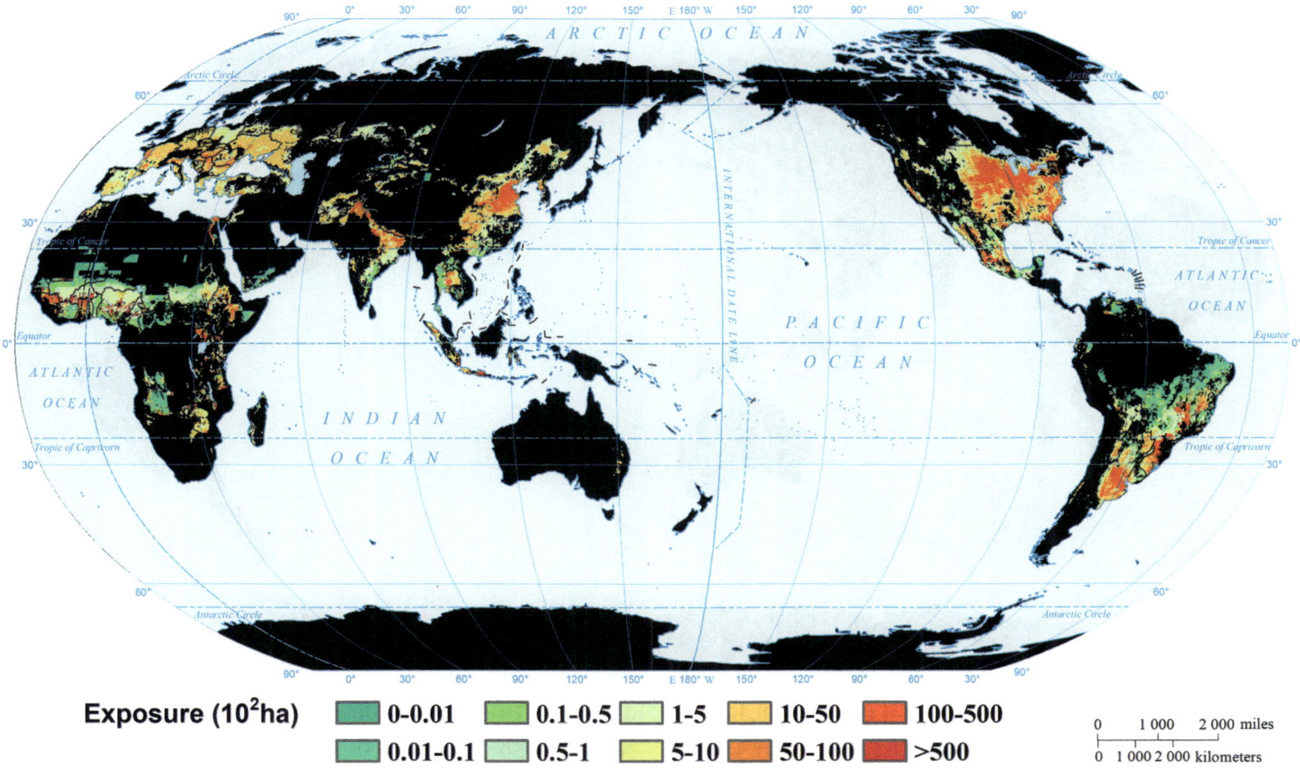

Global maize exposure to extremely high temperature (2050s, RCP4.5-SSP2)

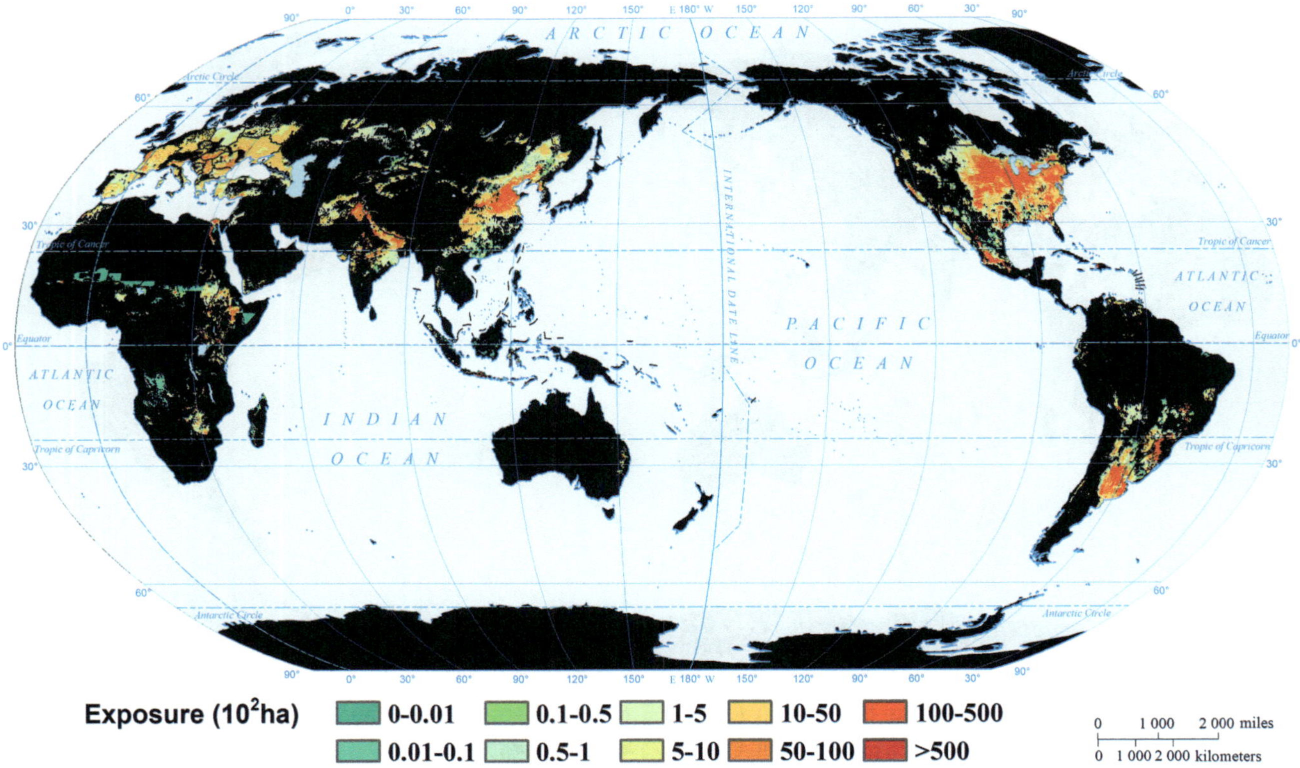

Global maize exposure to extremely high temperature (2050s, RCP8.5-SSP3)

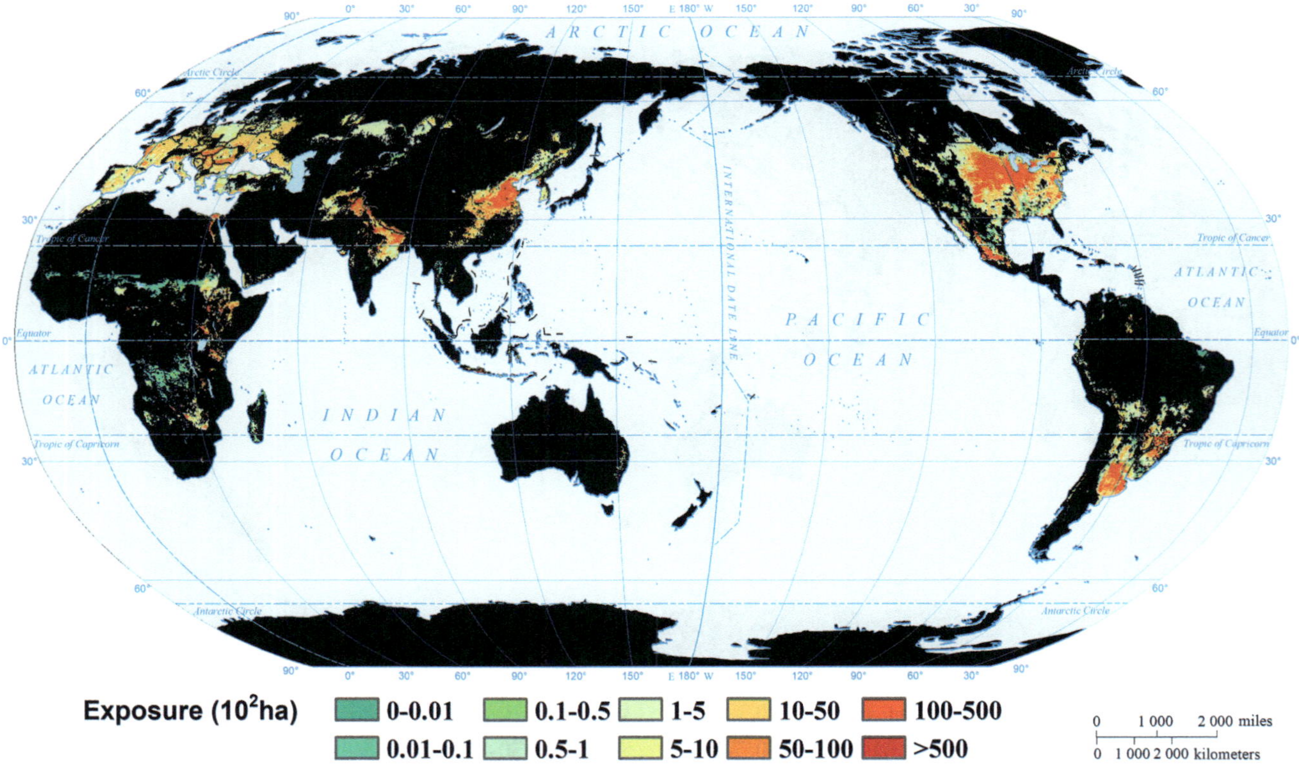

References

Change, P. C. (2018). Global wWarming of 1.5° C. Geneva: World Meteorological Organization, Geneva, Switzerland, 10.

Chen, Y., Z. Zhang, and Z., P. Wang, P., X. Song, X., X. Wei, X., & and F. Tao., F. . 2016. Identifying the impact of multi-hazards on crop yield—A case for heat stress and dry stress on winter wheat yield in northern China. *European Journal of Agronomy* 73: 55–63.

Deng, Z.Y., J.F, Xu., and L.N, Huang. (2009). Research Summary of the Damage Characteristics of the Wheat Dry Hot Wind in Northern China. Journal of Anhui Agricultural Sciences, 20.

Fahad, S., Adnan, M., Hassan, S., Saud, S., Hussain, S., Wu, C., Wang, D., Hakeem, K.R., Alharby, H.F., Turan, V. and Khan, M.A., 2019. Rice responses and tolerance to high temperature. In Advances in rice research for abiotic stress tolerance (pp. 201–224). Woodhead Publishing.

Holzkämper, A., P. Calanca, and J. Fuhrer. 2012. Statistical Crop Models: Predicting the Effects of Temperature and Precipitation Changes. *Climate Research* 51 (1): 11–21. https://doi.org/10.3354/cr01057.

Iizumi, Toshichika, and Navin Ramankutty. 2016. "Changes in Yield Variability of Major Crops for 1981–2010 Explained by Climate Change." Environmental Research Letters 11 (3). https://doi.org/10.1088/1748-9326/11/3/034003.

Janetos, A. C. (1997). Climate Cchange 1995: Impacts, Aadaptations and Mmitigation of Cclimate Cchange: Scientific-Ttechnical Aanalyses.: JSTOR.

Jiang, Q., Y. Yue, and Y., & and L. Gao., L. . 2019. The spatial-temporal patterns of heatwave hazard impacts on wheat in northern China under extreme climate scenarios. *Geomatics, Natural Hazards and Risk* 10 (1): 2346–2367.

Lobell, D.B., and M.B. Burke. 2010. On the use of statistical models to predict crop yield responses to climate change. *Agricultural and Forest Meteorology* 150 (11): 1443–1452.

Melillo, J., J. Borchers, J., J. Chaney, J., H. Fisher, H., S. Fox, S., A. Haxeltine, A. Janetos, D.W. Klicklighter, A. et al., . . . McGuire, A. (1995). Vegetation/ecosystem modeling and analysis project: Comparing biogeography and biogeochemistry models in a continental-scale study of terrestrial ecosystem responses to climate change and CO_2 doubling. *Global Biogeochemical Cycles* 9(4): 407--438.

Osborne, T.M., and T.R. Wheeler. 2013. Evidence for a Climate Signal in Trends of Global Crop Yield Variability over the Past 50 Years. *Environmental Research Letters* 8 (2): 024001. https://doi.org/10.1088/1748-9326/8/2/024001.

Oyebamiji, Oluwole K., Neil R. Edwards, Philip B. Holden, Paul H. Garthwaite, Sibyll Schaphoff, and Dieter Gerten. 2015. Emulating Global Climate Change Impacts on Crop Yields. *Statistical Modelling* 15 (6): 499–525. https://doi.org/10.1177/1471082X14568248.

Peng, S., J. Huang, J. E., Sheehy, J. E., R.C. Laza, R. C., R.M. Visperas, R. M., X. Zhong, G.S. Centeno, G.S. Khush, and K. G. Cassman. 2004. Rice yields decline with higher night temperature from global warming. *Proceedings of the National Academy of Sciences* 101 (27): 9971–9975.

Raimondo, Maria, Concetta Nazzaro, Giuseppe Marotta, and Francesco Caracciolo. 2020. Land Degradation and Climate Change: Global Impact on Wheat Yields. *Land Degradation & Development*. https://doi.org/10.1002/ldr.3699.

Mapping Global Industrial Value Added

Wei Song, Huiyi Zhu, Han Li, Qian Xue, and Yuanzhe Liu

1 Background

In the research of identifying the impact of climate change on the industrial economic system, the core step is to overlay climate data and industrial economic data with the same spatiotemporal resolution and perform spatial analysis (Zhao et al. 2017). However, the risk assessment of the industrial economic system is hampered by the lack of spatialized datasets of global and Chinese industrial economic system output value, especially under future climate scenarios, because it is difficult to accurately identify the output value of secondary and tertiary industries by conventional remote sensing methods. The existing spatial data of industrial economic system output value are mostly at the provincial, city, and county levels, with administrative areas as the smallest spatial units, which cannot represent the difference and spatial distribution of industrial output value within a province or a city. Therefore, it is difficult to carry out overlay analysis with gridded climate data in risk assessment. Although there are some spatialized data on the output value of a particular industry from research (Dong et al. 2016), in general, there is a lack of large-scale, high-resolution, and comprehensive spatial data of industrial output value.

Currently, the research methods of mapping economic data can be divided into three categories: spatial interpolation model (Tobler 1979), multi-source data fusion model (Li et al. 2018), and remote sensing inverse model (Wang et al. 2018). Compared with other spatial models, the nighttime light remote sensing data inversion model is characterized by simple implementation and high precision and is expected to solve the problem of large-scale data localization. However, the model is mainly used to analyze gross domestic product (GDP) and population data. Based on the existing applications, this study developed a method to spatialize the industrial output on a global scale, and the random forest algorithm in machine learning was used to map the industrial value added under different climate change scenarios in the future.

The definition of industrial value added is based on the World Bank's statistical standards. The industrial value added covers mining, manufacturing, construction, electricity, water, and gas sectors. Industrial value added statistics data are from the World Bank, and the data are in current U. S. dollars.

2 Method

The method for mapping industrial value added for future climate scenarios includes the following steps: (1) Mapping the current industrial value added; (2) Simulating the spatial boundary of future industrial value added; and (3) Estimating

Authors: Wei Song, Huiyi Zhu, Han Li, Qian Xue, Yuanzhe Liu.
Map Designers: Yuanyuan Jing, Jing'ai Wang, Ying Wang.
Language Editor: Song Wei.

W. Song (✉) · H. Zhu · H. Li · Q. Xue · Y. Liu
Key Laboratory of Land Surface Pattern and Simulation, Institute of Geographic Sciences and Natural Resources Research, Chinese Academy of Sciences, Beijing, 100101, China
e-mail: songw@igsnrr.ac.cn

H. Li
University of Chinese Academy of Sciences, Beijing, 100049, China

Q. Xue
Chongqing Jiaotong University, Chongqing, 400074, China

Y. Liu
Shandong Normal University, Jinan, 250358, China

P. Shi, *Atlas of Global Change Risk of Population and Economic Systems*, IHDP/Future Earth-Integrated Risk Governance Project Series, https://doi.org/10.1007/978-981-16-6691-9_12

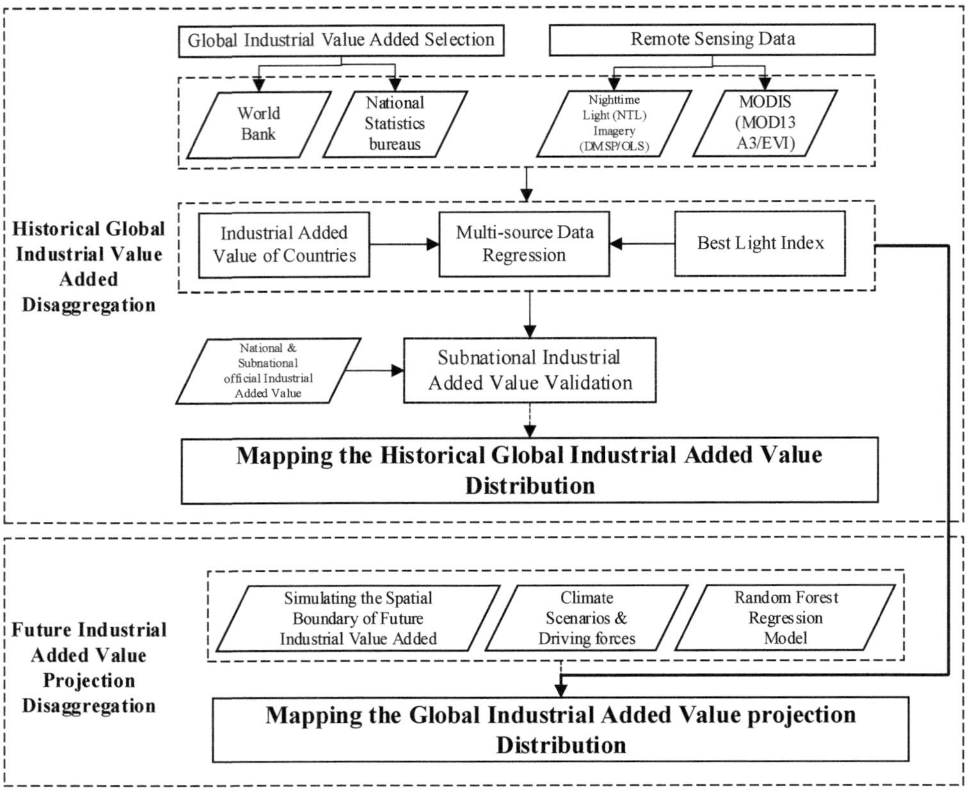

Fig. 1 Technical flowchart of mapping global industrial value added

the future industrial value added. Figure 1 shows the technical flowchart for mapping industrial value added.

2.1 Mapping the Current Industrial Value Added

Based on the 2010 global vegetation index data (Enhanced Vegetation Index, EVI, from MODIS) and the 2010 nighttime light remote sensing data (from DMSP/OLS), the adjustment nighttime light index (ANLTI) was constructed to preprocess the light saturation and overflow phenomena of nighttime light data, and the best light data in the world was obtained. The calculation formula is as follows (Zhuo et al. 2015):

$$\text{ANLTI} = \frac{2}{1 - \text{NT}_\text{n} + \text{EVI}} \times \text{NT} \quad (1)$$

ANLTI is the EVI adjusted nighttime light index, NT_n is the normalized nighttime light value, and NT is the original nighttime light value.

A regression model of the industrial value added was constructed using the adjusted nighttime light index (ANLTI) and industrial value added statistical data of countries from the World Bank. The data of global industrial value added in 2010 is generated using the following formula:

$$I = \frac{I_i \times \text{ANLTI}}{\text{ANLTI}_i} \quad (2)$$

I is the industrial value added of each pixel. I_i is the total industrial added value of the country in which the pixel is located, and ANLTI_i is the total EVI adjusted nighttime light index of the country in which the pixel is located.

In the global scope, statistical industrial value added in 178 provincial (state) regions were randomly selected for the correlation test, and the correlation coefficient is 0.93. Taking statistical data as the true values, the average accuracy of industrial value added in the 178 regions is 80.14% (Xue et al. 2018).

2.2 Simulating the Spatial Boundary of Future Industrial Value Added

Future industrial value added changes can be approximated by starting with industrial land use change. According to the principle of logistic–cellular automata (CA)–Markov simulation, the global land use data from ESA (European Space Agency, https://maps.elie.ucl.ac.be/CCI/viewer/index.php) in 2010 and 2015 were used to simulate the land use change in 2030 and 2050. The urban land was extracted as the spatial boundary of the future industrial added value. In the selection of driving factors, it is necessary to comprehensively describe the impact of different driving factors on land use change, and consider the research scale. Elevation, slope, population, GDP, and distance from river and road are selected as driving factors of land use. The data are from NOAA (National Oceanic and Atmospheric Administration, https://www.noaa.gov/), SEDA (Socioeconomic Data and Applications Center, https://sedac.ciesin.columbia.edu/), Global Risk Data Platform (https://preview.grid.unep.ch/), and Natural Earth (https://www.naturalearthdata.com/). The accuracy of the simulated land use in 2015 is verified by using the global land use data in 2015. The global average accuracy is 91.89%.

The logistic–CA–Markov model is mainly used in the simulation of land use change by combining the characteristics of the logistic regression model, the CA model, and the Markov model (Jamal et al. 2013). The logistic regression model can analyze the relationship between land use types and driving forces. The CA model can effectively simulate the spatial changes of the land use system and the Markov model can predict the quantitative changes of land use types so as to simulate land use changes more comprehensively and accurately.

2.3 Estimating Future Industrial Value Added

For the spatialization of industrial added value under different climate change scenarios in the future, appropriate factors of industrial added value change need to be selected. In order to determine the influencing factors of spatial change of industrial value added, land use, population density, and accessibility of the study area should be comprehensively considered. In addition, the distribution of rivers and lakes, and topographic features such as elevation and slope should be considered. Due to the many influencing factors, the ordinary regression model is difficult to comprehensively and accurately reflect the spatial distribution characteristics of the industrial added value under different climate change scenarios. Therefore, this study is based on machine learning, combined with the random forest model to build a spatial model of industrial added value under different climate change scenarios and the model is as follows (Xue and Song 2020):

$$F = (T, P, \mathrm{GDP}, \mathrm{Land}, \mathrm{Slope}, \mathrm{Pop}\ldots) \quad (3)$$

where F represents simulation results under different climate change scenarios, T represents average annual temperature, P represents average annual precipitation, GDP represents the gross domestic product, *Land* represents the spatial boundary of future industrial value added, and Pop represents the density of population.

The driving factors used to build the random forest model include industrial value added in 2010, the spatial distribution of urban land, elevation, slope, distance from rivers, distance from roads, distance from railroads, distance from residential settlements, and air temperature and precipitation (representing different climate change scenarios). When mapping the future industrial added value under different climate change scenarios, these driving factors are constant except for temperature and precipitation.

Based on the comprehensive consideration of the fitting speed and accuracy of the model, the parameters of the random forest model were tested. Finally, we built a total of 100 decision trees. In each decision tree, 90% of the samples were randomly selected to build the sample model, and the remaining 10% were used as test data. The simulation results show that the sample accuracy was 0.94 and the test sample accuracy was 0.81. The overall sample accuracy was relatively high, which can well explain the influence of various factors on the industrial value added, so it can be used for the simulation and prediction of industrial added value.

Through the statistics of the proportion of industrial value added in GDP (from the World Bank) and the GDP data under the Shared Socioeconomic Pathways (SSPs), the industrial added value of each country under different SSP scenarios in the future was obtained. According to the estimated proportion of industrial value added of each country and the proportion of industrial value added of each country under different SSP scenarios in the future, the regression model was constructed. Finally, the distribution of industrial added value under SSP1, SSP2, and SSP3 scenarios in 2030 and 2050 was obtained.

3 Results

The industrial value added under different future climate change scenarios was calculated by continent. Then we derived the statistical value of industrial value added of each continent under SSP1, SSP2, and SSP3 scenarios in 2010, 2030, and 2050 (Fig. 2). Overall, in the future, the industrial value added will increase obviously. Compared with other regions, Asia has the largest industrial value added, followed by North America, Europe, Africa, and South America, and Oceania has the smallest industrial value added. The industrial value added of SSP1 was the biggest, followed by SSP2, and the smallest was SSP3.

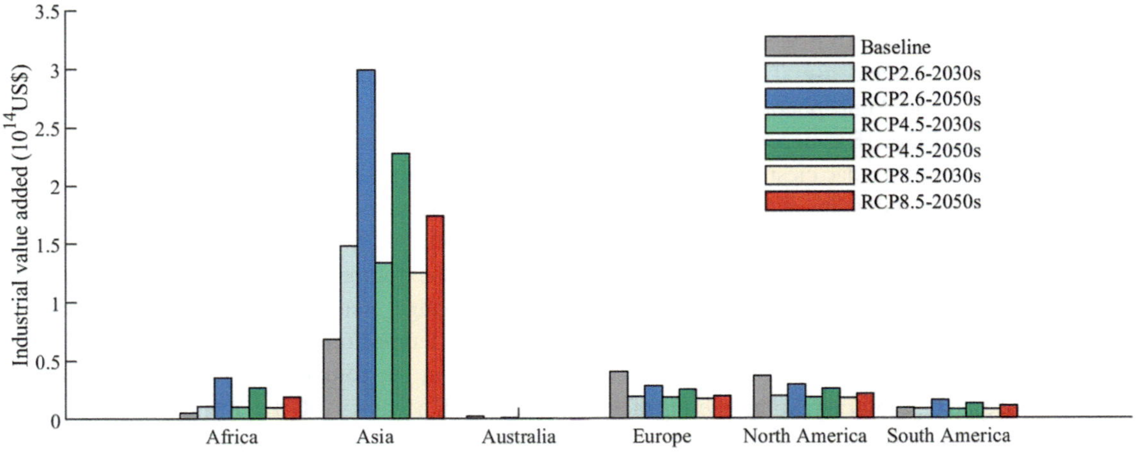

Fig. 2 Industrial value added under the Shared Socioeconomic Pathway (SSP) 1, SSP2, and SSP3 scenarios in 2030 and 2050

4 Maps

Global industrial value added (Baseline)

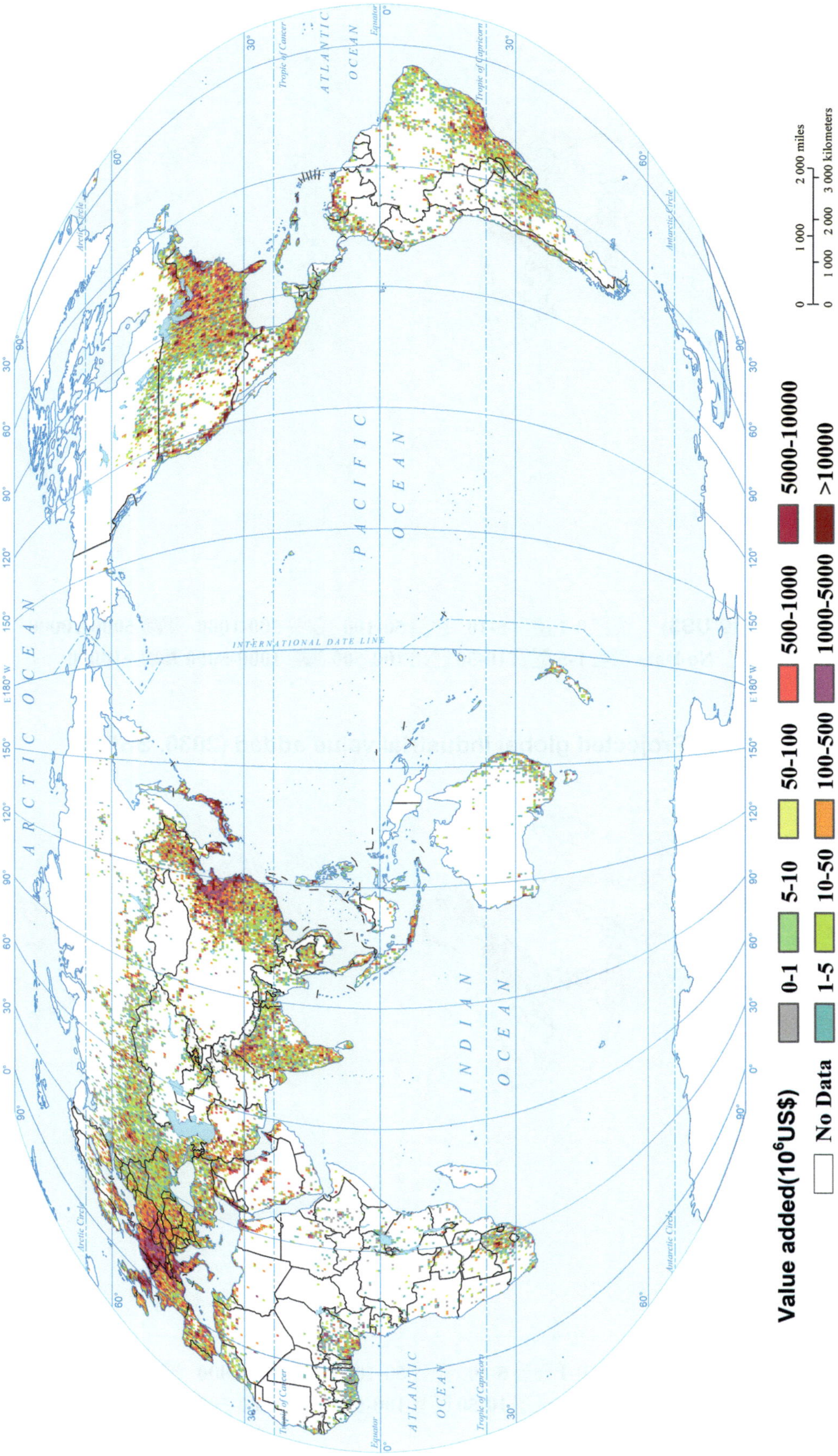

Projected global industrial value added (2030, SSP1)

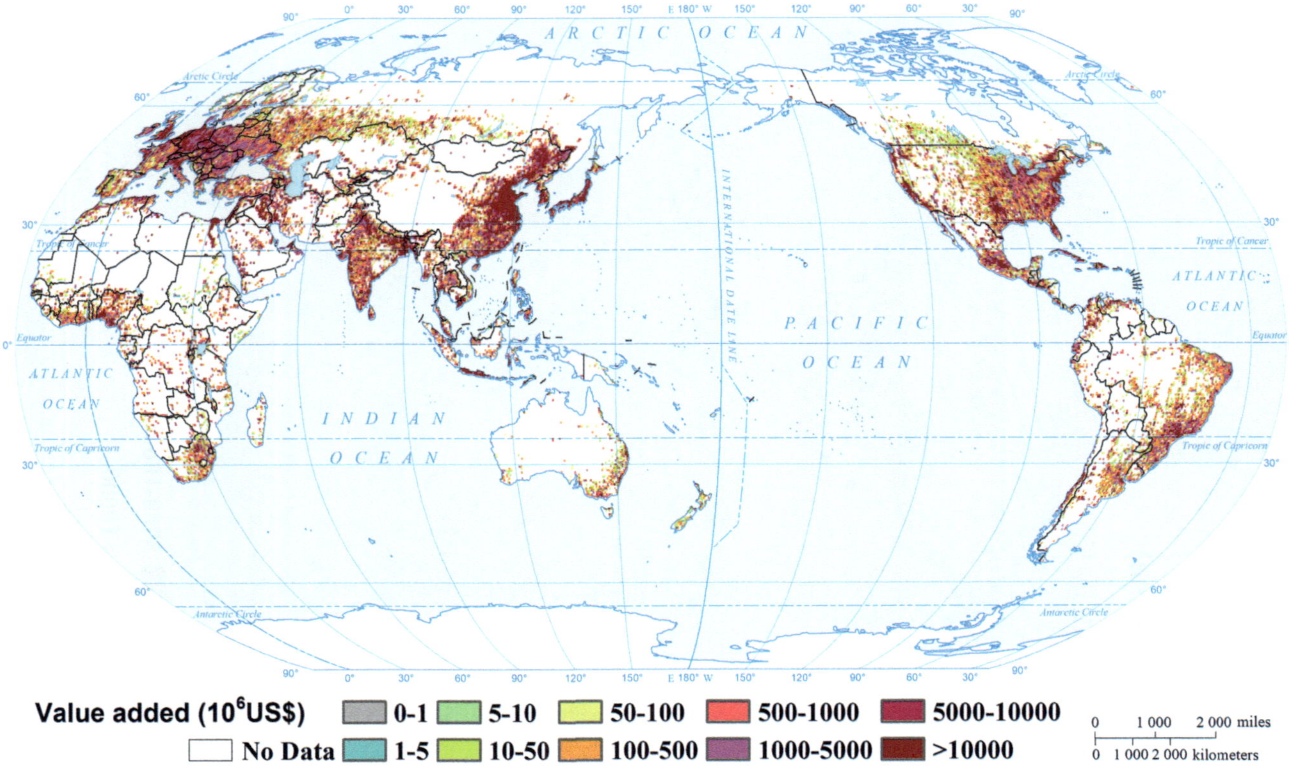

Projected global industrial value added (2030, SSP2)

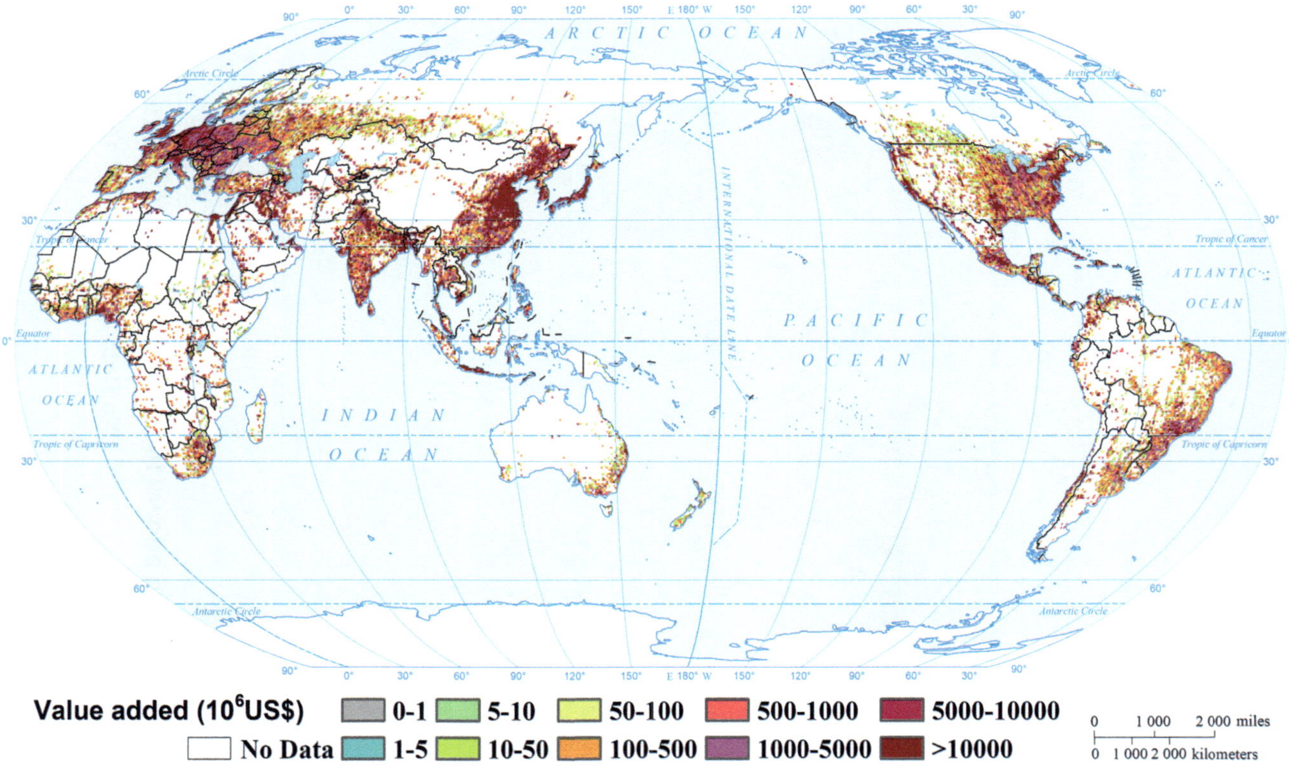

Projected global industrial value added (2030, SSP3)

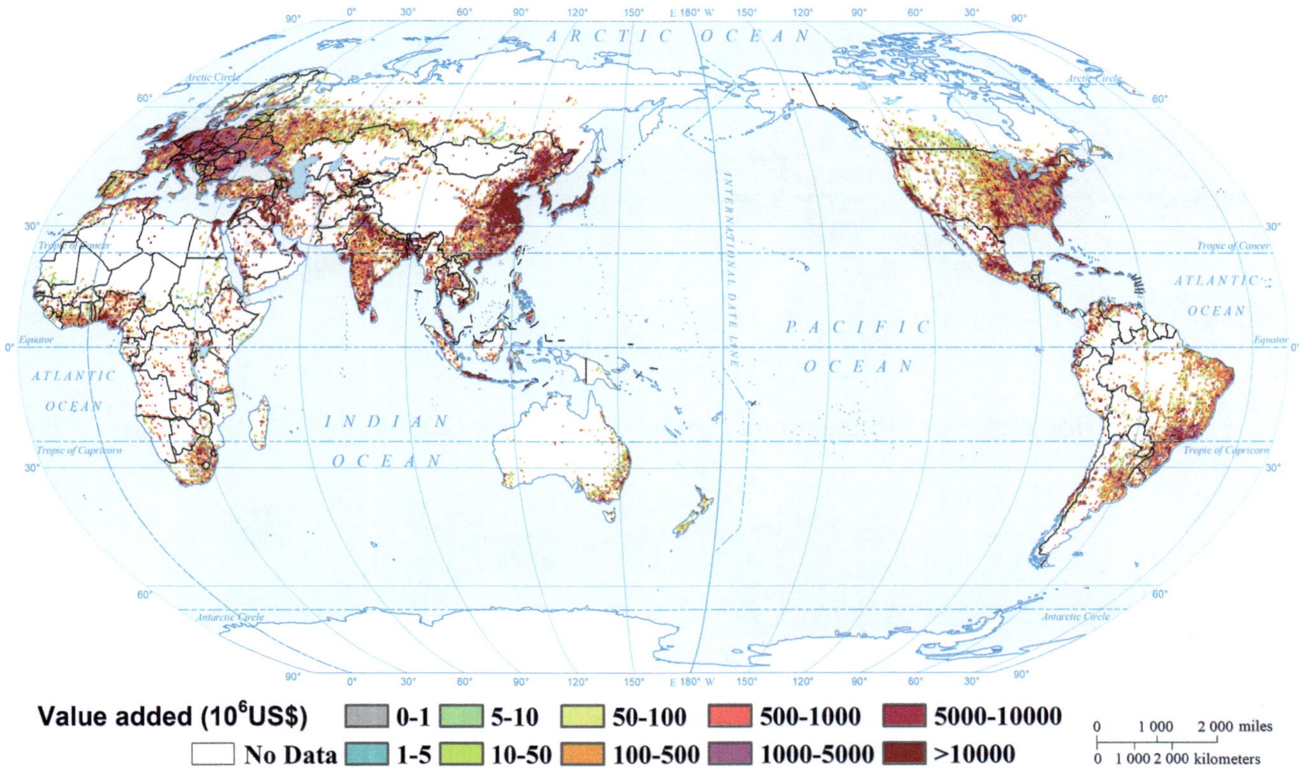

Projected global industrial value added (2050, SSP1)

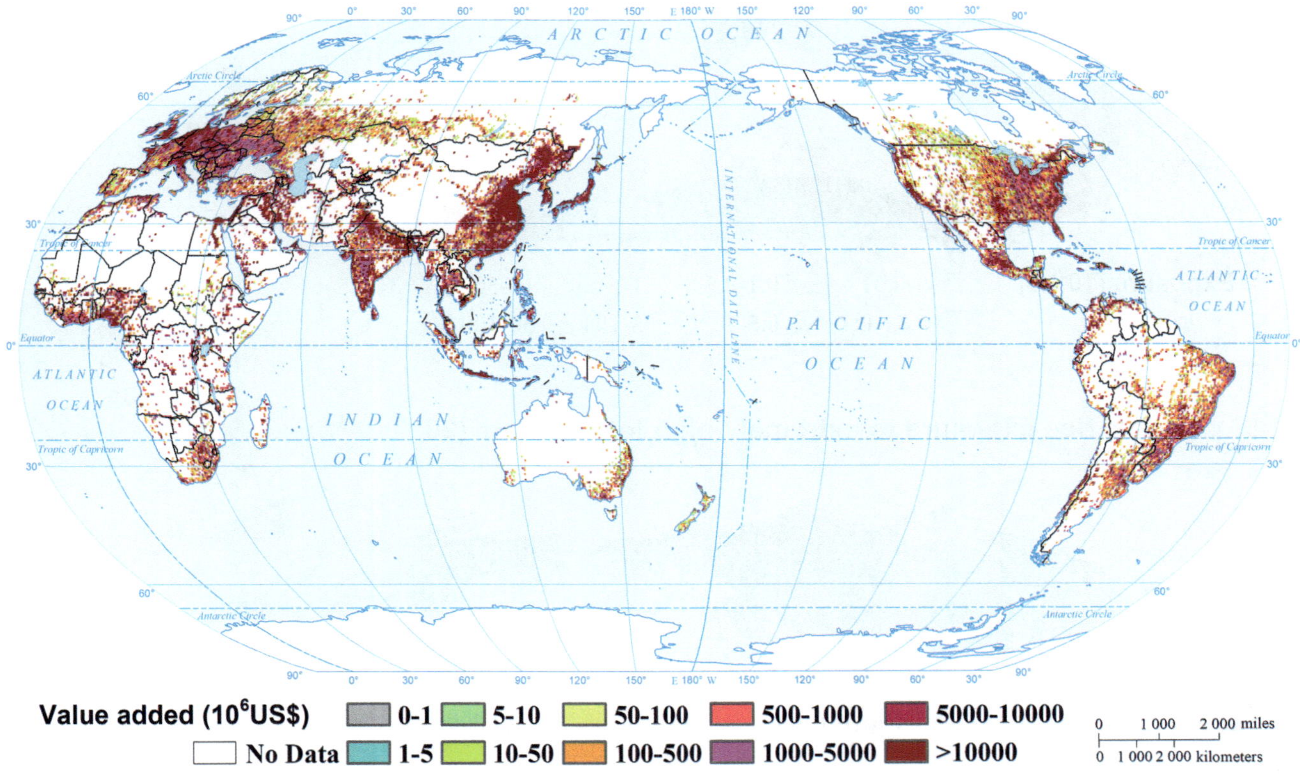

Projected global industrial value added (2050, SSP2)

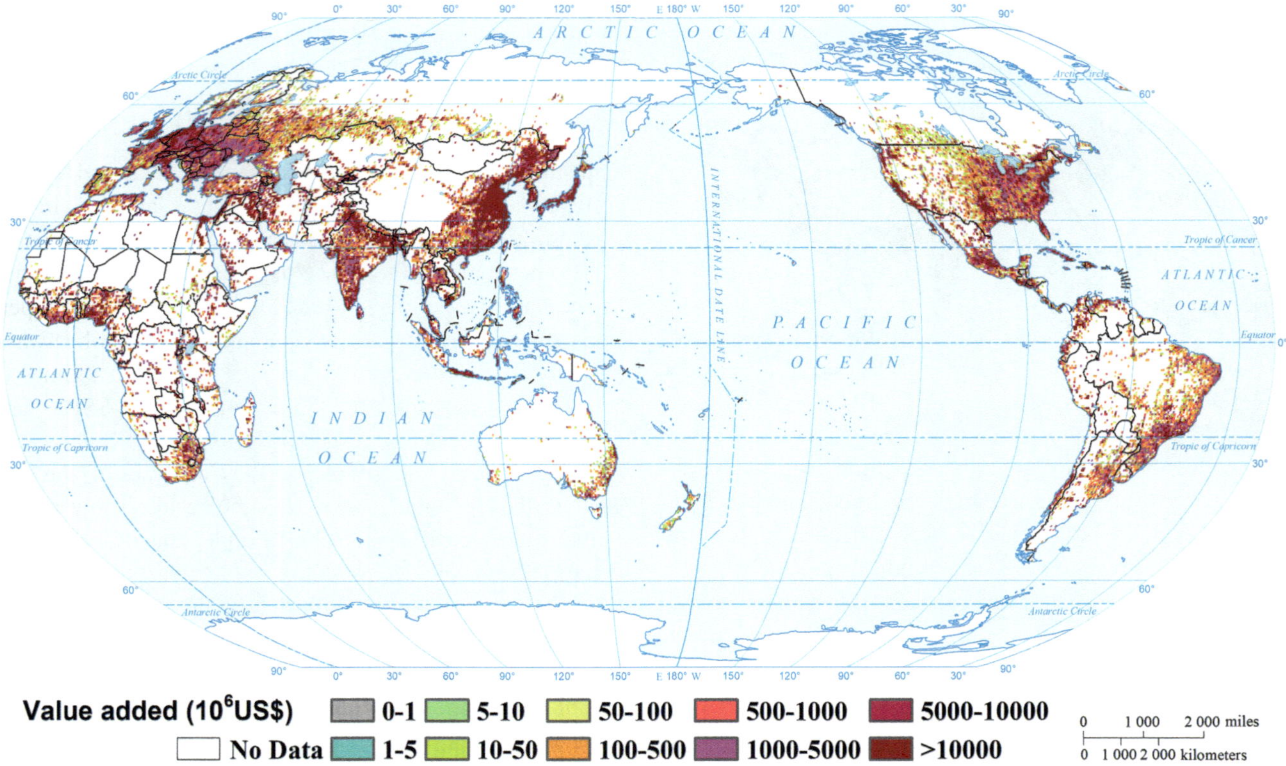

Projected global industrial value added (2050, SSP3)

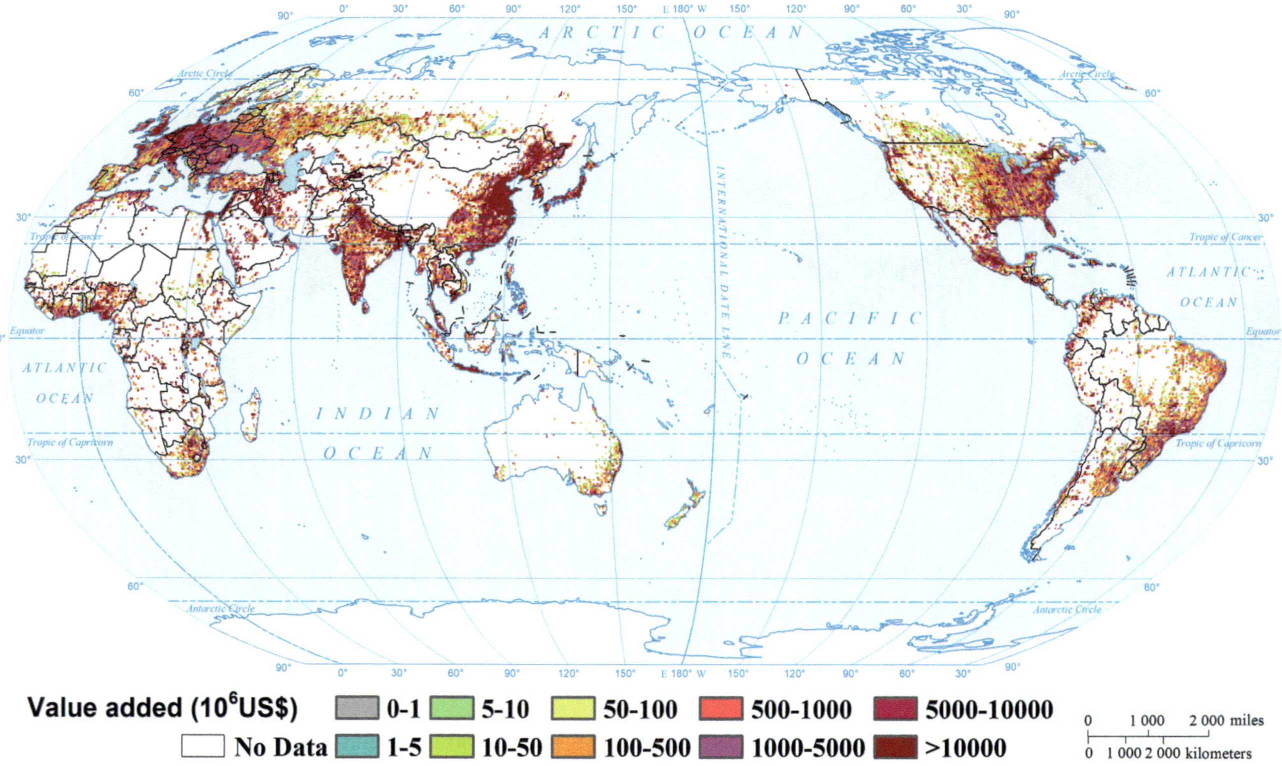

References

Dong, L., H. Liang, Z. Gao, and X. Luo. 2016. *J. Ren. Spatial Distribution of China's Renewable Energy Industry: Regional Features and Implications for a Harmonious Development Future. Renewable and Sustainable Energy Reviews* 58: 1521–1531.

Jamal, J.A., H. Marco, K. Wolfgang, and D. Ali. 2013. Integration of logistic regression, Markov chain and cellular automata models to simulate urban expansion. *International Journal of Applied Earth Observation and Geoinformation* 21: 265–275.

Li, L., J. Li, Z. Jiang, L. Zhao, and P. Zhao. 2018. Methods of population spatialization based on the classification information of buildings from China's first national geoinformation survey in urban area: A case study of Wuchang District, Wuhan City China. *Sensors* 18 (8): 2558.

Tobler, W.R. 1979. Smooth pycnophylactic interpolation for geographical regions. *Journal of the American Statistical Association* 74 (367): 519–530.

Wang, L., S. Wang, Y. Zhou, W. Liu, Y. Hou, J. Zhu, and F. Wang. 2018. Mapping population density in China between 1990 and 2010 using remote sensing. *Remote Sensing of Environment* 210: 269–281.

Xue, Q., and W. Song. 2020. Spatial distribution of China's industrial output values under global warming scenarios RCP4.5 and RCP8.5. *ISPRS International Journal of Geo-Information* 9(12): 724.

Xue, Q., W. Song, and H. Zhu. 2018. Global industrial added value 1 km square grid dataset. *Journal of Global Change Data & Discovery* 2 (1): 9–17.

Zhao, M., W. Cheng, C. Zhou, M. Li, N. Wang, and Q. Liu. 2017. GDP spatialization and economic differences in South China based on NPP-VIIRS nighttime light imagery. *Remote Sensing* 9 (7): 673.

Zhuo, L., X. Zhang, J. Zheng, H. Tao, and Y. Guo. 2015. An EVI-based method to reduce saturation of DMSP/OLS nighttime light data. *Acta Geographica Sinica* 70 (8): 1339–1350.

Mapping Global Road Networks

Wenxiang Wu and Lingyun Hou

1 Introduction

Various studies have put forward different methods for reasonably determining the development scale of regional road networks according to the economic development and traffic conditions of each country, and formed some conventional methods for projecting the future scale of road networks, such as land coefficient method, elastic coefficient method, time series method, connectivity analysis method, and generalized cost method. Most of the methods are based on the analysis of the existing data to derive the development pattern so as to determine the road network scale in a future year. The factors related to road network scale are land use type, economic development, and population growth. The object of this study is the global road system. In view of the large spatial scale of the research object, we selected the land coefficient method that is suitable for large-scale projection for future global road network scale.

In order to evaluate the response of the global road system to climate change and accurately explain the current global road spatial distribution characteristics, we spatially processed the current global road length data (2010) and produced the global road length raster data with 0.5° resolution (i.e., roads in kilometers per 0.5° × 0.5° grid).

In order to accurately assess the impact of climate change on the road system in the future, we used the collected data of total road length, population, area, and per capita GDP of countries over the years to project the global road lengths under different Shared Socioeconomic Pathways (SSPs) in 2030 and 2050 by using the land coefficient method and finally obtained the global road length data with 0.5° resolution under different SSPs in 2030 and 2050.

The dataset is about the global road network distribution, which is represented by the road length index, that is, the total length of roads in an area. It covers 166 countries/regions, reflecting the detailed spatial distribution of current and future roads in the world and improving the resolution from the national scale of the current projected global road data to 0.5°.

2 Data

In this study, the global road spatial distribution vector data were used to spatialize the current global roads; the historical road mileage, population, per capita GDP, and land area of various countries were used to fit the national land coefficient model, and the future global road spatial distribution was simulated and calculated by combining the population and economic data predicted by SSPs.

2.1 Global Road Distribution Data

We compared the global road dataset published by the Center for International Earth Science Information Network of Columbia University (https://sedac.ciesin.columbia.edu), DIVA-GIS road data system hosted by Robert J. Hijmans (http://dwww.diva-gis.org), and OpenStreetMap (https://www.openstreetmap.org). For regions of China, we also

Authors: Wenxiang Wu, Lingyun Hou
Map Designers: Qingyuan Ma, Jing'ai Wang, Ying Wang
Language Editor: Wenxiang Wu, Lingyun Hou

W. Wu (✉) · L. Hou
Key Laboratory of Land Surface Pattern and Simulation, Institute of Geographic Sciences and Natural Resources Research, Chinese Academy of Sciences, Beijing, 100101, China
e-mail: wuwx@igsnrr.ac.cn

P. Shi, *Atlas of Global Change Risk of Population and Economic Systems*, IHDP/Future Earth-Integrated Risk Governance Project Series, https://doi.org/10.1007/978-981-16-6691-9_13

considered the data of road distribution provided by the Resource and Environment Sciences and Data Center of the Institute of Geographic Sciences and Natural Resources Research, Chinese Academy of Sciences (http://www.resdc.cn). Considering the accuracy, comprehensiveness, and coverage of road data, and combined with online electronic maps such as Google Earth maps, the current global road distribution data were integrated as the basic data of global road data spatialization.

2.2 Historical Data of Population, per Capita GDP, and Regional Area and Road Length

The historical data of China's population, per capita GDP, and regional area and road length used in this study are from the provincial data of *China Statistical Yearbook* published by the National Bureau of statistics. Some of the statistical data are at the prefecture level, published in the provincial statistical yearbooks, and the time span is 1986–2016. The data of population, per capita GDP, and land area of other countries are from the world development indicators released by the World Bank (https://databank.worldbank.org/home.aspx). The total road length data are mainly from the World Bank (https://data.worldbank.org.cn) and the International Road Federation (https://www.irf.global) and supplemented by ASEAN statistical yearbook (https://www.aseansec.org). The time span is 1989–2014.

2.3 Shared Socioeconomic Pathways Data

This study projected the global road distribution in grids under SSP1, SSP2, and SSP3 scenarios in 2030 and 2050. The population and GDP forecast data used are derived from the global population and GDP data with 0.5° resolution under the SSP1–SSP5 scenarios produced by Jiang et al. (2017). The original 0.5° resolution population forecast data are from the global and national population data released by the International Institute for Applied Systems Analysis (IIASA), and the economic forecast data are from the GDP data provided by the Potsdam Institute for Climate Impact Research(PIK).

3 Method

For the production of the current global road length data, the current global road spatial distribution pattern was taken as the basis, and the global road length data with administrative regions as the basic statistical unit were disaggregated into the grids so as to realize the spatialization of global roads and obtain the global road spatial distribution data of 0.5° grid.

This study mainly estimated the future global road network distribution based on the land coefficient method, which considers that road length is directly proportional to the square root of population and area and the economic index coefficient of the area where the road network is located. The method can be used to calculate the reasonable theoretical road length in an area (Guo 2005). The calculation formula is as follows:

$$L = K \cdot \sqrt{P \cdot A} \tag{1}$$

where K is road network coefficient/economic coefficient; P is population (unit: 10,000 people); A is area (unit: 100 km^2); and L is road length.

The economic coefficient, population, and area in the model directly reflect the three most important influencing factors of the road network scale, which has strong practicability, and the results can reflect the actual demand of the projected area. Among the three parameters, the economic coefficient K is calculated as follows.

First, the data of road length (L_n), population (P_n), and area (A_n) over the years were used to calculate the economic coefficient K_n:

$$K_n = L_n / \sqrt{P_n \cdot A_n} \tag{2}$$

The relationship between per capita GDP (G_n) and the calculated K_n is obtained through regression:

$$K_n = a + b \cdot G_n \tag{3}$$

where a and b are regression coefficients.

Using the land coefficient method and the historical data of road length, population, per capita GDP (economic coefficient K), and (regional) land area, we simulated the relationship between road length and socioeconomic factors. Combined with different SSPs scenarios in the future, the global road length data in 2030 and 2050 were obtained (Fig. 1).

4 Results

A comparison of the current and future road lengths of the 10 countries with the longest total road mileage in the world in 2010 indicates that the change of road length in the developed countries represented by the United States is very small and that in the developing countries such as India and China is relatively large (Fig. 2).

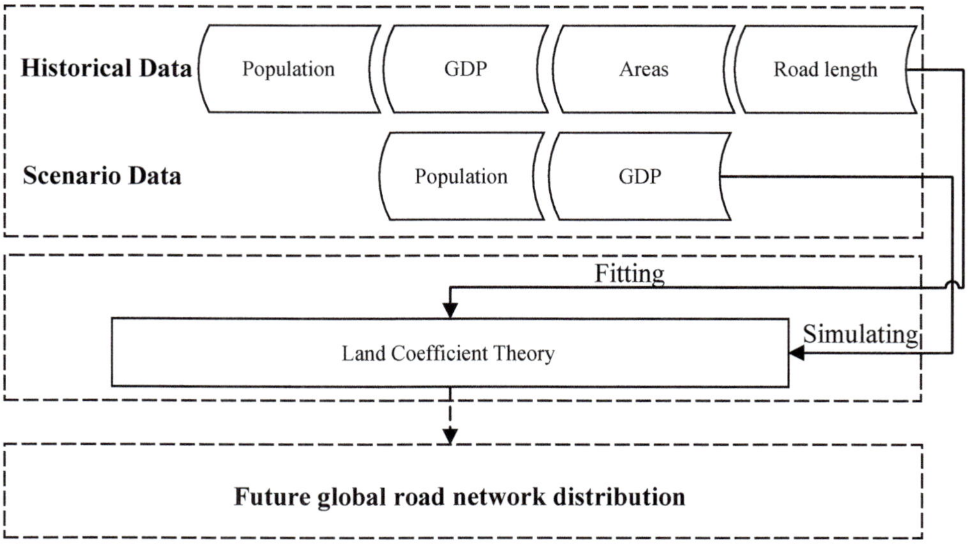

Fig. 1 Technical flowchart of mapping future road network

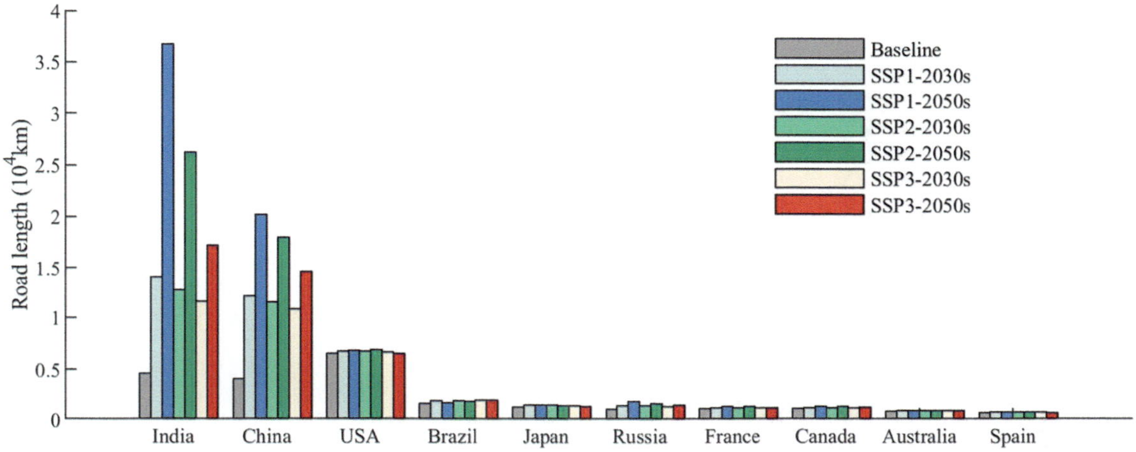

Fig. 2 Growth of road length of the top 10 countries with the longest total road mileage in 2010 under different Shared Socioeconomic Pathway (SSP) scenarios by 2030 and 2050

A comparison between Western Europe, northeastern United States, and the Bohai Rim of China indicates that the growth rate of road length in the Bohai Rim area is the highest, with clear differences under different SSPs and by different future years; while the growth rate of road length in Western Europe and the northeastern United States is not obvious, especially in the northeastern United States, which is similar to the current (2010) road length. The growth rate and trend of China's roads are similar to that of Asia as a whole, showing a high growth rate, and the growth rate varies significantly under different SSPs and by different years.

The total length of roads in all countries will increase significantly by 2030 and 2050. By 2030, the global total length of road will be about three times of the current (2010) total length, and by 2050, it will be about four to five times of the current level; the growth rate of the United States and Western Europe is the lowest, while that in Asia is the highest, especially in India and China. Under the sustainable development pathway SSP1 scenario, the growth of road network scale is the highest; under the regional competition pathway SSP3 scenario with higher challenges of climate change, the growth of road network scale is the smallest (Table 1).

Table 1 Growth scale of global roads under the Shared Socioeconomic Pathway (SSP) 1, SSP2, and SSP3 scenarios by 2030 and 2050

	SSP1	SSP2	SSP3
2030	3.02	2.88	2.71
2050	5.48	4.52	3.77

5 Maps

Global road network (Baseline)

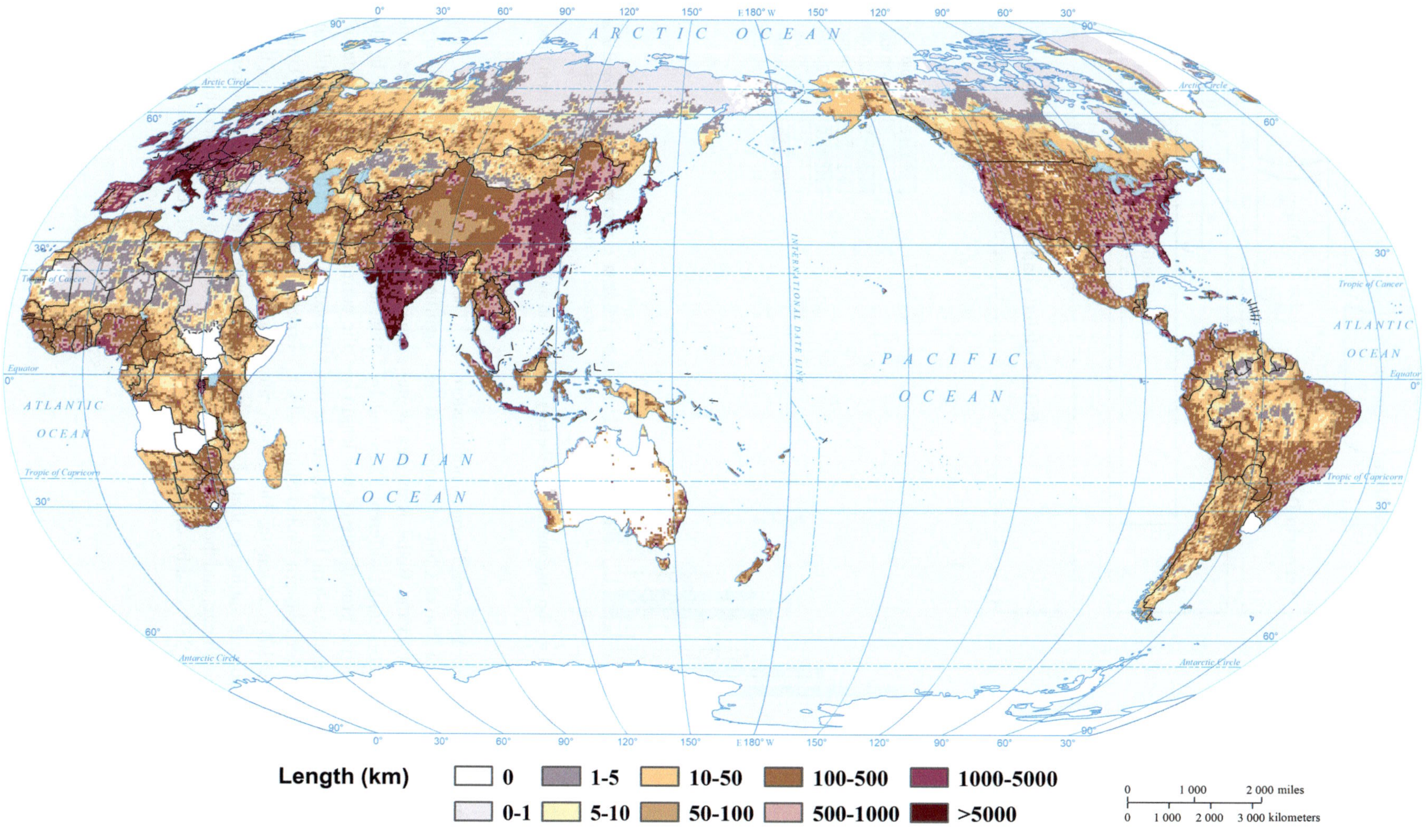

Projected global road network (2030, SSP1)

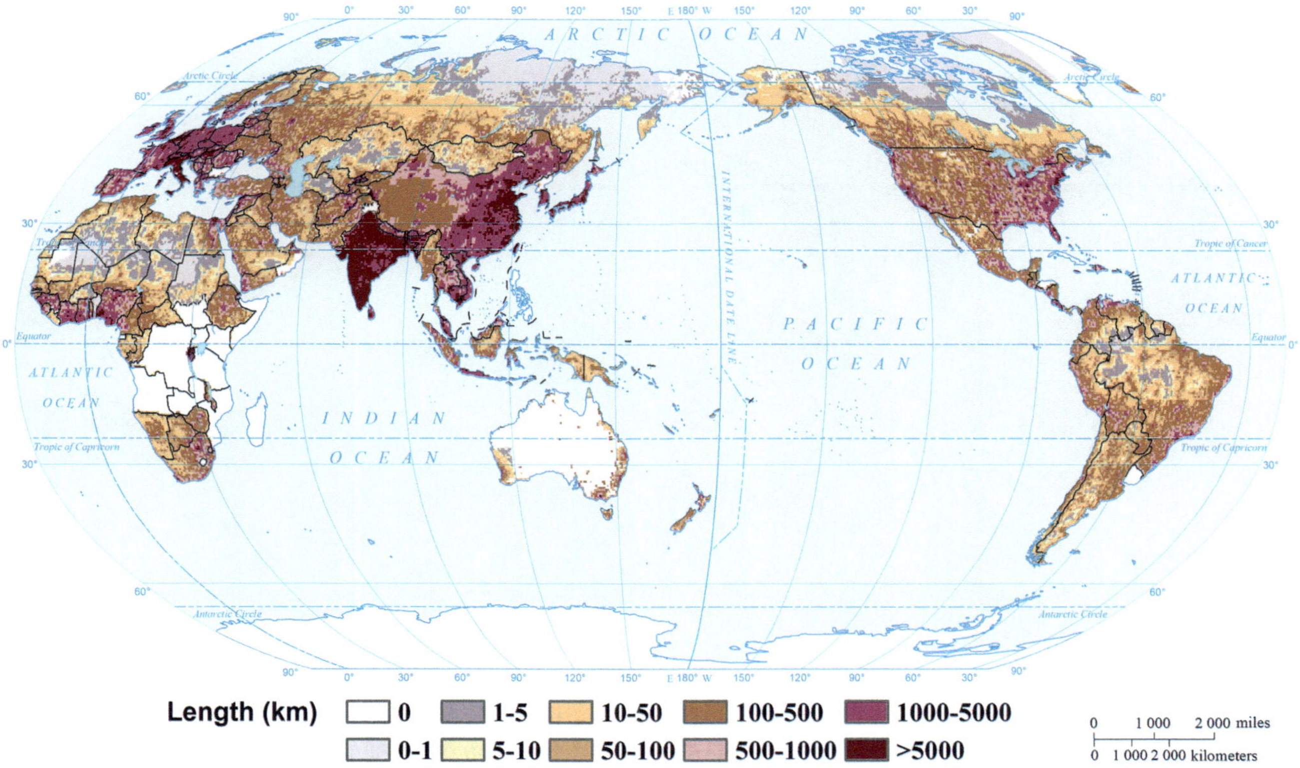

Projected global road network (2030, SSP2)

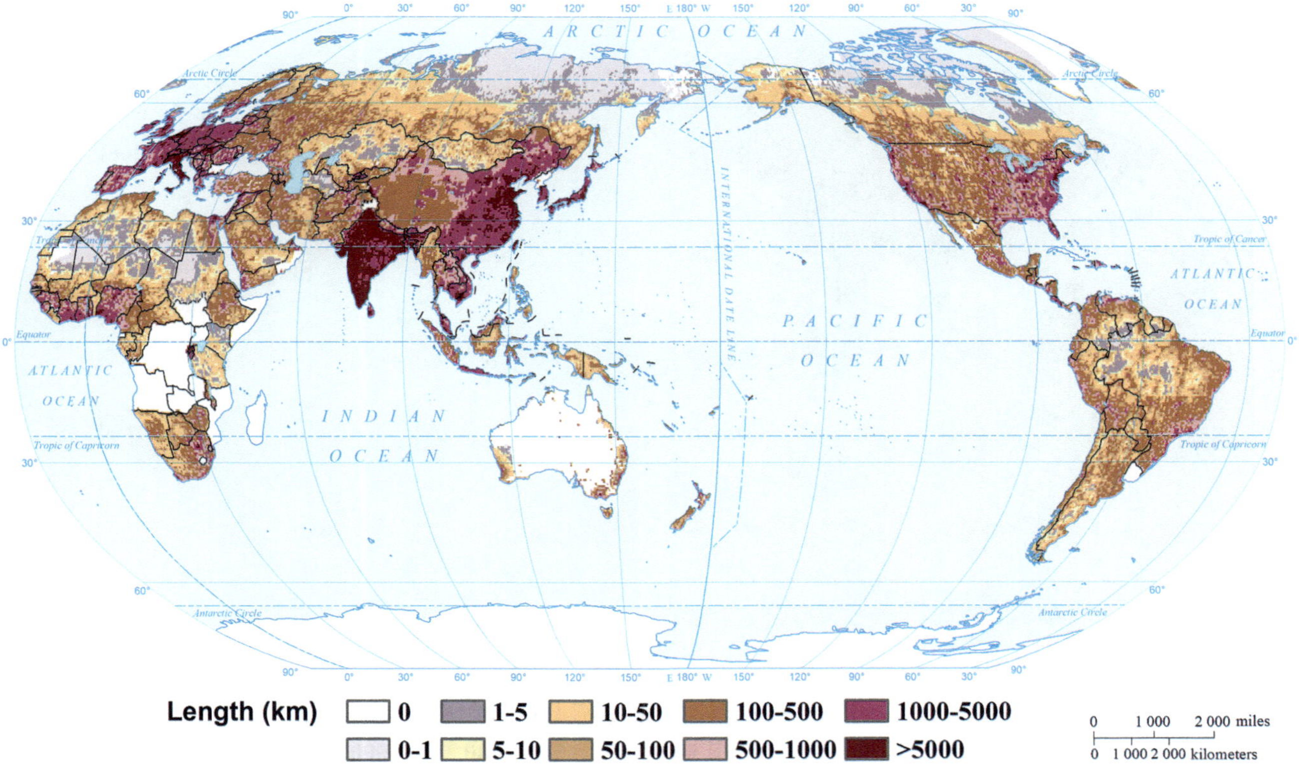

Projected global road network (2030, SSP3)

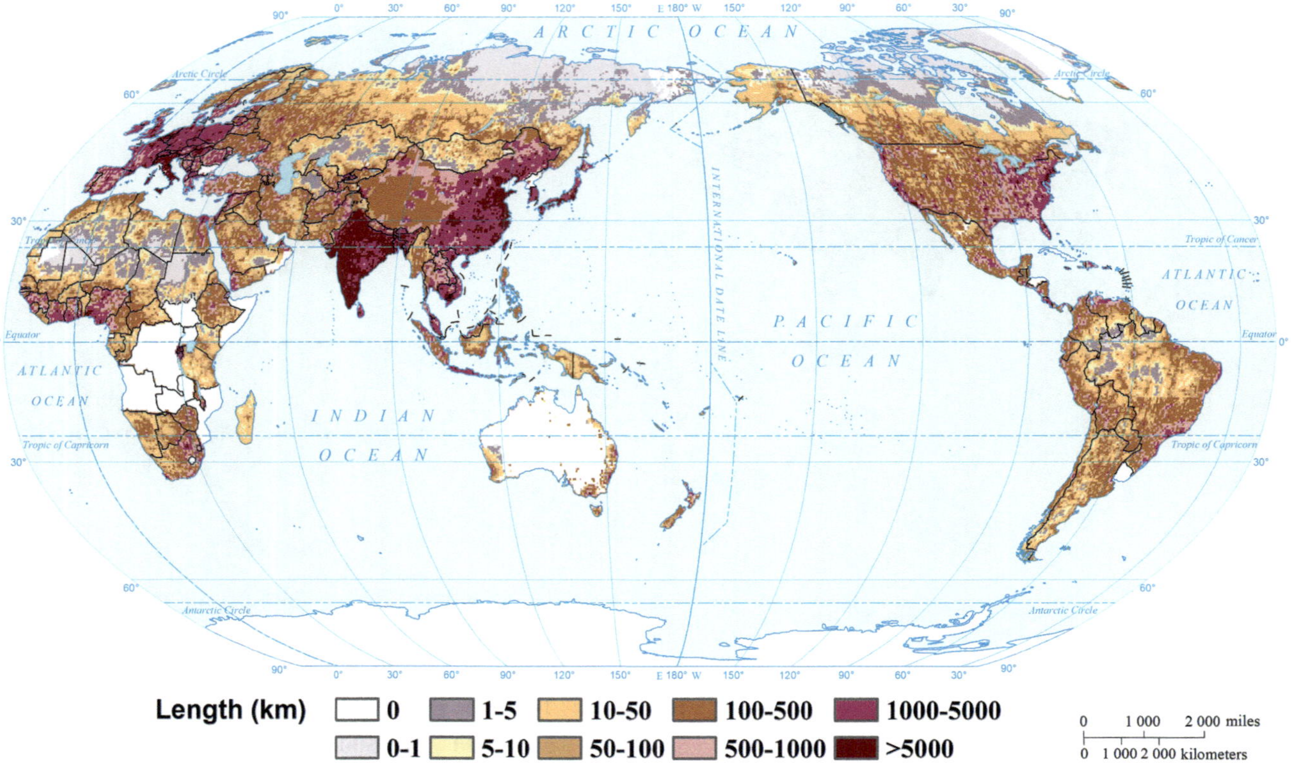

Projected global road network (2050, SSP1)

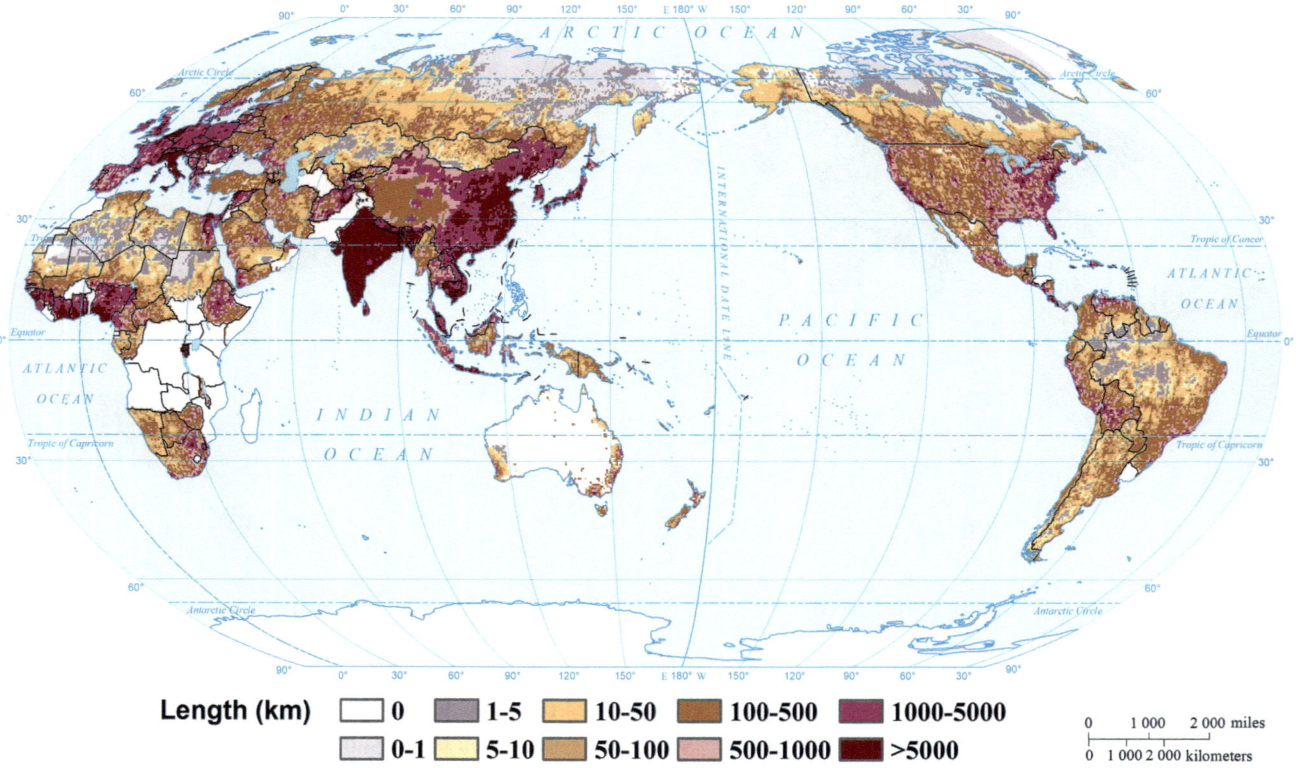

Projected global road network (2050, SSP2)

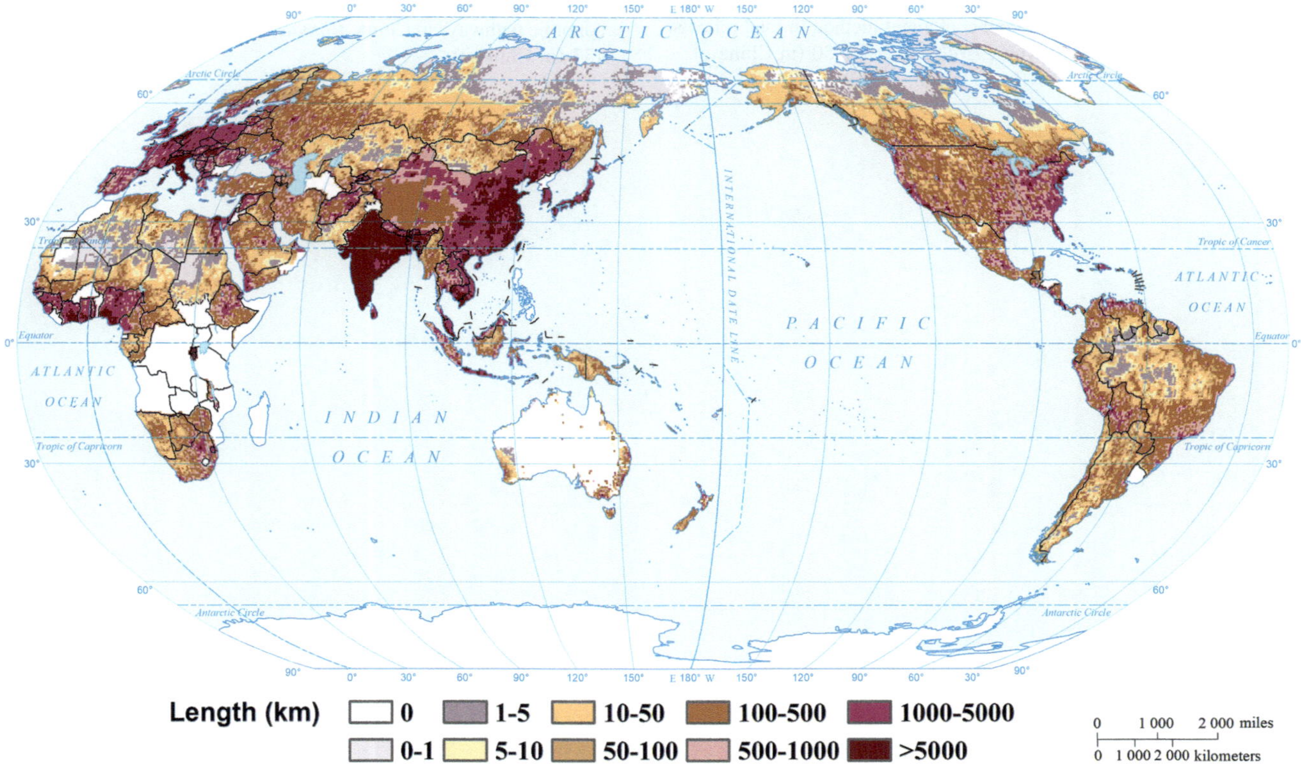

Projected global road network (2050, SSP3)

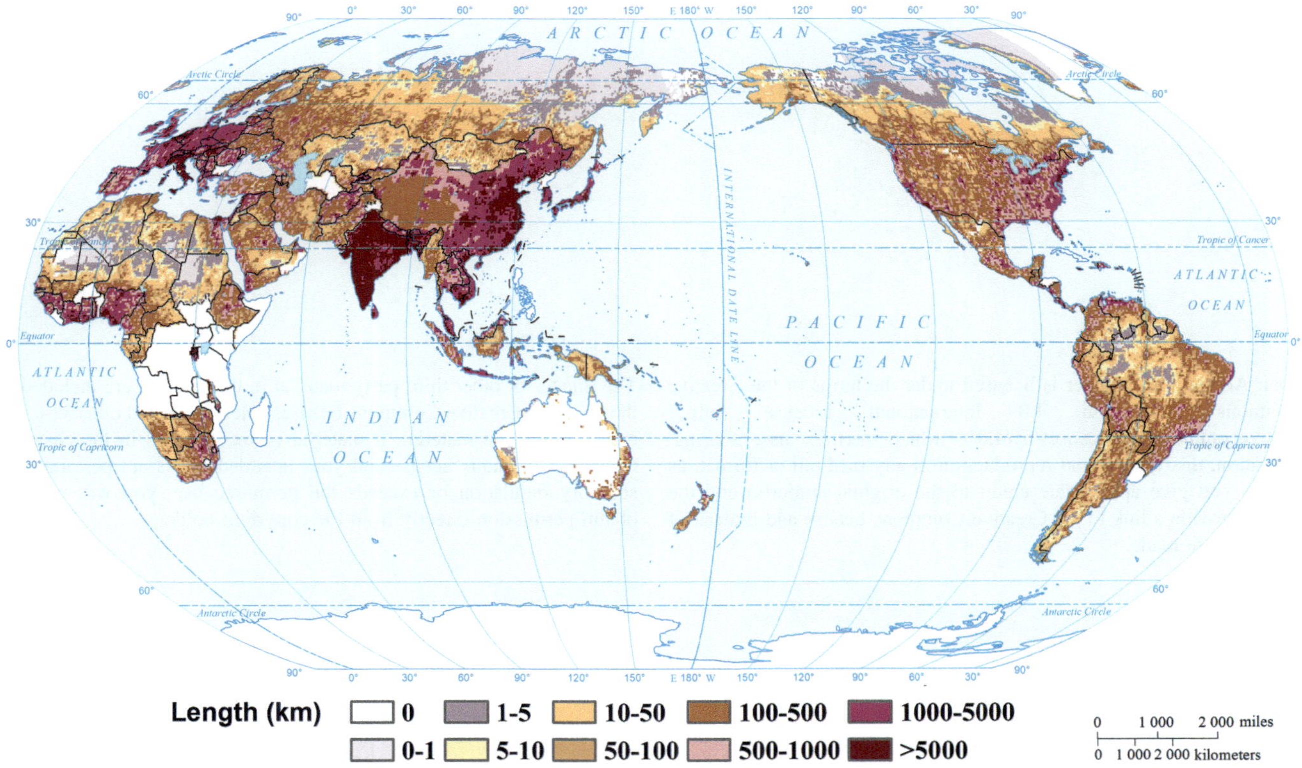

References

Guo, X.F. 2005. Application of land coefficient method in total mileage prediction of highway network. *Highway* 2: 77–80 (in Chinese).

Jiang, T., Zhao, J., Jing, C., Cao, L.G., Wang, Y.J., Sun, H.M., Wang, A.Q., Huang, J.L., Su, B.D., Wang, R. 2017. National and provincial population projected to 2100 under the shared socioeconomic pathways in China. *Climate Change Research* 13 (2): 128–137 (in Chinese).

Global Change Risks

Mapping Global Risk of Heatwave Mortality Under Climate Change

Qinmei Han, Weihang Liu, Wei Xu, and Peijun Shi

1 Background

Global warming has become a severe problem worldwide, where the average global temperature has steadily increased over recent decades accompanied by the abnormally hot weather (IPCC 2013). Since the 1950s, heatwave events have increased in frequency, intensity, and duration and their impact on human health will also increase under enhanced global warming (Perkins-Kirkpatrick and Lewis 2020). Heatwaves have become one of the most serious climate events in the world. Exposed to heatwave is associated with increasing mortality—for instance, the European heatwave of 2003 induced more than 70,000 additional deaths in France, Germany, Italy, Spain, and other countries (Robine et al. 2008). For Russia as a whole, the death toll of the 2010 summer heatwave totaled 55,000 people (Barriopedro et al. 2011).

Considering the ever-worsening impact of heatwaves, future projection of heat-related mortality under climate change has been widely studied in recent decades. Heatwave mortality shows an increasing trend under high-emission scenarios (Huang et al. 2011), especially in temperate areas (Gasparrini et al. 2017). Existed studies mainly focus on regional-, country-, or city-level risk assessments, but the distribution and variation of heatwave mortality risk at a global scale need to be further quantitatively evaluated. Mapping heat-related mortality and finding the hotspots will help provide the policy recommendations at both global and national levels.

This study examined future heatwave mortality risk of the world at the grid level (0.25° × 0.25°) and country level, respectively, based on the disaster system theory (Shi 1991, 1996, 2002), using the data of temperature and population changes. We also evaluated the decadal mortality risk change in the 2030s (2016–2035) and the 2050s (2046–2065) as compared to the baseline period (1986–2005) using high spatial resolution climate and population data under different Representative Concentration Pathway (RCP) and Shared Socioeconomic Pathway (SSP) scenarios, namely RCP2.6-SSP1, RCP4.5-SSP2, and RCP8.5-SSP3. To estimate the regional change, we adopted the regionalization recommended by the Intergovernmental Panel on Climate Change (IPCC), which divides the world into 26 regions.

2 Method

In this study, we used daily maximum temperature as the metric to calculate heatwave. The daily maximum temperature data were from the NEX-GDDP dataset, which was released by the National Aeronautics and Space Administration (NASA) in June 2015 (https://dataserver.nccs.nasa.gov/thredds/catalog/bypass/NEX-GDDP/catalog.html). The spatial resolution of the dataset is 0.25° × 0.25°. There are 21 general circulation models in this dataset that contain two RCPs—RCP4.5 and RCP8.5—for the period from 1950 to 2100. The temperature data of lower emissions scenario RCP2.6 was computed by sub-project 1. The population projection data used in this study contain SSP1 − 3 computed by sub-project 2. We calculated heatwave intensity and mortality risk for the periods 2016–2035 and 2046–2065 in the future compared to 1986–2005 under the

Authors: Qinmei Han, Weihang Liu, Wei Xu, Peijun Shi.
Map Designers: Xinli Liao, Jing'ai Wang, Ying Wang.
Language Editor: Wei Xu.

Q. Han · W. Liu · W. Xu · P. Shi (✉)
Faculty of Geographical Science, Beijing Normal University, Beijing, 100875, China
e-mail: spj@bnu.edu.cn

W. Xu · P. Shi
State Key Laboratory of Earth Surface Processes and Resource Ecology, Beijing Normal University, Beijing, 100875, China

P. Shi, *Atlas of Global Change Risk of Population and Economic Systems*, IHDP/Future Earth-Integrated Risk Governance Project Series, https://doi.org/10.1007/978-981-16-6691-9_14

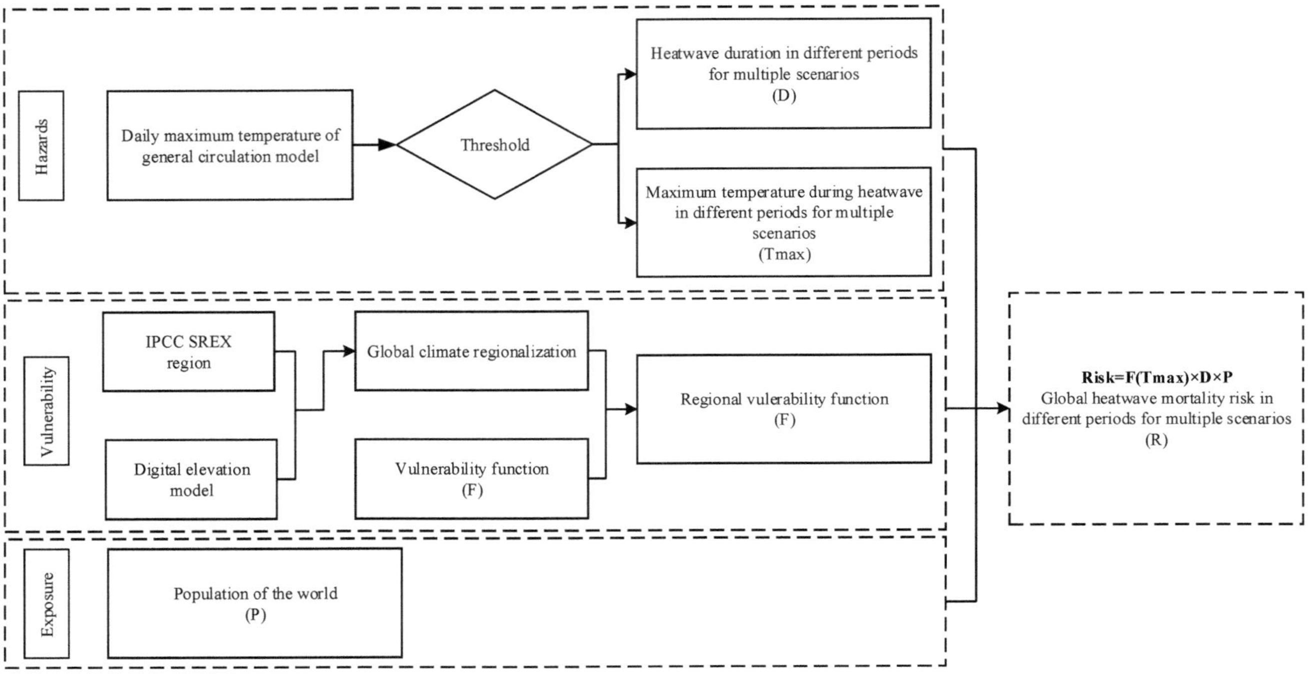

Fig. 1 Technical flowchart for mapping global heatwave mortality risk under climate change

RCP2.6-SSP1, RCP4.5-SSP2, and RCP8.5-SSP3 scenarios (Fig. 1). The maps shown in section 4 are multi-model ensemble results.

2.1 Heatwave Intensity Metrics

To better estimate regional changes of heatwave mortality risk in the world, we chose a relative threshold instead of an absolute threshold. A heatwave event here was defined as at least three consecutive days exceeding the given threshold, which is the 95th percentile value of the daily maximum temperature series over the baseline period 1986–2005 at the grid level, and if the 95th percentile value was lower than 25 °C, we set 25 °C as the threshold in this grid. The annual heatwave duration was defined as the total heatwave days in a given year. The annual heatwave maximum temperature was defined as the maximum temperature of all the heatwave events in a year.

2.2 Population Vulnerability

In this study, six mortality vulnerability curves for typical cities in the world were adopted (Gosling et al. 2007). In the IPCC-SREX reports, the world is divided into 26 regions (Seneviratne et al. 2012). We regrouped the 26 regions into 6 groups according to climate types, latitude zones, and terrain elevation. Each vulnerability curve was applied to a group of the IPCC-SREX regions to map heatwave mortality risk of the world.

2.3 Mortality Risk Function

Heatwave mortality risk of the world is assessed with formula (1):

$$R = F(T_{max}) \times D \times P \tag{1}$$

where R is the annual heatwave mortality risk, F represents the regional vulnerability function, $T_{\max}$ refers to the annual maximum temperature during heatwaves, P refers to the annual total population of each grid, and D is the annual heatwave duration (days). We computed the annual heatwave mortality risk for the selected periods—1986–2005, 2016–2035, and 2046–2065. To map the spatial distribution, we then computed the 20-year average value for each period. All calculations, analyses, and the figures of the meteorological metrics and heatwave mortality risk were performed and mapped at the grid level (0.25° × 0.25°).

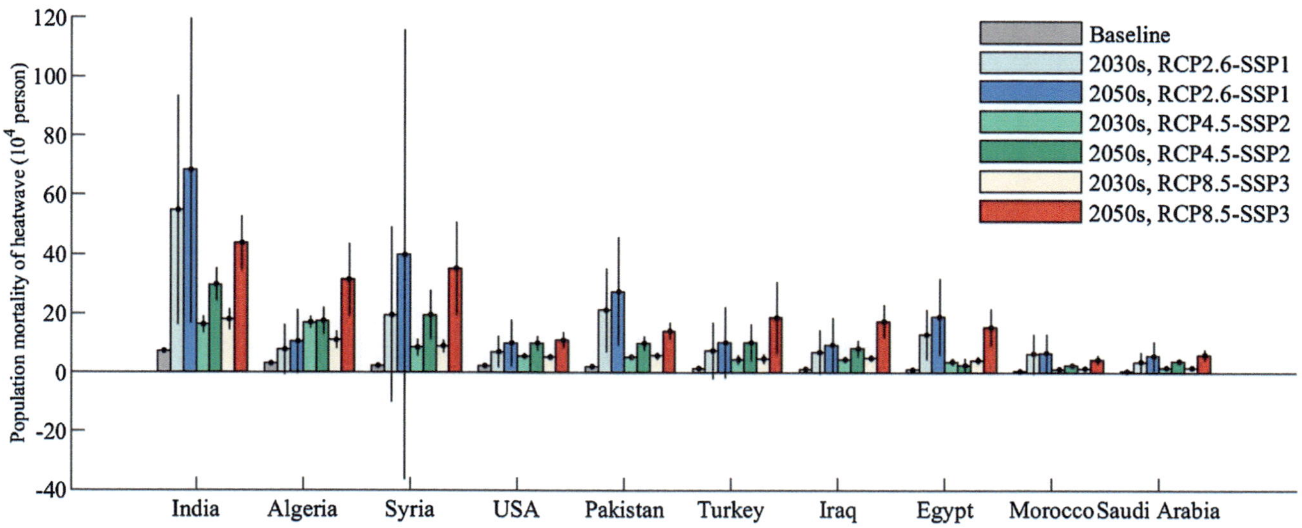

Fig. 2 Heatwave mortality risk of the top 10 countries. The error bar represents the one standard deviation across the 21 (13 for Representative Concentration Pathway (RCP) 2.6) general circulation models and 3 shared socioeconomic pathways (SSPs)

3 Results

3.1 Heatwave Intensity

The results show that the spatial distribution of heatwave mortality risk for the two periods (2030s and 2050s) under the three scenarios (RCP2.6-SSP1, RCP4.5-SSP2, and RCP8.5-SSP3) are very similar. For the maximum temperature during heatwaves, regions with high values are distributed in North Africa, West Asia, India, and Oceania, mainly near 23°N, and the temperature decreases from the high value areas to the north and south, respectively.

Heatwave duration decreases from the equator to the poles. The highest duration areas include Central Africa, West Asia, Central Asia, Central South America, and Oceania. The longest heatwave days are in Central Africa, Indonesia, and northern South America. The areas with high values in the 2050s are significantly larger than that in the 2030s under different scenarios.

Generally, global averaged heatwave duration during 1986–2005 are under 10 days per year, whereas it increases to 28 days per year in the 2050s under the high-emission scenarios.

3.2 Mortality Risk

High mortality risk areas for heatwaves are mainly distributed in the Northern Hemisphere including India peninsula, West Asia, the Mediterranean area, and eastern North America at the grid level. High latitudes in the Northern Hemisphere are mainly of lower risk as compared to the other regions. Overall, global heatwave mortality risk for the baseline period (1986–2005) is about 289,576 persons per year. In comparison, the annual average heatwave mortality risk increases by 8 times, 5 times, and 8 times in the 2050s under the RCP8.5-SSP3, RCP4.5-SSP2, and RCP2.6-SSP1scenarios, respectively.

We performed the zonal statistics analysis of the risk result at the continental level and the results show a substantial increase during the 2050s under the RCP8.5-SSP3 scenario. The heatwave mortality risk increases by 10 times in Africa followed by 8 times in Asia and 6 times in South America. There are 4-times, 4-times, and 3.5-times increase of heatwave mortality risk in Europe, North America, and Oceania, respectively, in the 2050s under the RCP8.5-SSP3 scenario.

The heatwave mortality risk at the country scale was also derived and ranked. The top ten countries with high heatwave mortality are showed in Fig. 2. These countries are mainly located in North Africa and West Asia, distributed around 30°N. Compared to the baseline period, the heatwave mortality risks of all countries increase significantly under different scenario combinations, especially in the 2050s (Fig. 2).

4 Maps

Global heatwave mortality risk (Baseline)

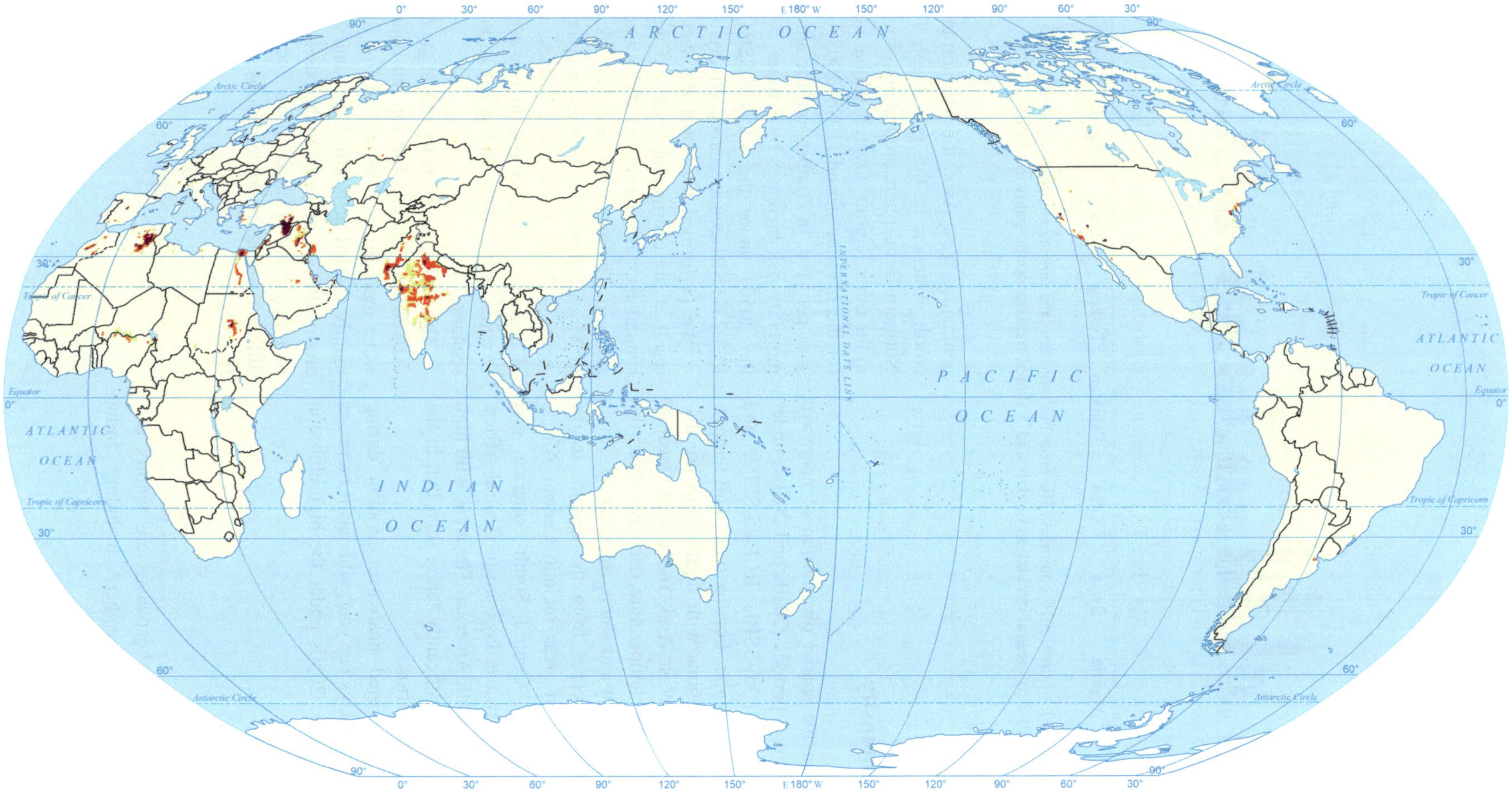

Mortality (person) >100 30-100 20-30 10-20 ≤10 No Data

Global heatwave mortality risk (2030s, RCP2.6-SSP1)

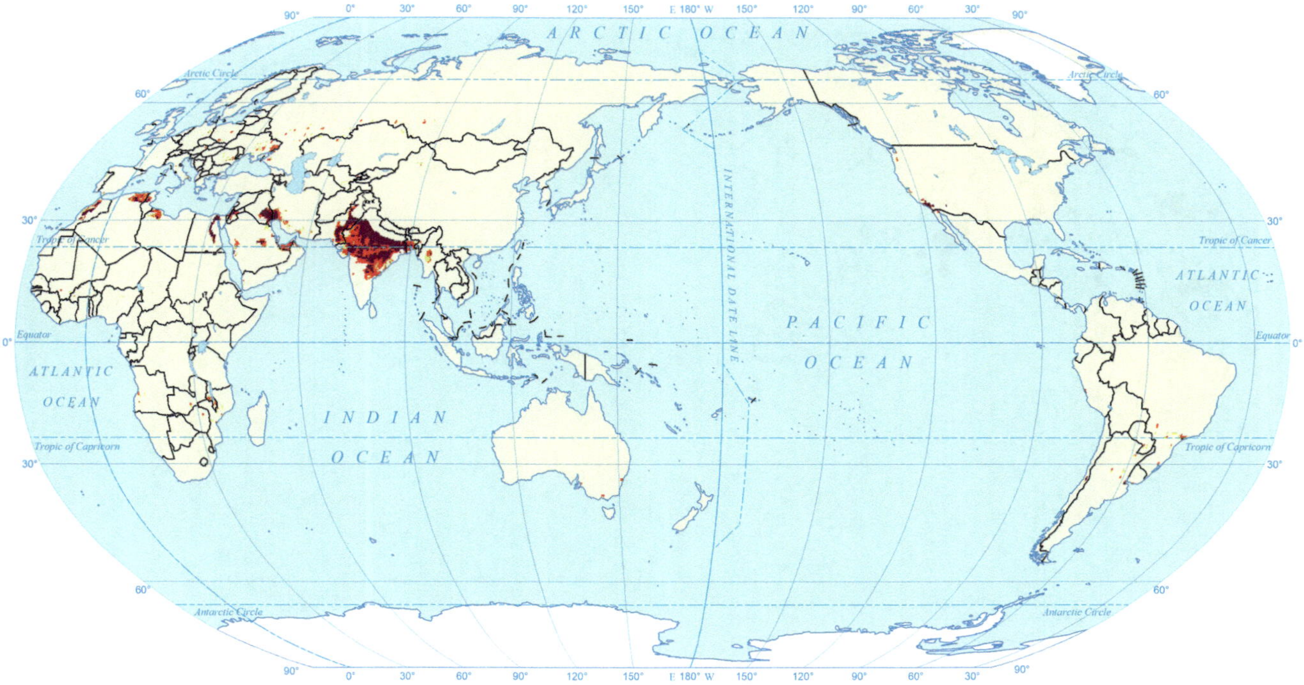

Mortality (person) >100 30-100 20-30 10-20 ≤10 No Data

0 1 000 2 000 miles
0 1 000 2 000 kilometers

Global heatwave mortality risk (2030s, RCP4.5-SSP2)

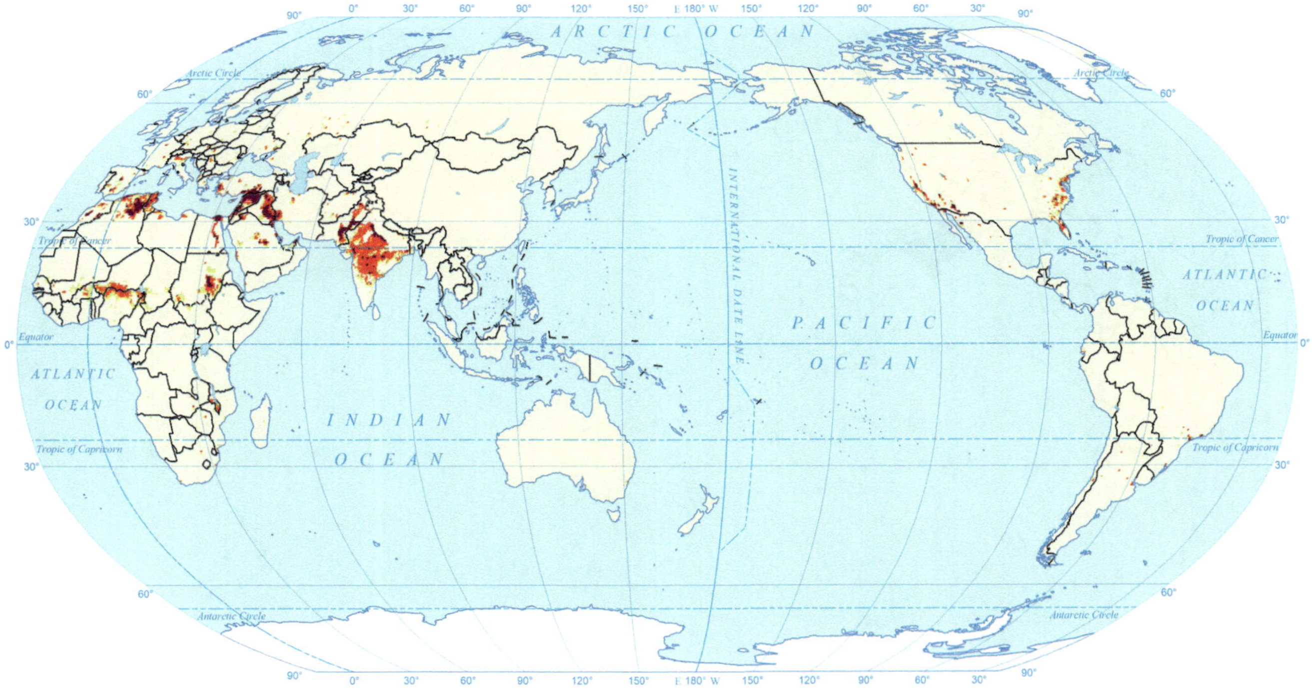

Mortality (person) >100 30-100 20-30 10-20 ≤10 No Data

0 1 000 2 000 miles
0 1 000 2 000 kilometers

Global heatwave mortality risk (2030s, RCP8.5-SSP3)

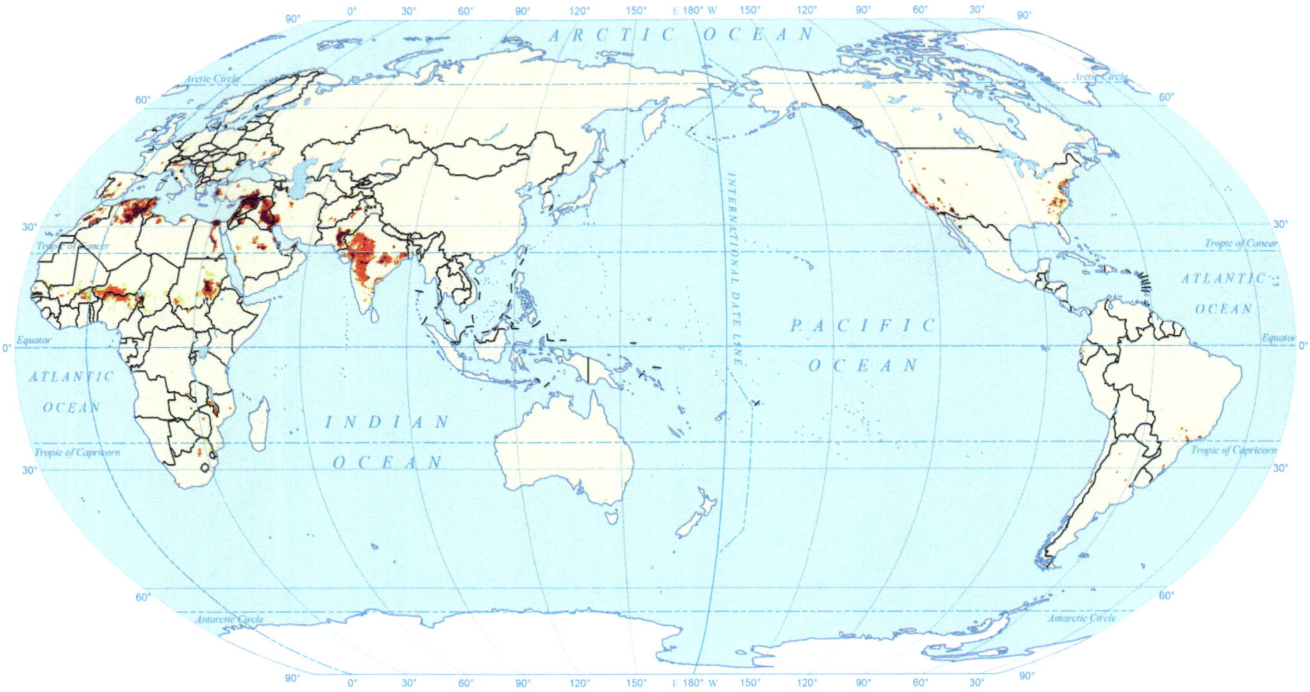

Mortality (person) >100 30-100 20-30 10-20 ≤10 No Data

0 1 000 2 000 miles
0 1 000 2 000 kilometers

Global heatwave mortality risk (2050s, RCP2.6-SSP1)

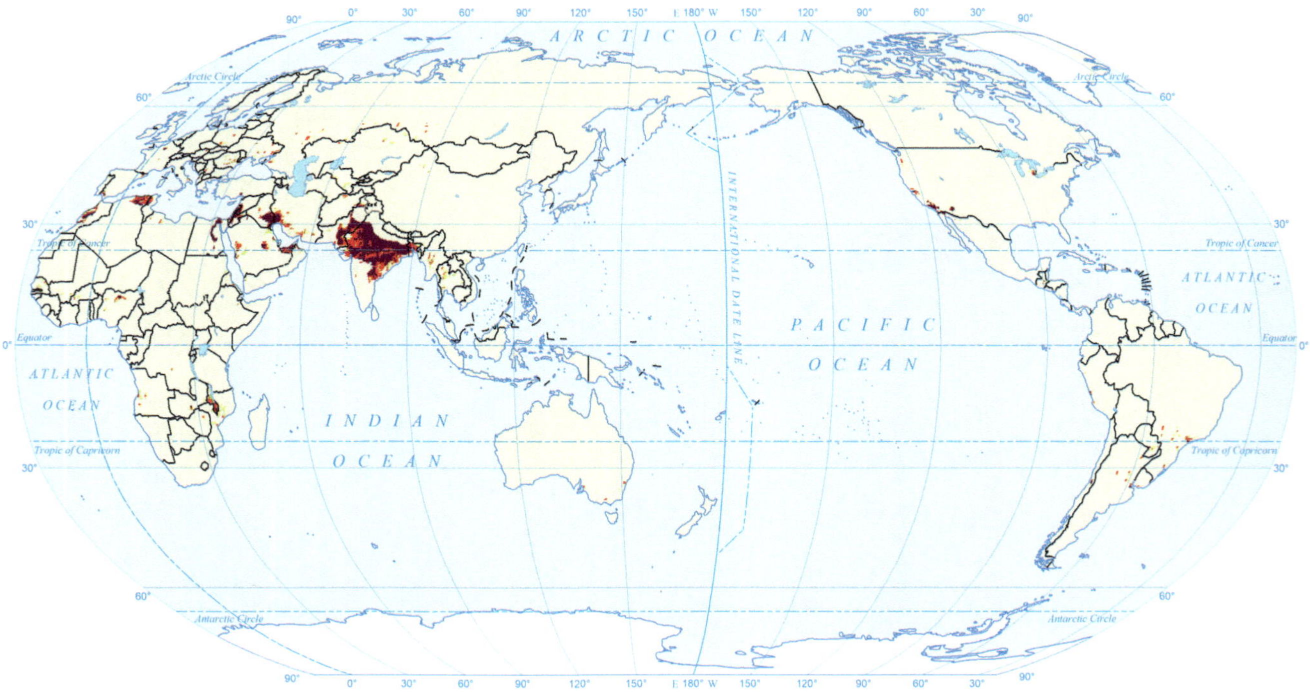

Mortality (person) >100 30-100 20-30 10-20 ≤10 No Data

0 1 000 2 000 miles
0 1 000 2 000 kilometers

Global heatwave mortality risk (2050s, RCP4.5-SSP2)

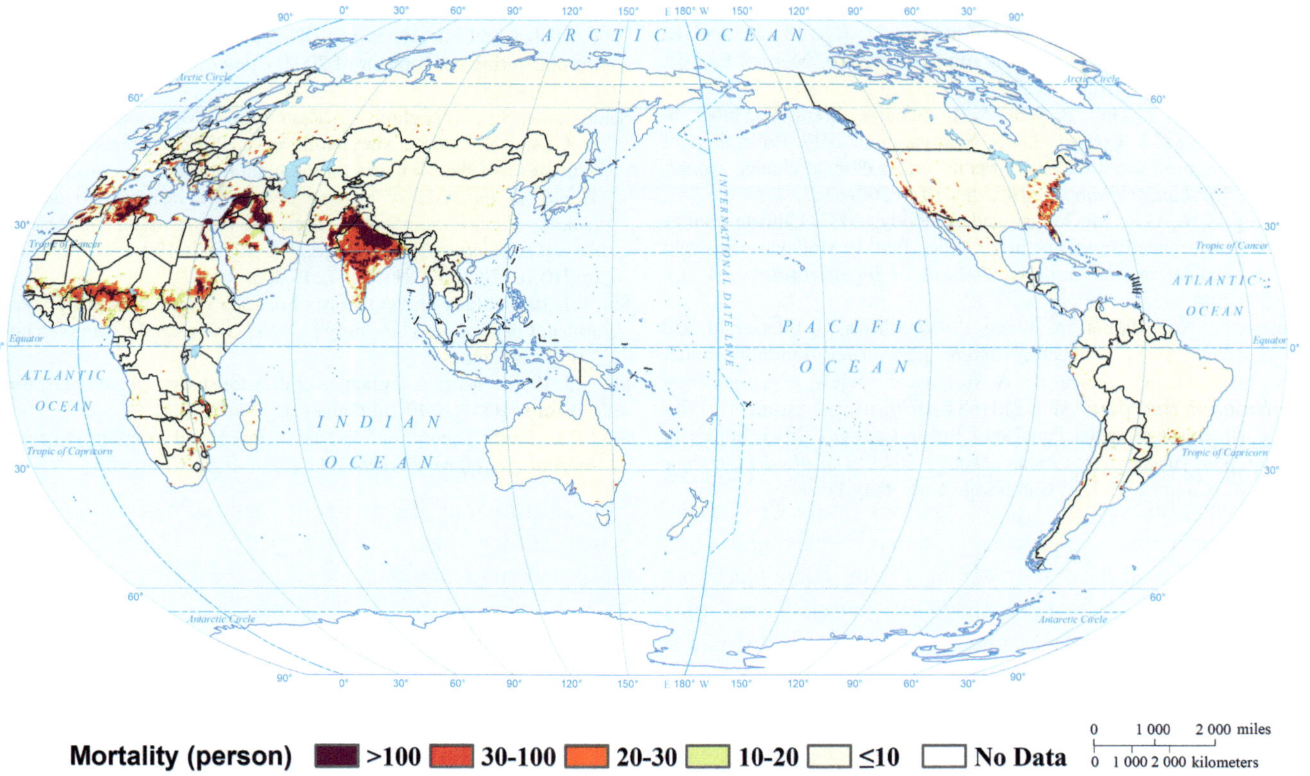

Global heatwave mortality risk (2050s, RCP8.5-SSP3)

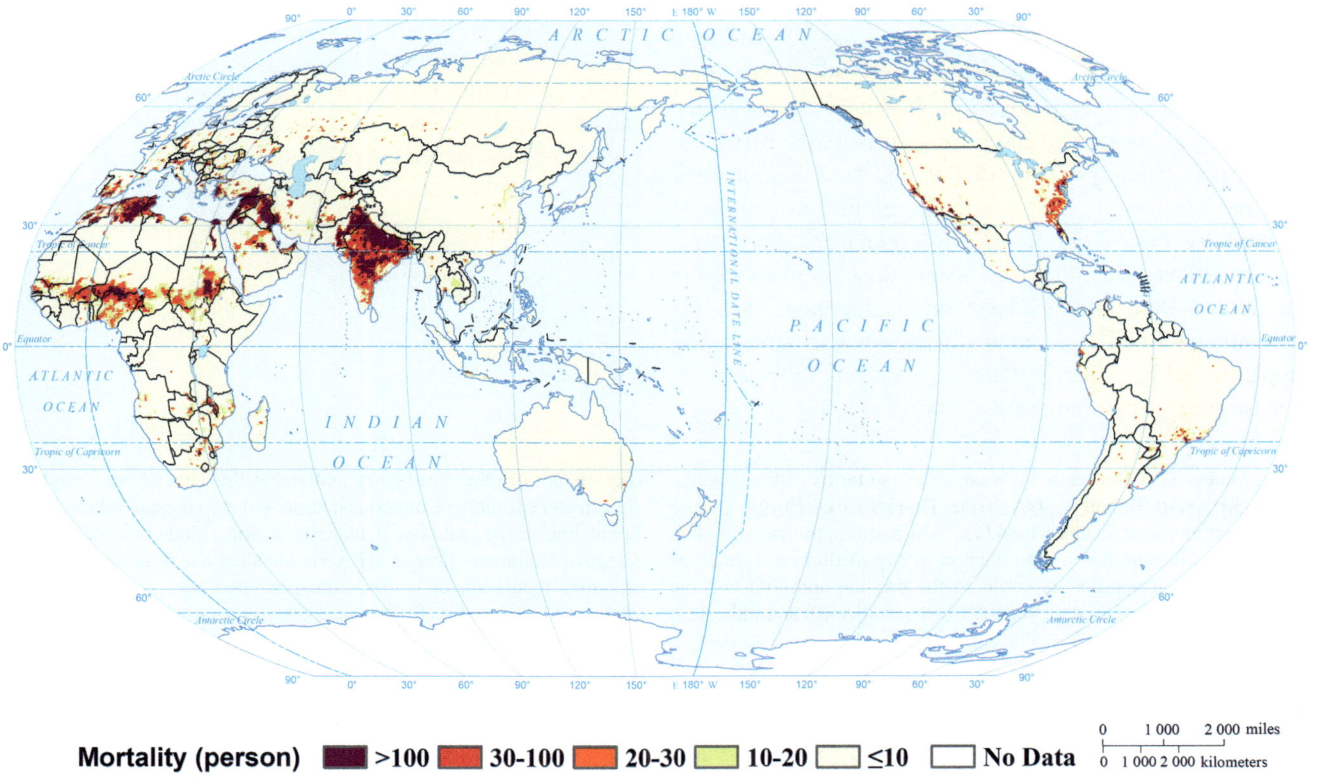

References

Barriopedro, D., E.M. Fischer, J. Luterbacher, R.M. Trigo, and R. Garcia-Herrera. 2011. The hot summer of 2010: Map of Europe. *Science* 332 (April): 220–224.

Gasparrini, A., Y. Guo, F. Sera, A.M. Vicedo-Cabrera, V. Huber, S. Tong, M.S.Z.S. Coelho, P.H.N. Saldiva, et al. 2017. Projections of temperature-related excess mortality under climate change scenarios. *The Lancet Planetary Health* 1 (9): e360–e367.

Gosling, S.N., G.R. McGregor, and A. Páldy. 2007. Climate change and heat-related mortality in six cities Part 1: Model construction and validation. *International Journal of Biometeorology* 51 (6): 525–540.

Huang, C., A.G. Barnett, X. Wang, P. Vaneckova, G. FitzGerald, and S. Tong. 2011. Projecting future heat-related mortality under climate change scenarios: A systematic review. *Environmental Health Perspectives* 119 (12): 1681–1690.

IPCC (Intergovernmental Panel on Climate Change). 2013. Summary for policymakers. In *Climate change 2013: The physical science basis*. Cambridge, UK: Cambridge University Press.

Perkins-Kirkpatrick, S.E., and S.C. Lewis. 2020. Increasing trends in regional heatwaves. *Nature Communications* 11 (1): 1–8.

Robine, J.M., S.L.K. Cheung, S. Le Roy, H. Van Oyen, C. Griffiths, J. P. Michel, and F.R. Herrmann. 2008. Death toll exceeded 70,000 in Europe during the summer of 2003. *Comptes Rendus Biologies* 331 (2): 171–178.

Seneviratne, S.I., N. Nicholls, D. Easterling, C.M. Goodess, S. Kanae, J. Kossin, Y. Luo, J. Marengom, et al. 2012. Changes in climate extremes and their impacts on the natural physical environment. In *Managing the risks of extreme events and disasters to advance climate change adaptation: A special report of the intergovernmental panel on climate change*, 9781107025, 109–230. https://doi.org/10.1017/CBO9781139177245.006.

Shi, P.J. 1991. Study on the theory of disaster research and its practice. *Journal of Nanjing University (natural Sciences)* 11 (Supplement): 37–42 (in Chinese).

Shi, P.J. 1996. Theory and practice of disaster study. *Journal of Natural Disasters* 5 (4): 6–17 (in Chinese).

Shi, P.J. 2002. Theory on disaster science and disaster dynamics. *Journal of Natural Disasters* 11 (3): 1–9 (in Chinese).

Mapping Global Risk of River Flood Mortality

Junlin Zhang, Xinli Liao, and Wei Xu

1 Background

Globally, river flooding induced by heavy rainfall frequently causes fatalities every year (Jongman et al. 2015; CRED and UNISDR 2018; CRED 2019). Particularly, heavy rainfall will increase in the future with climate warming (Liao et al. 2019). This could lead to greater rain-induced local flooding in some watersheds or regions (IPCC 2012). Besides, exposed populations to floods are increasing with the socioeconomic development (Jongman et al. 2015; Winsemius et al. 2018; Liao et al. 2019).

Generally, river flooding risk assessment has two steps. The first is to simulate river flooding hazard using hydrological or hydrodynamic model and inundation model, and the second is to calculate affected populations by overlaying the population data and flood hazard maps, and the results are used to assess the risk of affected population by floods (Arnell and Gosling 2016; Lim et al. 2018). Projected future precipitation and social-economic datasets provide a basis for these studies. However, when assessing risks in the population, most studies are concerned with the affected population (Alfieri et al. 2015; Dottori et al. 2016; Wing et al. 2018) and few studies concern mortality risks. Shi and Kasperson (2015) and Jongman et al. (2015) assessed the mortality risk using baseline mortality rate. Kinoshita et al. (2018) simulated changing mortality vulnerability with time to assess risks, but the vulnerability is measured by an index rather than vulnerability functions that involve certain physical processes. The difficulty of mortality risk research is lacking proven vulnerability functions. In addition, many studies focus on the risk at the country and regional levels, lacking grid level high spatial resolution results.

In order to address the issues, this study assessed the global mortality risks from floods by two main steps. The first was to develop mortality vulnerability functions for all countries by revising an existing vulnerability function. Then future potential death tolls of the 2030s and the 2050s were estimated at the 2.5' grid level under the Representative Concentration Pathway (RCP) and Shared Socioeconomic Pathway (SSP) scenarios of RCP4.5-SSP2 and RCP8.5-SSP3. The results were compiled to produce the risk maps at the 0.25° grid level and country level.

2 Method

Figure 1 shows the technical flowchart for mapping flood mortality risk of the world. The study revised the existing vulnerability functions by adjustment coefficient that is calculated based on recorded death tolls. Then future death tolls were estimated using predicted inundation data and population data, and the adjusted vulnerability functions. Future death tolls at the grid level were then aggregated to other geographic units. Finally, the risk size and model uncertainty are analyzed.

2.1 Estimation of Risks

2.1.1 Estimation of Losses for the Baseline Period

The mortality vulnerability function is represented by Eq. (1) (Jonkman et al. 2008). Using the function, historical death tolls were estimated by Eqs. (2) and (3).

Authors: Junlin Zhang, Xinli Liao, Wei Xu.**Map Designers**: Junlin Zhang, Jing'ai Wang, Ying Wang.**Language Editor**: Wei Xu.

J. Zhang · X. Liao · W. Xu (✉)
Faculty of Geographical Science, Beijing Normal University, Beijing, 100875, China
e-mail: xuwei@bnu.edu.cn

W. Xu
State Key Laboratory of Earth Surface Processes and Resource Ecology, Beijing Normal University, Beijing, 100875, China

P. Shi, *Atlas of Global Change Risk of Population and Economic Systems*, IHDP/Future Earth-Integrated Risk Governance Project Series, https://doi.org/10.1007/978-981-16-6691-9_15

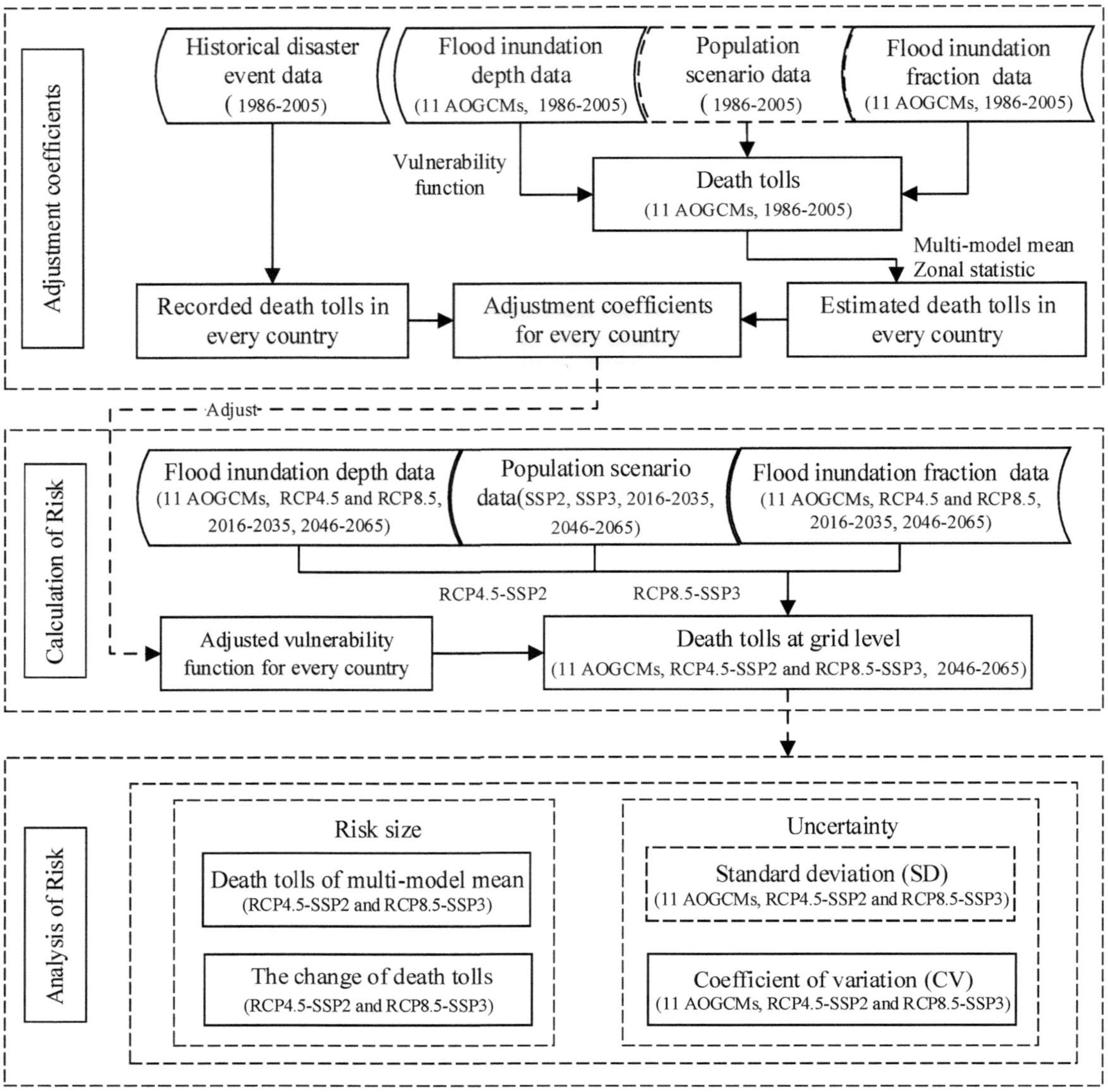

Fig.1 Technical flowchart for mapping global risk of river flood mortality. AOGCM = Atmosphere–Ocean General Circulation Model; RCP = Representative Concentration Pathway; SSP = Shared Socioeconomic Pathway

$$V(d) = \Phi\left[\frac{\ln(d) - 7.60}{2.75}\right] \quad (1)$$

where Φ is the cumulative normal distribution. d is the water depth.

$$L_{his\_i\_j} = V(d_{his\_i\_j}) \times S_{his\_j} \times f_{his\_i\_j} \quad (2)$$

$$L_{his} = \frac{1}{11 \times 20}\sum_{j=1}^{20}\sum_{i=1}^{11} L_{his\_i\_j} \quad (3)$$

where i is the order of the 11 Atmosphere–Ocean General Circulation Models (AOGCMs); j is the sequence of the 20 years; his is the baseline period (1986–2005); $L_{his\_i\_j}$, $d_{his\_i\_j}$, and $f_{his\_i\_j}$ are the estimated death tolls, water depth, and inundation fraction for the ith AOGCM in the jth year for the baseline period, respectively; S_{his_j} is the population size in the jth year of the baseline period; L_{his} is the annual average death tolls, that is, the multi-model ensemble for the baseline period; $V(d)$ is the vulnerability function.

2.1.2 Calculation of Adjustment Coefficients

The study revised the mortality vulnerability function (Eq. 1) for countries to reduce the diversity of vulnerability functions in different areas.

Using Eq. (4), the adjustment coefficients (K_c values) were calculated for the countries with total recorded deaths and estimated deaths both greater than zero during the baseline period. On this basis, the adjustment coefficient is the minimum of above calculated K_c values for the countries with total recorded death tolls equal to zero during the baseline period; and the adjustment coefficient is the average of above calculated K_c values for the countries with total recorded death tolls greater than zero but total estimated deaths equal to zero.

$$K_c = \frac{\sum_{j=1}^{20} SL_{his\_c\_j}}{L_{his\_c}} \quad (4)$$

where K_c is the adjustment coefficient of country c; j represents the sequential number of the 20 years in the baseline period; $SL_{his\_c\_j}$ is the recorded death tolls of country c.

The adjusted vulnerability function is shown in Eq. (5).

$$AdjV_c(d) = K_c \times V(d) \quad (5)$$

where $AdjV_c(d)$ is the adjusted vulnerability function of country c.

2.1.3 Calculation of Future Losses and Change

Future death tolls of a grid in country c were estimated according to Eq. (6), based on future predicted inundation and population data and adjusted vulnerability function. Next, the study averaged the results of 20 years for all AOGCMs (Eq. 7) to compute uncertainties; then we averaged the results of the 11 AOGCMs as the death tolls of the 2030s or the 2050s to reduce model uncertainties (Eq. 8). Finally, the changes of death tolls from the baseline period to future periods were calculated by Eq. (9).

$$L_{fut\_i\_j} = AdjV_c(d_{fut\_i\_j}) \times S_{fut\_j} \times f_{fut\_i\_j} \quad (6)$$

where $L_{fut\_i\_j}$, $d_{fut\_i\_j}$, and $f_{fut\_i\_j}$ are the estimated death tolls, water depth, and inundation fraction for the ith AOGCM in the jth year for a future period, respectively; S_{fut_j} is the population data in the jth year of a future period.

$$L_{fut\_i} = \frac{1}{20}\sum_{j=1}^{20} L_{fut\_i\_j} \quad (7)$$

where L_{fut_i} is the average death tolls of 20 years (2016–2035 or 2046–2065) for the ith AOGCM.

$$L_{fut} = \frac{1}{11}\sum_{i=1}^{11} L_{fut\_i} \quad (8)$$

where L_{fut} is the death tolls of the 2030s or the 2050s for the multi-model ensemble, which averaged the results from the 11 AOGCMs.

$$\Delta L = L_{fut} - L_{his} \quad (9)$$

where ΔL is the change of death tolls for the 2030s or the 2050s relative to the baseline period.

2.2 Model Uncertainty

The result uncertainty of multi-models is measured by standard deviation (Eq. 10) and coefficient of variation CV, the ratio of SD to L in Eq. 10.

$$SD = \sqrt{\frac{\sum_{i=1}^{11}(L_i - L)^2}{11}} \quad (10)$$

where L_i is the average death tolls of 20 years for the ith AOGCM (L_{his_i} or L_{fut_i}). L is the average death tolls of 20 years for the multi-model ensemble (L_{his} or L_{fut}).

3 Results

Globally, the annual average death tolls of the 2030s are approximately 21 thousand persons for the RCP4.5-SSP2 scenario and 23 thousand persons for the RCP8.5-SSP3 scenario; they increase 0.57 and 0.69 times relative to the baseline period, respectively. The annual average death tolls of the 2050s are approximately 26 thousand persons for the RCP4.5-SSP2 scenario and 32 thousand persons for the RCP8.5-SSP3 scenario; they increase 0.88 and 1.31 times relative to the baseline period. The patterns of spatial distribution are similar for different scenarios; high-risk areas are located in East Asia, South Asia, and Southeast Asia, particularly in eastern coastal China and the Ganges River Basin.

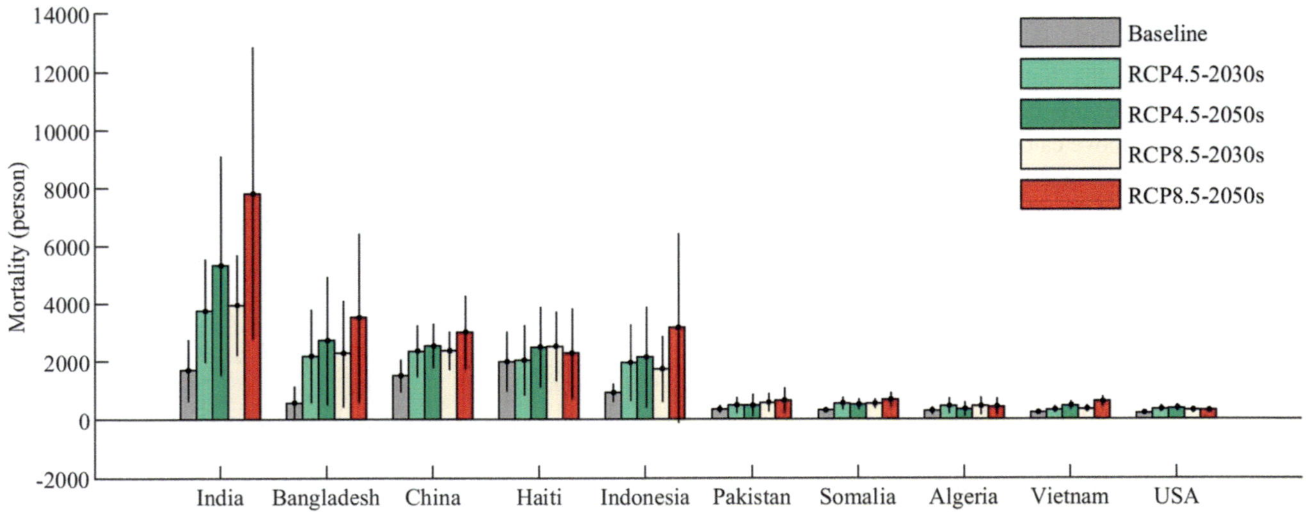

Fig.2 Annual average death tolls of the top 10 high-risk countries (in descending order by death toll). **The error bar represents** the standard deviations across the 11 Atmosphere–Ocean General Circulation Models (AOGCMs)

Using zonal statistics of the death toll results, we calculated the annual average death tolls at the national level. Figure 2 shows the annual average death tolls and errors (measured by standard deviation) of the top ten high-risk countries. The death tolls are higher for India, Bangladesh, China, Haiti, and Indonesia, and lower in Pakistan, Somalia, Algeria, Viet Nam, and the United States. For most countries, the death tolls of the 2050s are higher than that of the 2030s; and the death tolls of the RCP8.5-SSP3 scenario are higher than that of the RCP4.5-SSP2 scenario. The changes of death tolls are higher in India and Bangladesh, increasing 1.22–3.63 times for India and 2.82–5.16 times for Bangladesh. The risk changes of Haiti are the lowest, about 0.03–0.27 times (Fig. 2).

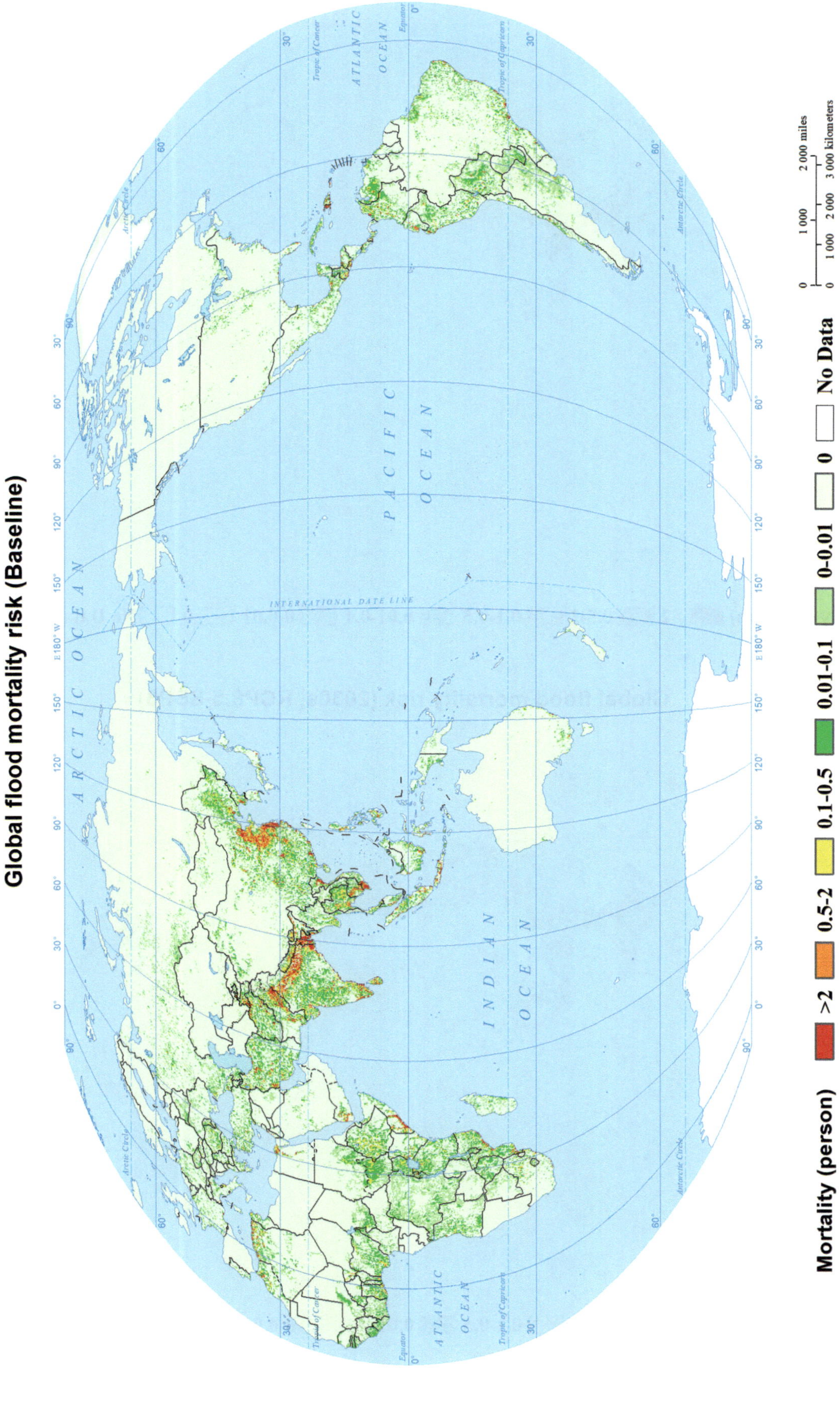
Global flood mortality risk (Baseline)
Mortality (person)
>2
0.5-2
0.1-0.5
0.01-0.1
0-0.01
0
No Data
ARCTIC OCEAN
PACIFIC OCEAN
INDIAN OCEAN
ATLANTIC OCEAN
INTERNATIONAL DATE LINE
Tropic of Cancer
Tropic of Capricorn
Equator
Arctic Circle
Antarctic Circle
0 1000 2000 miles
0 1000 2000 3000 kilometers

Global flood mortality risk (2030s, RCP4.5-SSP2)

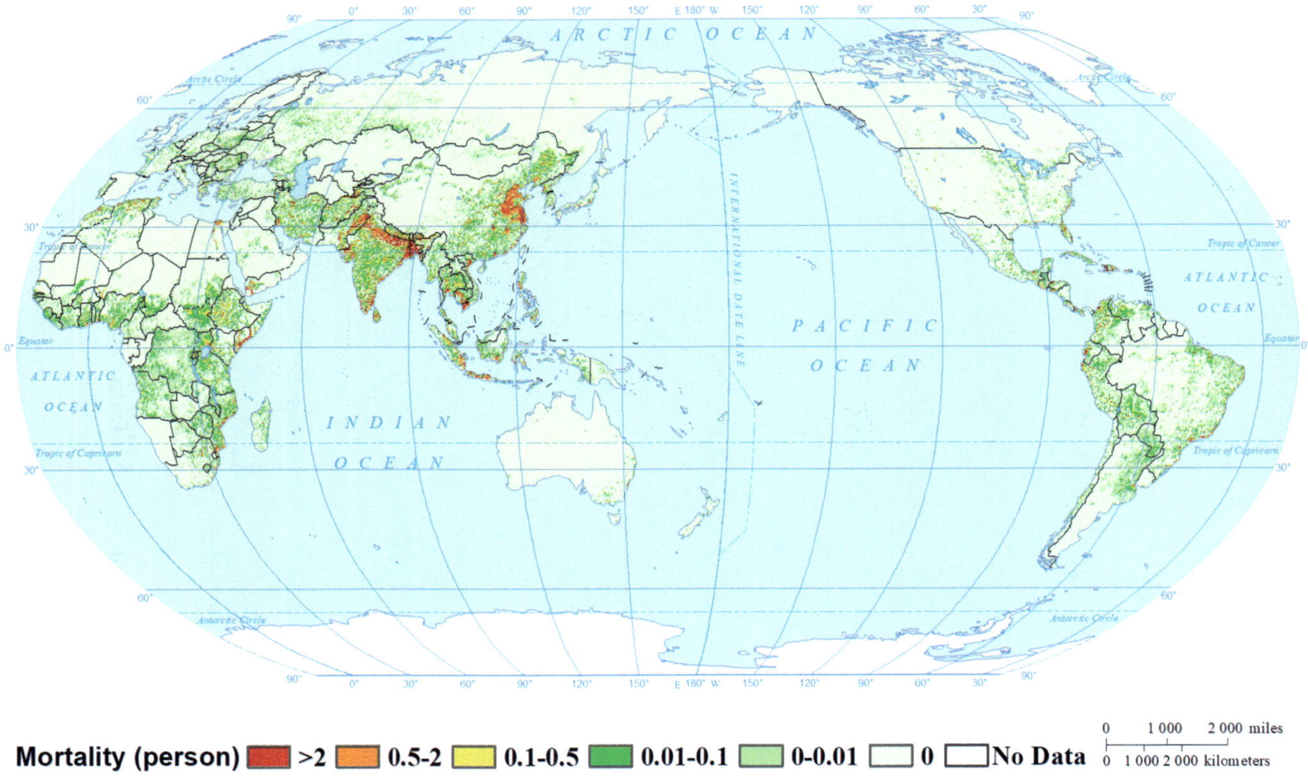

Global flood mortality risk (2030s, RCP8.5-SSP3)

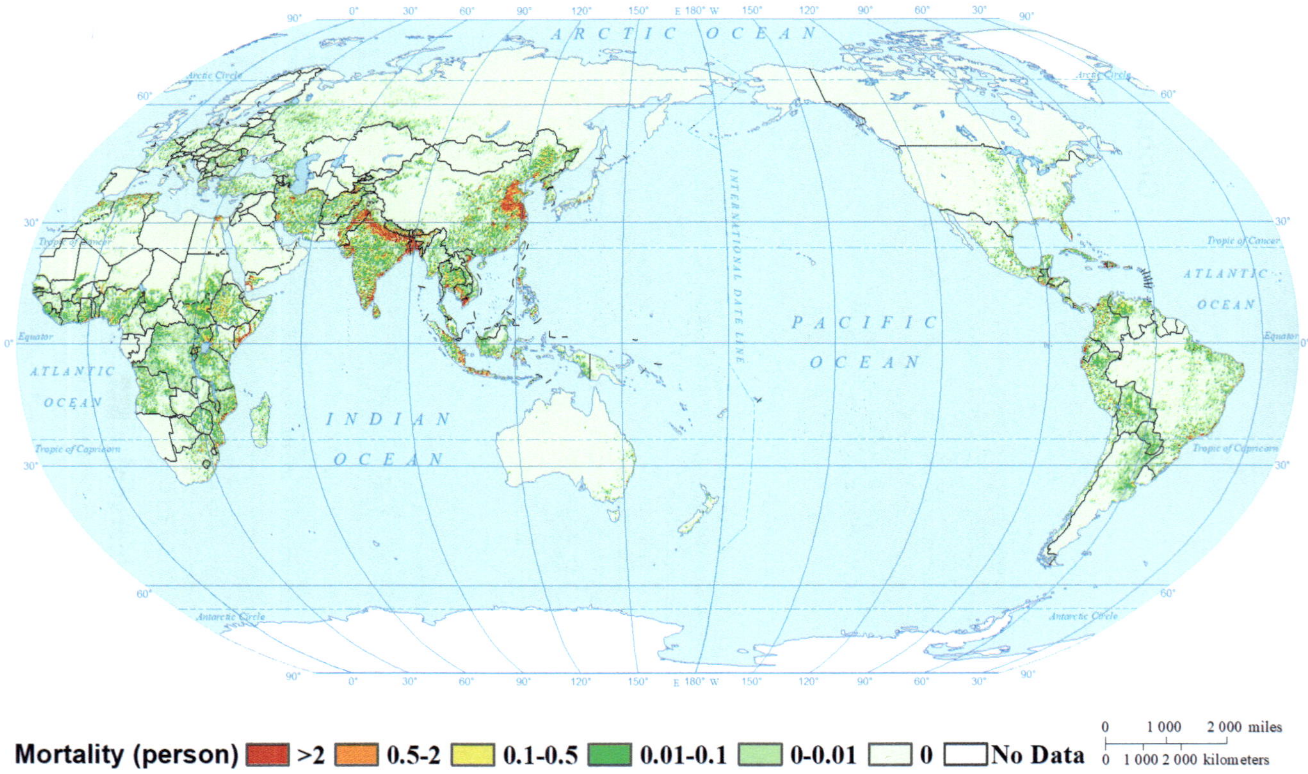

Global flood mortality risk (2050s, RCP4.5-SSP2)

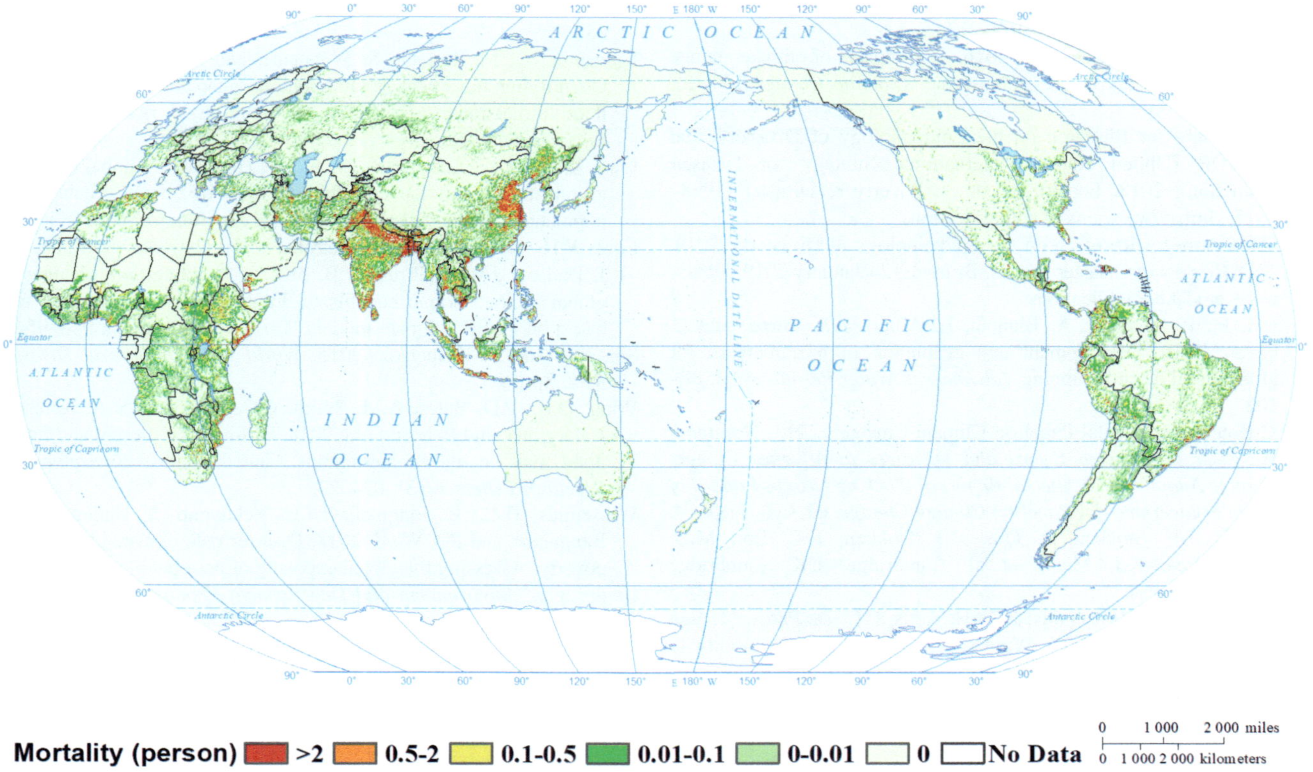

Global flood mortality risk (2050s, RCP8.5-SSP3)

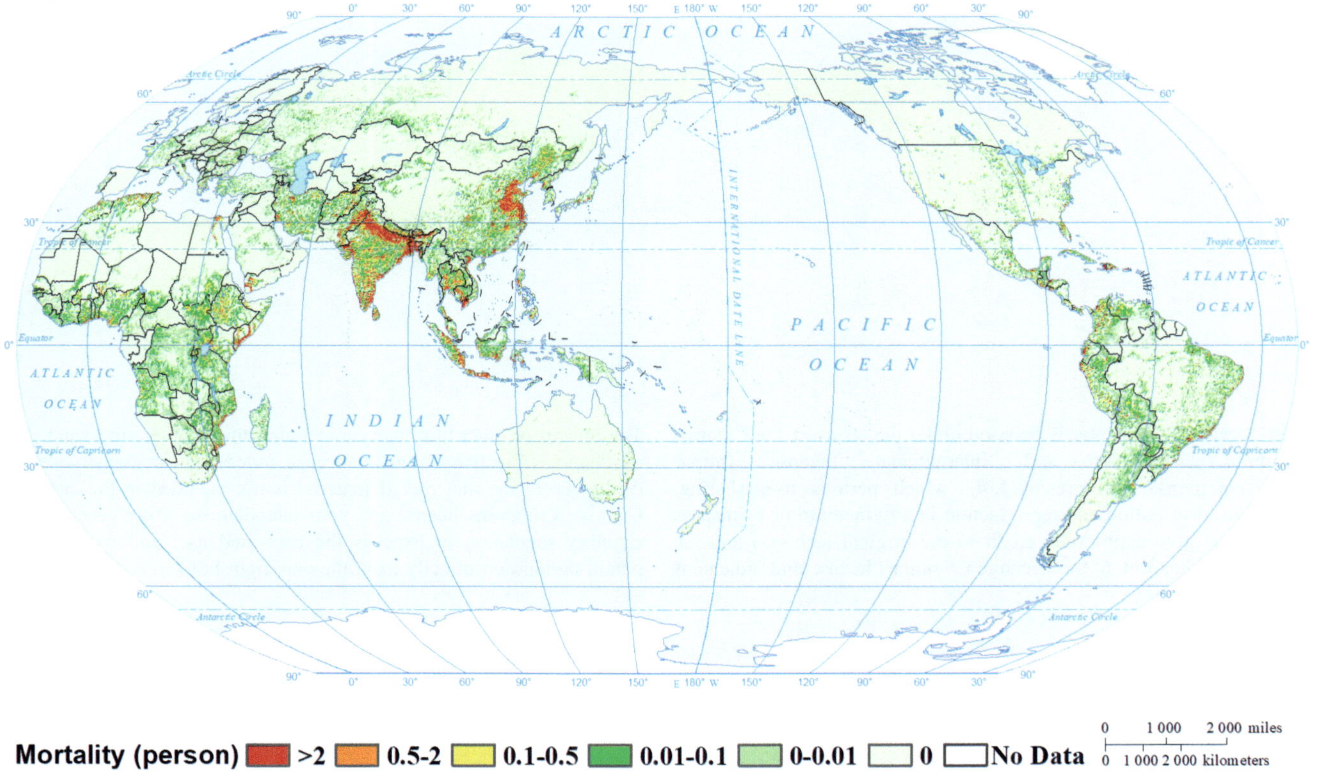

References

Alfieri, L., L. Feyen, F. Dottori, and A. Bianchi. 2015. Ensemble flood risk assessment in Europe under high end climate scenarios. *Global Environmental Change* 35: 199–212.

Arnell, N.W., and S.N. Gosling. 2016. The impacts of climate change on river flood risk at the global scale. *Climatic Change* 134 (3): 387–401.

CRED (Centre for Research on the Epidemiology of Disasters) and UNISDR (United Nations International Strategy for Disaster Reduction). 2018. Economic Losses, Poverty & Disaster (1998–2017). https://www.emdat.be/publications.

CRED (Centre for Research on the Epidemiology of Disasters). 2019. 2018 Review of Disaster Events. Brussels, 24 January 2019. https://www.emdat.be/publications.

Dottori, F., P. Salamon, A. Bianchi, L. Alfieri, F.A. Hirpa, and L. Feyen. 2016. Development and evaluation of a framework for global flood hazard mapping. *Advances in Water Resources* 94: 87–102.

IPCC (Intergovernmental Panel on Climate Change). 2012. *Managing the Risks of Extreme Events and Disasters to Advance Climate Change Adaptation. A Special Report of Working Groups I and II of the Intergovernmental Panel on Climate Change*, ed. C.B. Field, V. Barros, T.F. Stocker, D. Qin, D.J. Dokken, K.L. Ebi, M.D. Mastrandrea, K.J. Mach, et al. Cambridge, UK: Cambridge University Press.

Jongman, B., H.C. Winsemius, J.C.J.H. Aerts, E.C. De Perez, M. Van Aalst, W. Kron, and P.J. Ward. 2015. Declining vulnerability to river floods and the global benefits of adaptation. *Proceedings of the National Academy of Sciences of the United States of America* 112 (18): 201414439.

Jonkman, S.N., J.K. Vrijling, and A.C.W.M. Vrouwenvelder. 2008. Methods for the estimation of loss of life due to floods: A literature review and a proposal for a new method. *Natural Hazards* 46: 353–389.

Kinoshita, Y., M. Tanoue, S. Watanabe, and Y. Hirabayashi. 2018. Quantifying the effect of autonomous adaptation to global river flood projections: Application to future flood risk assessments. *Environmental Research Letters* 13(1): 014006.

Liao, X., W. Xu, J. Zhang, Y. Li, and Y. Tian. 2019. Global exposure to rainstorms and the contribution rates of climate change and population change. *Science of the Total Environment* 663: 644–653.

Lim, W.H., D. Yamazaki, S. Koirala, Y. Hirabayashi, S. Kanae, S. J. Dadson, J.W. Hall, and F.B. Sun. 2018. Long-term changes in global socioeconomic benefits of flood defenses and residual risk based on CMIP5 climate models. *Earth's Future* 6 (7): 938–954.

Shi, P.J., and R. Kasperson. 2015. *World Altas of Natural Disaster Risk*. Springer.

Wing, O.E., P.D. Bates, A.M. Smith, C.C. Sampson, K.A. Johnson, J. Fargione, and P. Morefield. 2018. Estimates of present and future flood risk in the conterminous United States. *Environmental Research Letters* 13(3): 034023.

Winsemius, H.C., B. Jongman, T.I.E. Veldkamp, S. Hallegatte, M. Bangalore, and P.J. Ward. 2018. Disaster risk, climate change, and poverty: Assessing the global exposure of poor people to floods and droughts. *Environment and Development Economics* 23 (3): 328–348.

Mapping Global Risk of GDP Loss to River Floods

Junlin Zhang, Xinli Liao, and Wei Xu

1 Background

Globally, river flooding induced by heavy rainfall frequently causes serious economic losses every year (Jongman et al. 2015; CRED and UNISDR 2018; CRED 2019). Particularly, heavy rainfall will increase in the future with climate warming (Liao et al. 2019). This could lead to greater rain-induced local flooding in some watersheds or regions (IPCC 2012). Besides, exposed assets to floods are increasing with the socioeconomic development (Jongman et al. 2015; Winsemius et al. 2018; Liao et al. 2019).

Generally, river flooding risk assessment has two steps. The first is to simulate river flooding hazard using hydrological or hydrodynamic model and inundation model, and the second is to calculate economic losses by overlaying the economic data and flood hazard maps, and the results are used to assess the economic loss risks from floods (Arnell and Gosling 2016; Lim et al. 2018). Projected future precipitation and social-economic datasets provide a basis for these studies. Vulnerability functions of economic loss have been developed for the risk assessment (Alfieri et al. 2015; Muis et al. 2015; Dottori et al. 2016; Sarhadi et al. 2016; Wing et al. 2018). But most studies are concerned with asset losses and there are few studies on GDP loss risks. In addition, many studies focus on the risk at the country and region levels, lacking grid-level high spatial resolution results.

In order to address the issues, this study assessed the GDP loss risks from floods by two main steps. The first was to develop vulnerability functions of GDP losses for the countries by revising an existing vulnerability function. Then future potential GDP losses of the 2030s and the 2050s were estimated at the 2.5' grid level under the Representative Concentration Pathway (RCP) and Shared Socioeconomic Pathway (SSP) scenarios of RCP4.5-SSP2 and RCP8.5-SSP3. The results were compiled to produce the risk maps at the 0.25° grid level and country level.

2 Method

Figure 1 shows the technical flowchart for mapping the risk of GDP loss to floods of the world. This study revised the existing vulnerability functions by adjustment coefficient that is calculated based on recorded GDP losses. Then future losses were estimated using predicted inundation data and GDP data, and the adjusted vulnerability functions. Future losses at the grid level were then aggregated to other geographic units. Finally, the risk size and model uncertainty are analyzed.

2.1 Estimation of Risks

2.1.1 Estimation of Losses for the Baseline Period

The vulnerability function of asset losses is represented by Eq. (1) (Wing et al. 2018). Using the function, historical asset losses were estimated by Eqs. (2) and (3).

Authors: Junlin Zhang, Xinli Liao, Wei Xu.
Map Designers: Junlin Zhang, Jing'ai Wang, Ying Wang.
Language Editor: Wei Xu.

J. Zhang · X. Liao · W. Xu (✉)
Institute of Disaster Risk Science, Beijing Normal University, Beijing, 100875, China
e-mail: xuwei@bnu.edu.cn

W. Xu
State Key Laboratory of Earth Surface Processes and Resource Ecology, Beijing Normal University, Beijing, 100875, China

P. Shi, *Atlas of Global Change Risk of Population and Economic Systems*, IHDP/Future Earth-Integrated Risk Governance Project Series, https://doi.org/10.1007/978-981-16-6691-9_16

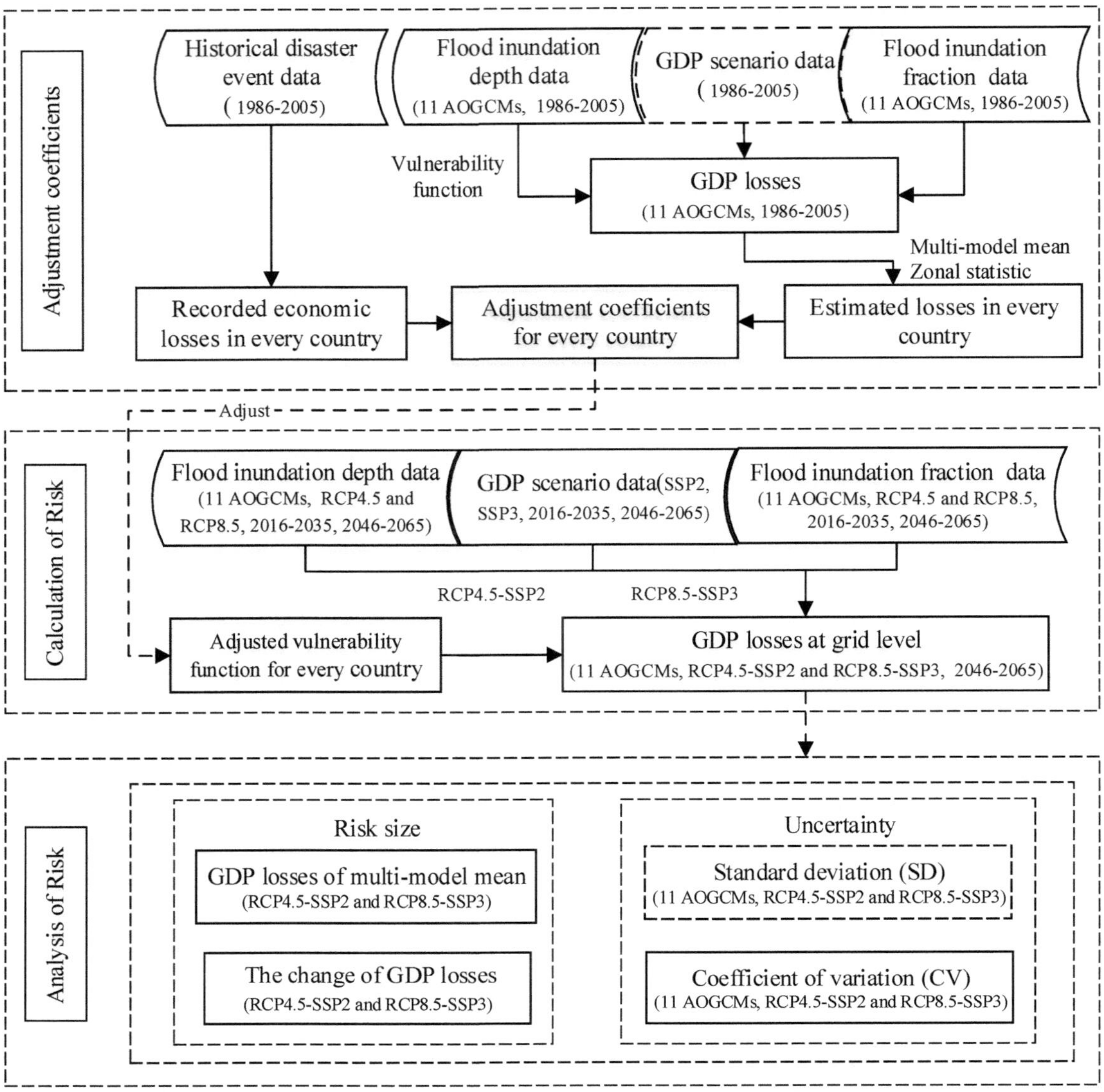

Fig. 1 Technical flowchart for mapping global risk of GDP loss to river floods. AOGCM = Atmosphere–Ocean General Circulation Model; RCP = Representative Concentration Pathway; SSP = Shared Socioeconomic Pathway

$$V(d) = \begin{cases} -0.0067d^4 + 0.0723d^3 - 0.233d^2 + 0.3953d, d < 5 \\ 0.0038d + 0.962, d \geqslant 5 \end{cases} \tag{1}$$

$$L_{his\_i\_j} = V(d_{his\_i\_j}) \times S_{his\_j} \times f_{his\_i\_j} \tag{2}$$

$$L_{his} = \frac{1}{11 \times 20} \sum_{j=1}^{20} \sum_{i=1}^{11} L_{his\_i\_j} \tag{3}$$

where i is the order of the 11 Atmosphere–Ocean General Circulation Models (AOGCMs); j is the sequence of the 20 years; his is the baseline period (1986–2005); $L_{\text{his}\_i\_j}$, $d_{\text{his}\_i\_j}$, and $f_{\text{his}\_i\_j}$ are the estimated losses, water depth, and inundation fraction for the ith AOGCM in the jth year for the baseline period, respectively; S_{his_j} is the GDP data in the jth year of the baseline period; L_{his} is the annual average loss, that is multi-model ensemble for the baseline period; $V(d)$ is the vulnerability function.

2.1.2 Calculation of Adjustment Coefficients

The study revised the vulnerability function of asset loss (Eq. 1) for countries to build the vulnerability function of GDP loss. Using Eq. (4), the adjustment coefficients (K_c values) were calculated for the countries with total recorded losses and estimated losses both greater than zero during the baseline period. On this basis, the adjustment coefficient is the minimum of above calculated K_c values for the countries with total recorded losses equal to zero during the baseline period; and the adjustment coefficient is the average of above calculated K_c values for the countries with total recorded losses greater than zero but total estimated losses equal to zero.

$$K_c = \frac{\sum_{j=1}^{20} SL_{his\_c\_j}}{L_{his\_c}} \quad (4)$$

where K_c is the adjustment coefficient of country c; j represents the sequential number of the 20 years in the baseline period; $SL_{\mathrm{his}\_c\_j}$ is the recorded direct economic losses of country c.

The adjusted vulnerability function is shown in Eq. (5).

$$AdjV_c(d) = K_c \times V(d) \quad (5)$$

where $AdjV_c(d)$ is the adjusted vulnerability function of country c.

2.1.3 Calculation of Future Losses and Change

Future losses of a grid in country c were estimated according to Eq. (6), based on future predicted inundation and GDP data and the adjusted vulnerability function. Next, the study averaged the results of 20 years for all AOGCMs (Eq. 7) to compute uncertainties; then we averaged the results of the 11 AOGCMs as the losses of the 2030s or the 2050s to reduce model uncertainties (Eq. 8). Finally, the changes of losses from the baseline period to the future period were calculated by Eq. (9).

$$L_{fut\_i\_j} = AdjV_c(d_{fut\_i\_j}) \times S_{fut\_j} \times f_{fut\_i\_j} \quad (6)$$

where $L_{\mathrm{fut}\_i\_j}$, $d_{\mathrm{fut}\_i\_j}$, and $f_{\mathrm{fut}\_i\_j}$ are the estimated GDP losses, water depth, and inundation fraction for the ith AOGCM in the jth year for a future period, respectively; S_{fut_j} is the GDP data in the jth year of a future period.

$$L_{fut\_i} = \frac{1}{20}\sum_{j=1}^{20} L_{fut\_i\_j} \quad (7)$$

where L_{fut_i} is the average losses of 20 years (2016–2035 or 2046–2065) for the ith AOGCM.

$$L_{fut} = \frac{1}{11}\sum_{i=1}^{11} L_{fut\_i} \quad (8)$$

where L_{fut} is the losses of the 2030s or the 2050s for the multi-model ensemble, which averaged the results from the 11 AOGCMs.

$$\Delta L = L_{fut} - L_{his} \quad (9)$$

where ΔL is the change of losses for the 2030s or the 2050s relative to the baseline period.

2.2 Model Uncertainty

The result uncertainty of multi-models is measured by standard deviation (Eq. 10) and coefficient of variation CV, the ratio of SD to L in Eq. 10.

$$SD = \sqrt{\frac{\sum_{i=1}^{11} (L_i - L)^2}{11}} \quad (10)$$

where L_i is the average losses of 20 years for the ith AOGCM (L_{his_i} or L_{fut_i}). L is the average losses of 20 years for the multi-model ensemble (L_{his} or L_{fut}).

3 Results

Globally, the annual average GDP losses of the 2030s are approximately USD 223 billion for the RCP4.5-SSP2 scenario and USD 199 billion for the RCP8.5-SSP3 scenario; they increase 4.82 and 4.19 times relative to the baseline period, respectively. The annual average GDP losses of the 2050s are approximately USD 447 billion for the RCP4.5-SSP2 scenario and USD 429 billion for the RCP8.5-SSP3 scenario; they increase 10.69 and 10.21 times relative to the baseline period. The patterns of spatial distribution are similar for different scenarios. High-risk areas are mainly located in East Asia, South Asia, and Southeast Asia. The risks of GDP loss are the highest in the eastern coastal areas of China, the vicinity of the Ganges River Basin, and the coastal areas of the Indo-China Peninsula. Overall, the risk of the 2050s is higher than that of the 2030s. For Africa, the risks are higher in the western, southern, and eastern Africa. For South America, the risks are higher in Liano Orinoco Plain and La Plata Plain. The high-risk areas in Europe are spread out. The high-risk areas in North America are distributed in Atlantic Coastal Plain, along Saint Lawrence River, and in the southern coastal area.

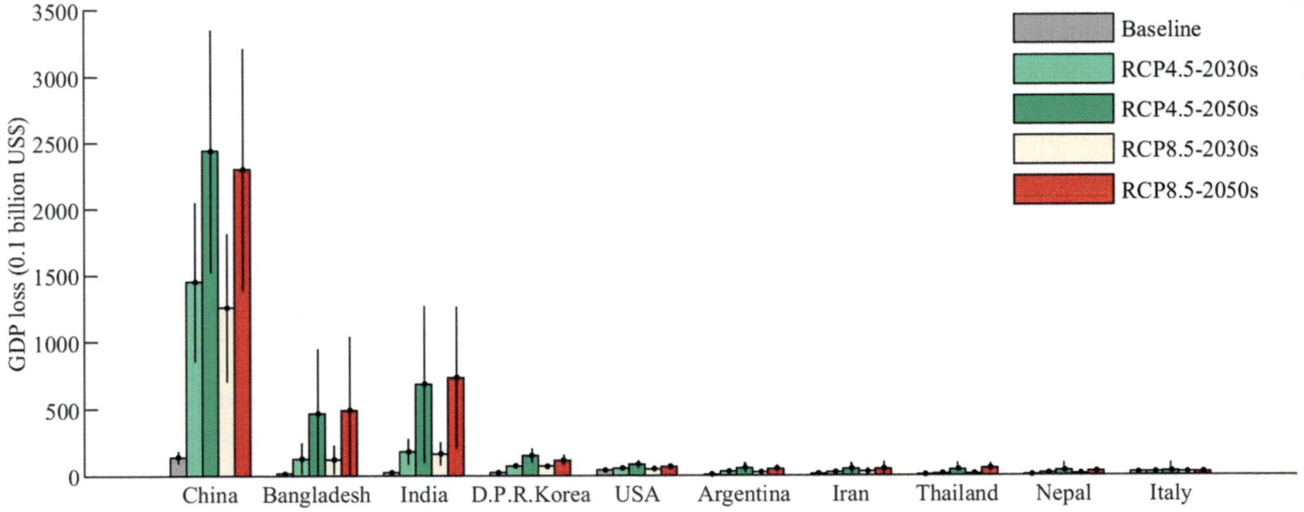

Fig. 2 Annual average GDP losses of the top 10 high-risk countries (in descending order by GDP loss). The error bar represents the standard deviations across 11 Atmosphere–Ocean General Circulation Models (AOGCMs)

The high-risk areas in Australia are mainly in the southeast coastal area.

By zonal statistics of the GDP loss results, we calculated the annual average GDP losses at the national level. Figure 2 shows the annual average GDP losses and errors (measured by standard deviation) of the top ten high-risk countries. The GDP loss risks are the highest for China, higher than USD 100 billion for the 2030s and the 2050s, increasing more than eight times relative to the baseline period. The average annual GDP losses of the remaining countries are all lower than USD 100 billion. The changes are greater for India and Bangladesh relative to the baseline period. In most countries, the risks are higher in the 2050s than in the 2030s.

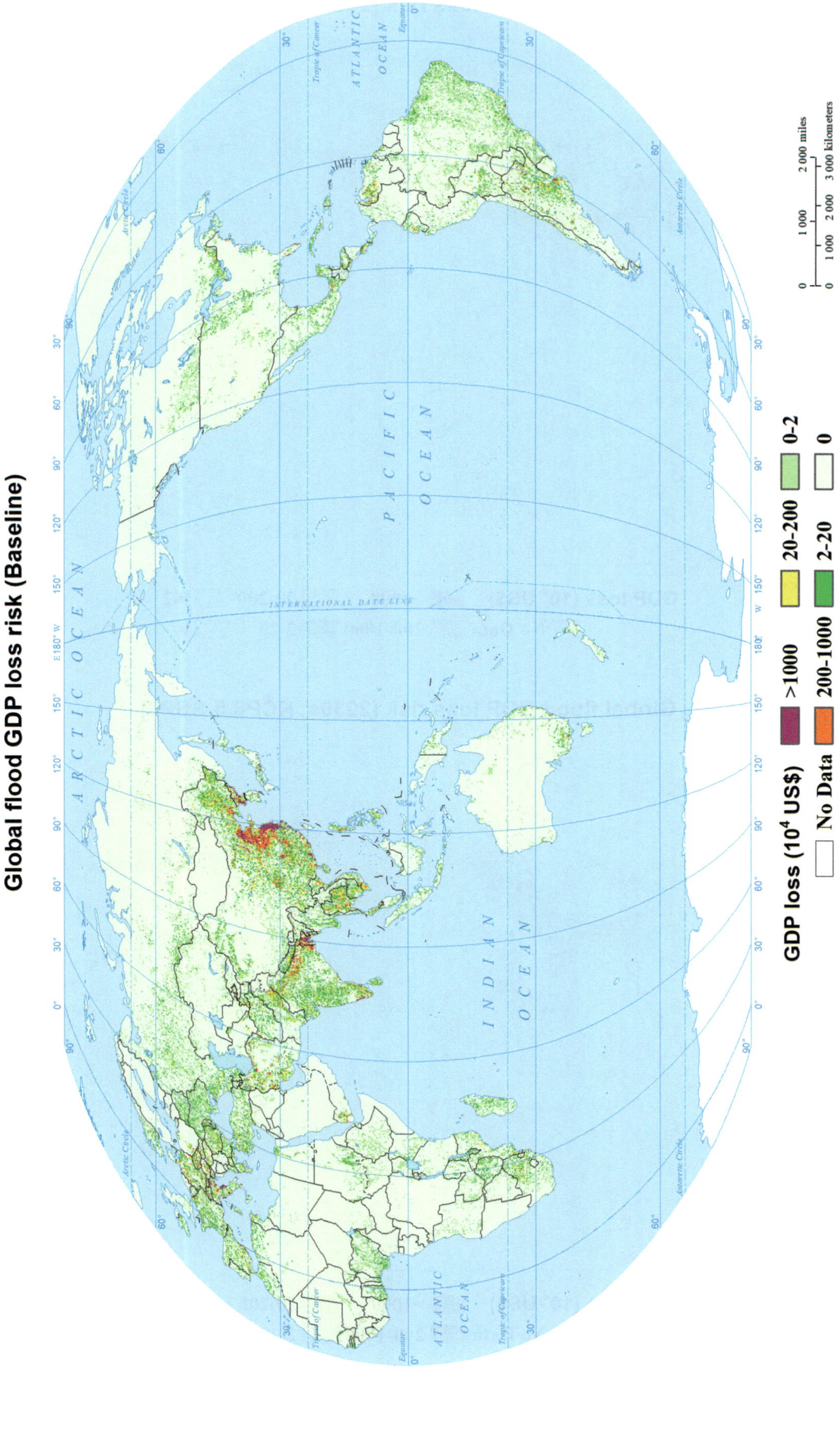
Global flood GDP loss risk (Baseline)
GDP loss (10^4 US$)
No Data
>1000
200-1000
20-200
2-20
0-2
0
ARCTIC OCEAN
PACIFIC OCEAN
INDIAN OCEAN
ATLANTIC OCEAN
INTERNATIONAL DATE LINE
Arctic Circle
Tropic of Cancer
Equator
Tropic of Capricorn
Antarctic Circle
0 1 000 2 000 miles
0 1 000 2 000 3 000 kilometers

Global flood GDP loss risk (2030s, RCP4.5-SSP2)

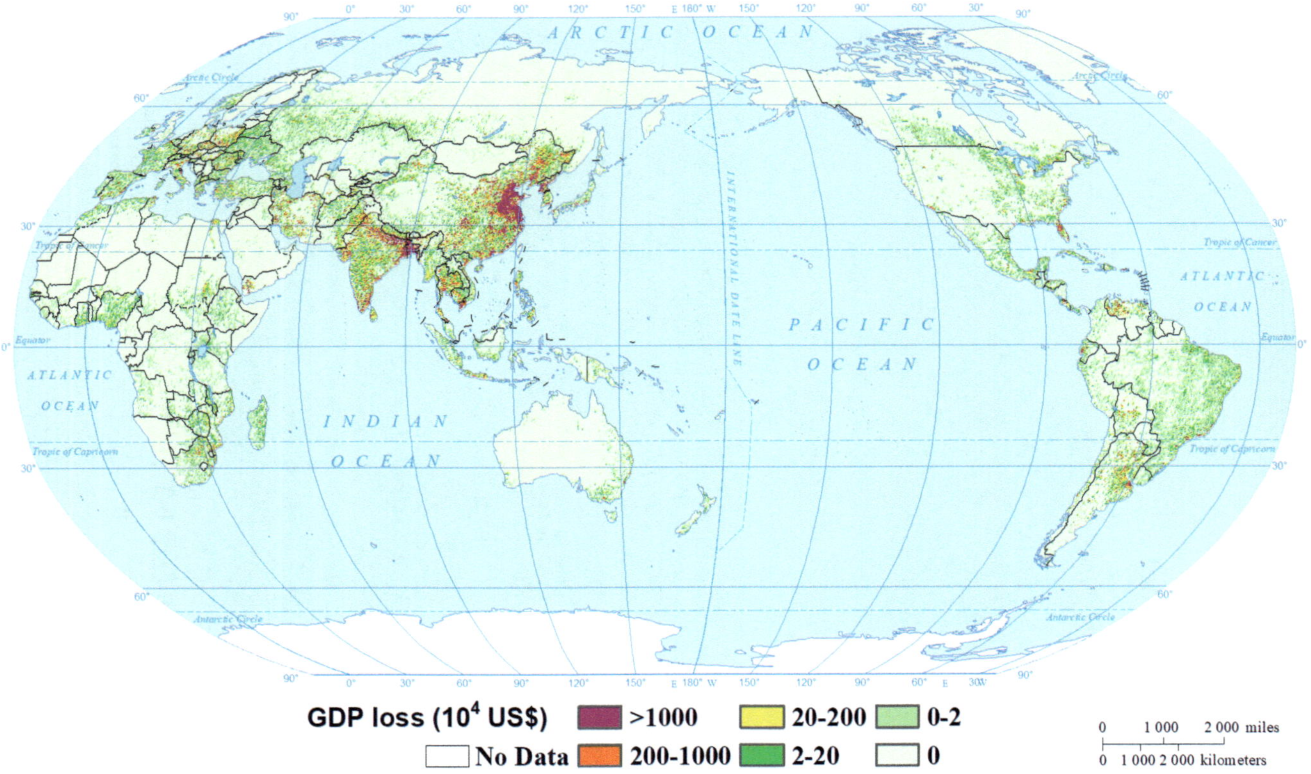

Global flood GDP loss risk (2030s, RCP8.5-SSP3)

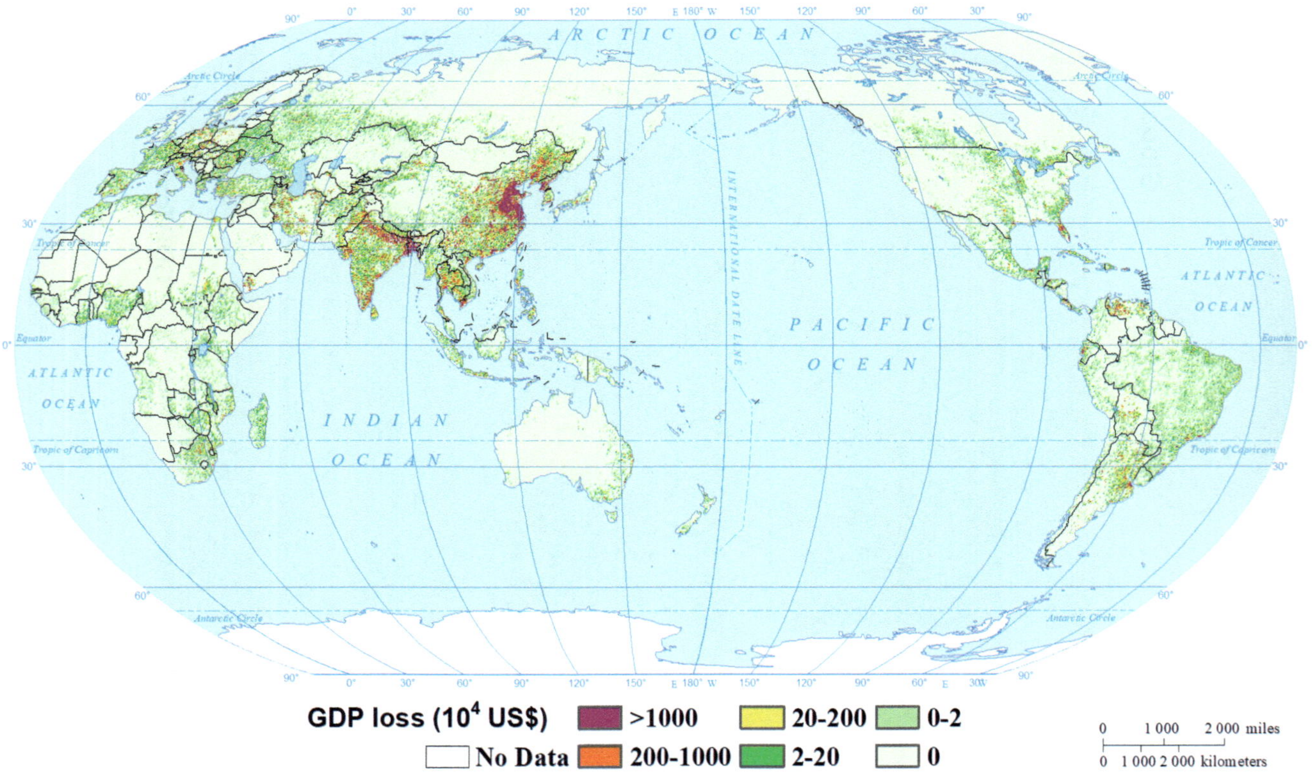

Global flood GDP loss risk (2050s, RCP4.5-SSP2)

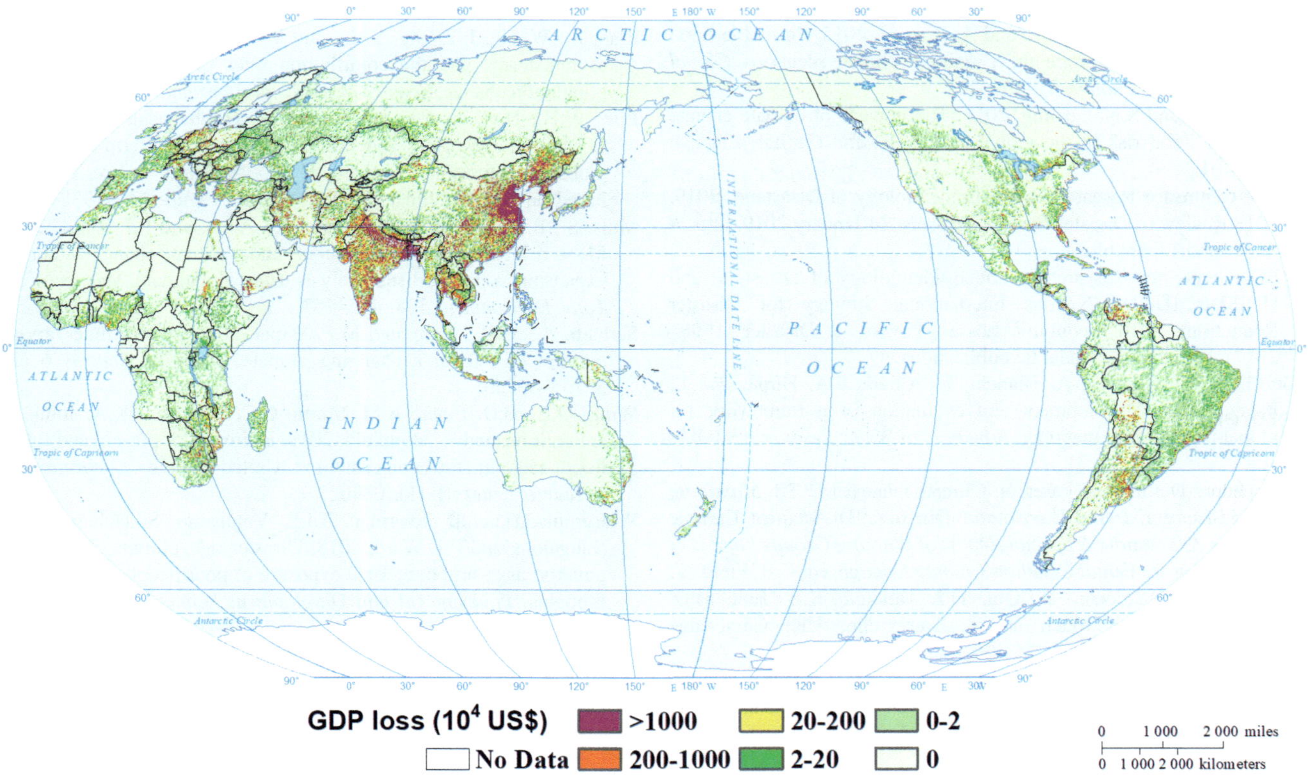

Global flood GDP loss risk (2050s, RCP8.5-SSP3)

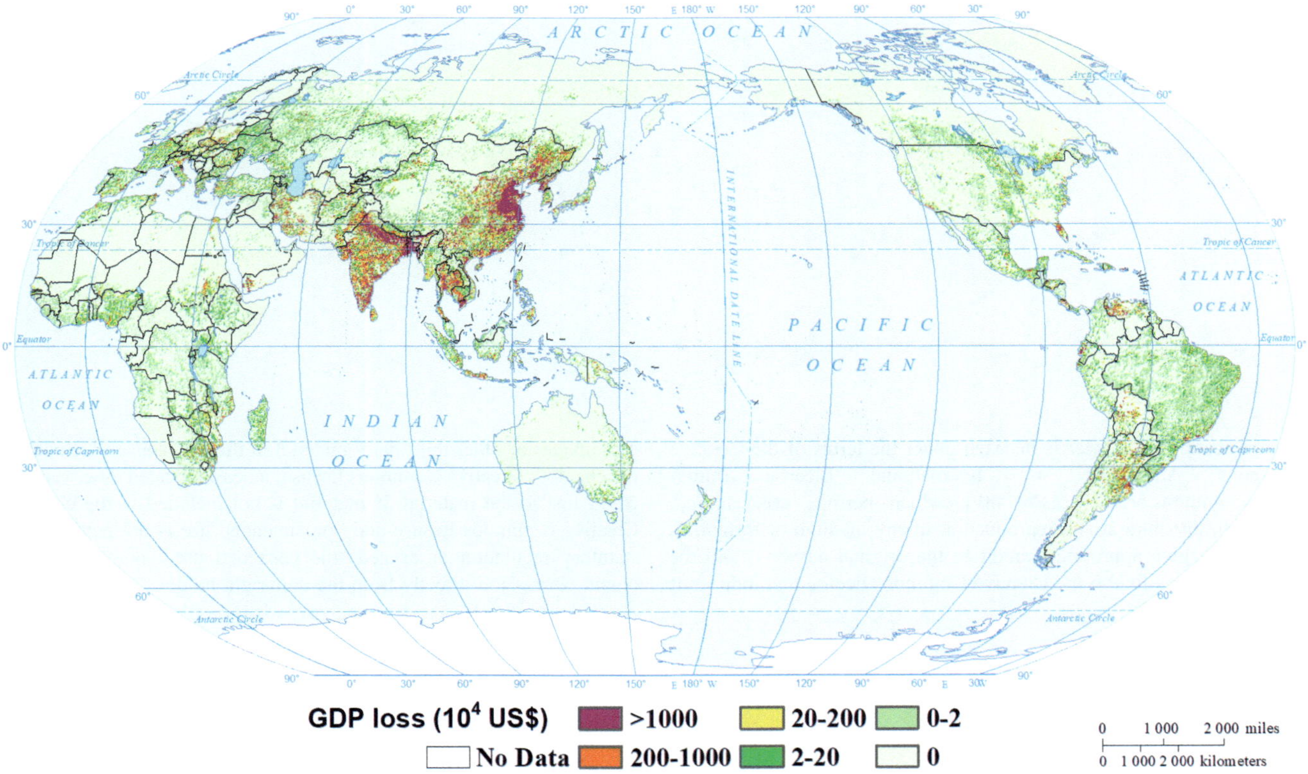

References

Alfieri, L., L. Feyen, F. Dottori, and A. Bianchi. 2015. Ensemble flood risk assessment in Europe under high end climate scenarios. *Global Environmental Change* 35: 199–212.

Arnell, N.W., and S.N. Gosling. 2016. The impacts of climate change on river flood risk at the global scale. *Climatic Change* 134 (3): 387–401.

CRED (Centre for Research on the Epidemiology of Disasters). 2019. 2018 Review of Disaster Events. Brussels, 24 January 2019. https://www.emdat.be/publications.

CRED (Centre for Research on the Epidemiology of Disasters) and UNISDR (United Nations International Strategy for Disaster Reduction). 2018. Economic Losses, Poverty & Disaster (1998–2017). https://www.emdat.be/publications.

Dottori, F., P. Salamon, A. Bianchi, L. Alfieri, F.A. Hirpa, and L. Feyen. 2016. Development and evaluation of a framework for global flood hazard mapping. *Advances in Water Resources* 94: 87–102.

IPCC (Intergovernmental Panel on Climate Change). 2012. *Managing the Risks of Extreme Events and Disasters to Advance Climate Change Adaptation. A Special Report of Working Groups I and II of the Intergovernmental Panel on Climate Change*, ed. C.B. Field, V. Barros, T.F. Stocker, D. Qin, D.J. Dokken, K.L. Ebi, M.D. Mastrandrea, K.J. Mach, et al. Cambridge, UK: Cambridge University Press.

Jongman, B., H.C. Winsemius, J.C.J.H. Aerts, E.C. De Perez, M. Van Aalst, W. Kron, and P.J. Ward. 2015. Declining vulnerability to river floods and the global benefits of adaptation. *Proceedings of the National Academy of Sciences of the United States of America* 112 (18): 201414439.

Liao, X., W. Xu, J. Zhang, Y. Li, and Y. Tian. 2019. Global exposure to rainstorms and the contribution rates of climate change and population change. *Science of the Total Environment* 663: 644–653.

Lim, W.H., D. Yamazaki, S. Koirala, Y. Hirabayashi, S. Kanae, S. J. Dadson, J.W. Hall, and F.B. Sun. 2018. Long-term changes in global socioeconomic benefits of flood defenses and residual risk based on CMIP5 climate models. *Earth's Future* 6 (7): 938–954.

Muis, S., B. Guneralp, B. Jongman, J.C. Aerts, and P.J. Ward. 2015. Flood risk and adaptation strategies under climate change and urban expansion: a probabilistic analysis using global data. *Science of the Total Environment* 538: 445–457.

Sarhadi, A., M.C. Ausin, and M.P. Wiper. 2016. A new time-varying concept of risk in a changing climate. *Scientific Reports* 6 (1): 35755–35755.

Wing, O.E., P.D. Bates, A.M. Smith, C.C. Sampson, K.A. Johnson, J. Fargione, and P. Morefield. 2018. Estimates of present and future flood risk in the conterminous United States. *Environmental Research Letters* 13(3): 034023.

Winsemius, H.C., B. Jongman, T.I.E. Veldkamp, S. Hallegatte, M. Bangalore, and P.J. Ward. 2018. Disaster risk, climate change, and poverty: assessing the global exposure of poor people to floods and droughts. *Environment and Development Economics* 23 (3): 328–348.

Mapping Global Risk of Crop Yield Under Climate Change

Weihang Liu, Shuo Chen, Qingyang Mu, Tao Ye, and Peijun Shi

1 Introduction

Risk of crop yield under climate change refers to the potential changes in crop yield (mean yield, interannual yield variability, and lower extreme yield) caused by climate change. Increases in the interannual variability of yield and the likelihood of lower yield extremes can affect the livelihoods of farmers (Morton 2007), increase pressure on inter-temporal food reserves (Bobenrieth et al. 2013), induce large price changes in the global market or even destabilize regions of the world (Sternberg 2011). The risk of crop yield is one of the primary drivers of food system instability (IPCC 2017), and its response to climate change is critical to understanding the impact of climate change on food security (FAO 2019). Previous studies have mostly focused on the variation in the mean yield in response to climate change (Ray et al. 2015). Climate change impacts on crop yield risk, however, remains a key research gap (IPCC 2017).

Recent studies have provided evidence for changes in the yield risk of major cereal crops and identified significant global impacts of climate change, either at the country level or the 0.5° grid level (Osborne and Wheeler 2013; Iizumi and Ramankutty 2016). These studies have been followed up by regional county-level analysis of the interannual yield variability of maize. Most recently, efforts have been concerned with predicting the impact of future climate change on the interannual yield variability, focusing on wheat and maize at both the global and regional scales, using site-based process-based crop models (Liu et al. 2019) and statistical models (Urban et al. 2012; Ben-Ari et al. 2018; Tigchelaar et al. 2018).

This study evaluated the changes in the global staple crop yield risk from three aspects—mean yield, interannual yield variability, and lower extreme yield—under the RCP2.6-2030s, RCP2.6-2050s, RCP4.5-2030s, RCP4.5-2050s, RCP8.5-2030s, and RCP8.5-2050s scenarios. The crop yield risk was evaluated by using multi-model ensemble (MME) simulation with global high spatial resolution (0.25°) climate forcing data. Emulators of global gridded crop models (GGCM) were developed to ensure efficient prediction (Lobell and Burke 2010; Holzkämper et al. 2012; Oyebamiji et al. 2015; Raimondo et al. 2020). The present results may provide crucial information for climate risk assessment and effective adaptations.

Authors: Weihang Liu, Shuo Chen, Qingyang Mu, Tao Ye, Peijun Shi.
Map Designers:
Wheat: Weihang Liu, Jing'ai Wang, Ying Wang.
Maize: Qingyang Mu, Jing'ai Wang, Ying Wang.
Rice: Shuo Chen, Jing'ai Wang, Ying Wang.
Language Editor: Tao Ye.

W. Liu · S. Chen · Q. Mu · T. Ye (✉) · P. Shi
Faculty of Geographical Science, Beijing Normal University, Beijing, 100875, China
e-mail: yetao@bnu.edu.cn

T. Ye · P. Shi
State Key Laboratory of Earth Surface Processes and Resource Ecology, Beijing Normal University, Beijing, 100875, China

P. Shi, *Atlas of Global Change Risk of Population and Economic Systems*, IHDP/Future Earth-Integrated Risk Governance Project Series, https://doi.org/10.1007/978-981-16-6691-9_17

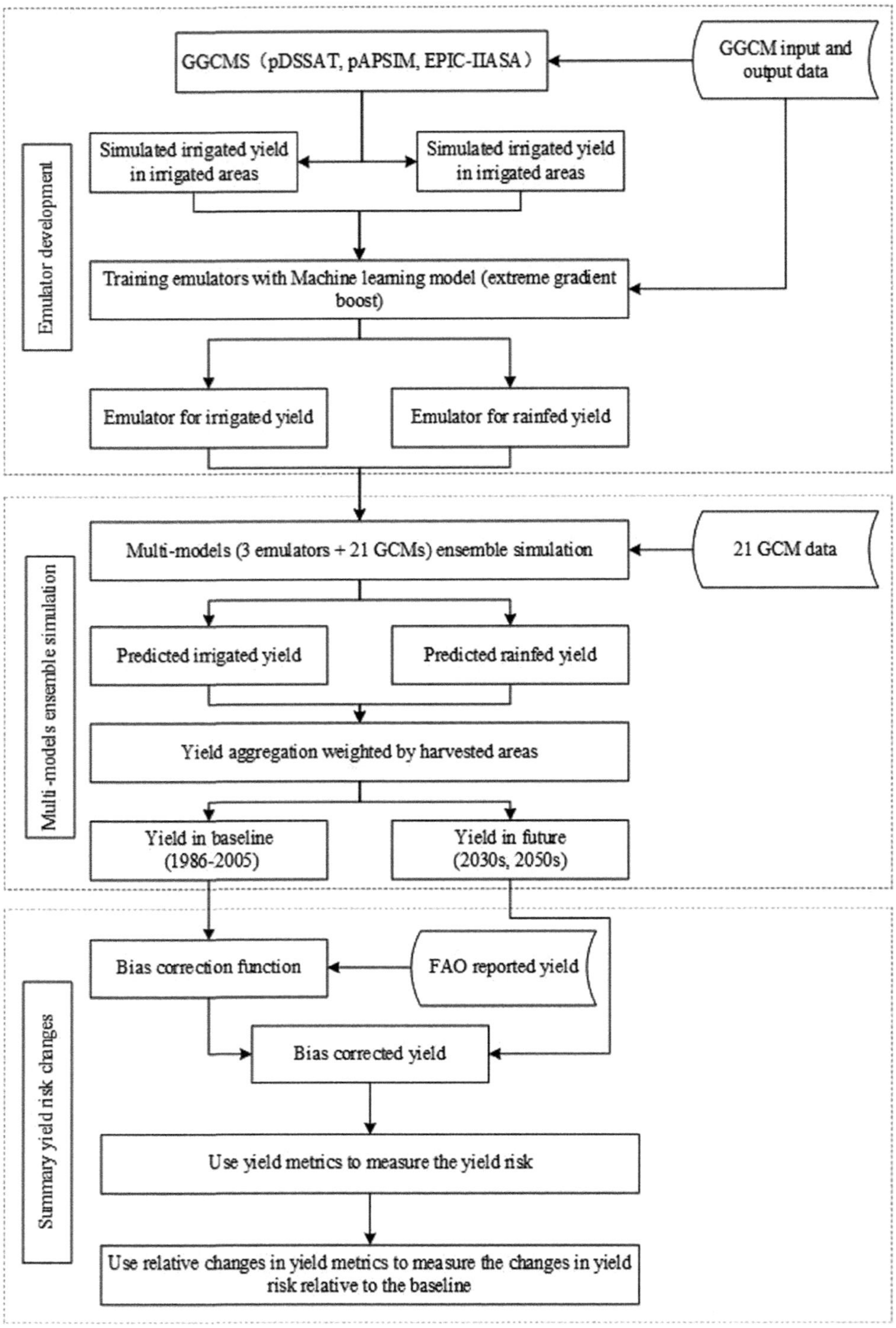

Fig. 1 Technical flowchart for mapping crop yield risk of the world under climate change, taking wheat as an example. GGCM = Global Gridded Crop Model; MME = Multi-model ensemble; GCM = General Circulation Model

2 Method

The method for detecting changes in the wheat yield variability for future climate scenarios includes the following steps (Fig. 1): (1) Develop emulators for the process-based models in the GGCMs; (2) Conduct MME prediction of the global wheat yield at fine spatial resolution; (3) Correct the bias in the predicted global yield using country-level reported yield to best capture the actual interannual yield variability; (4) Summarize yield risk change. Figure 1 shows the technical flowchart for mapping the risk of crop yield of the world under climate change.

2.1 Development of the Global Gridded Crop Model (GGCM) Emulators

Emulator training via extreme gradient boosting

The development of the emulators consists of training machine learning models on specific GGCM input and output datasets so that the models may replicate the complex process of yield simulation within the crop model. Variables that have been frequently reported to significantly influence wheat yield were prepared as the predicting variables (Folberth et al. 2019), and they cover climate, soil type, length of vegetation growth period, and management practices. All the data for training were computed/adapted from the GGCMs' input and output datasets. For climate data, the monthly variables and the growing season variables were considered. Soil properties, the length of vegetation growth period, and management practices were site-specific variables.

In total, six emulators were trained for the three GGCMs (pDSSAT, pAPSIM, and EPIC-IIASA) each with two cultivation possibilities (rainfed and irrigation). An extreme gradient boosting (XGB) algorithm was used due to its better performance in terms of goodness-of-fit, cross-validation errors, and computation efficiency than the random forest algorithm (Folberth et al. 2019). The predicting variables and the simulated yield in the GGCMs were randomly split into training and validation sets that contained 75% and 25% of the samples (Yue et al. 2019), respectively.

Predicting global crop yield by MME simulation

The MME yield prediction was performed at the global level for the 0.25° grids for the years 1986–2005 (baseline period), 2016–2035 (the 2030s), and 2046–2065 (the 2050s) under the RCP2.6, RCP4.5, and RCP8.5 scenarios. The MME approach has been proven to be a reliable method in reproducing the main effects anticipated for climate change when simulations are compared with observation (Asseng et al. 2015; Frieler et al. 2017). The large climate model-crop model emulator setup in our framework enabled a robust MME estimate as well as analysis of spatial heterogeneity and inter-model uncertainty (Martre et al. 2015). There were 432 treatments (13 or 21 general circulation models (GCMs) × 3 emulators for the baseline period, RCP2.6-2030s, RCP2.6-2050s, RCP4.5-2030s, RCP4.5-2050s, RCP8.5-2030s, and RCP8.5-2050s) each simulated for 20-year periods. For future yield predictions, planting dates, soil properties, and management practices were assumed to remain constant through time. The MME median across the 3 emulators and the 13 or 21 GCMs was taken as the final estimate.

Bias correction of the predicted yield in the baseline period and in future applications

The emulator-predicted global wheat yields were corrected for bias using the national yield reports of the FAO. Rainfed and irrigated yield were first aggregated to the grid and national levels using an area-weighted average (Müller et al. 2017).

2.2 Measuring the Yield Risk

To fully describe the yield risk, the conventional approach was to derive the probability distribution or the cumulative distribution of the yield (Coble et al. 2010; Ye et al. 2015). In this study, we computed the mean yield (multi-year average), yield interannual variability (multi-year standard deviation), and the lower yield extremes (multi-year 10th percentile) across different 20-year periods. To account for the uncertainty associated with the projected input climate data, the computation was conducted for simulated yield series derived independently with each GCM and emulator, and the results of the MME median were reported.

3 Results

3.1 Wheat Yield Risk

Globally, countries with higher mean wheat yield are mainly located in Europe, while those with lower yield are mainly located in East Africa, the Middle East, and the northern United States. The change of yield risk shows a complex connection with warming. For higher emissions and warming scenarios, although countries at higher latitudes would be more likely to experience higher interannual variability of yield, their mean yield and lower extreme yield tend to increase, compared with the baseline period. Countries at low latitudes, e.g., India, would suffer more, that is, its lower extreme yield will be much worse in the future while its yield interannual variability slightly declines.

From the perspective of mean wheat yield, among the top 10 wheat-producing countries in the world (Fig. 2), the mean yields of France and Germany stand out, exceeding 6 t/ha. The mean yields of the other wheat-producing countries are lower than 3 t/ha. There are two patterns of variation of the mean yields under warming conditions. On the one hand, the rising temperature will bring a favorable effect on mean wheat yield. For example, the mean yields in China, the United States, Russia, France, Canada, Germany, and Turkey will gradually climb up with increasing warming. On the

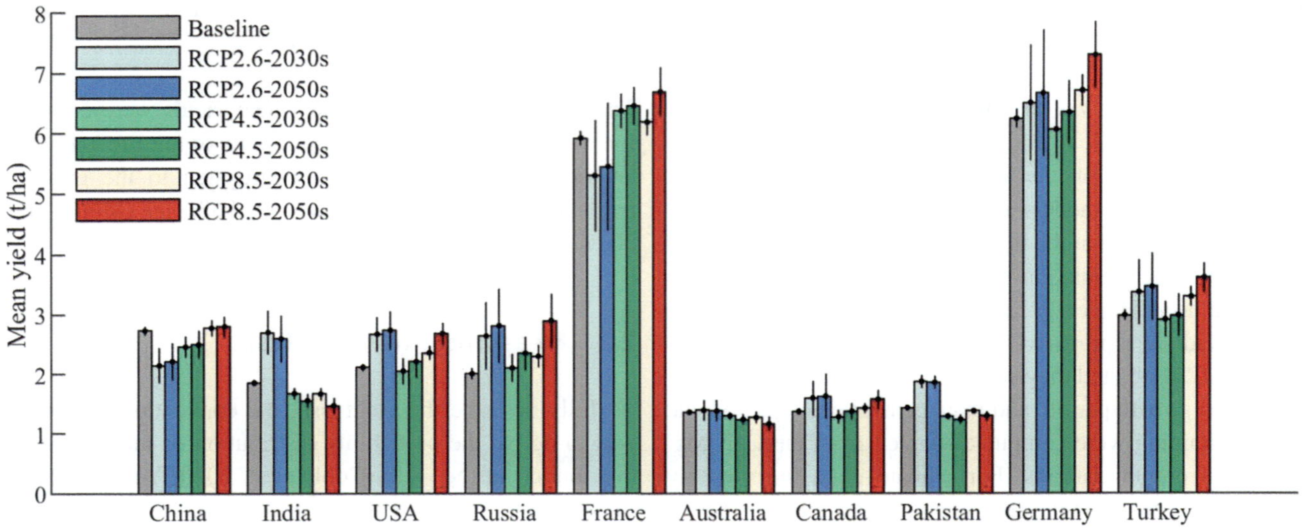

Fig. 2 Mean wheat yield of the top 10 major wheat-producing countries (in descending order by total production). The error bar represents the one standard deviation across the 21 general circulation models (GCMs) and three emulators

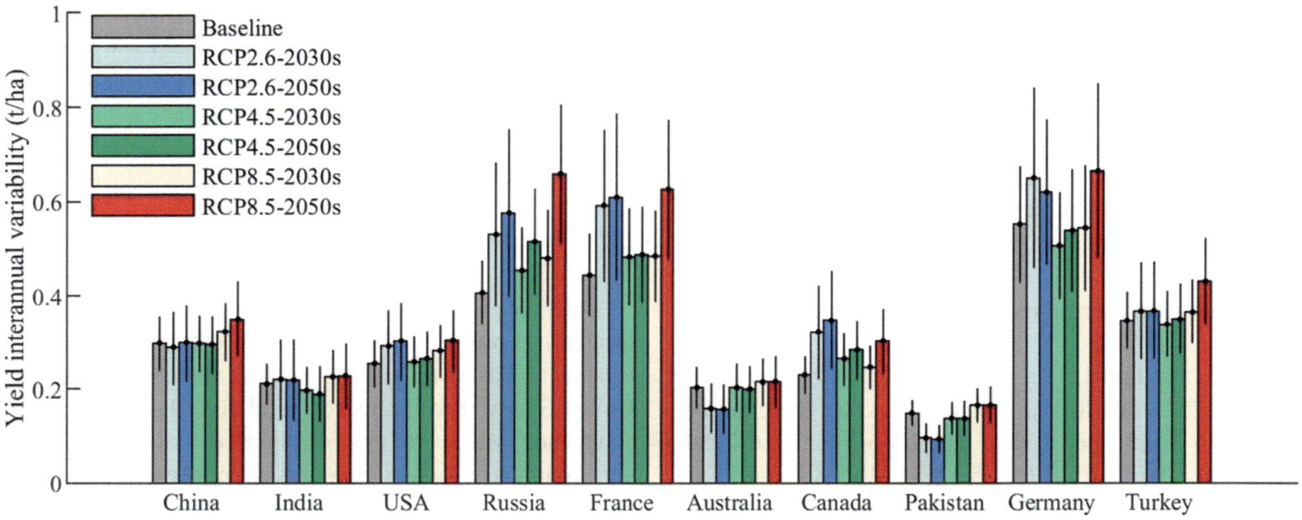

Fig. 3 Wheat yield interannual variability of the top 10 major wheat-producing countries (in descending order by total production). The error bar represents the one standard deviation across the 21 general circulation models (GCMs) and three emulators

other hand, warming will impose an adverse influence on the other wheat-producing countries such as India, Australia, and Pakistan, whose mean yields will fall during the future periods.

From the perspective of wheat yield interannual variability, among the top ten wheat-producing countries in the world (Fig. 3), Russia, France, and Germany have higher wheat yield interannual variability. China, India, and the United States—the top three countries in wheat production—are at the middle level, while Pakistan has the lowest interannual variability. With increasing warming (e.g., higher emissions scenario or further into the future), the interannual variability will rise in all countries, but the degree will vary. In China, the interannual variability of yield will grow insignificantly under the medium emissions scenario (RCP4.5) but will increase slightly (less than 0.4 t/ha) under the high emissions scenario (RCP8.5). The interannual variability in France is flat under the three scenarios of RCP4.5-2030s, RCP4.5-2050s, and RCP8.5-2030s. However, when the temperature rises to a considerable degree (RCP8.5-2050s), the interannual variability will increase sharply.

From the perspective of lower extreme wheat yield, among the top ten wheat-producing countries in the world (Fig. 4), the countries with the highest lower extreme wheat yield are France and Germany, where the yields are as high

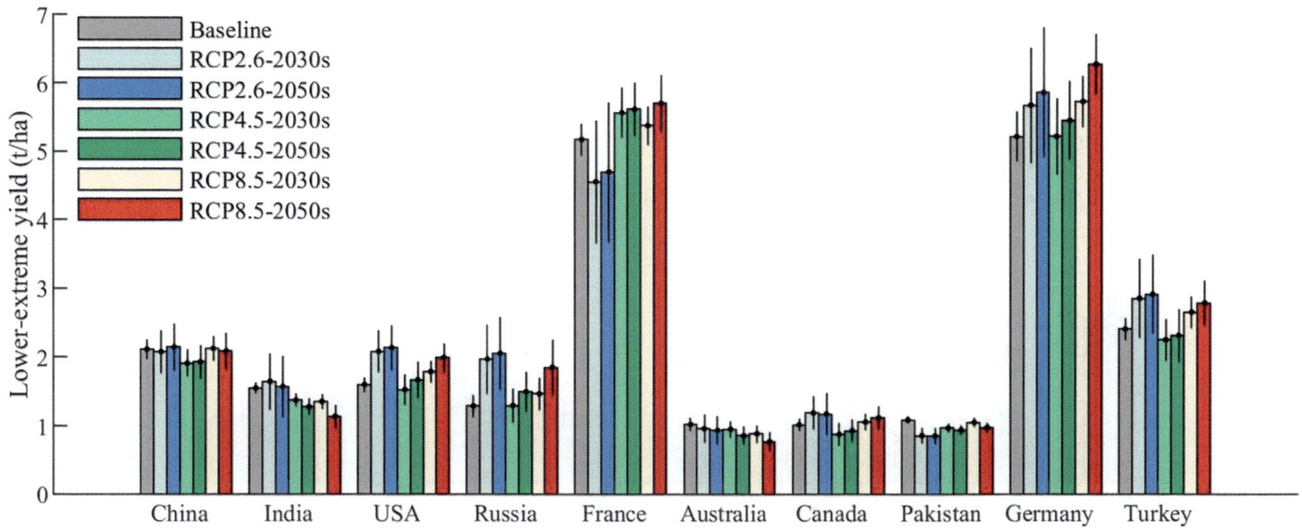

Fig. 4 Lower extreme wheat yield of the top 10 major wheat-producing countries (in descending order by total production). The error bar represents the one standard deviation across the 21 general circulation models (GCMs) and three emulators

as 4–6 t/ha, while the countries with the lowest wheat yield are Australia, Canada, and Pakistan, where the yields can be as low as 1 t/ha. The relationship between changes in lower extreme wheat yields and warming is not uniform across countries: with the increase of warming, most countries show an overall upward trend, but India and Australia show a downward trend. For example, a warmer world will cause lower extreme wheat yield in India to become even lower. However, it will have a favorable impact in high latitude areas, especially in the United States, where the lower extreme wheat yield will increase steadily under the four scenarios of RCP4.5-2030s, RCP4.5-2050s, RCP8.5-2030s, and RCP8.5-2050s.

3.2 Maize Yield Risk

Globally, countries with higher mean maize yield are mainly located in Europe, northern China, and the United States while those with lower yield are mainly located in central Africa and southern India. Climate change will have a two-sided effect on maize production in most of the regions. Warming will reduce the maize yield interannual variability risk (lower variability) but increase the lower extreme wheat yield risk (lower yield) in the regions between south of the Sahara and north of the equator in Africa and northern India. Similarly, it will increase the variability risk (higher variability) but reduce the extreme yield risk (higher yield) in

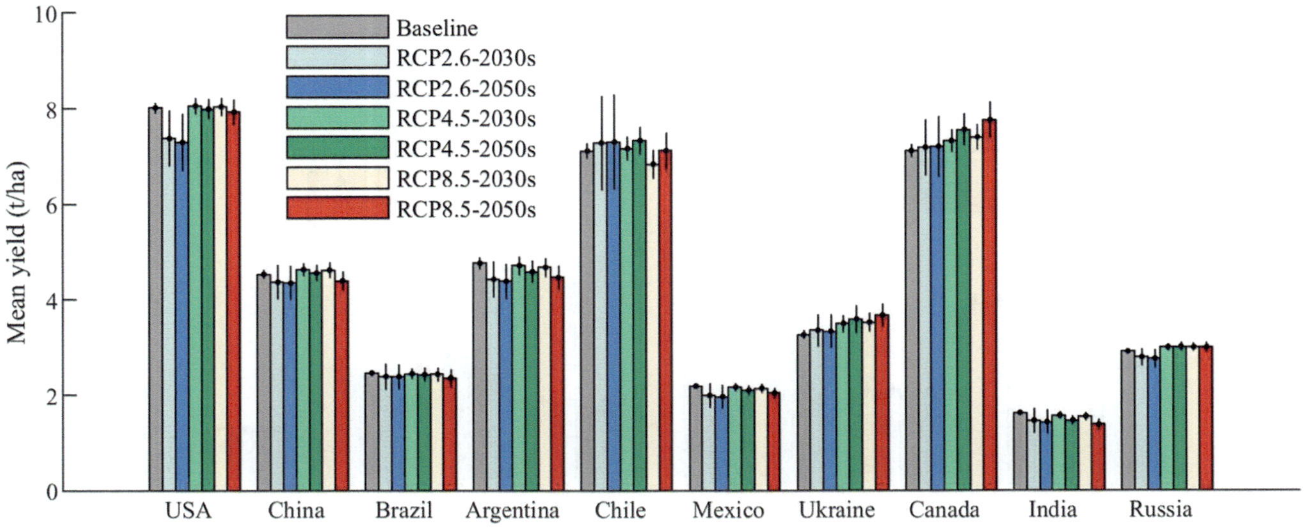

Fig. 5 Mean maize yield of the top 10 major maize-producing countries (in descending order by total production). The error bar represents the one standard deviation across the 21 general circulation models (GCMs) and three emulators

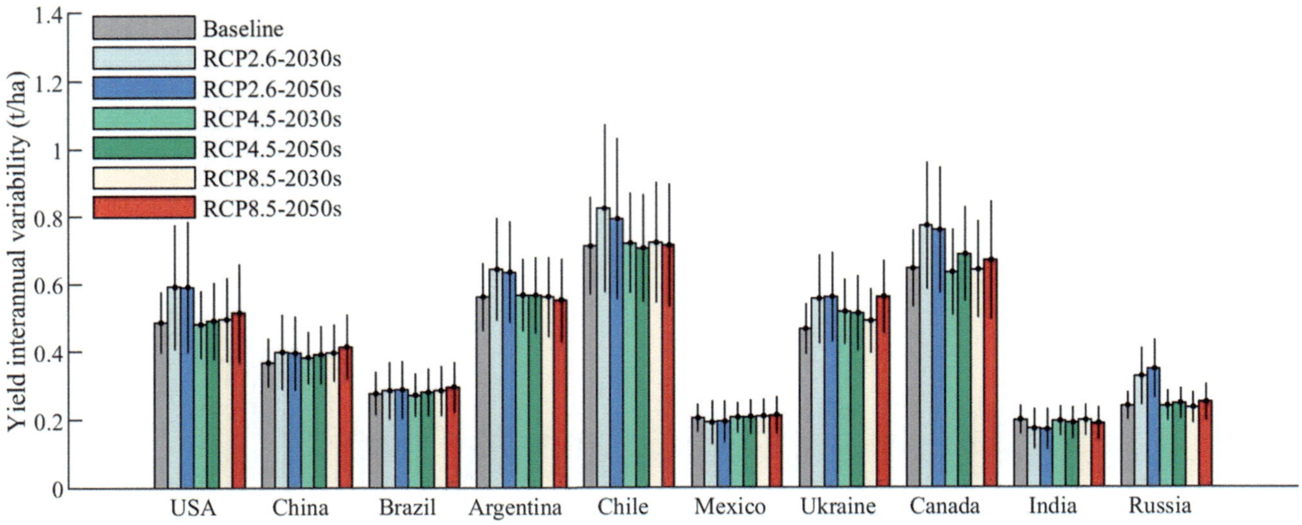

Fig. 6 Maize yield interannual variability of the top 10 major maize-producing countries (in descending order by total production). The error bar represents the one standard deviation across the 21 general circulation models (GCMs) and three emulators

Europe and southeast Africa. On the contrary, the North China Plain and the central United States will be surely harmed from warming, while the Great Lakes region of the United States and central South America will benefit from it.

From the perspective of mean maize yield, among the top ten maize-producing countries in the world (Fig. 5), the United States, Chile, and Canada have the highest mean maize yields, with values higher than 7 t/ha. The mean maize yields of China and Argentina are moderate (around 4.5 t/ha), while that of Brazil, Mexico, Ukraine, and India are lower, ranging from 1.5 t/ha to 3.5 t/ha. Few maize-producing countries will benefit from warming excluding Ukraine and Canada, whose mean maize yield will increase during the future periods. The mean maize yield of the United States, China, Brazil, Chile, and Russia in the future periods tend to maintain a relatively consistent level compared to that in the baseline period. In comparison, the mean maize yields of Argentina, Mexico, and India will decline slightly in the future.

From the perspective of maize yield interannual variability, among the top ten maize-producing countries in the world (Fig. 6), in Chile and Canada, the interannual variability is higher than 0.6 t/ha. Conversely, in Mexico, India, and Russia, it is low and about 0.2 t/ha. With the increase of warming, the interannual variability of the United States, China, Brazil, and Mexico will show a steady increasing

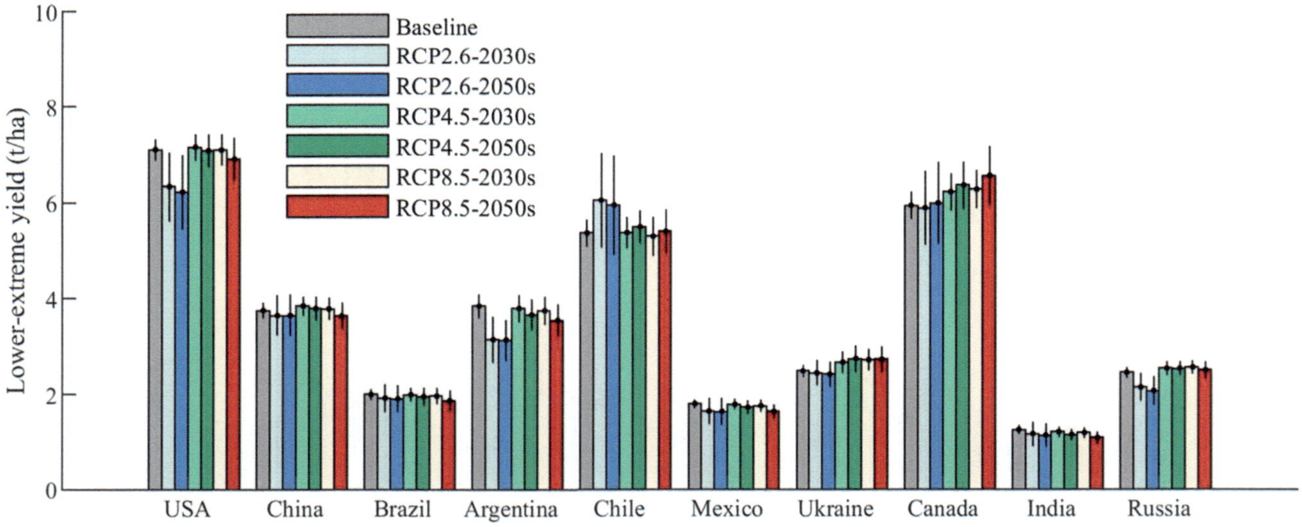

Fig. 7 Lower extreme maize yield of the top 10 major maize-producing countries (in descending order by total production). The error bar represents the one standard deviation across the 21 general circulation models (GCMs) and three emulators

trend—for example, under the RCP4.5 and RCP8.5 scenarios, the interannual variability will become higher over time in Canada and Russia. However, the interannual variability under the RCP4.5-2050s scenario will be slightly higher than that under the RCP8.5-2030s scenario. In contrast, the interannual variability in Argentina, Chile, and India will show a descending trend in general.

From the perspective of lower extreme maize yield, among the top ten maize-producing countries in the world (Fig. 7), lower extreme yields are higher in the United States, Chile, and Canada, with values greater than 5 t/ha during all the periods and under all the scenarios. There are large differences in the relationship between the changes of warming and lower extreme maize yield among the top ten maize-producing countries. The lower extreme maize yield in Canada tends to grow as temperature rise, but in most of the countries (such as the United States, China, Brazil, Chile, Ukraine, and Russia), it will fall after rising. The rising lower extreme maize yield in short term may be a result of the increasing precipitation. However, the lower extreme maize yields in Argentina, Mexico, and India will be generally on the decline.

3.3 Rice Yield Risk

Globally, countries with higher mean rice yield are mainly located in the Mediterranean region, China, the southeastern United States, and Peru while those with lower yield are mainly located in Africa. There is uncertainty about the risks in some areas, such as central Africa and coastal Asia, because they will have increasing yield interannual variability but higher lower extreme rice yield. However, in general, climate change will reduce the rice yield risk at mid-latitude regions, especially Ukraine, northeastern China, and southeastern South America, where yield interannual variability will reduce along with the increasing lower extreme rice yield. Conversely, most of the regions in the low-latitude area, namely those in eastern Africa, southern Asia, and northern South America, would experience higher risk in terms of yield interannual variability and lower extreme yield.

From the perspective of mean rice yield, among the top ten rice-producing countries in the world (Fig. 8), China, the United States, and Spain have the highest mean rice yields (more than 6 t/ha) and Peru and Iran have lower but still considerable mean yields (around 5.5 t/ha). Indonesia and Ukraine have the mean rice yields of 4 t/ha, while Brazil, India, and Nigeria have the lowest ones (less than 2.5 t/ha). Warming has little effect on mean rice yields, especially for Brazil, India, Peru, Spain, Nigeria, and Iran, where yields will remain largely unchanged under the RCP4.5-2030s, RCP4.5-2050s, and RCP8.5-2030s scenarios, although they will decrease under the RCP8.5-2050s scenario compared with that in the baseline period. However, the mean rice yields of China and Ukraine will increase slightly while those of Indonesia and the United States will decrease marginally.

From the perspective of rice yield interannual variability, among the top ten rice-producing countries in the world (Fig. 9), Ukraine has the greatest interannual variability (>0.8 t/ha), followed by Indonesia (>0.6 t/ha). However, as the top three rice-producing countries, China, Brazil, and India have relatively little interannual variability. In several major rice-producing countries, the increase of interannual variability of rice yield has a weak positive relationship with the increase of warming. In China, Brazil, India, Indonesia,

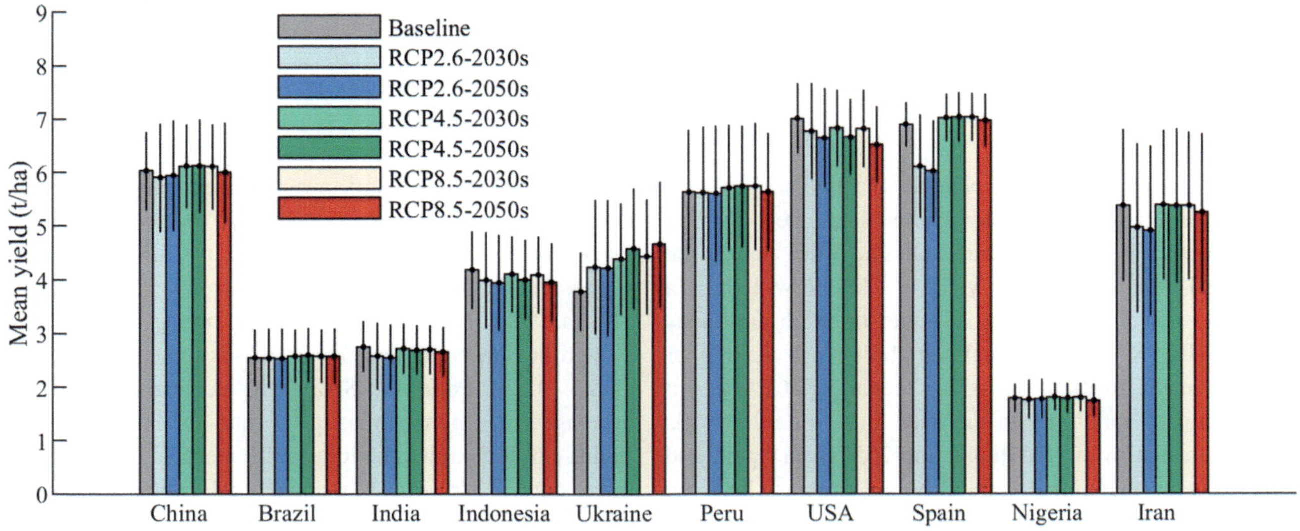

Fig. 8 Mean rice yield of the top 10 major rice-producing countries (in descending order by total production). The error bar represents the one standard deviation across the 21 general circulation models (GCMs) and three emulators

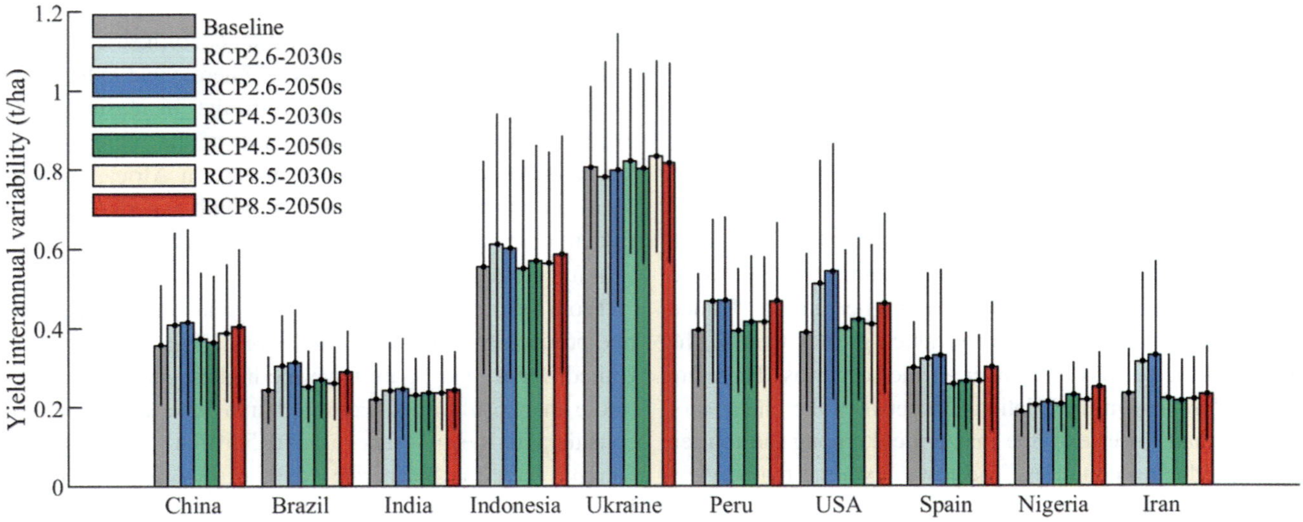

Fig. 9 Rice yield interannual variability of the top 10 major rice-producing countries (in descending order by total production). The error bar represents the one standard deviation across the 21 general circulation models (GCMs) and three emulators

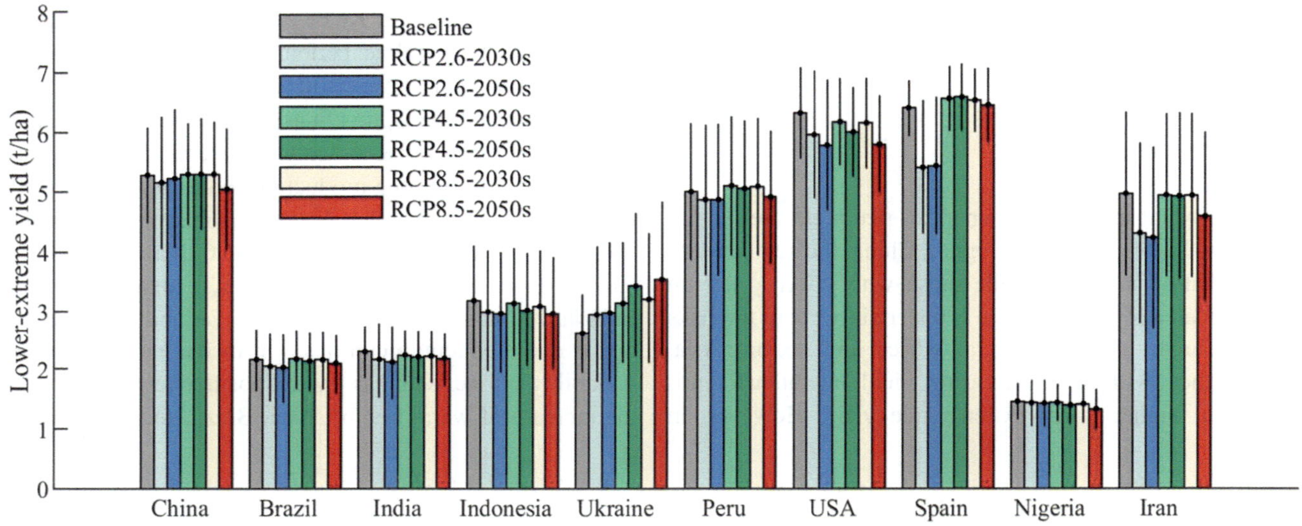

Fig. 10 Lower extreme rice yield of the top 10 major rice-producing countries (in descending order by total production). The error bar represents the one standard deviation across the 21 general circulation models (GCMs) and three emulators

Peru, the United States, and Nigeria, the interannual variability of rice yield will increase as the climate gets warmer, but the trend is not significant. On the other hand, the interannual variability of rice yield in Ukraine will show a fluctuating upward trend, that is, under the high emissions scenario (RCP8.5), interannual variability will be higher than that in the medium emissions scenario (RCP4.5), but the interannual variability in the near future (the 2030s) is higher than that in the far future (the 2050s). In Spain and Iran, the yield interannual variability will decrease first, but increase slightly when it is much warmer.

From the perspective of lower extreme rice yield, among the top ten rice-producing countries in the world (Fig. 10), the lower extreme rice yield is high in China, the United States, and Spain (>5 t/ha), followed by Brazil and India (2–3 t/ha) and Nigeria (1–2 t/ha). Among the top ten rice-producing countries, the lower -extreme rice yields in China, Brazil, Peru, and Spain will increase first and then decrease under higher emissions scenario and further into the future, but in Ukraine the lower extreme rice yield will keep increasing. The lower extreme rice yields in India, Indonesia, the United States, Nigeria, and Iran will generally decrease.

4 Maps

Global wheat mean yield (Baseline)

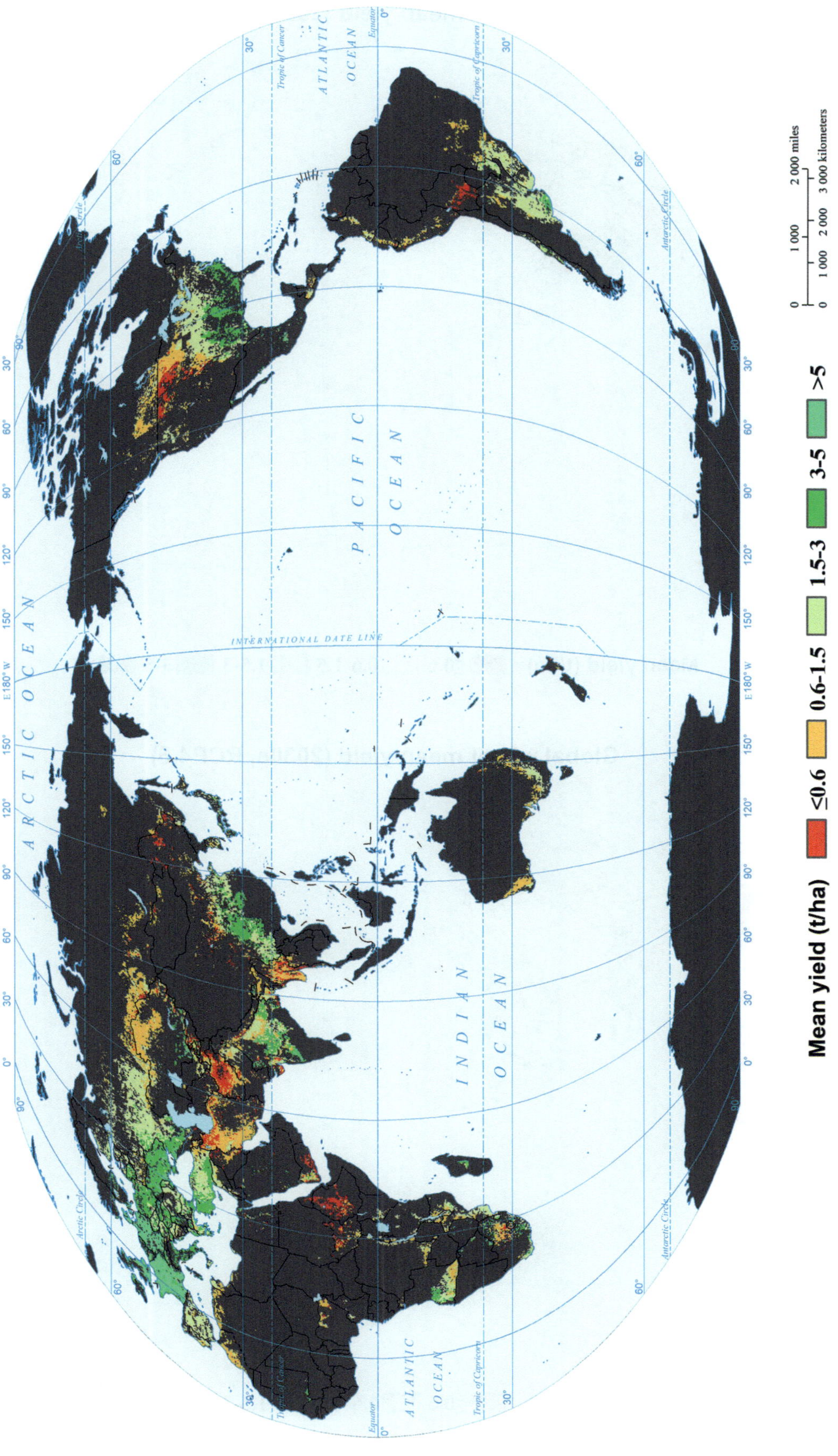

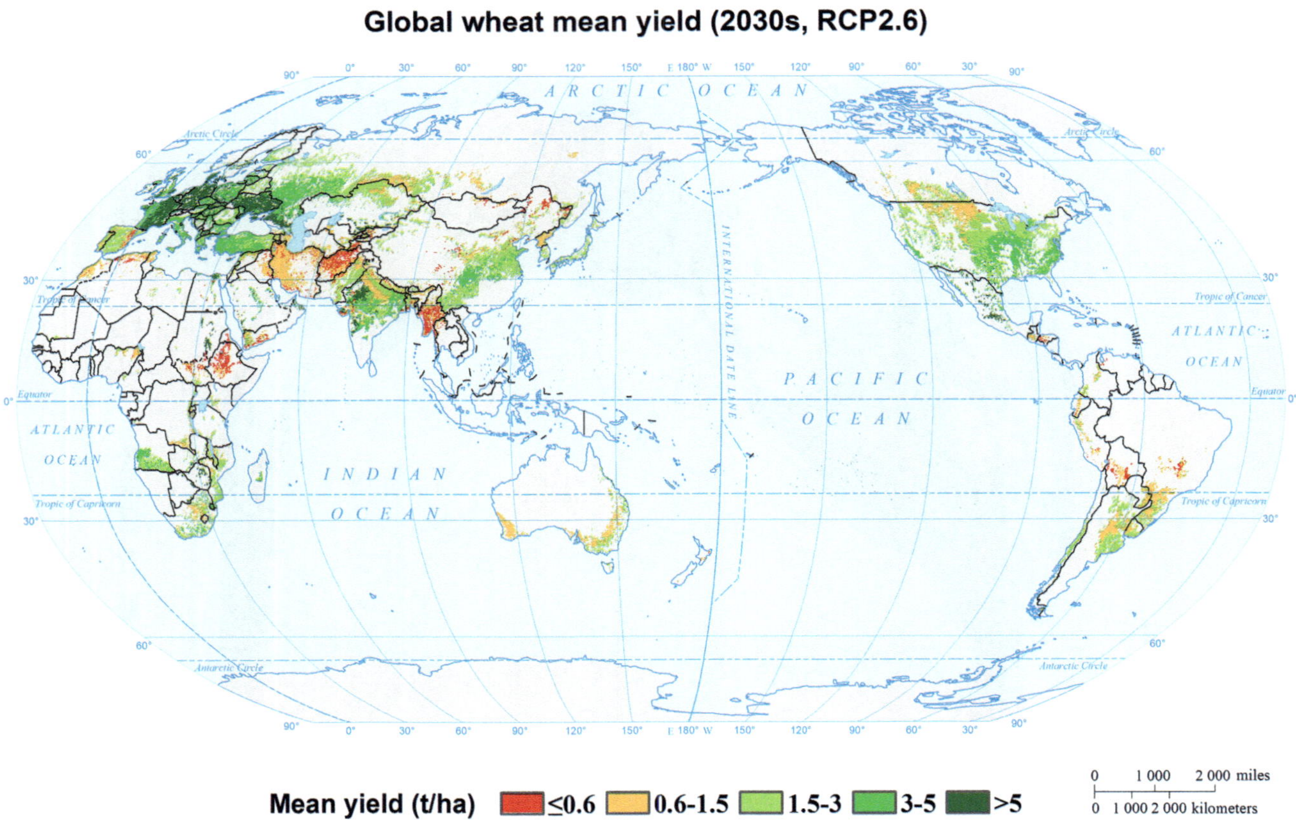
Global wheat mean yield (2030s, RCP2.6)
ARCTIC OCEAN
PACIFIC OCEAN
INDIAN OCEAN
ATLANTIC OCEAN
ATLANTIC OCEAN
INTERNATIONAL DATE LINE
Arctic Circle
Tropic of Cancer
Equator
Tropic of Capricorn
Antarctic Circle
Mean yield (t/ha) ≤0.6 0.6-1.5 1.5-3 3-5 >5
0 1 000 2 000 miles
0 1 000 2 000 kilometers

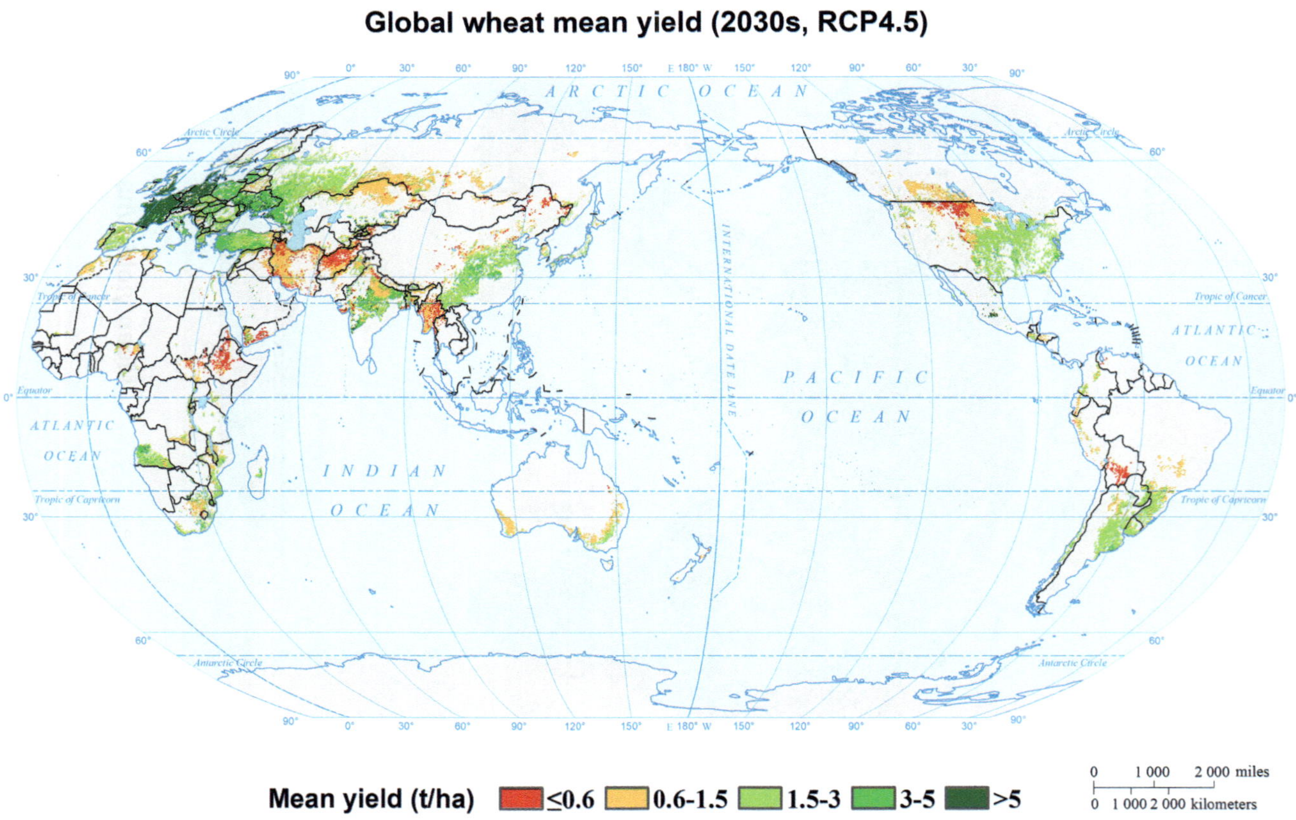
Global wheat mean yield (2030s, RCP4.5)
ARCTIC OCEAN
PACIFIC OCEAN
INDIAN OCEAN
ATLANTIC OCEAN
ATLANTIC OCEAN
INTERNATIONAL DATE LINE
Arctic Circle
Tropic of Cancer
Equator
Tropic of Capricorn
Antarctic Circle
Mean yield (t/ha) ≤0.6 0.6-1.5 1.5-3 3-5 >5
0 1 000 2 000 miles
0 1 000 2 000 kilometers

Global wheat mean yield (2030s, RCP8.5)

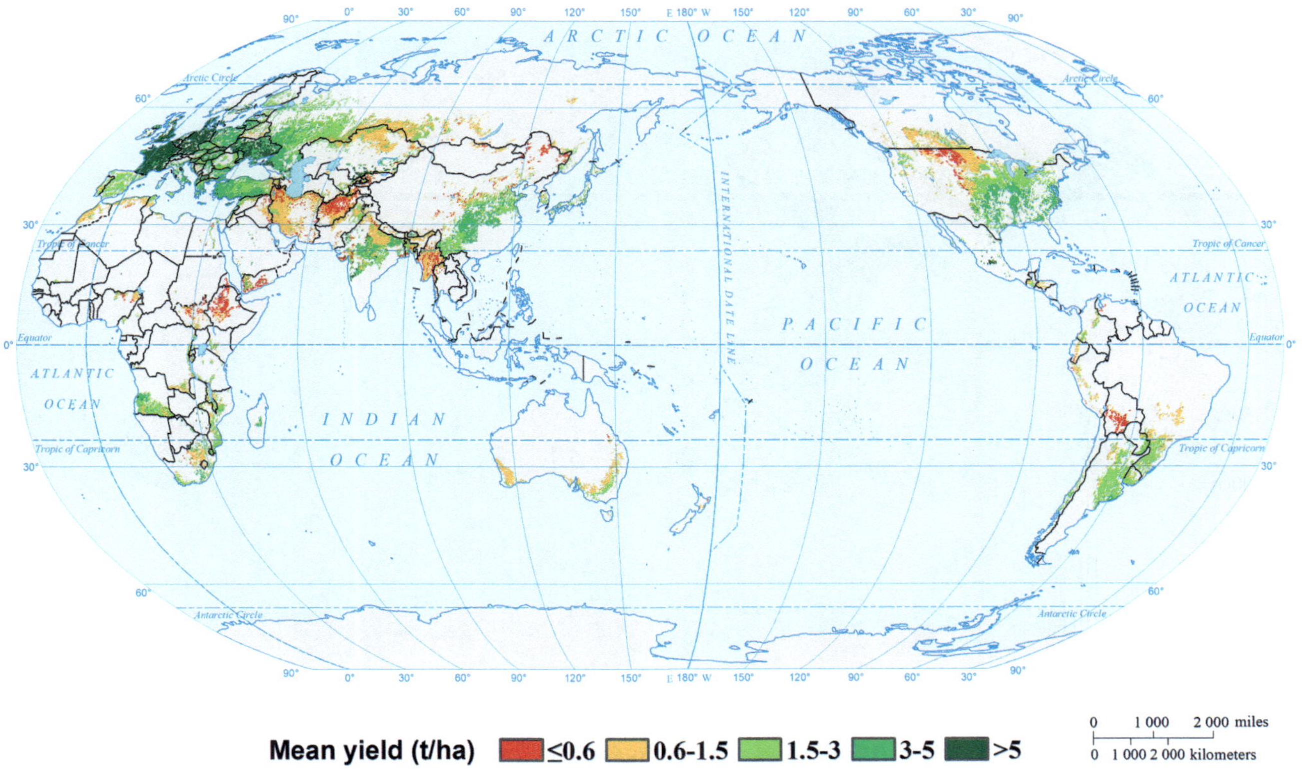

Mean yield (t/ha) ≤0.6 0.6-1.5 1.5-3 3-5 >5

Global wheat mean yield (2050s, RCP2.6)

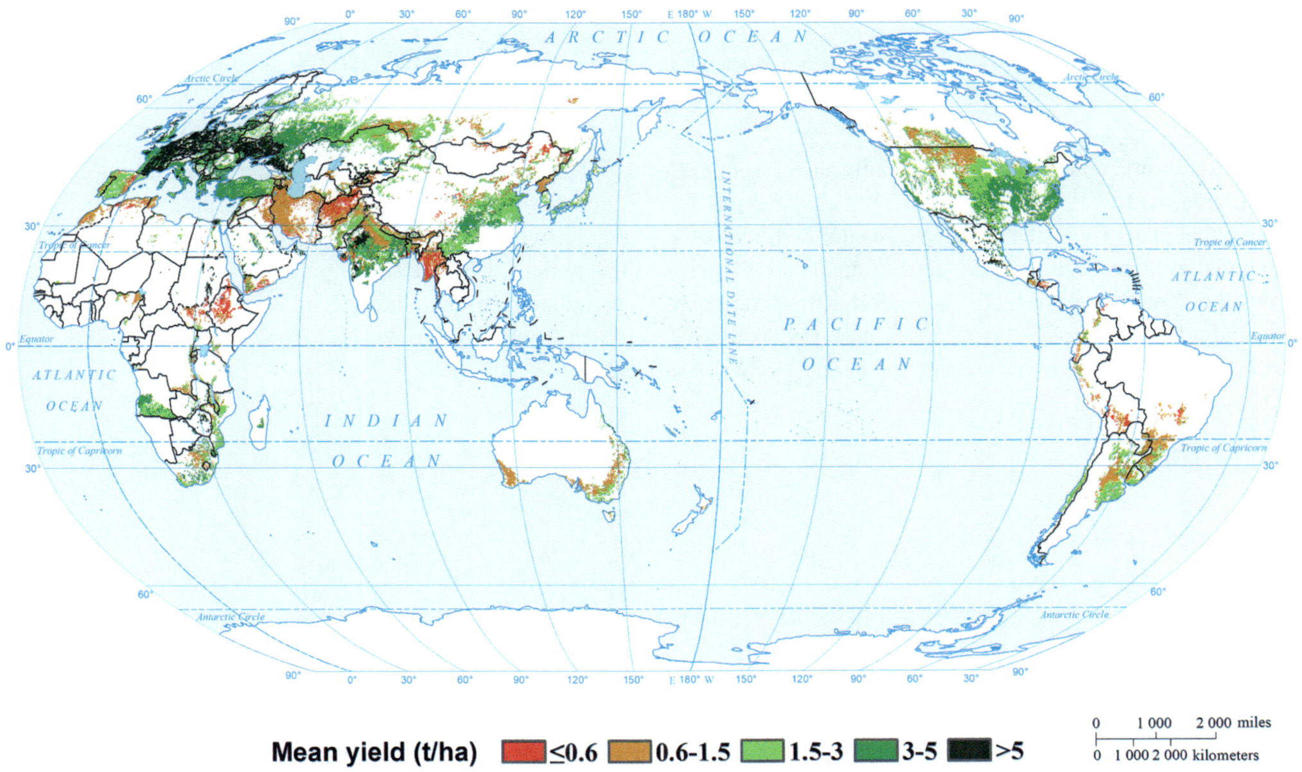

Mean yield (t/ha) ≤0.6 0.6-1.5 1.5-3 3-5 >5

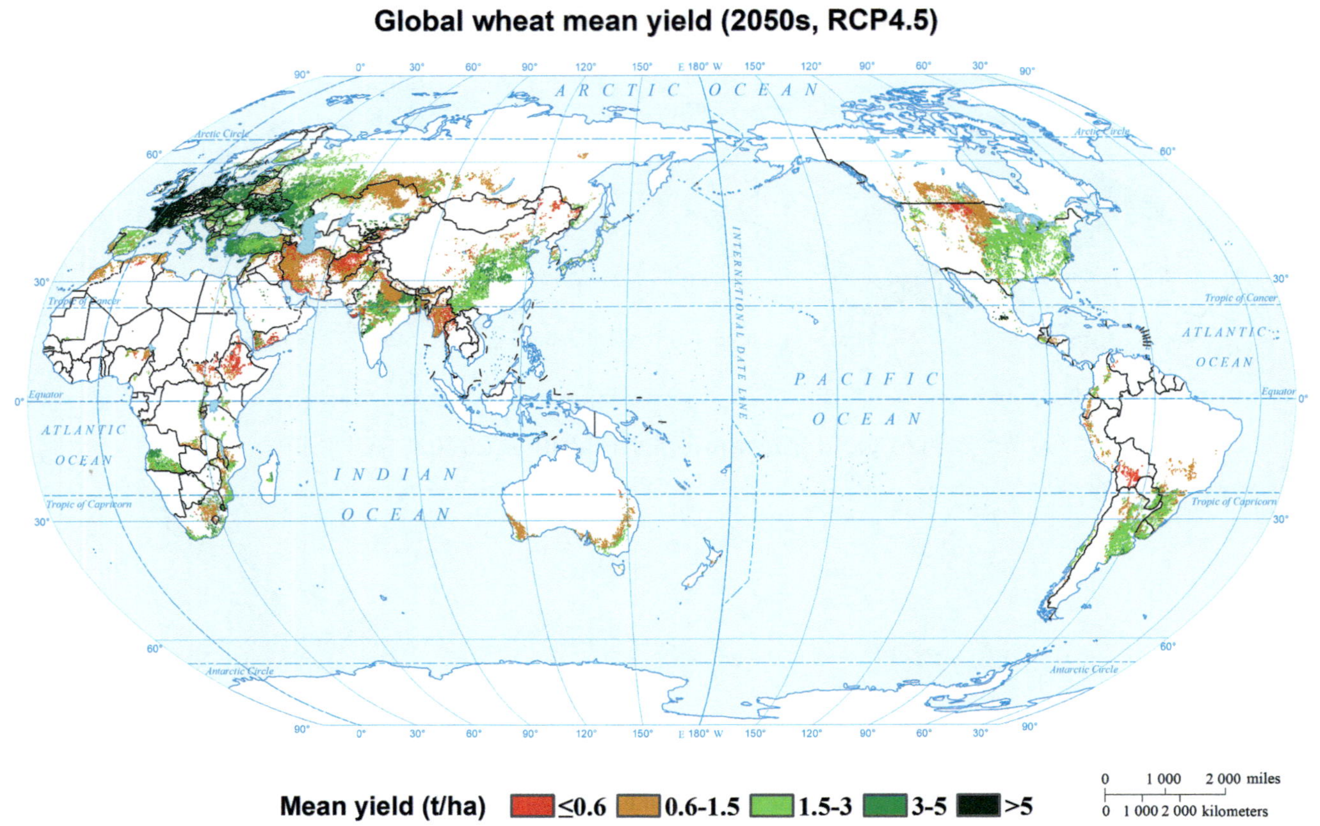
Global wheat mean yield (2050s, RCP4.5)
ARCTIC OCEAN
PACIFIC OCEAN
ATLANTIC OCEAN
ATLANTIC OCEAN
INDIAN OCEAN
INTERNATIONAL DATE LINE
Arctic Circle
Tropic of Cancer
Equator
Tropic of Capricorn
Antarctic Circle
Mean yield (t/ha)
≤0.6
0.6-1.5
1.5-3
3-5
>5
0 1 000 2 000 miles
0 1 000 2 000 kilometers

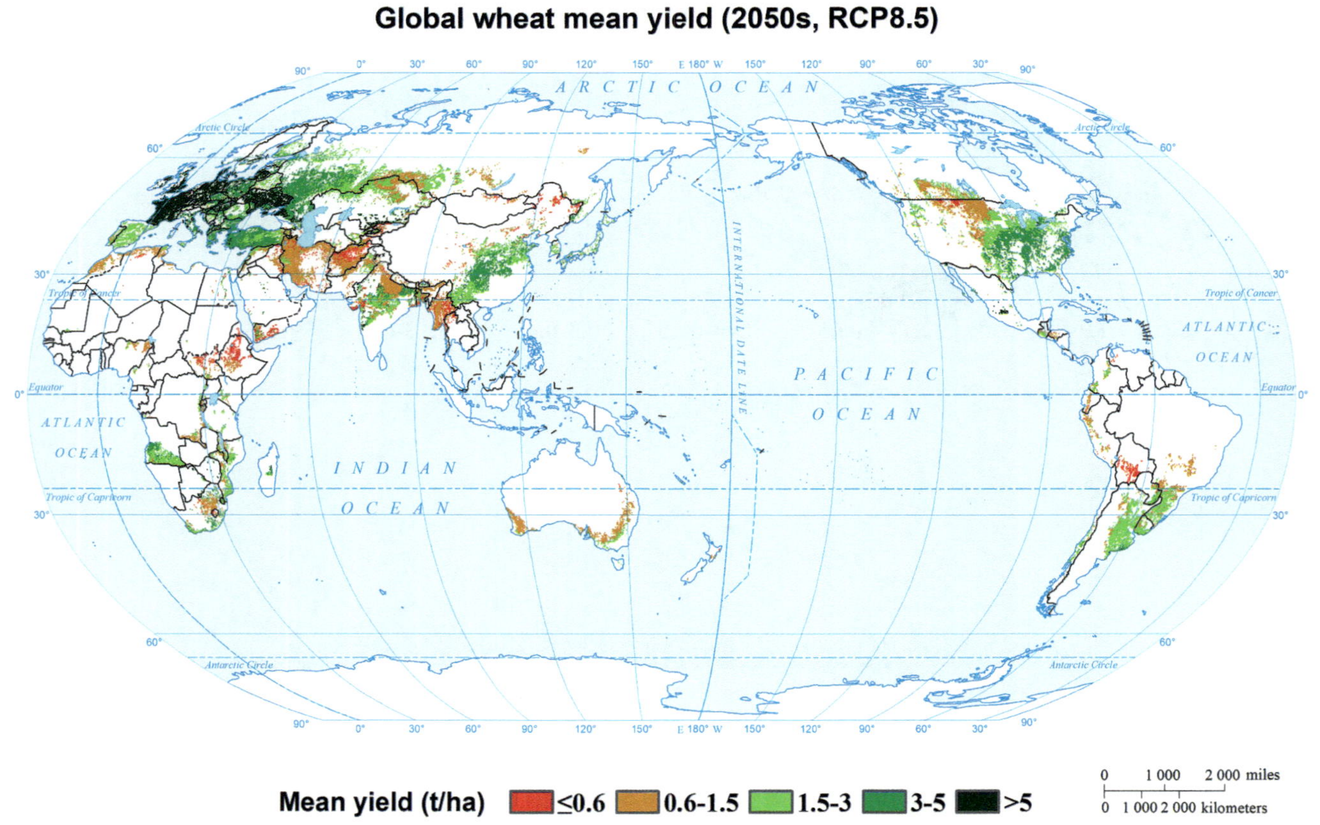
Global wheat mean yield (2050s, RCP8.5)
ARCTIC OCEAN
PACIFIC OCEAN
ATLANTIC OCEAN
ATLANTIC OCEAN
INDIAN OCEAN
INTERNATIONAL DATE LINE
Arctic Circle
Tropic of Cancer
Equator
Tropic of Capricorn
Antarctic Circle
Mean yield (t/ha)
≤0.6
0.6-1.5
1.5-3
3-5
>5
0 1 000 2 000 miles
0 1 000 2 000 kilometers

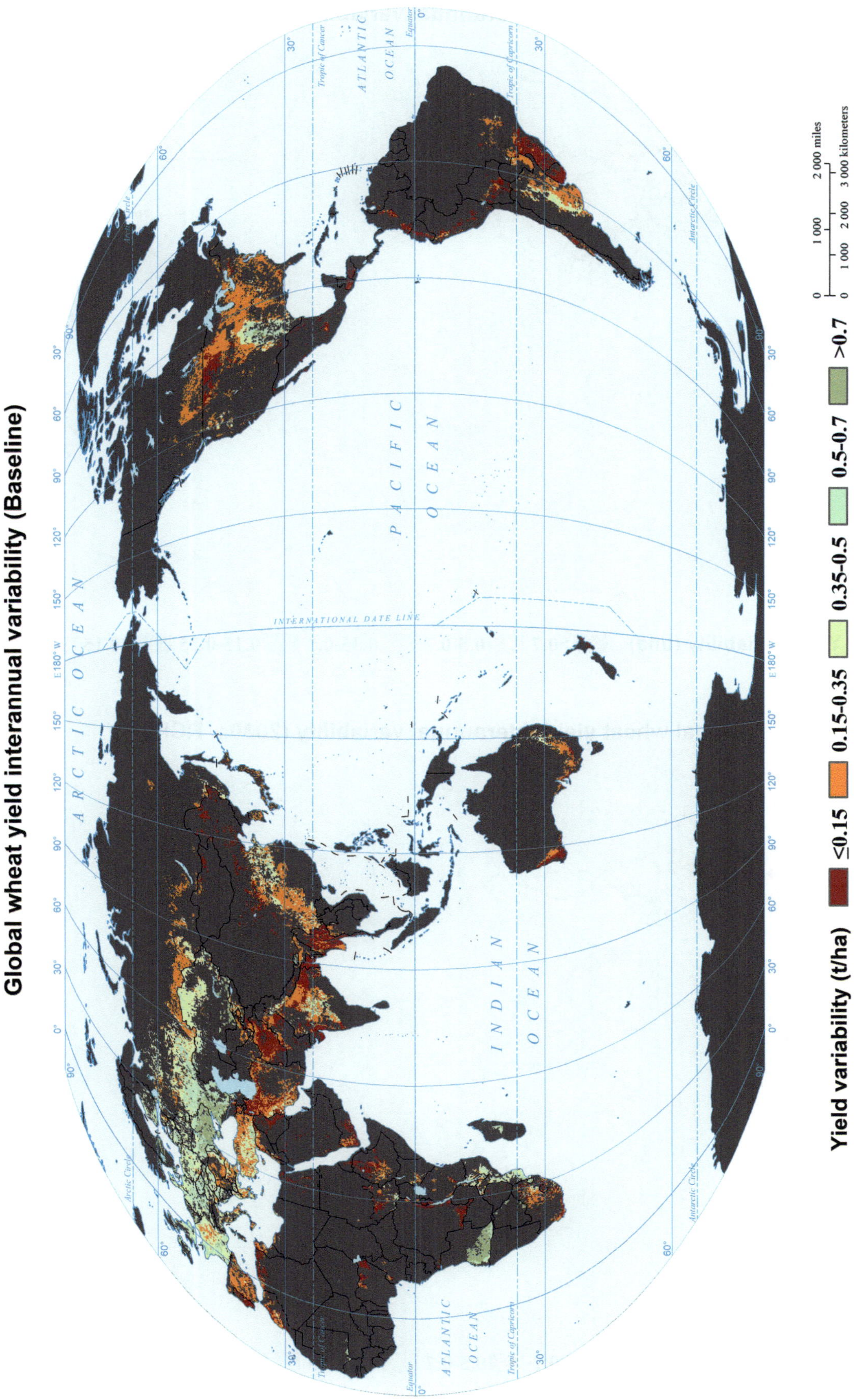
Global wheat yield interannual variability (Baseline)
Yield variability (t/ha)
≤0.15
0.15-0.35
0.35-0.5
0.5-0.7
>0.7
0 1 000 2 000 miles
0 1 000 2 000 3 000 kilometers
ARCTIC OCEAN
PACIFIC OCEAN
ATLANTIC OCEAN
INDIAN OCEAN
INTERNATIONAL DATE LINE
Tropic of Cancer
Tropic of Capricorn
Equator
Arctic Circle
Antarctic Circle
E 180° W

Global wheat yield interannual variability (2030s, RCP2.6)

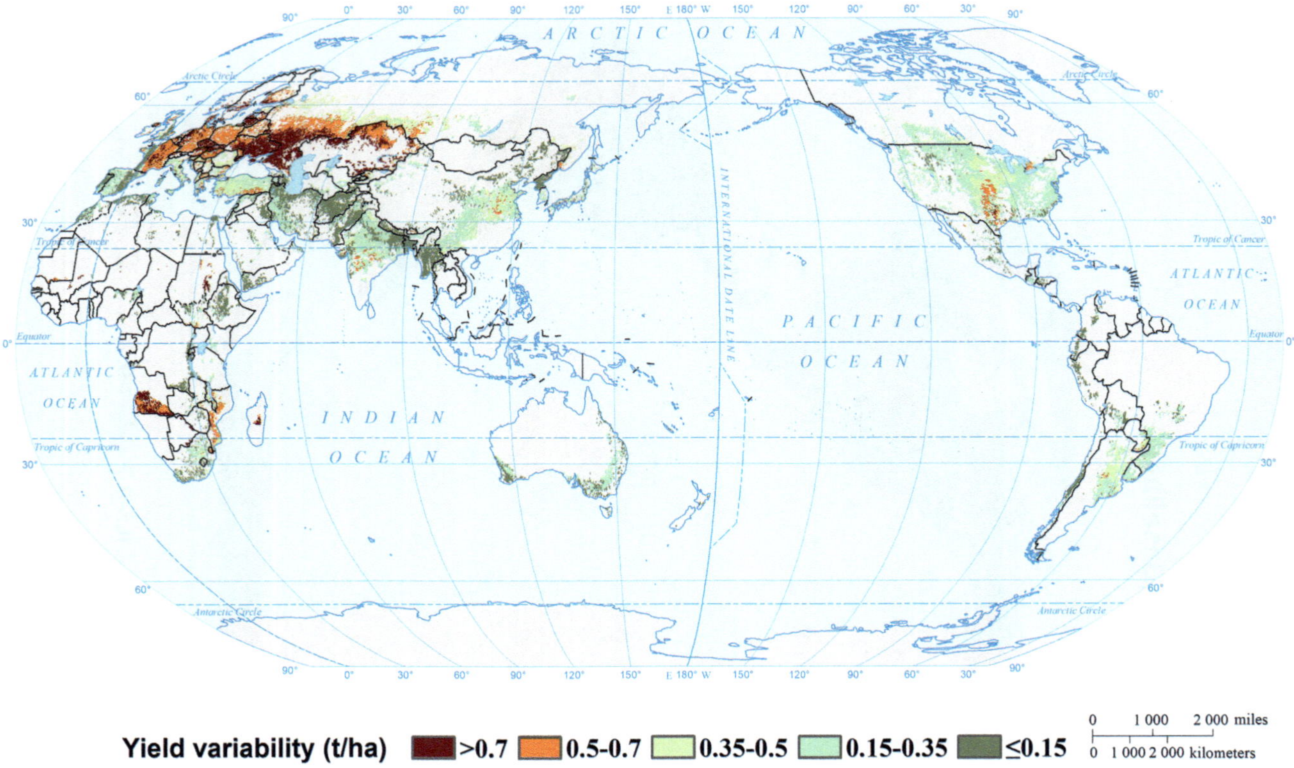

Global wheat yield interannual variability (2030s, RCP4.5)

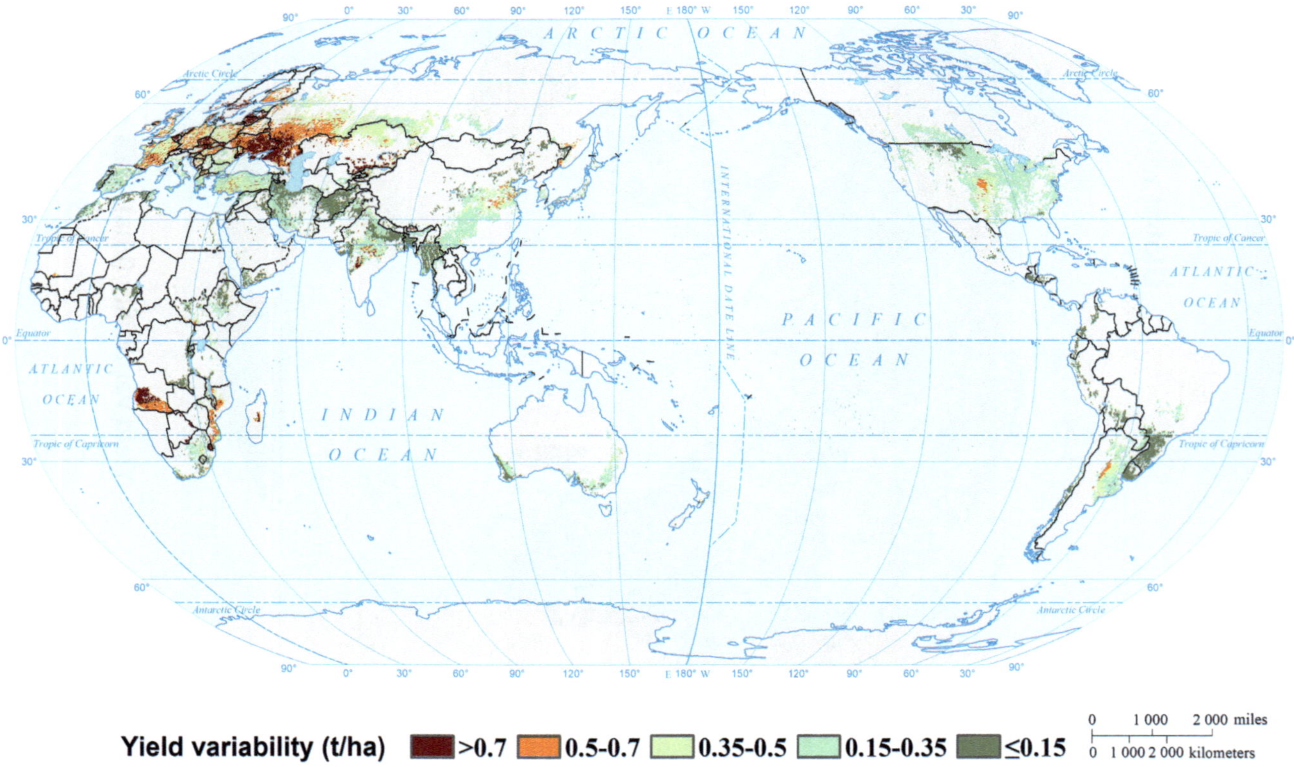

Global wheat yield interannual variability (2030s, RCP8.5)

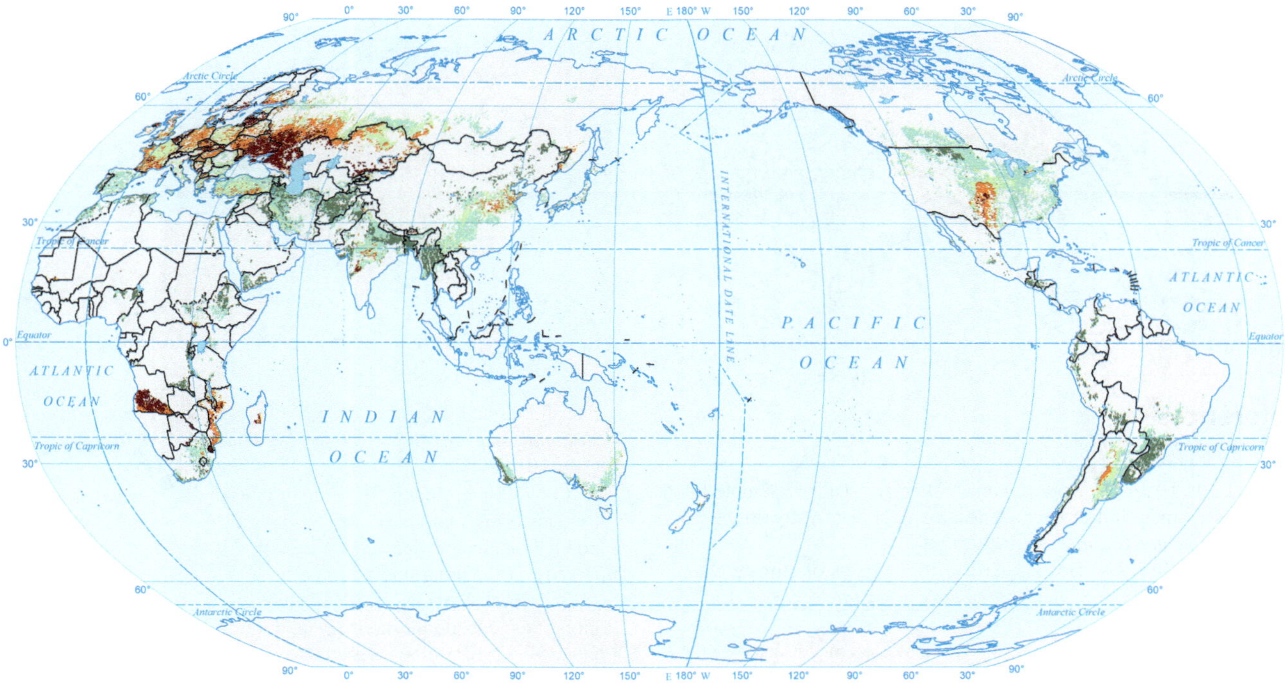

Yield variability (t/ha) >0.7 0.5-0.7 0.35-0.5 0.15-0.35 ≤0.15

0 1 000 2 000 miles
0 1 000 2 000 kilometers

Global wheat yield interannual variability (2050s, RCP2.6)

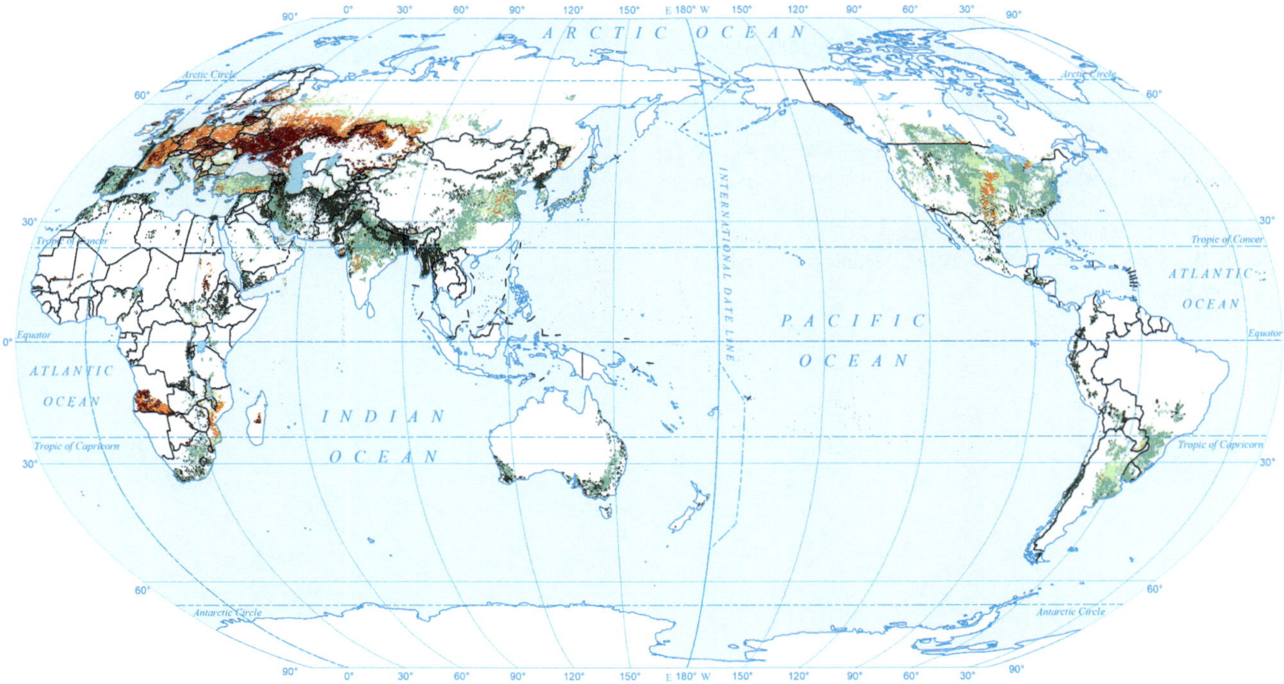

Yield variability (t/ha) >0.7 0.5-0.7 0.35-0.5 0.15-0.35 ≤0.15

0 1 000 2 000 miles
0 1 000 2 000 kilometers

Global wheat yield interannual variability (2050s, RCP4.5)

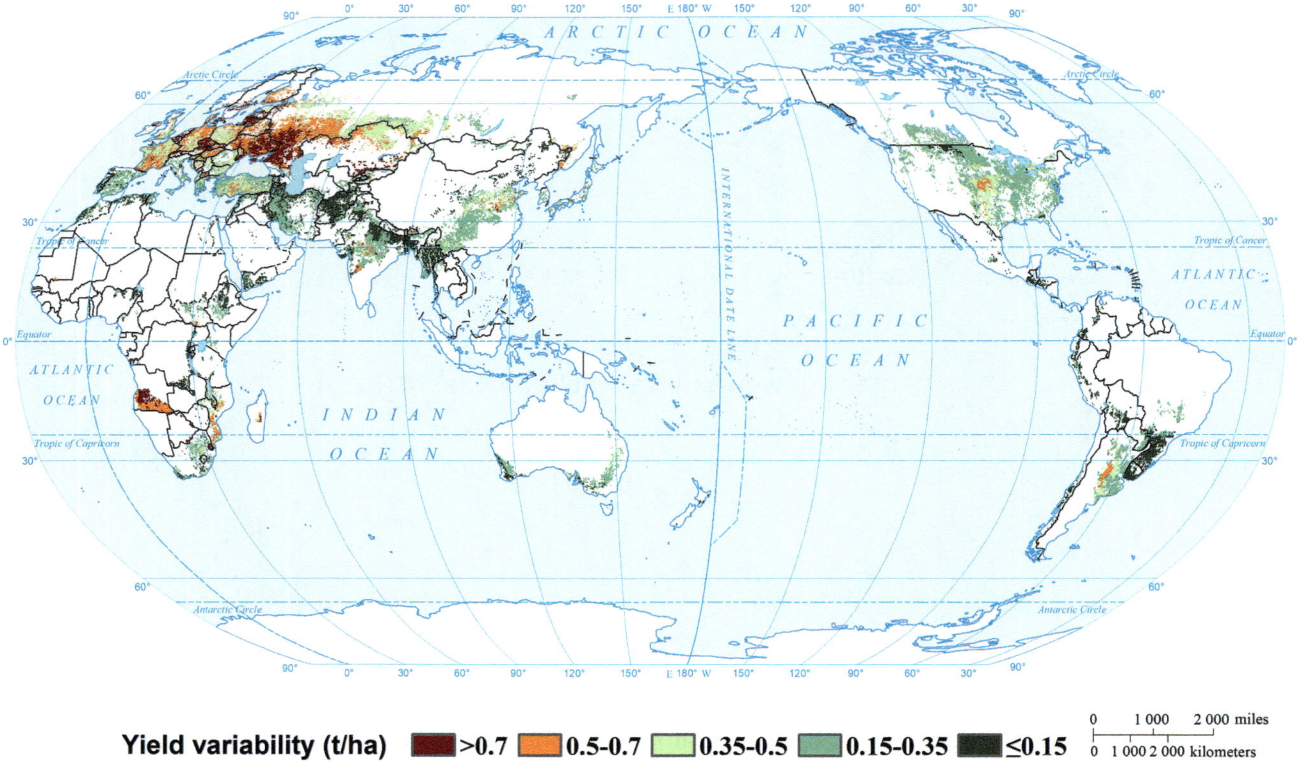

Global wheat yield interannual variability (2050s, RCP8.5)

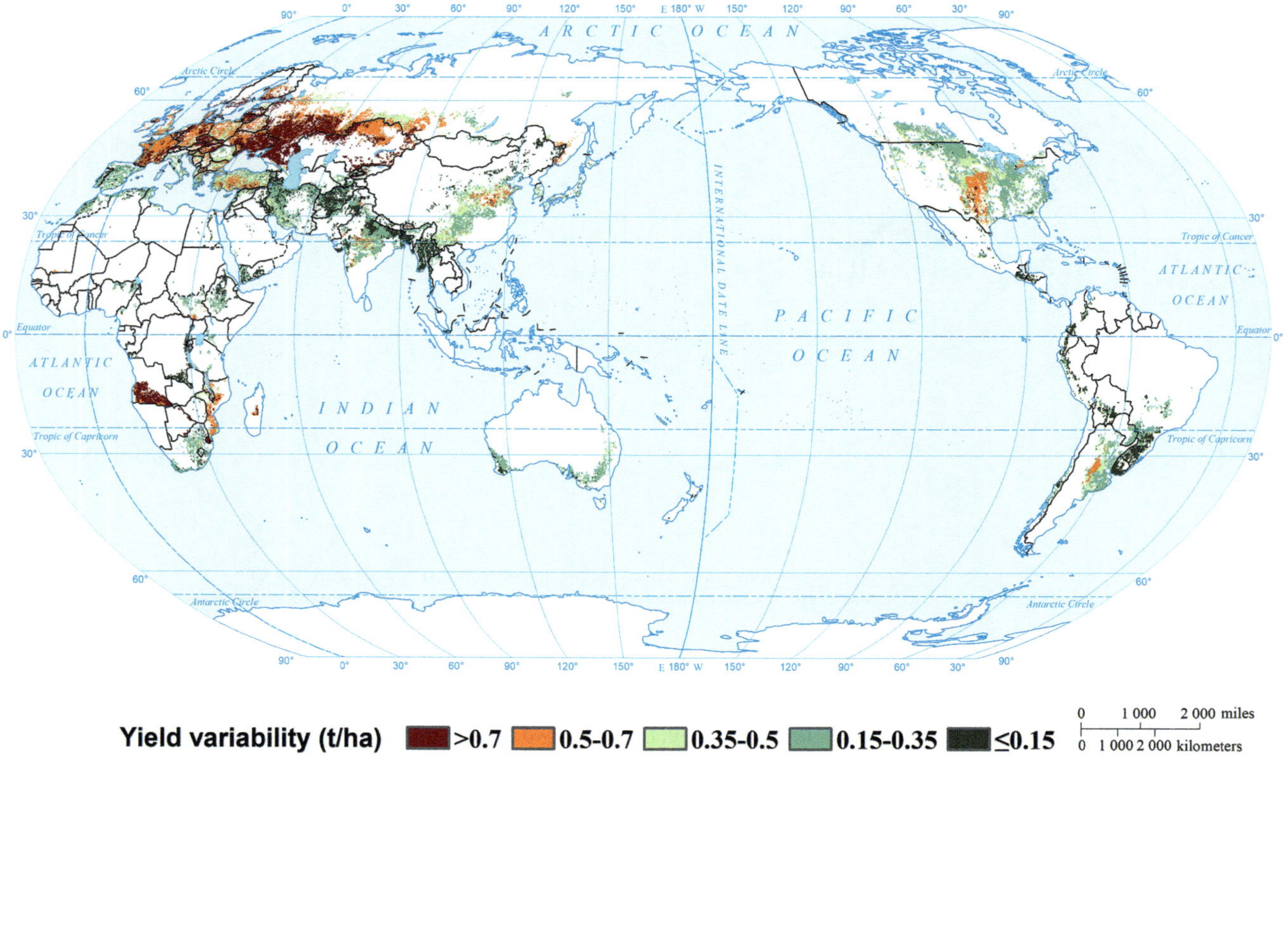

Global wheat lower-extreme yield (Baseline)

Yield extremes (t/ha): ≤0.35 | 0.35-0.6 | 0.6-1.8 | 1.8-3 | >3

0 1 000 2 000 miles
0 1 000 2 000 3 000 kilometers

Global wheat lower-extreme yield (2030s, RCP2.6)

Yield extremes (t/ha) ≤0.35 0.35-0.6 0.6-1.8 1.8-3 >3

Global wheat lower-extreme yield (2030s, RCP4.5)

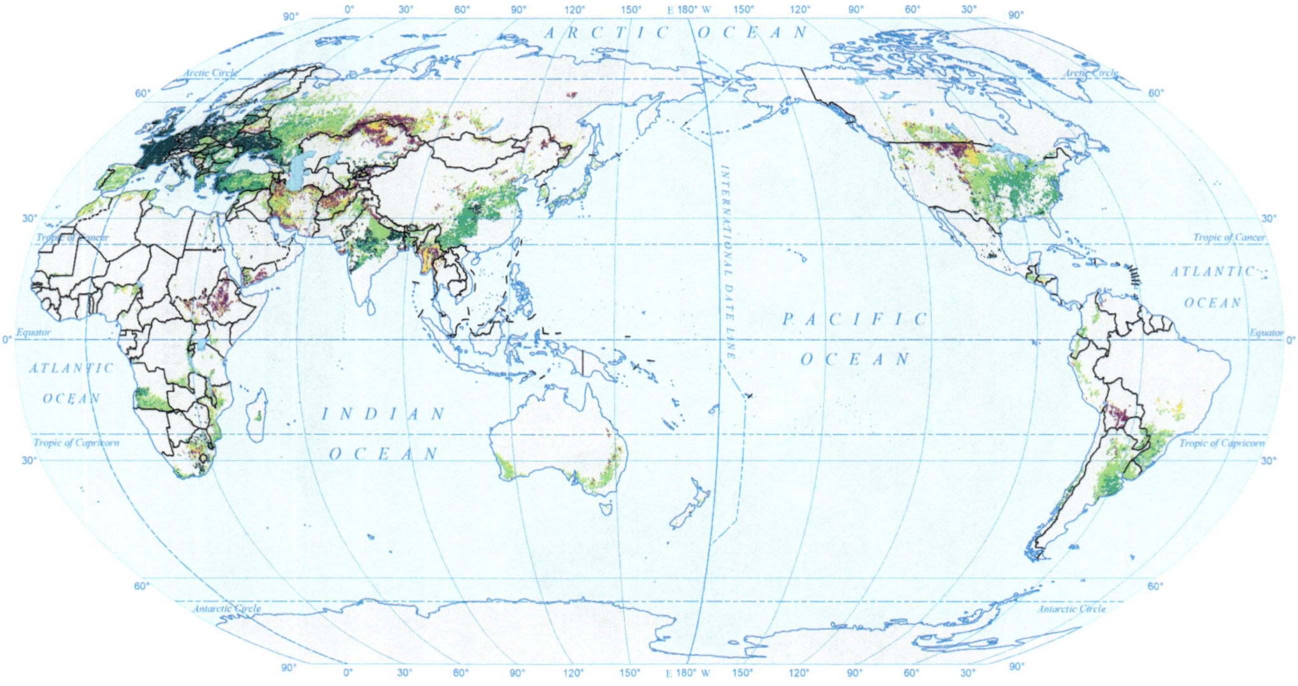

Yield extremes (t/ha) ≤0.35 0.35-0.6 0.6-1.8 1.8-3 >3

Global wheat lower-extreme yield (2030s, RCP8.5)

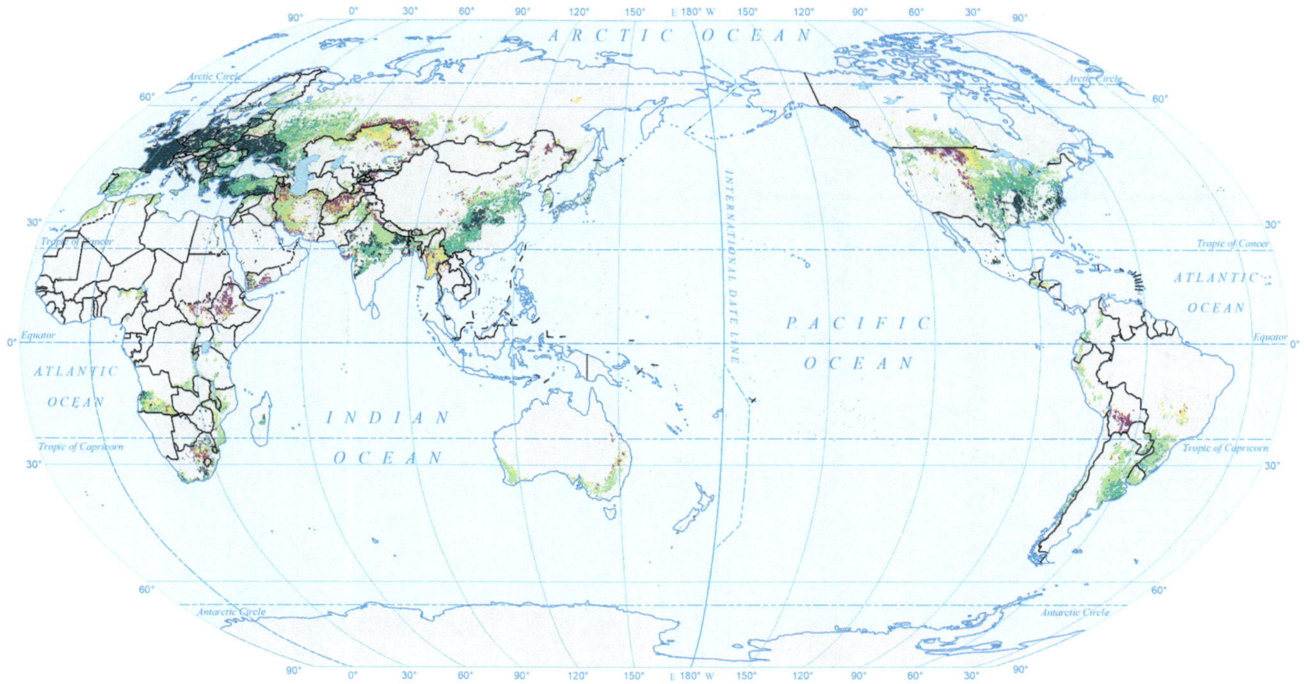

Yield extremes (t/ha) ≤0.35 0.35-0.6 0.6-1.8 1.8-3 >3

0 1 000 2 000 miles
0 1 000 2 000 kilometers

Global wheat lower-extreme yield (2050s, RCP2.6)

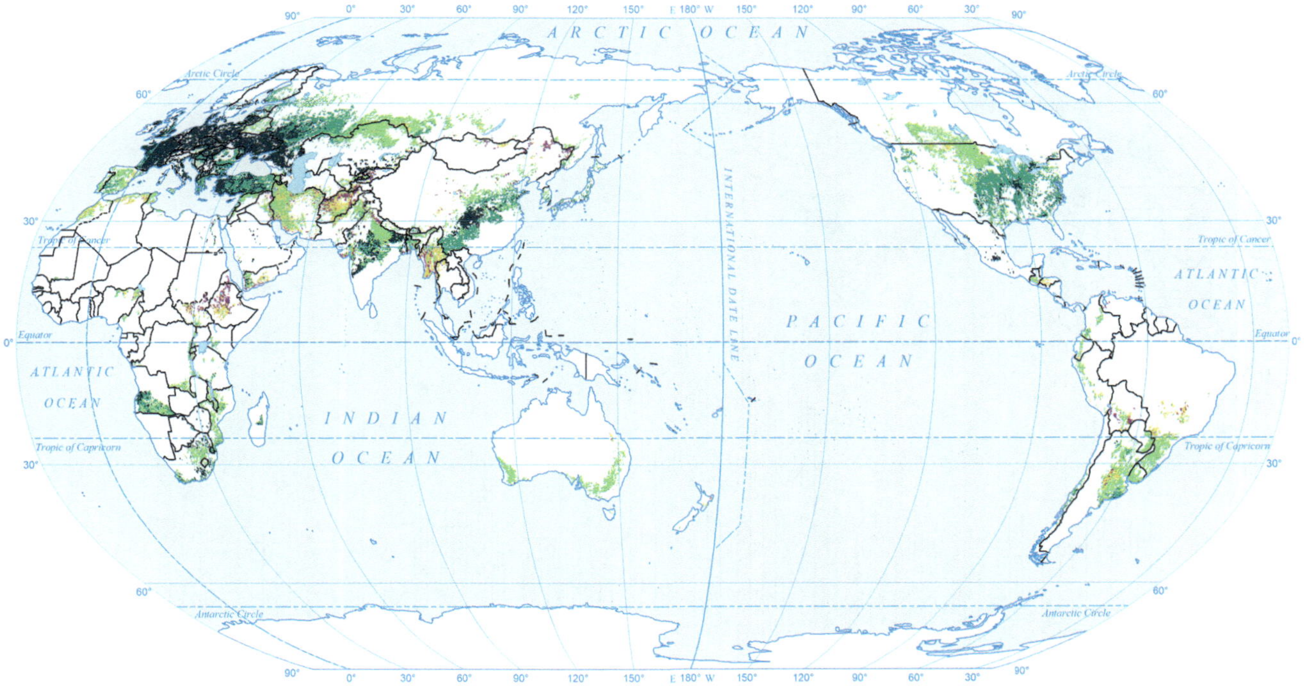

Yield extremes (t/ha) ≤0.35 0.35-0.6 0.6-1.8 1.8-3 >3

0 1 000 2 000 miles
0 1 000 2 000 kilometers

Global wheat lower-extreme yield (2050s, RCP4.5)

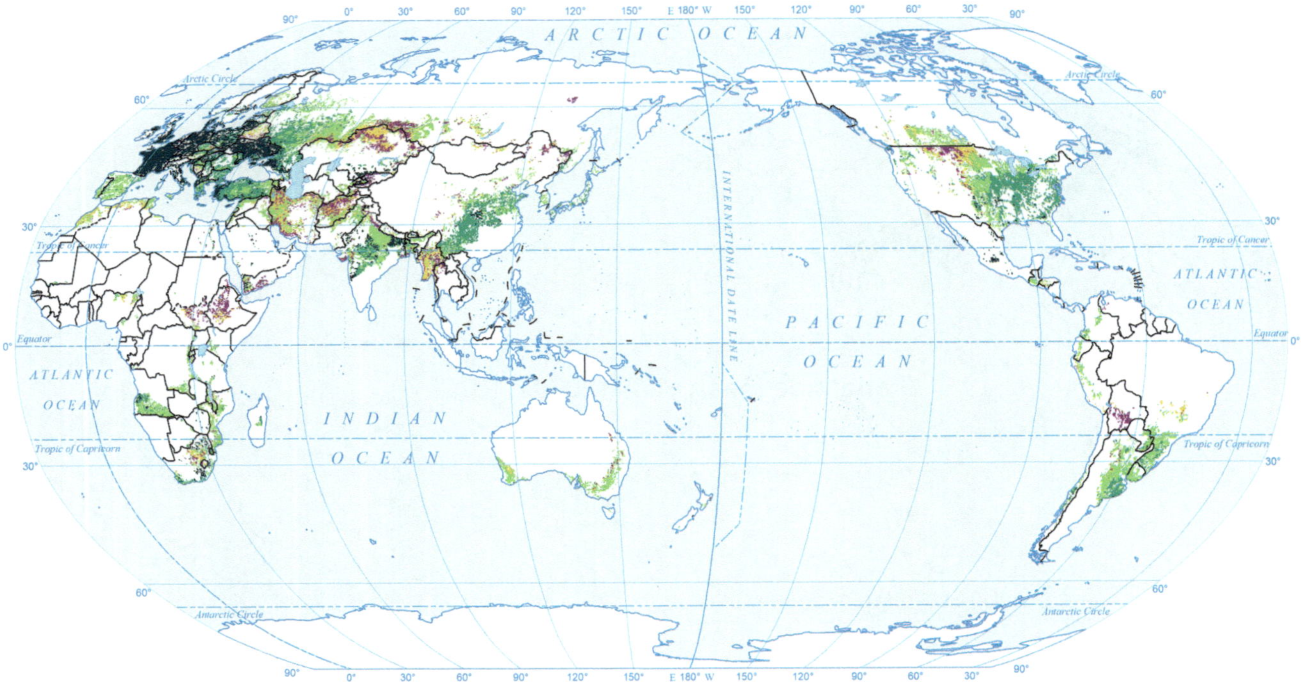

Yield extremes (t/ha) ≤0.35 0.35-0.6 0.6-1.8 1.8-3 >3

0 1 000 2 000 miles
0 1 000 2 000 kilometers

Global wheat lower-extreme yield (2050s, RCP8.5)

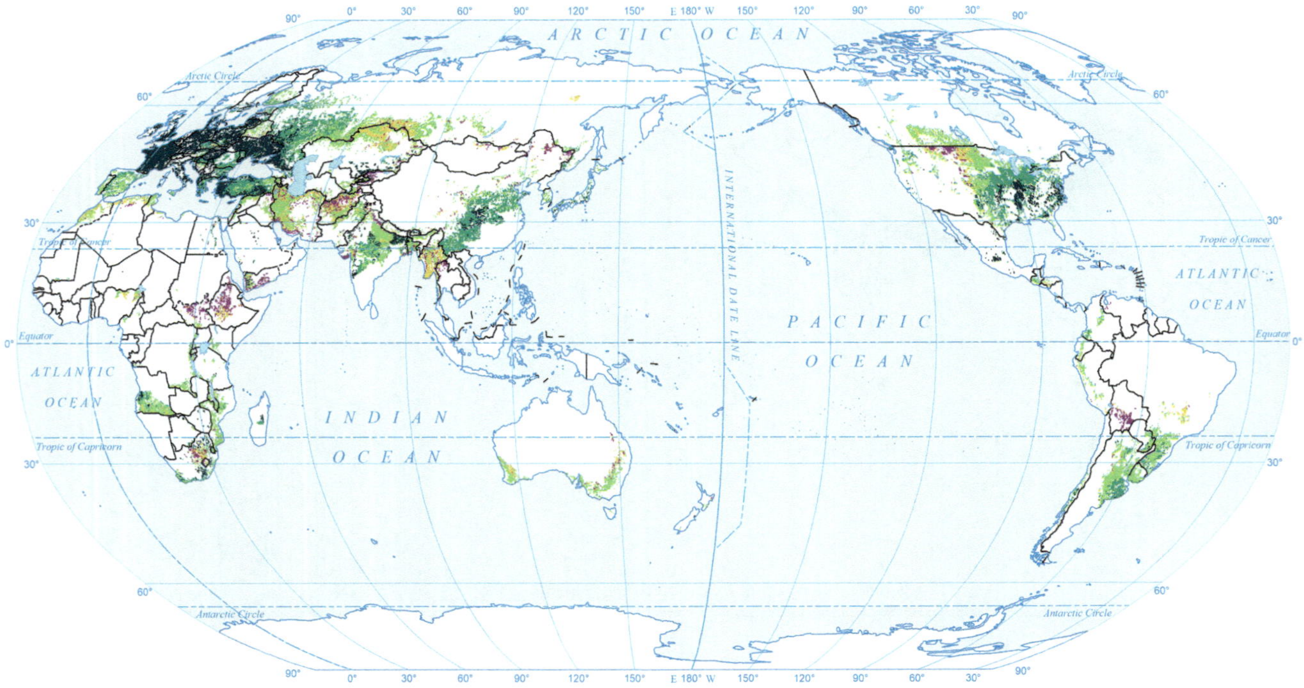

Yield extremes (t/ha) ≤0.35 0.35-0.6 0.6-1.8 1.8-3 >3

0 1 000 2 000 miles
0 1 000 2 000 kilometers

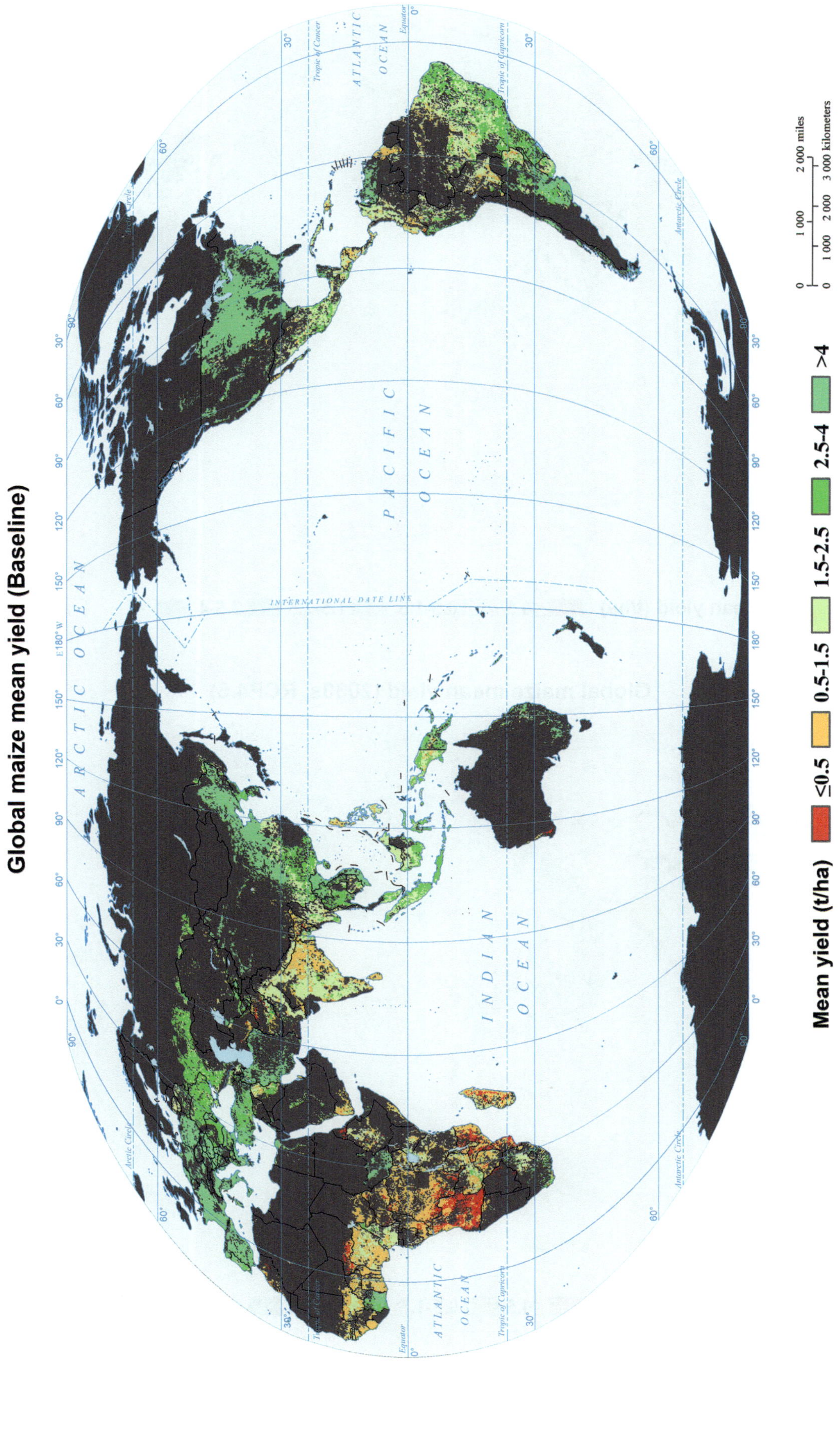
Global maize mean yield (Baseline)
ARCTIC OCEAN
ATLANTIC OCEAN
PACIFIC OCEAN
INDIAN OCEAN
INTERNATIONAL DATE LINE
Tropic of Cancer
Equator
Tropic of Capricorn
Arctic Circle
Antarctic Circle
Mean yield (t/ha)
≤0.5
0.5-1.5
1.5-2.5
2.5-4
>4
0 1 000 2 000 miles
0 1 000 2 000 3 000 kilometers

Global maize mean yield (2030s, RCP2.6)

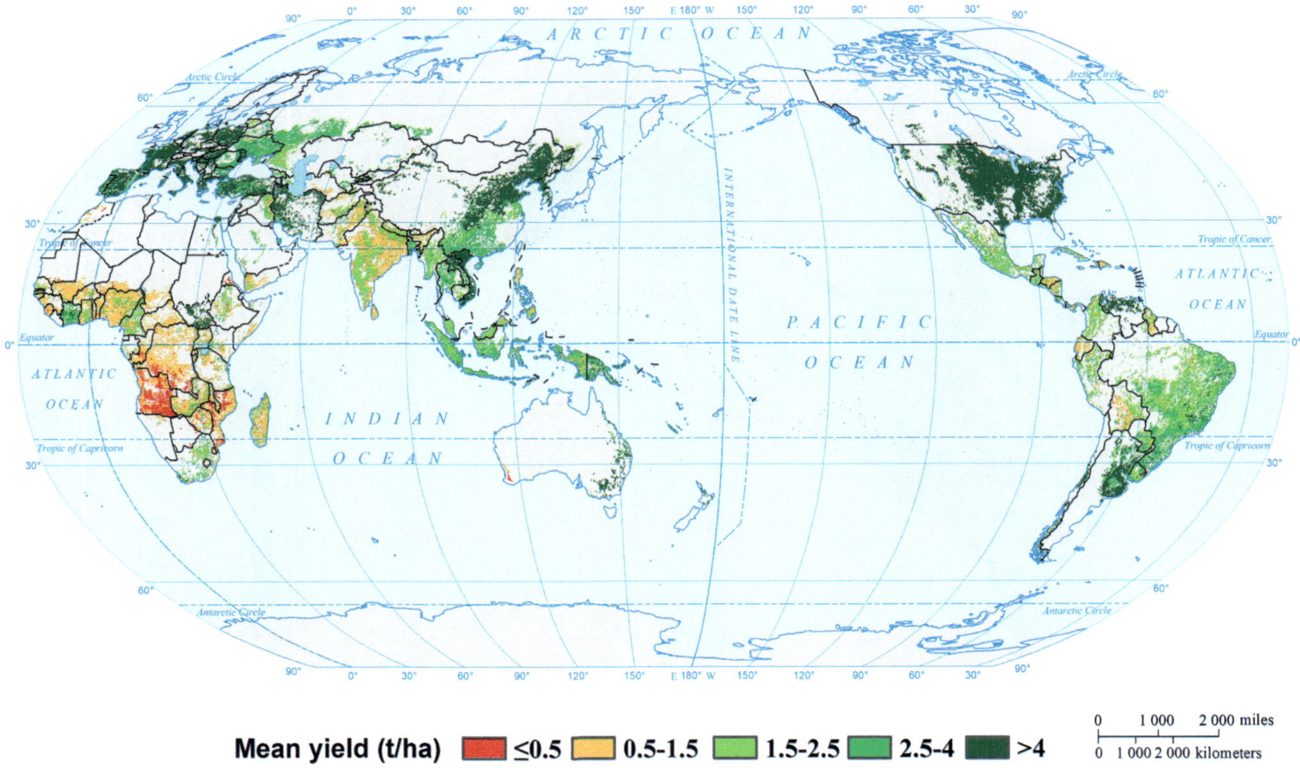

Mean yield (t/ha) ≤0.5 0.5-1.5 1.5-2.5 2.5-4 >4

Global maize mean yield (2030s, RCP4.5)

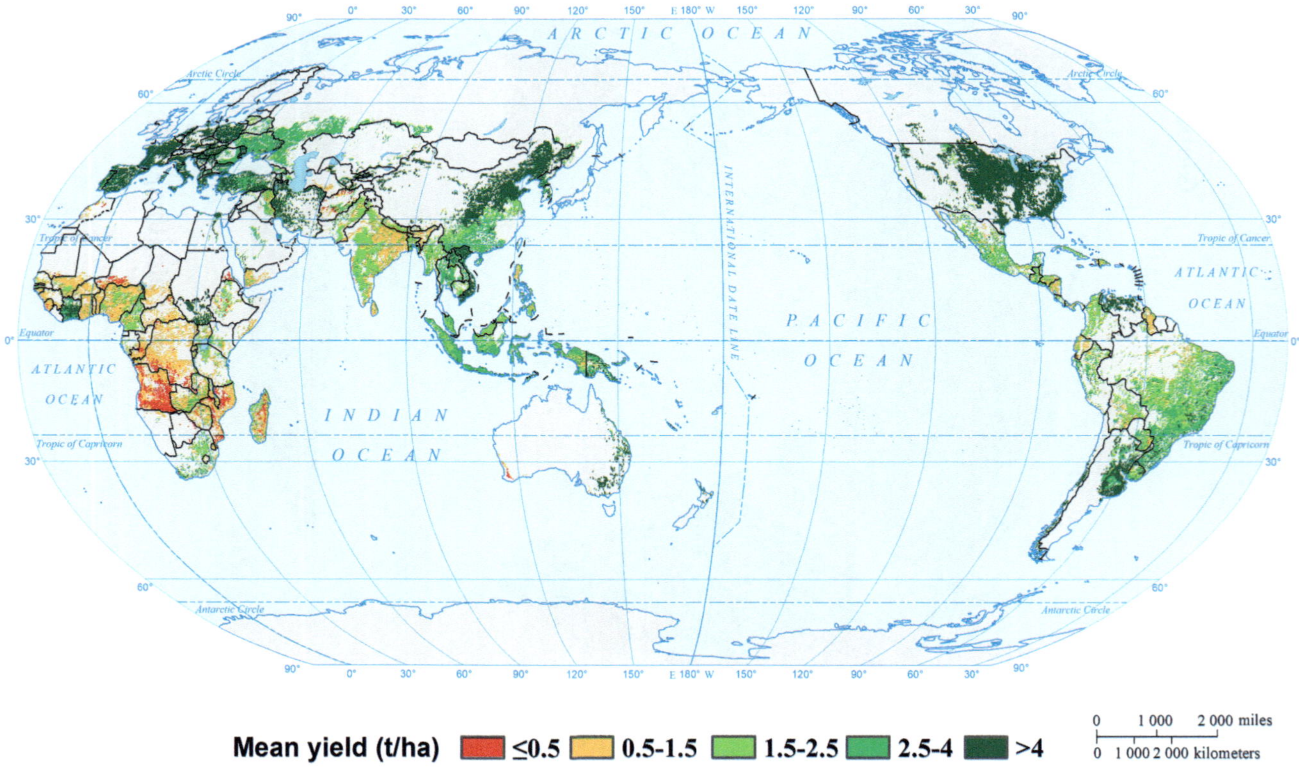

Mean yield (t/ha) ≤0.5 0.5-1.5 1.5-2.5 2.5-4 >4

Global maize mean yield (2030s, RCP8.5)

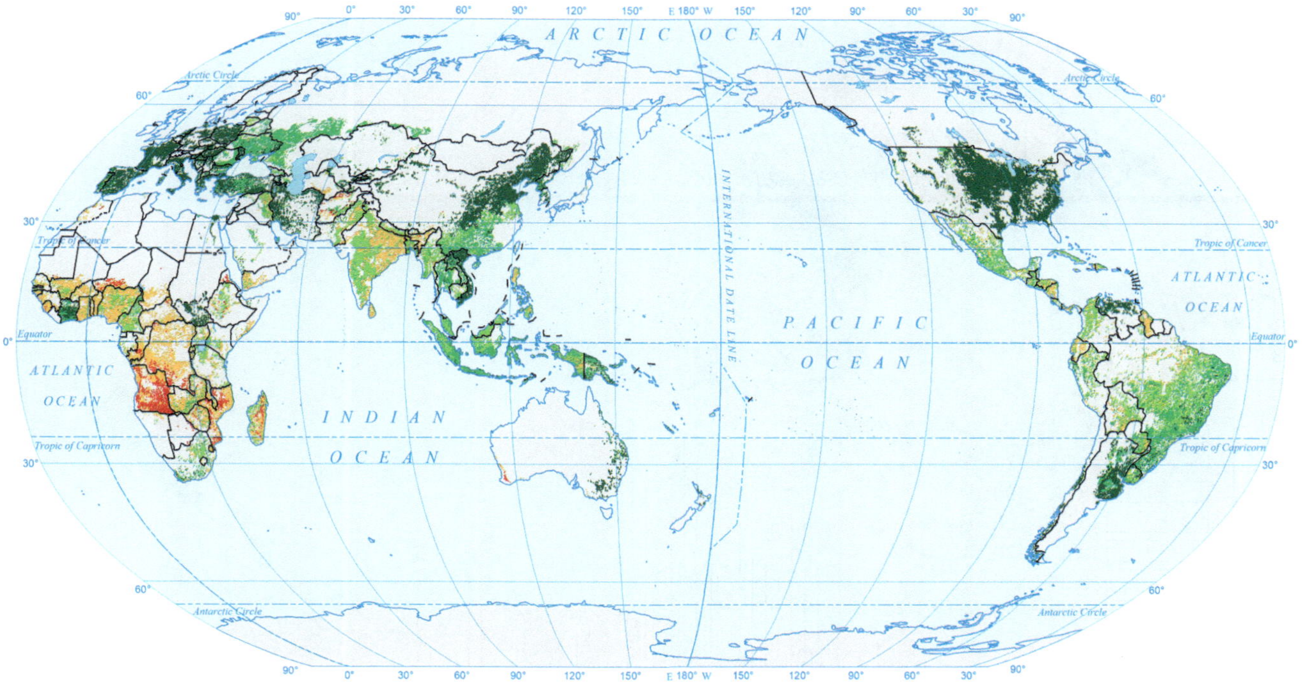

Mean yield (t/ha) ≤0.5 0.5-1.5 1.5-2.5 2.5-4 >4

0 1 000 2 000 miles
0 1 000 2 000 kilometers

Global maize mean yield (2050s, RCP2.6)

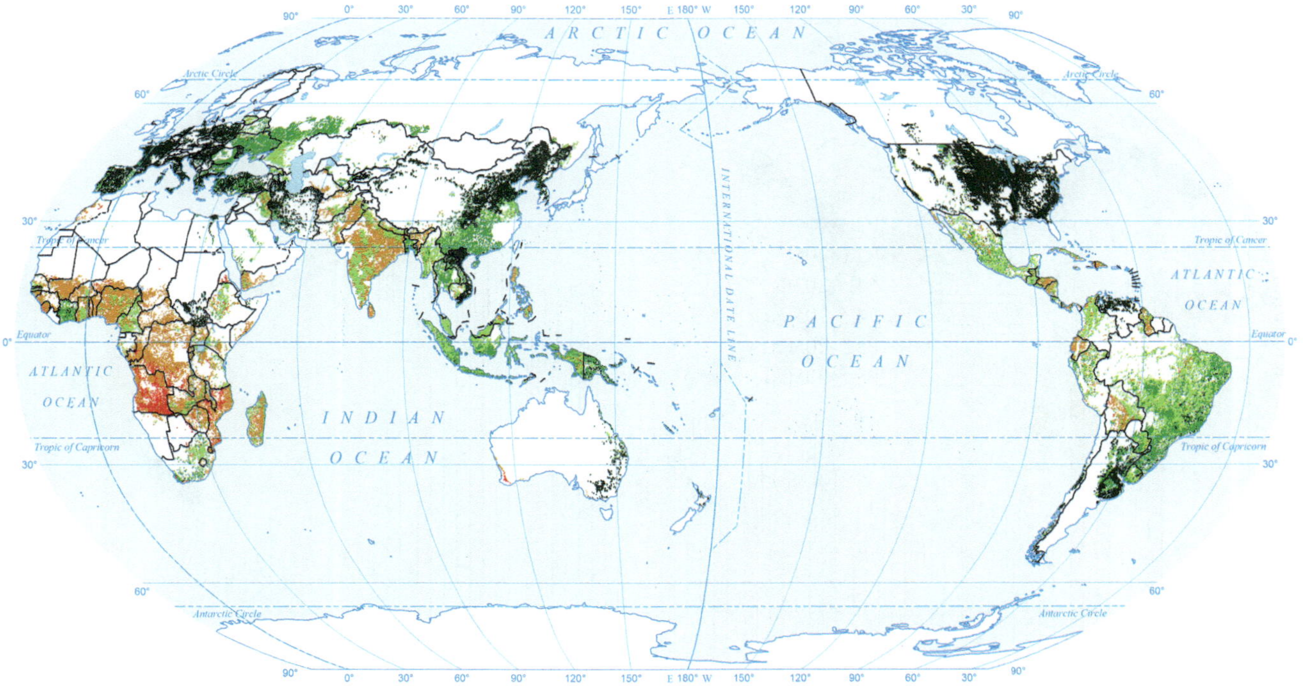

Mean yield (t/ha) ≤0.5 0.5-1.5 1.5-2.5 2.5-4 >4

0 1 000 2 000 miles
0 1 000 2 000 kilometers

Global maize mean yield (2050s, RCP4.5)

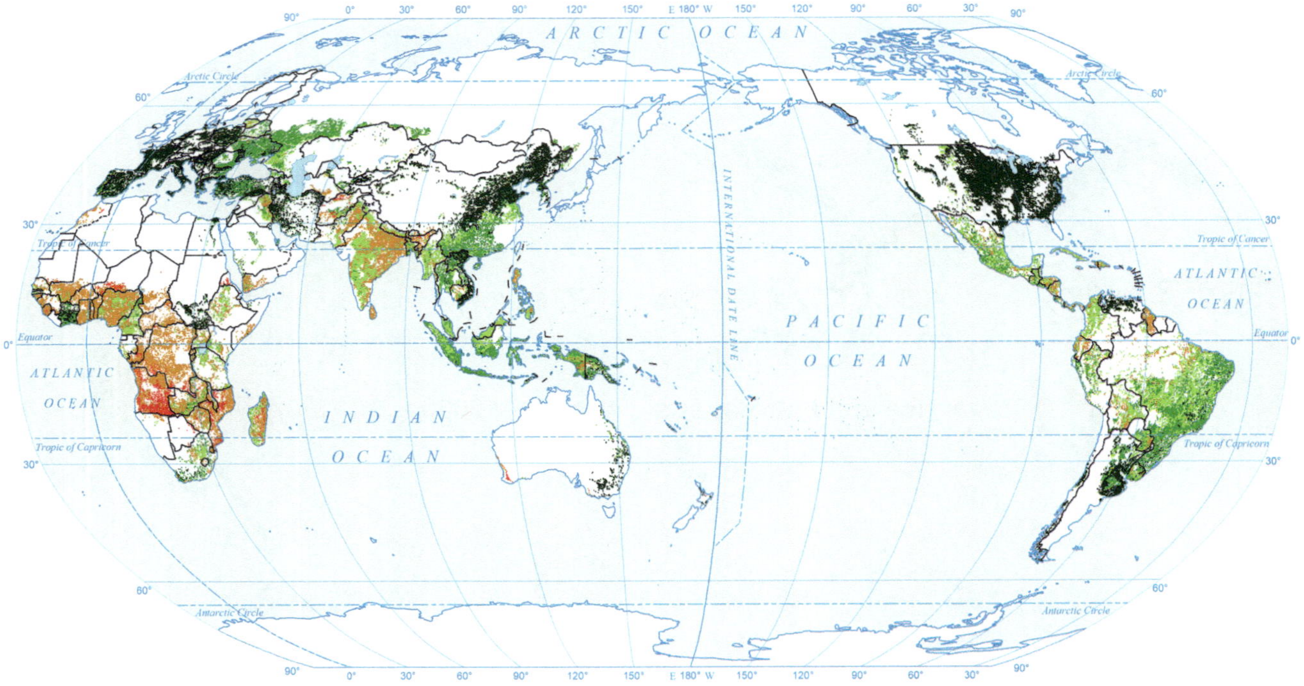

Mean yield (t/ha) ≤0.5 0.5-1.5 1.5-2.5 2.5-4 >4

0 1 000 2 000 miles
0 1 000 2 000 kilometers

Global maize mean yield (2050s, RCP8.5)

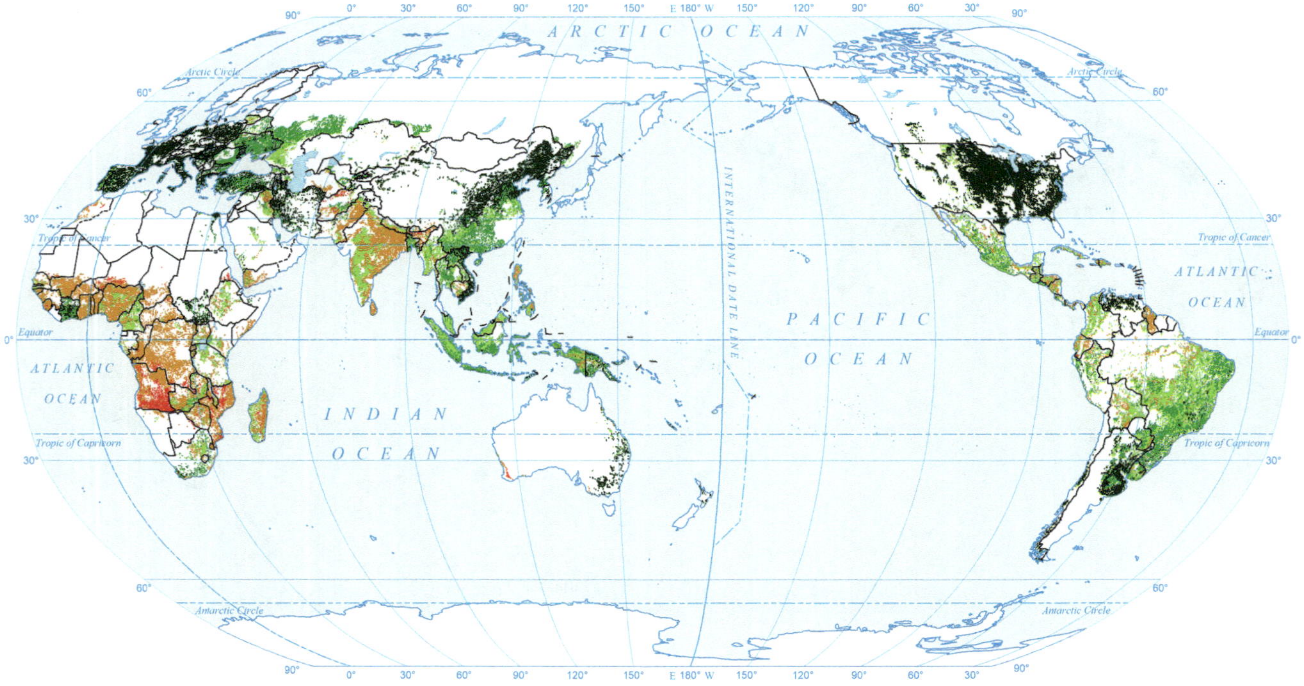

Mean yield (t/ha) ≤0.5 0.5-1.5 1.5-2.5 2.5-4 >4

0 1 000 2 000 miles
0 1 000 2 000 kilometers

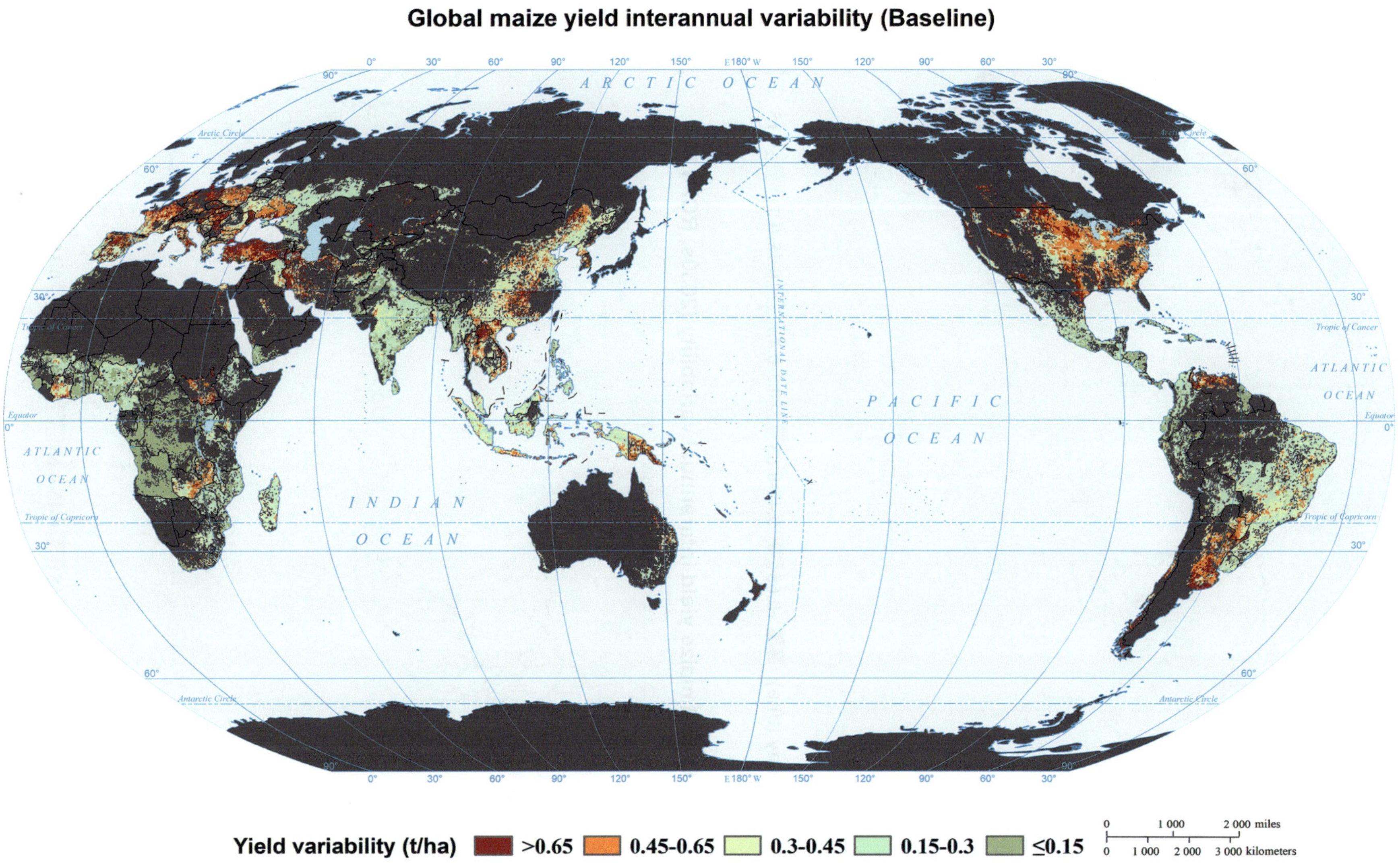
Global maize yield interannual variability (Baseline)
ARCTIC OCEAN
PACIFIC OCEAN
ATLANTIC OCEAN
INDIAN OCEAN
INTERNATIONAL DATE LINE
Arctic Circle
Tropic of Cancer
Equator
Tropic of Capricorn
Antarctic Circle
0° 30° 60° 90° 120° 150° E 180° W 150° 120° 90° 60° 30°
Yield variability (t/ha)
>0.65
0.45-0.65
0.3-0.45
0.15-0.3
≤0.15
0 1 000 2 000 miles
0 1 000 2 000 3 000 kilometers

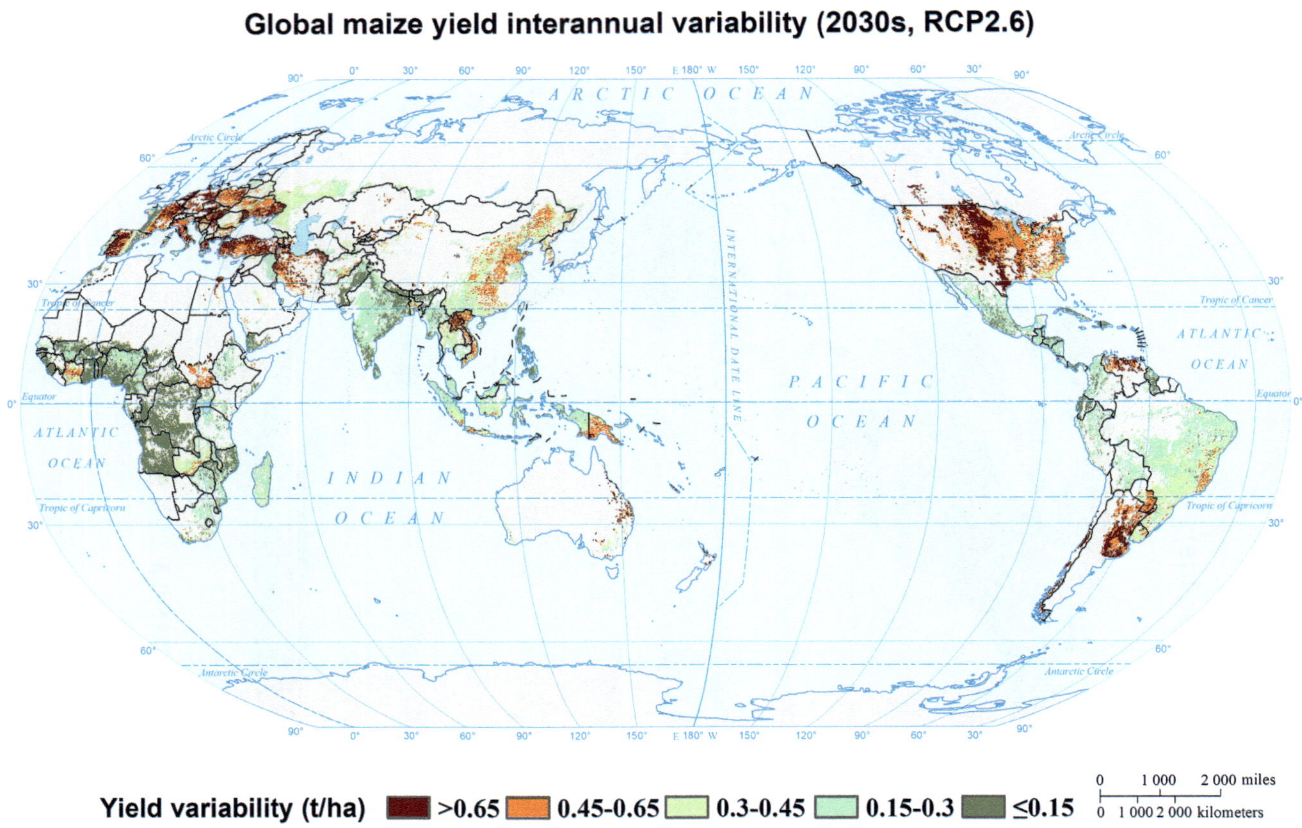
Global maize yield interannual variability (2030s, RCP2.6)
ARCTIC OCEAN
PACIFIC OCEAN
ATLANTIC OCEAN
INDIAN OCEAN
INTERNATIONAL DATE LINE
Arctic Circle
Tropic of Cancer
Equator
Tropic of Capricorn
Antarctic Circle
Yield variability (t/ha) >0.65 0.45-0.65 0.3-0.45 0.15-0.3 ≤0.15
0 1 000 2 000 miles
0 1 000 2 000 kilometers

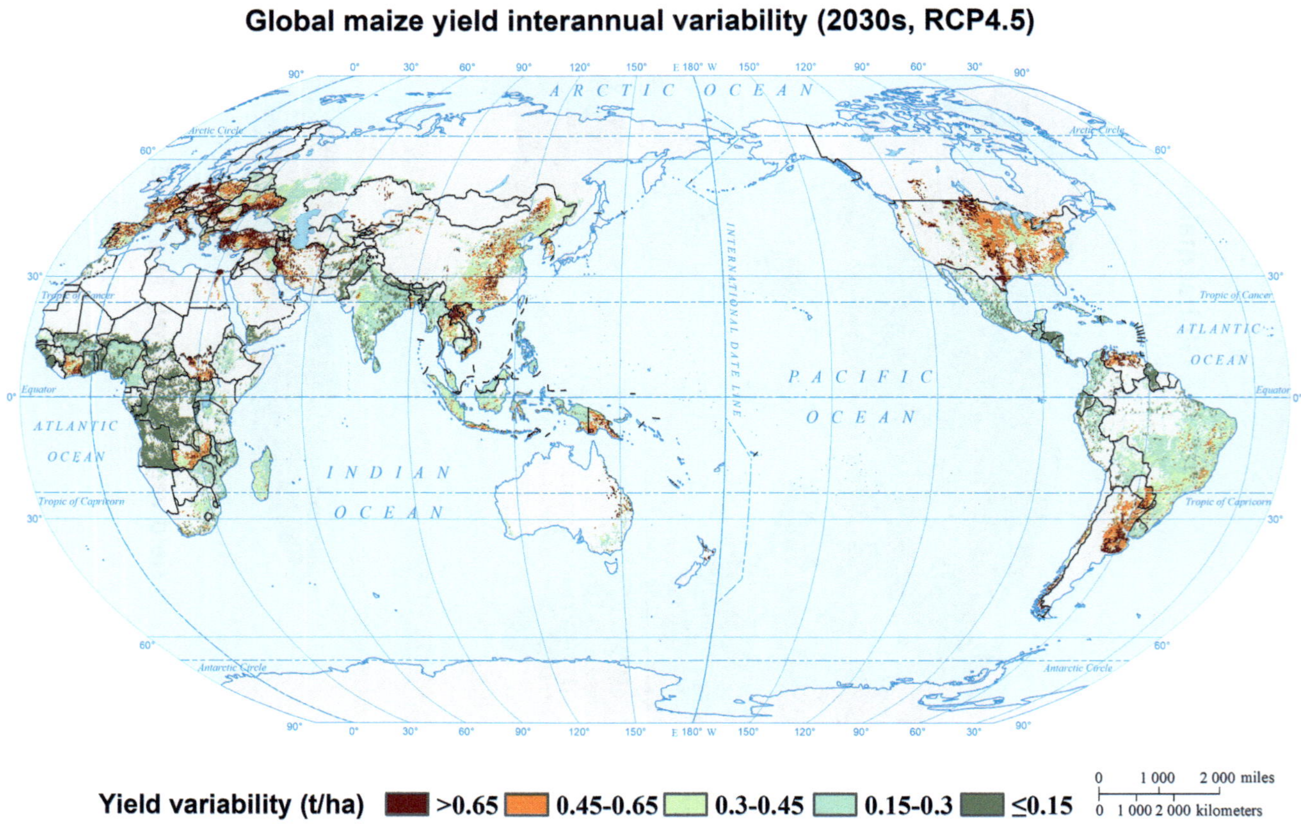
Global maize yield interannual variability (2030s, RCP4.5)
ARCTIC OCEAN
PACIFIC OCEAN
ATLANTIC OCEAN
INDIAN OCEAN
INTERNATIONAL DATE LINE
Arctic Circle
Tropic of Cancer
Equator
Tropic of Capricorn
Antarctic Circle
Yield variability (t/ha) >0.65 0.45-0.65 0.3-0.45 0.15-0.3 ≤0.15
0 1 000 2 000 miles
0 1 000 2 000 kilometers

Global maize yield interannual variability (2030s, RCP8.5)

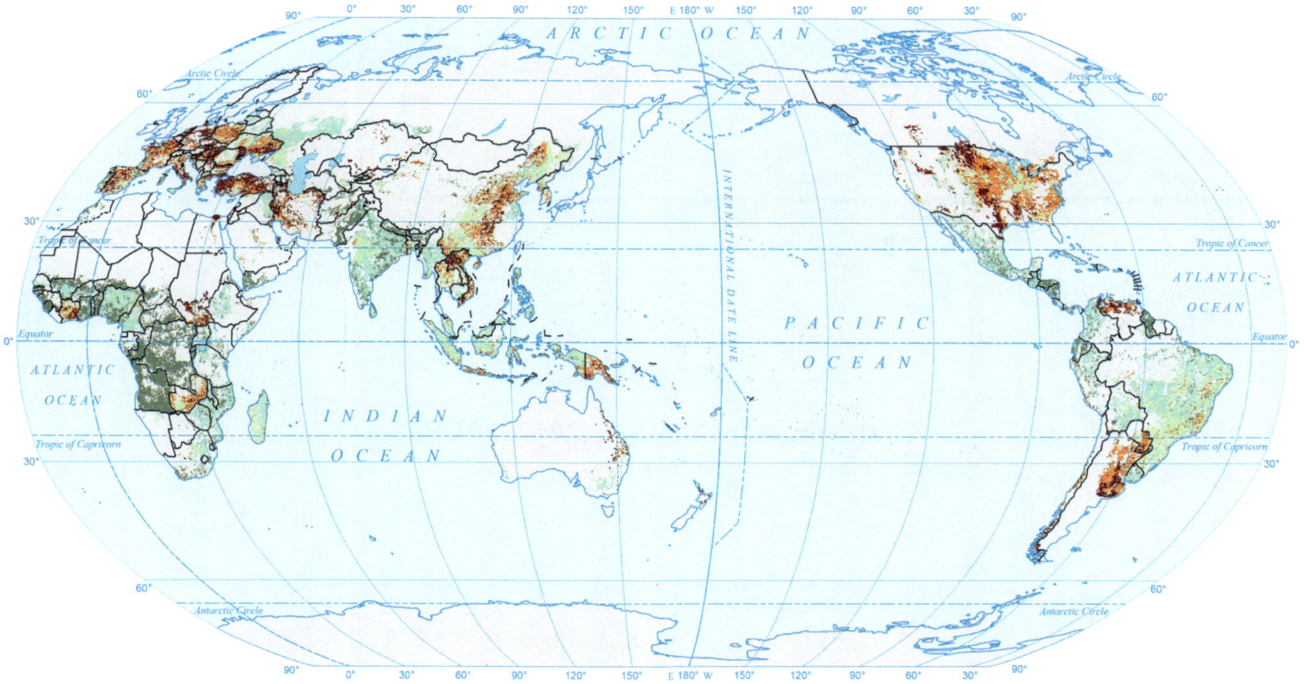

Yield variability (t/ha) >0.65 0.45-0.65 0.3-0.45 0.15-0.3 ≤0.15

Global maize yield interannual variability (2050s, RCP2.6)

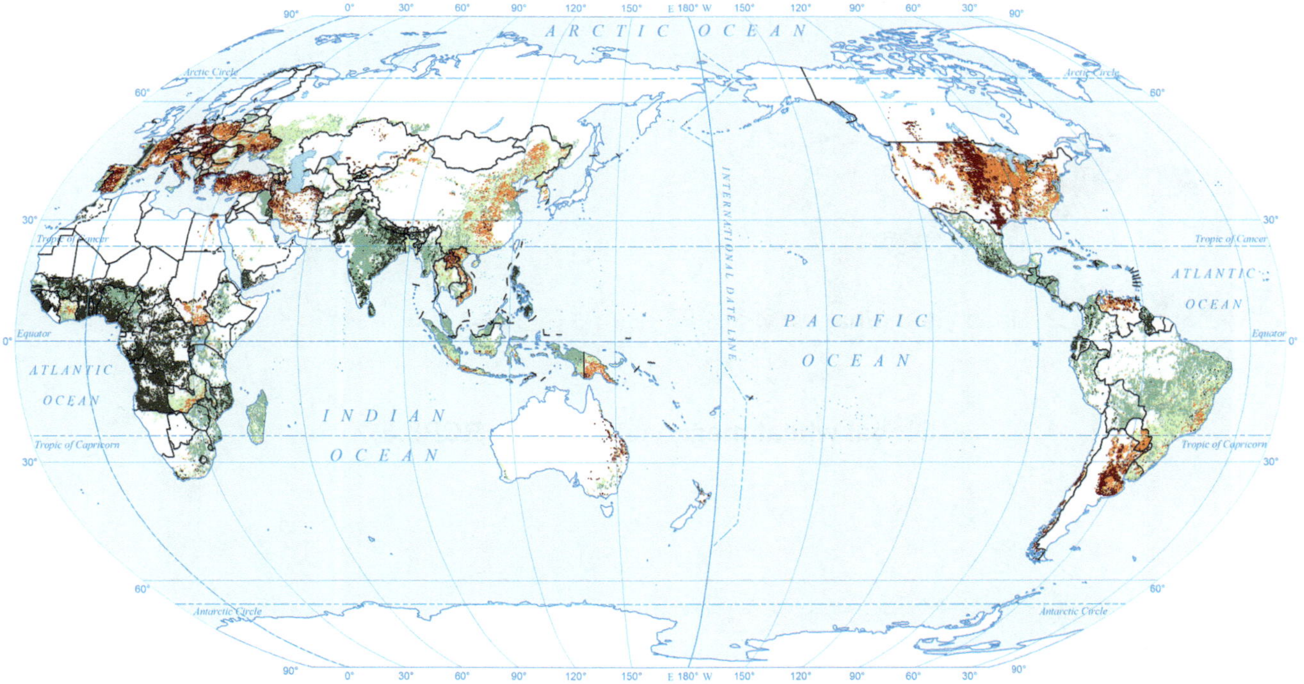

Yield variability (t/ha) >0.65 0.45-0.65 0.3-0.45 0.15-0.3 ≤0.15

Global maize yield interannual variability (2050s, RCP4.5)

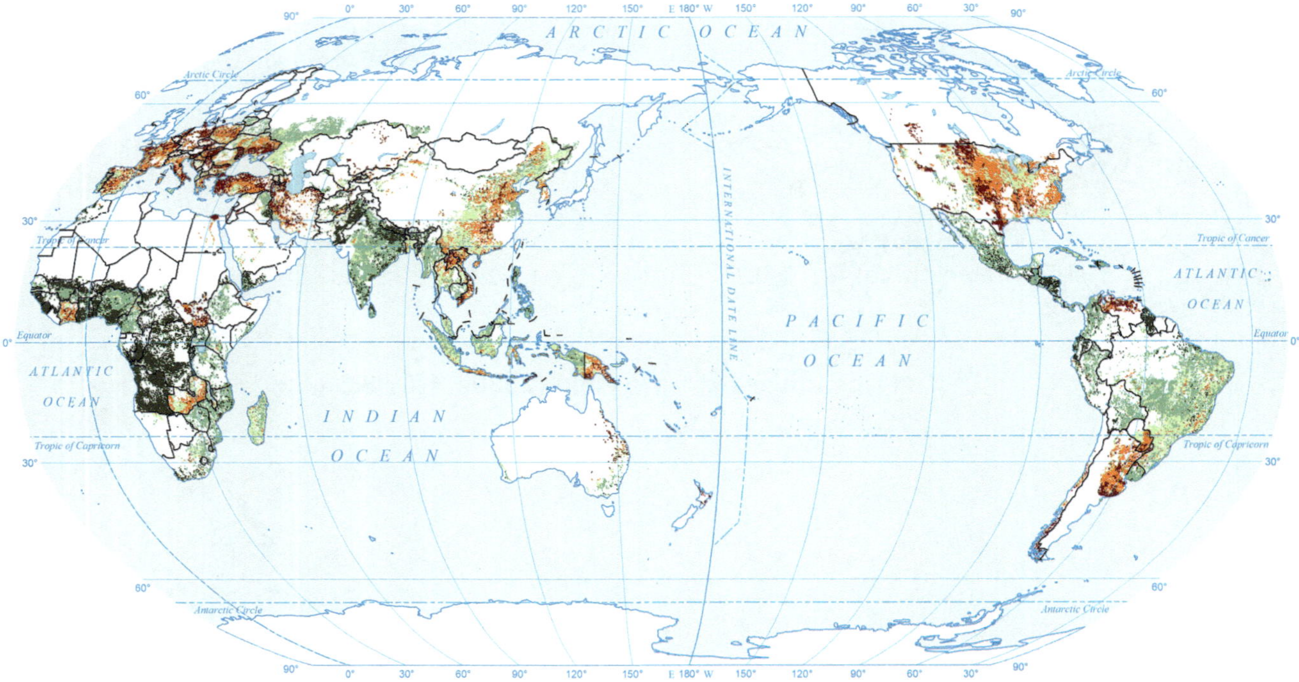

Yield variability (t/ha) >0.65 0.45-0.65 0.3-0.45 0.15-0.3 ≤0.15

0 1 000 2 000 miles
0 1 000 2 000 kilometers

Global maize yield interannual variability (2050s, RCP8.5)

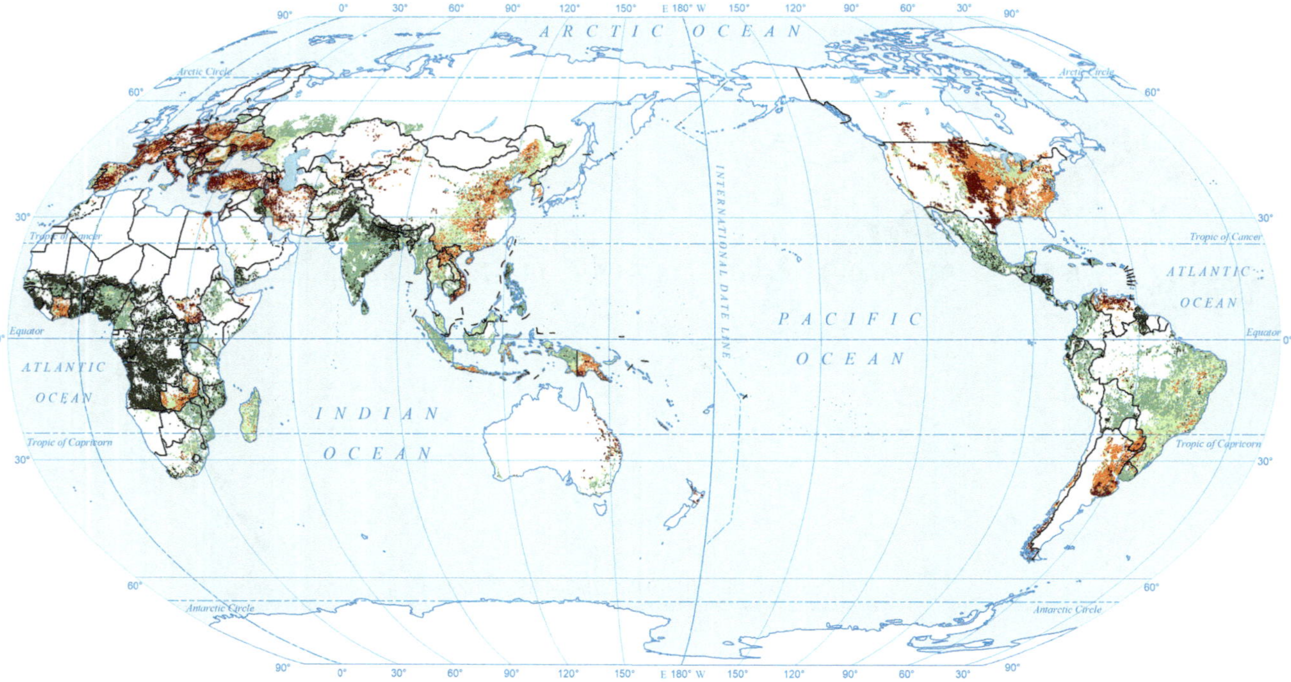

Yield variability (t/ha) >0.65 0.45-0.65 0.3-0.45 0.15-0.3 ≤0.15

0 1 000 2 000 miles
0 1 000 2 000 kilometers

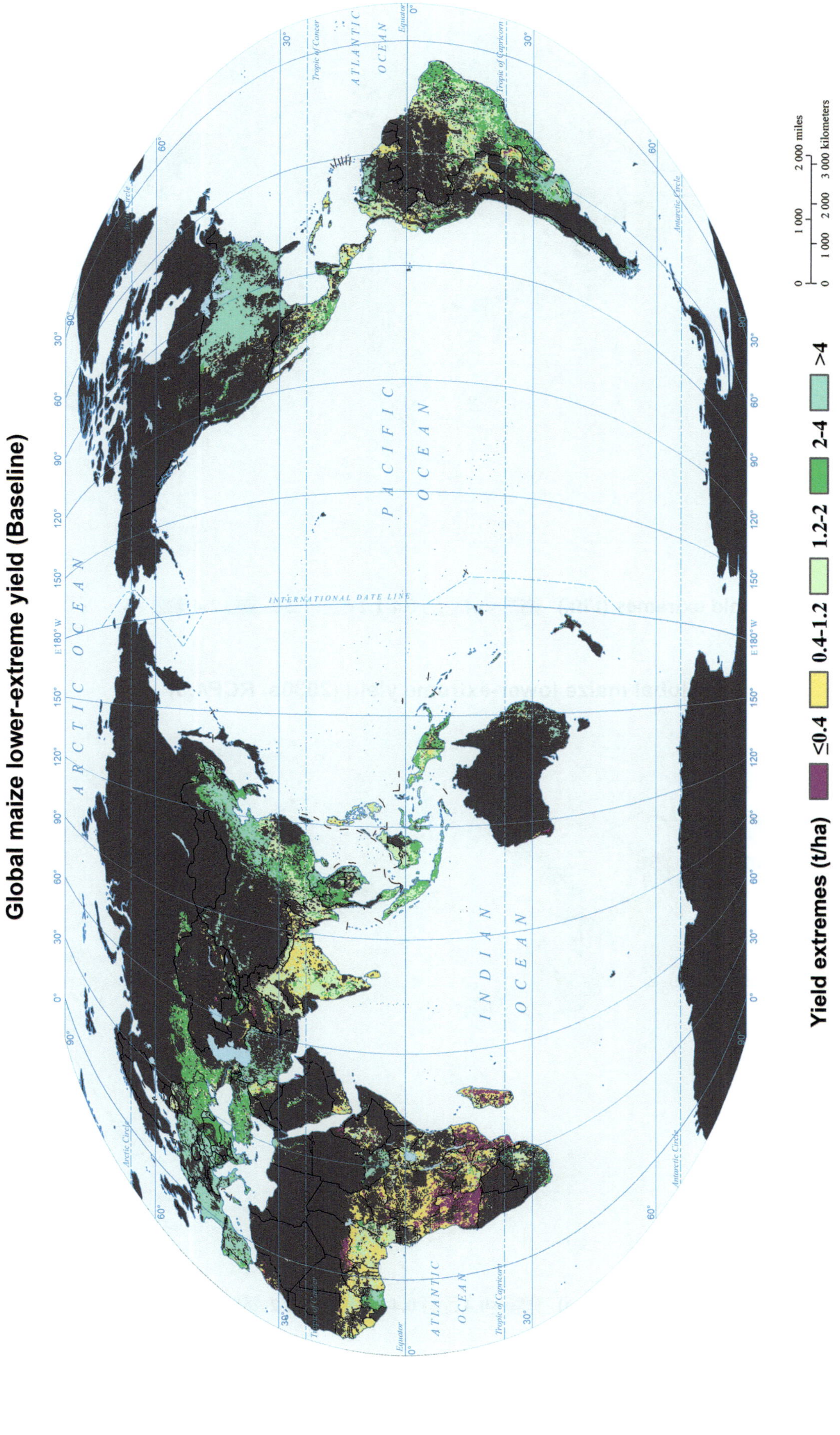
Global maize lower-extreme yield (Baseline)
ARCTIC OCEAN
PACIFIC OCEAN
ATLANTIC OCEAN
INDIAN OCEAN
INTERNATIONAL DATE LINE
Tropic of Cancer
Tropic of Capricorn
Equator
Arctic Circle
Antarctic Circle
Yield extremes (t/ha)
≤0.4
0.4-1.2
1.2-2
2-4
>4
0 1 000 2 000 miles
0 1 000 2 000 3 000 kilometers

Global maize lower-extreme yield (2030s, RCP2.6)

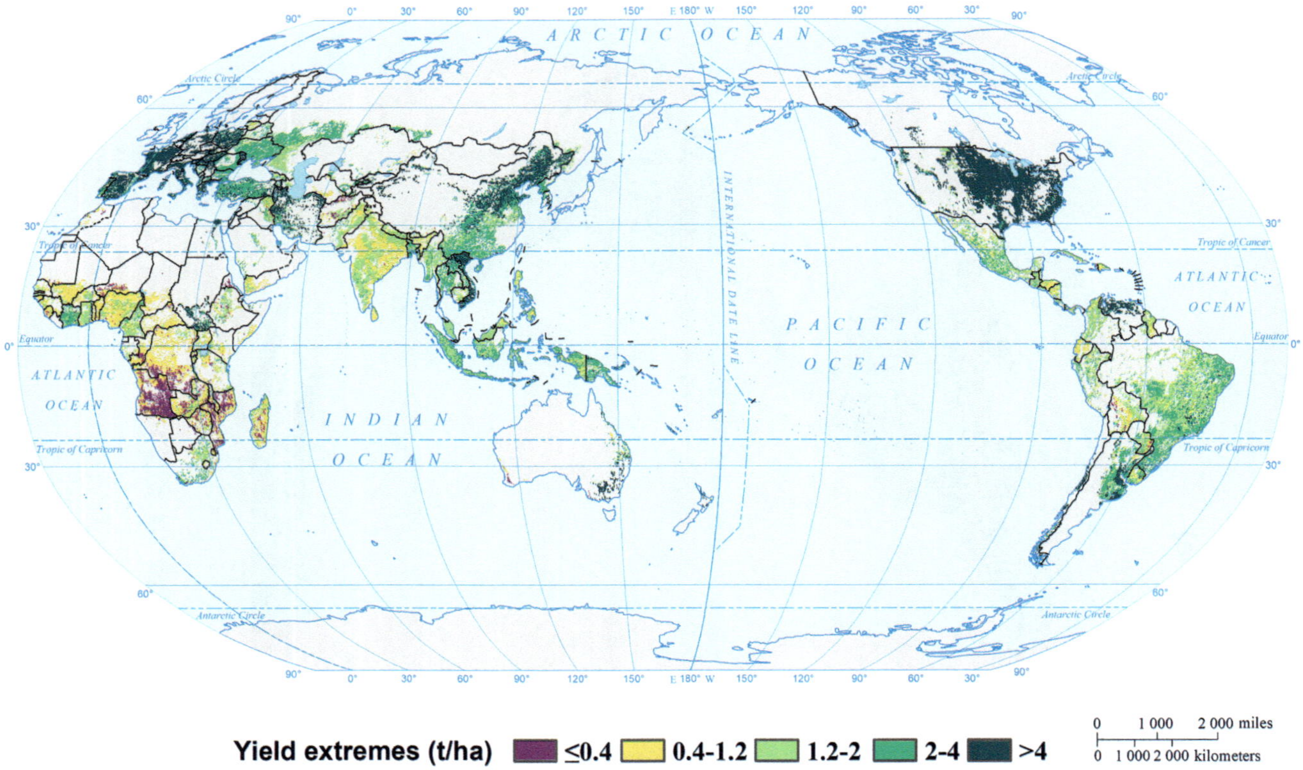

Global maize lower-extreme yield (2030s, RCP4.5)

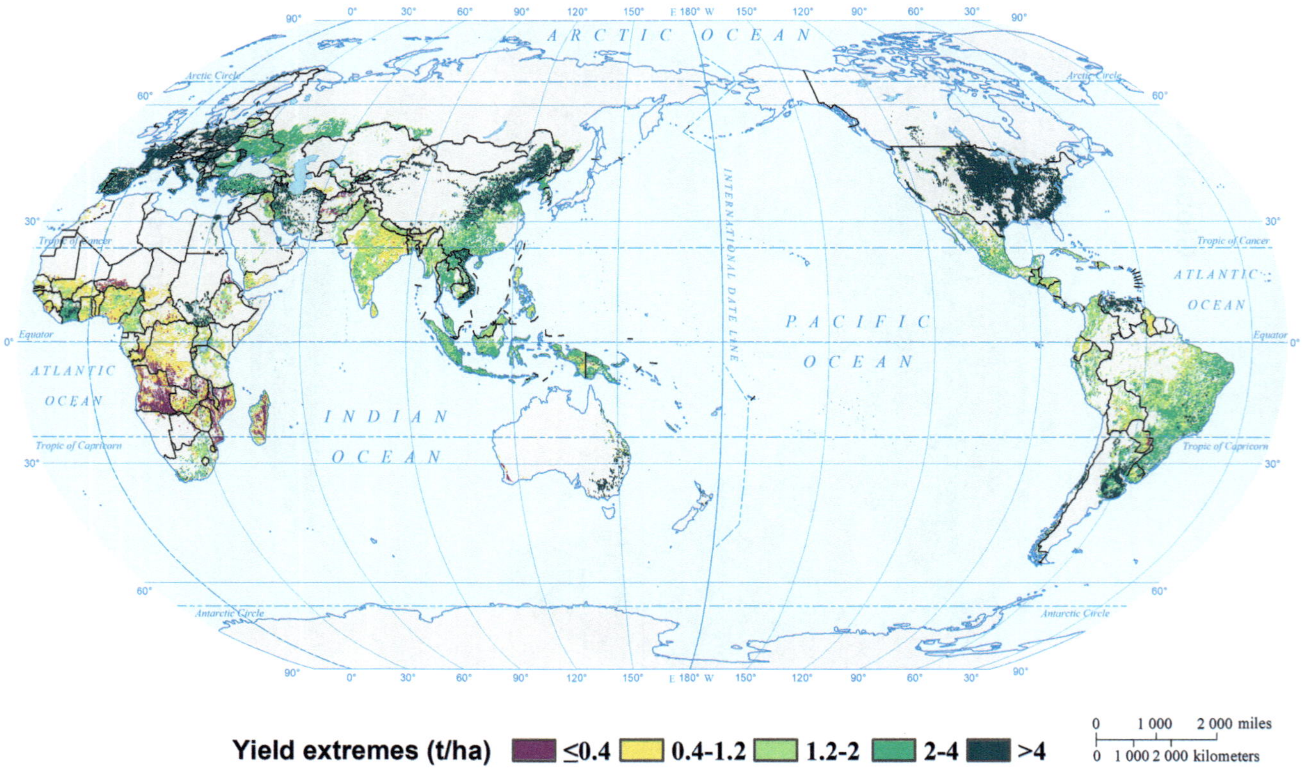

Global maize lower-extreme yield (2030s, RCP8.5)

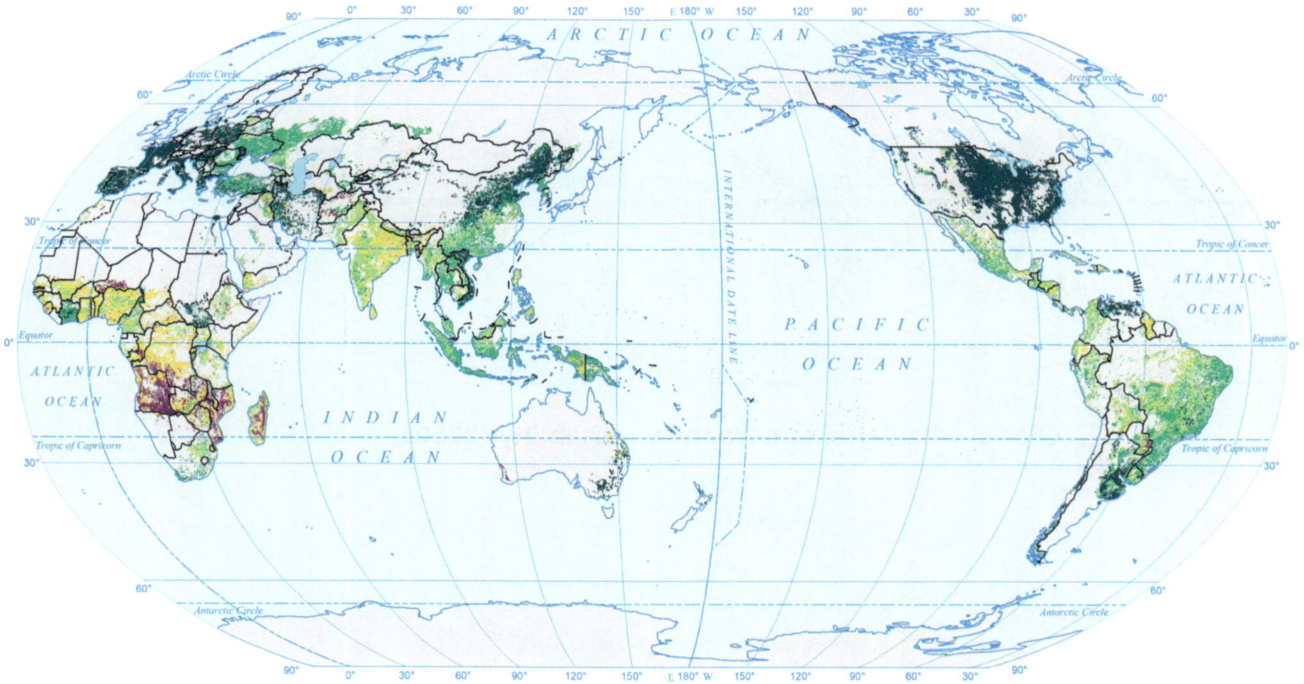

Yield extremes (t/ha) ≤0.4 0.4-1.2 1.2-2 2-4 >4

Global maize lower-extreme yield (2050s, RCP2.6)

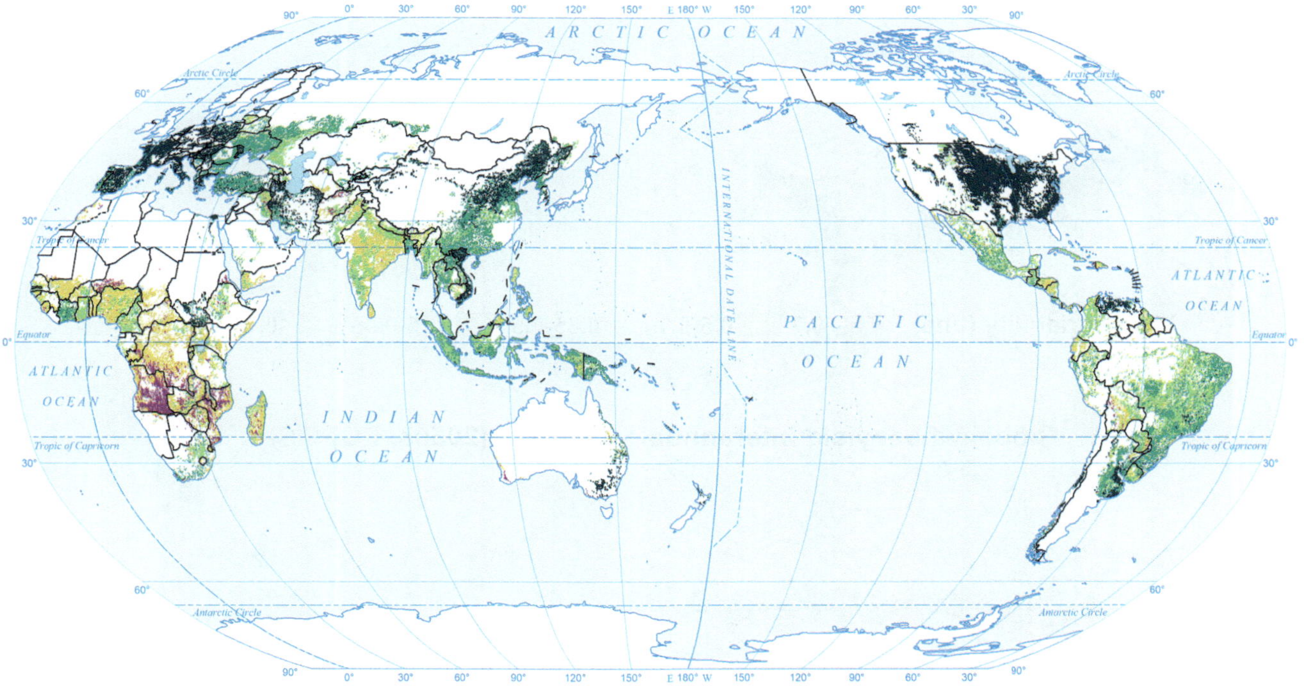

Yield extremes (t/ha) ≤0.4 0.4-1.2 1.2-2 2-4 >4

Global maize lower-extreme yield (2050s, RCP4.5)

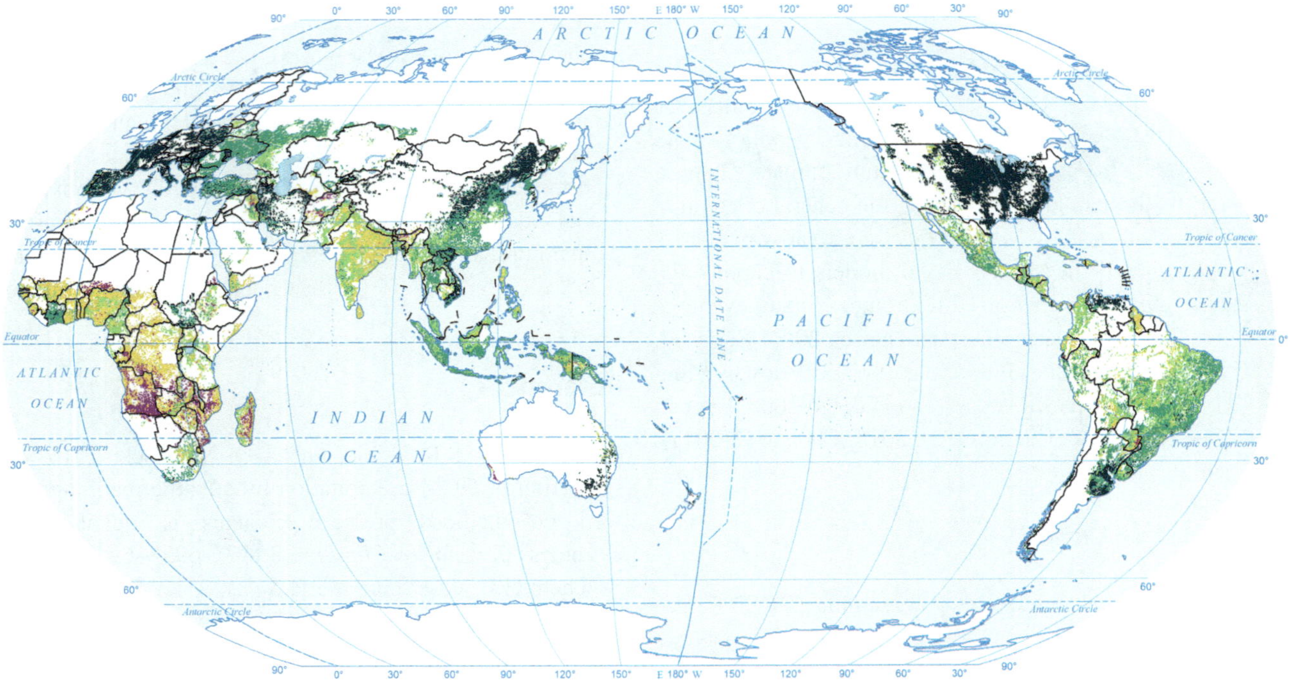

Yield extremes (t/ha) ≤0.4 0.4-1.2 1.2-2 2-4 >4

Global maize lower-extreme yield (2050s, RCP8.5)

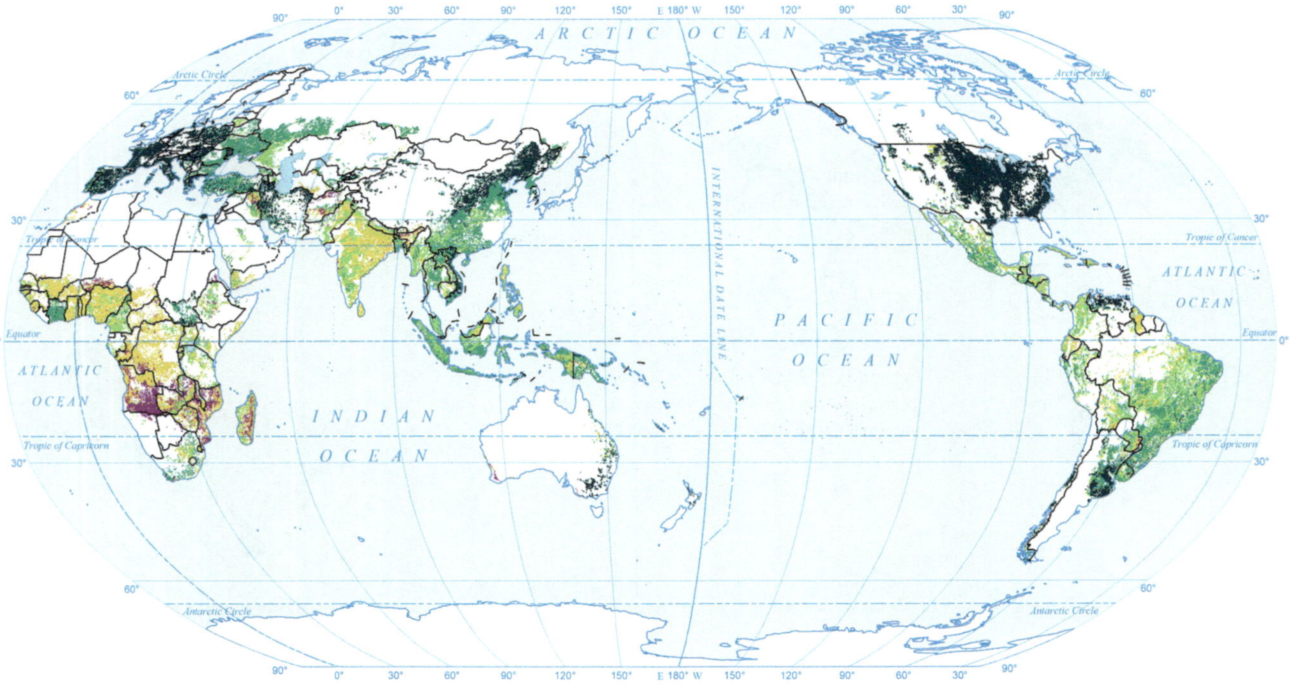

Yield extremes (t/ha) ≤0.4 0.4-1.2 1.2-2 2-4 >4

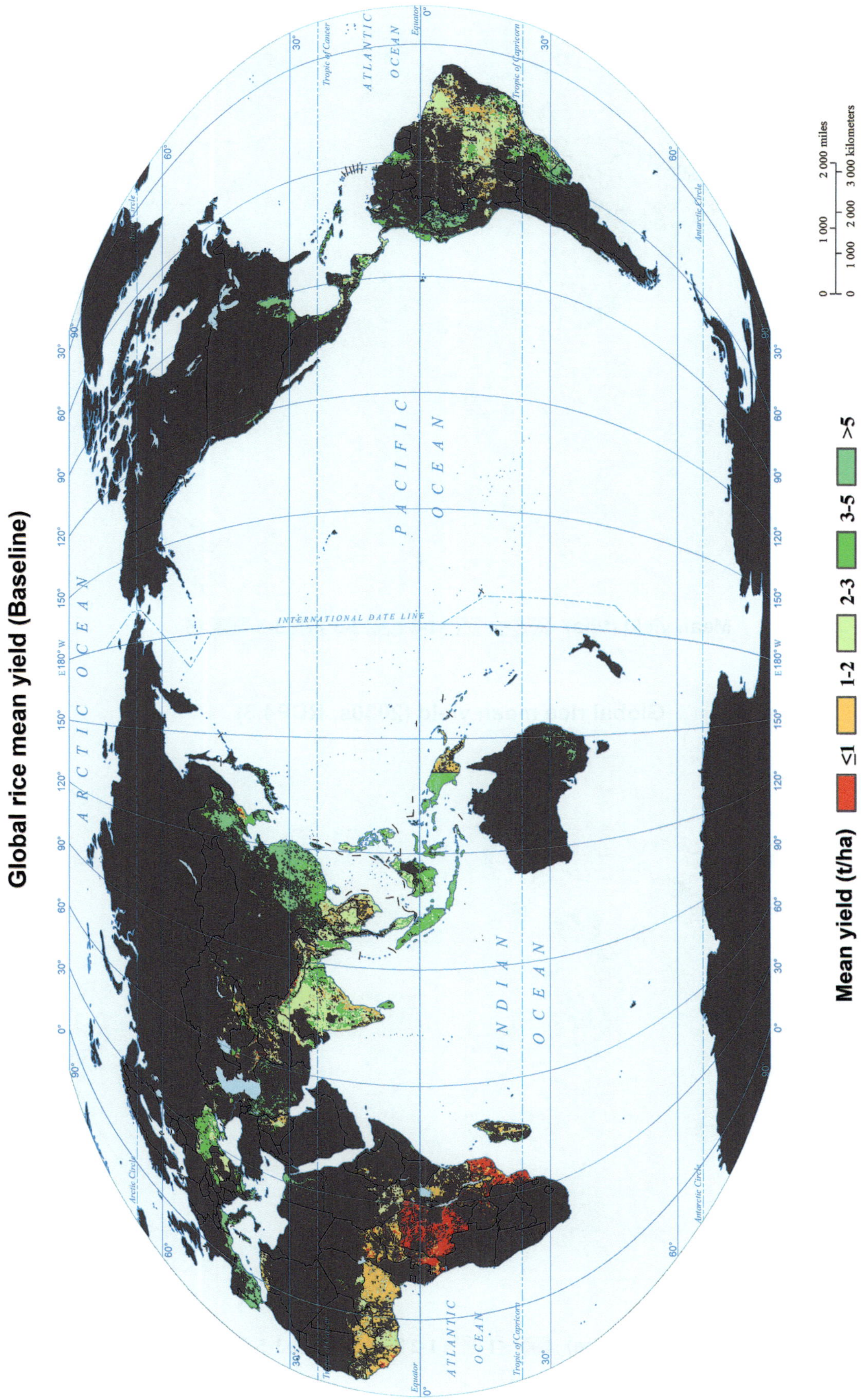
Global rice mean yield (Baseline)
Mean yield (t/ha)
≤1
1-2
2-3
3-5
>5
ARCTIC OCEAN
PACIFIC OCEAN
ATLANTIC OCEAN
INDIAN OCEAN
INTERNATIONAL DATE LINE
Equator
Tropic of Cancer
Tropic of Capricorn
Arctic Circle
Antarctic Circle
0 1 000 2 000 miles
0 1 000 2 000 3 000 kilometers

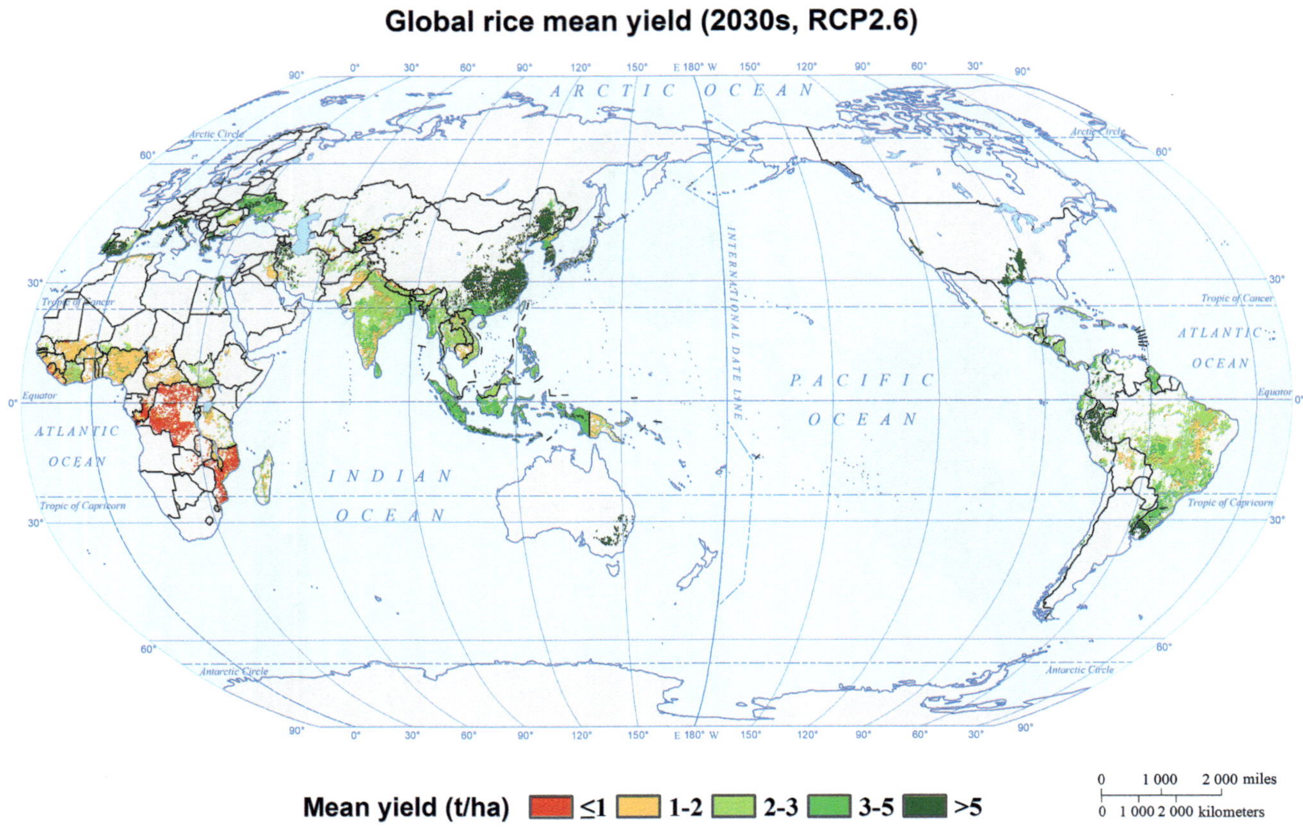
Global rice mean yield (2030s, RCP2.6)
ARCTIC OCEAN
PACIFIC OCEAN
INDIAN OCEAN
ATLANTIC OCEAN
INTERNATIONAL DATE LINE
Arctic Circle
Tropic of Cancer
Equator
Tropic of Capricorn
Antarctic Circle
Mean yield (t/ha) ≤1 1-2 2-3 3-5 >5
0 1 000 2 000 miles
0 1 000 2 000 kilometers

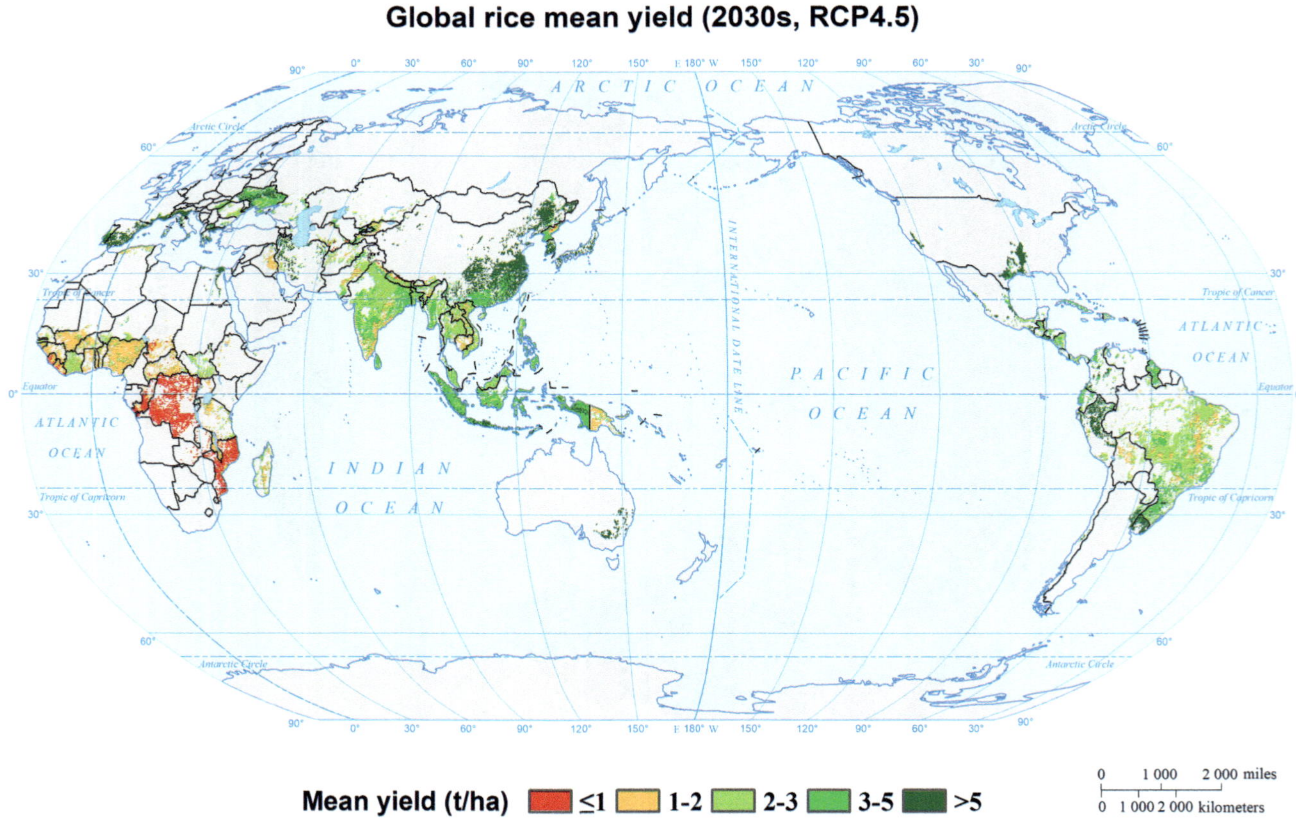
Global rice mean yield (2030s, RCP4.5)
ARCTIC OCEAN
PACIFIC OCEAN
INDIAN OCEAN
ATLANTIC OCEAN
INTERNATIONAL DATE LINE
Arctic Circle
Tropic of Cancer
Equator
Tropic of Capricorn
Antarctic Circle
Mean yield (t/ha) ≤1 1-2 2-3 3-5 >5
0 1 000 2 000 miles
0 1 000 2 000 kilometers

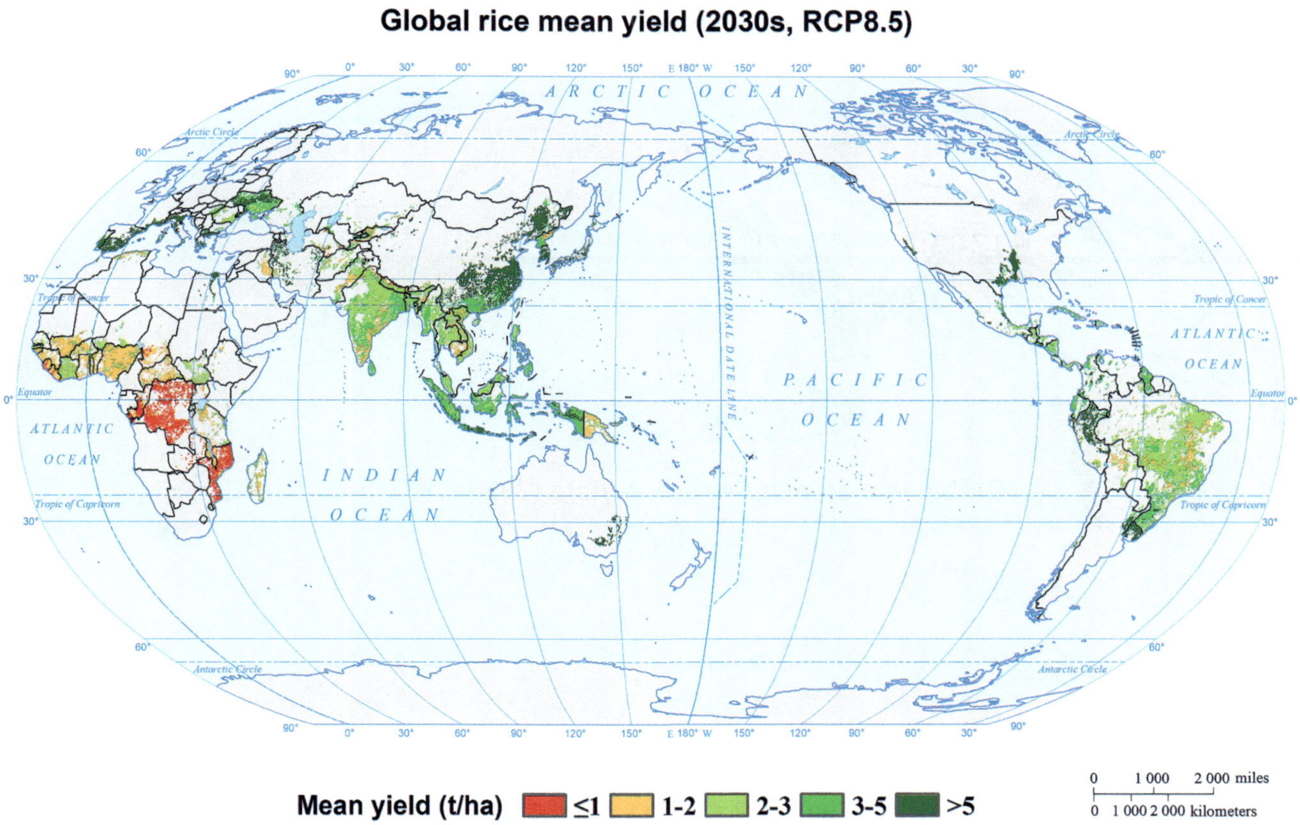
Global rice mean yield (2030s, RCP8.5)
ARCTIC OCEAN
PACIFIC OCEAN
INDIAN OCEAN
ATLANTIC OCEAN
INTERNATIONAL DATE LINE
Arctic Circle
Tropic of Cancer
Equator
Tropic of Capricorn
Antarctic Circle
Mean yield (t/ha) ≤1 1-2 2-3 3-5 >5
0 1 000 2 000 miles
0 1 000 2 000 kilometers

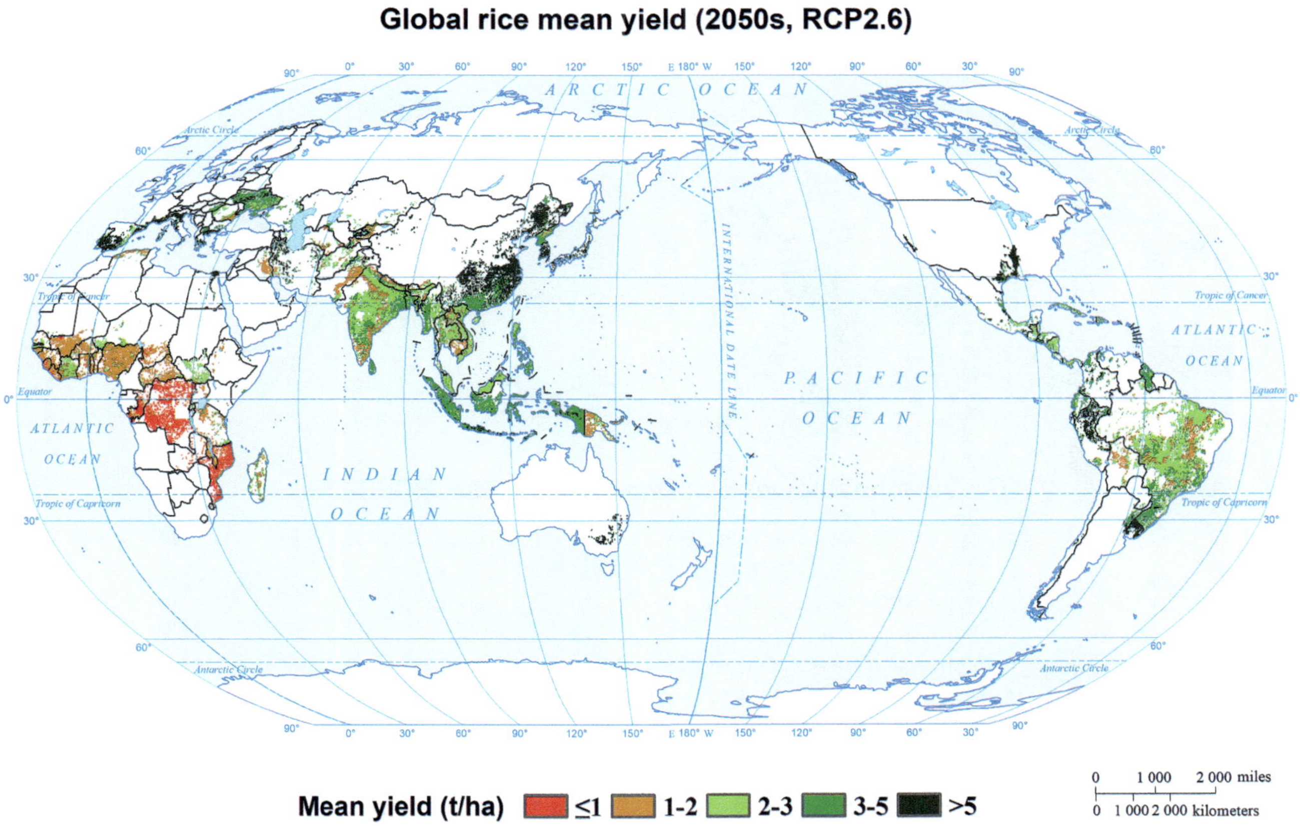
Global rice mean yield (2050s, RCP2.6)
ARCTIC OCEAN
PACIFIC OCEAN
INDIAN OCEAN
ATLANTIC OCEAN
INTERNATIONAL DATE LINE
Arctic Circle
Tropic of Cancer
Equator
Tropic of Capricorn
Antarctic Circle
Mean yield (t/ha) ≤1 1-2 2-3 3-5 >5
0 1 000 2 000 miles
0 1 000 2 000 kilometers

Global rice mean yield (2050s, RCP4.5)

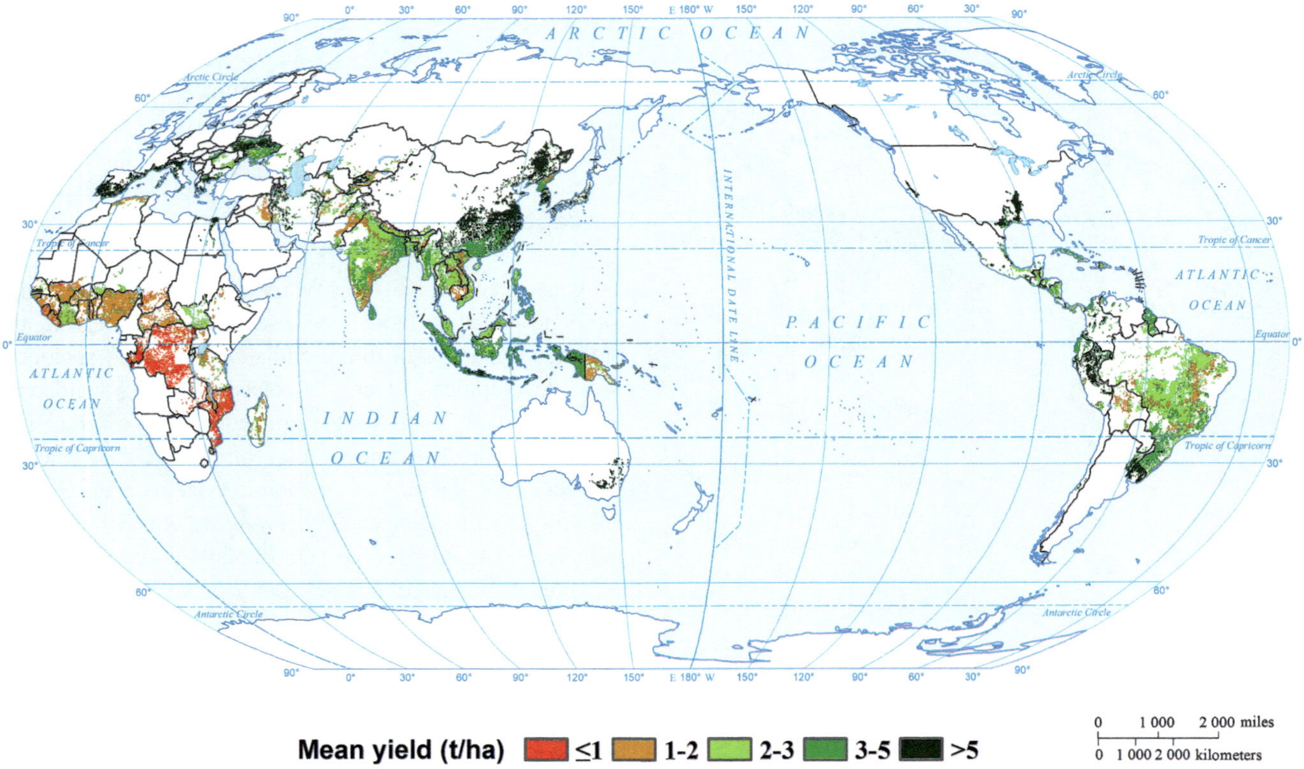

Global rice mean yield (2050s, RCP8.5)

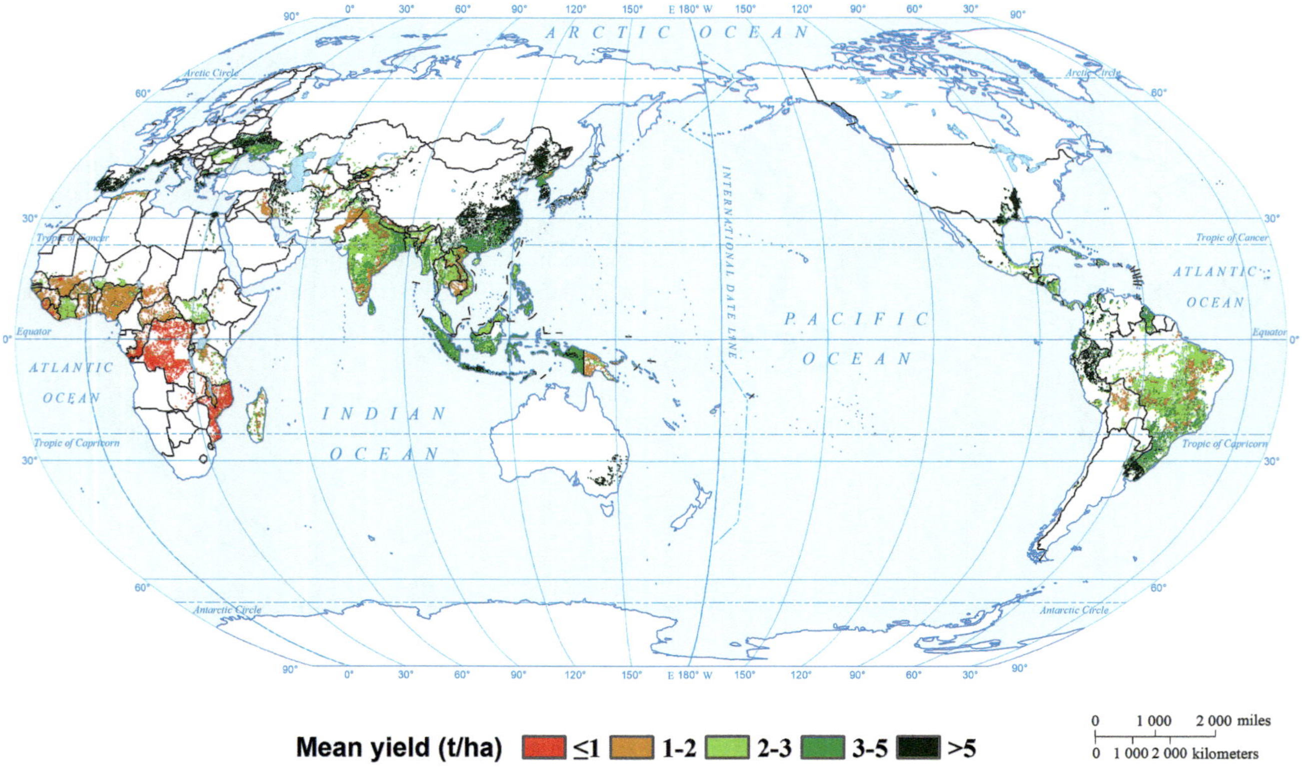

Global rice yield interannual variability (2030s, RCP2.6)

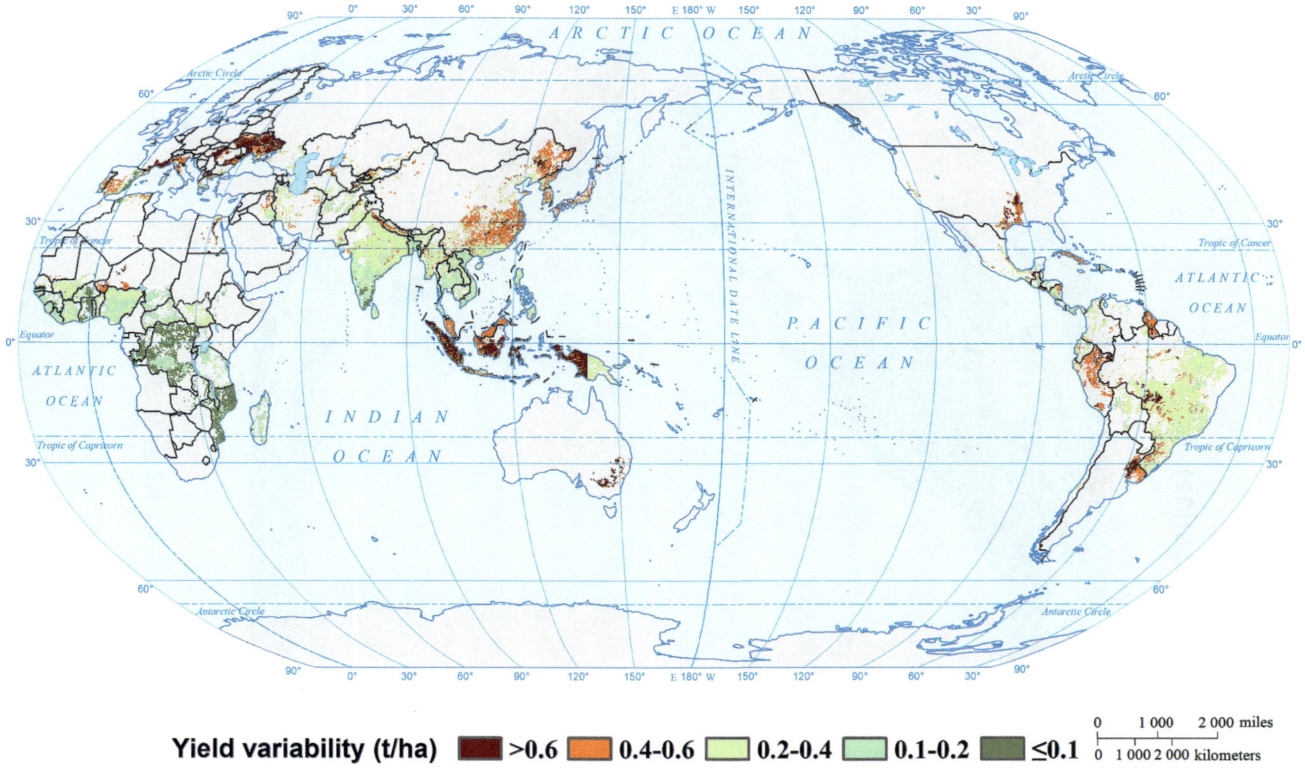

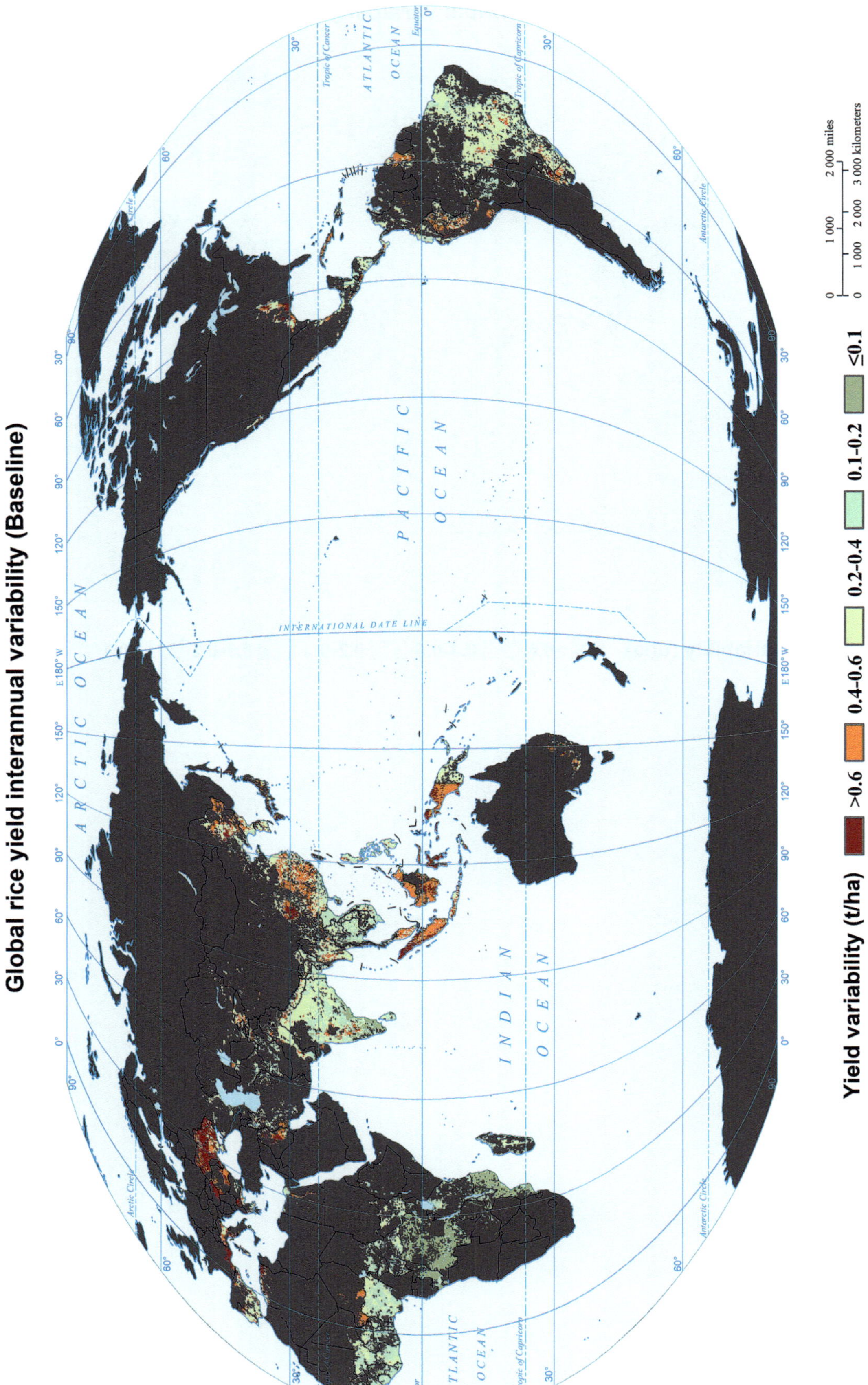
Global rice yield interannual variability (Baseline)
ARCTIC OCEAN
PACIFIC OCEAN
ATLANTIC OCEAN
INDIAN OCEAN
INTERNATIONAL DATE LINE
Tropic of Cancer
Tropic of Capricorn
Equator
Arctic Circle
Antarctic Circle
Yield variability (t/ha)
>0.6
0.4-0.6
0.2-0.4
0.1-0.2
≤0.1
0 1 000 2 000 miles
0 1 000 2 000 3 000 kilometers

Global rice yield interannual variability (2030s, RCP4.5)

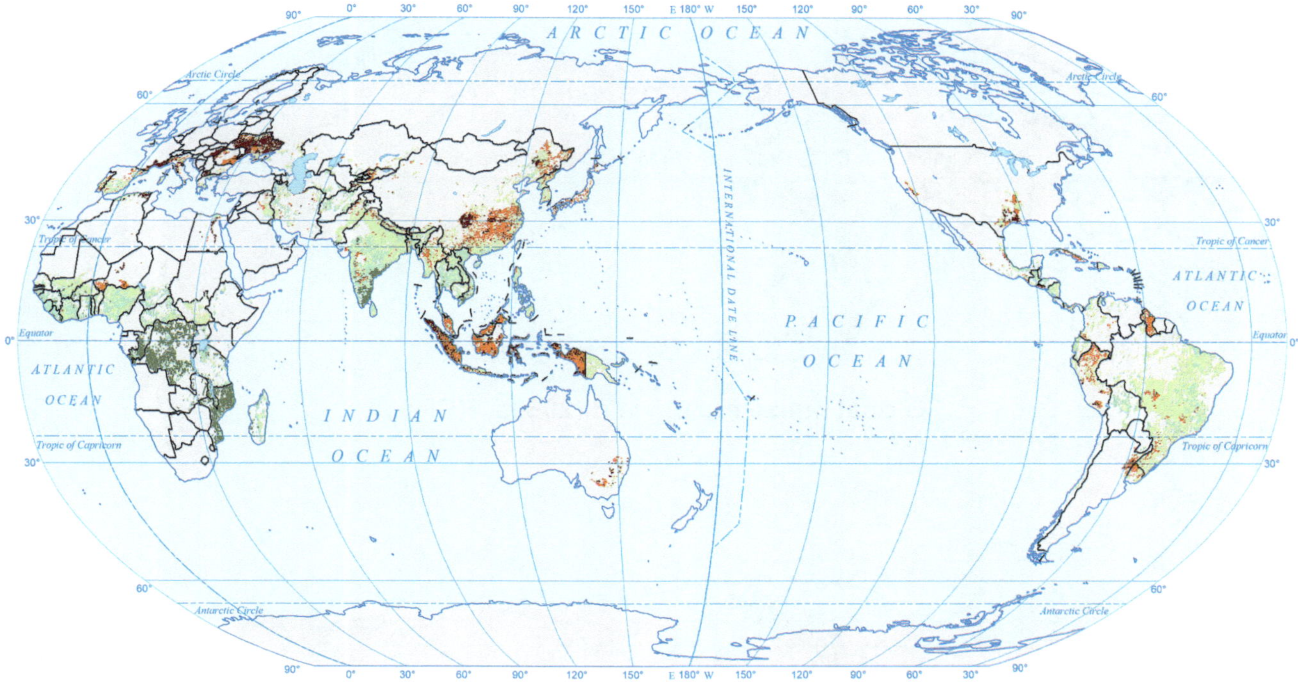

Yield variability (t/ha) >0.6 0.4-0.6 0.2-0.4 0.1-0.2 ≤0.1 0 1 000 2 000 miles 0 1 000 2 000 kilometers

Global rice yield interannual variability (2030s, RCP8.5)

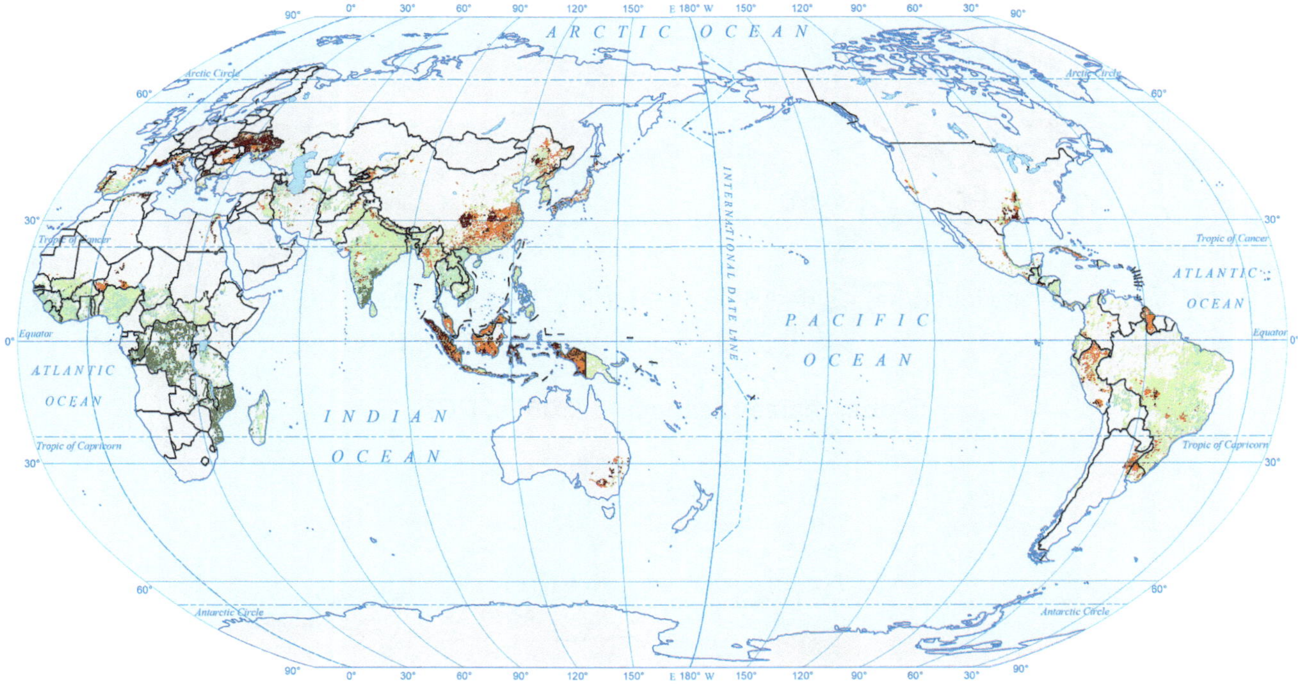

Yield variability (t/ha) >0.6 0.4-0.6 0.2-0.4 0.1-0.2 ≤0.1 0 1 000 2 000 miles 0 1 000 2 000 kilometers

Global rice yield interannual variability (2050s, RCP2.6)

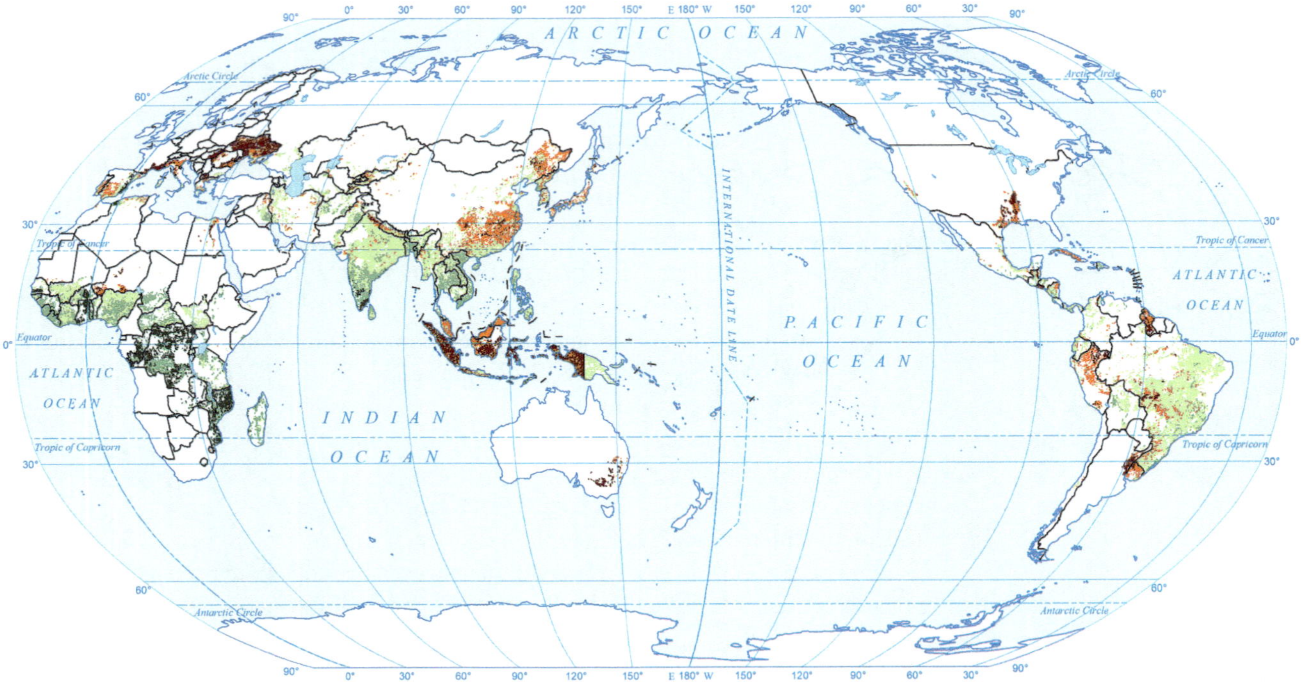

Yield variability (t/ha) >0.6 0.4-0.6 0.2-0.4 0.1-0.2 ≤0.1

0 1 000 2 000 miles
0 1 000 2 000 kilometers

Global rice yield interannual variability (2050s, RCP4.5)

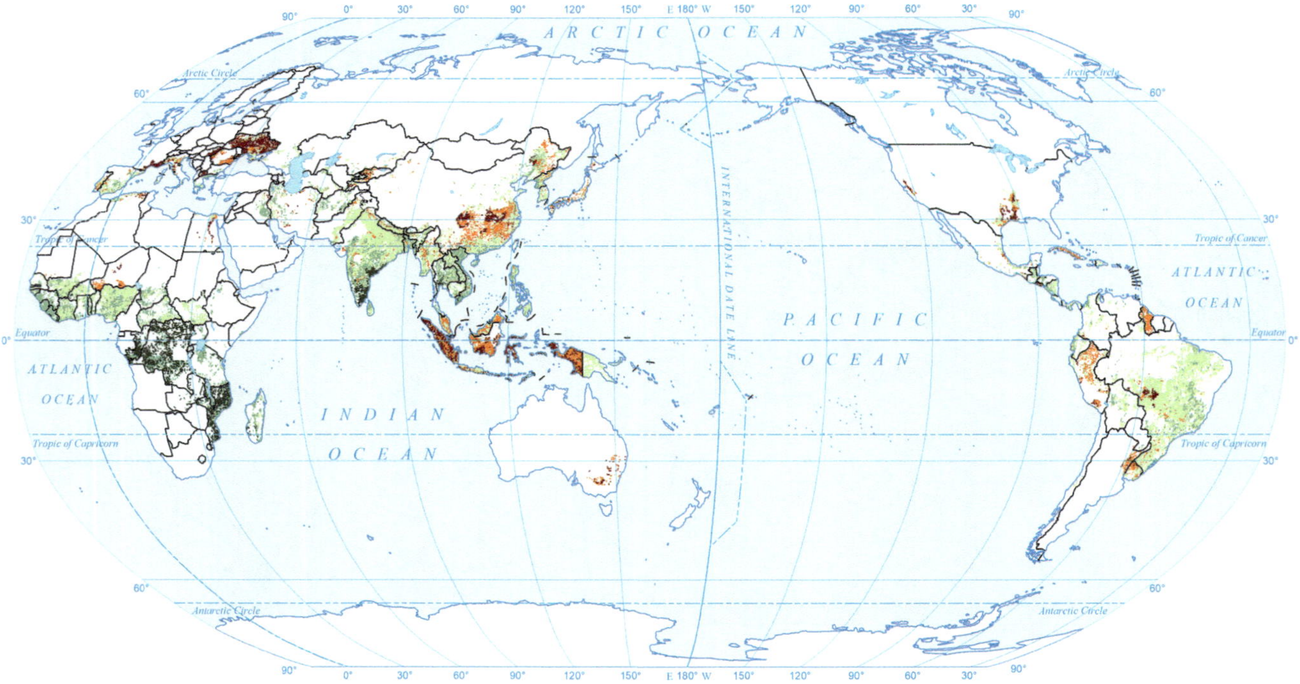

Yield variability (t/ha) >0.6 0.4-0.6 0.2-0.4 0.1-0.2 ≤0.1

0 1 000 2 000 miles
0 1 000 2 000 kilometers

Global rice yield interannual variability (2050s, RCP8.5)

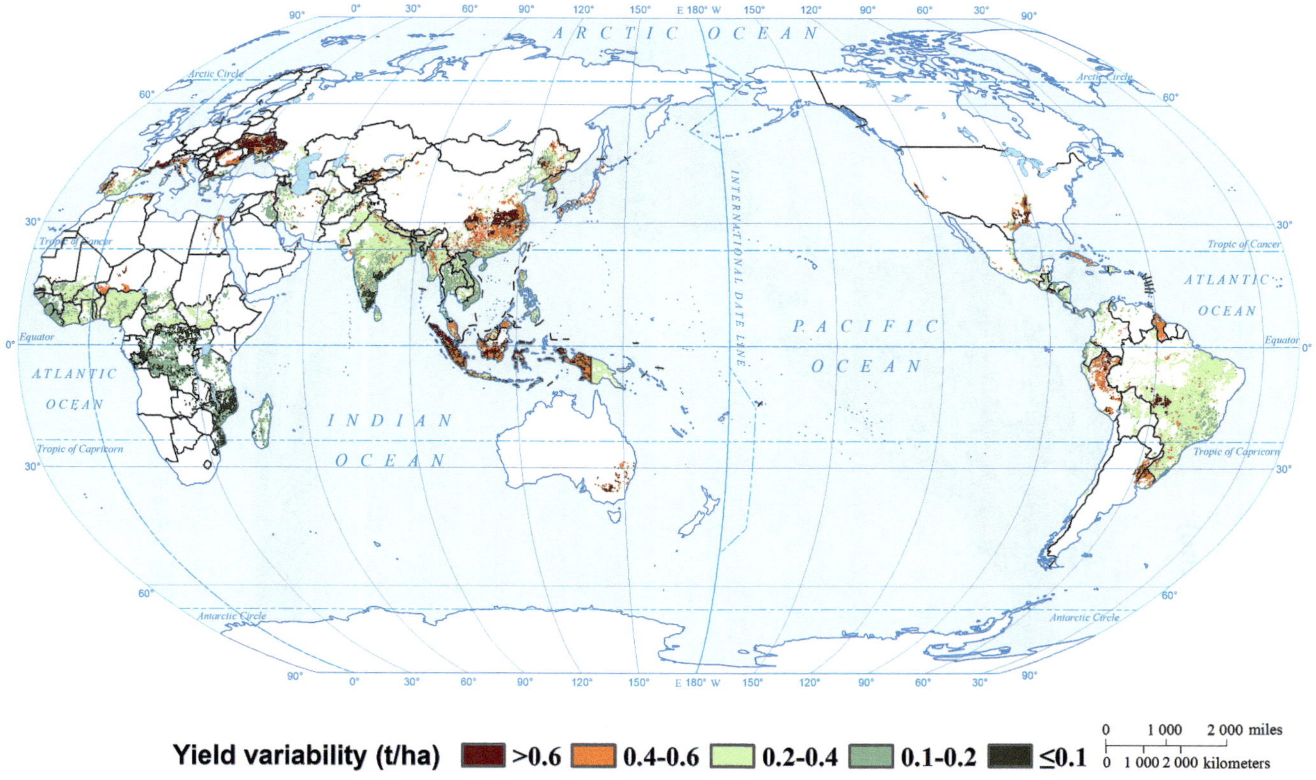

Yield variability (t/ha) >0.6 0.4-0.6 0.2-0.4 0.1-0.2 ≤0.1

Global rice lower-extreme yield (2030s, RCP2.6)

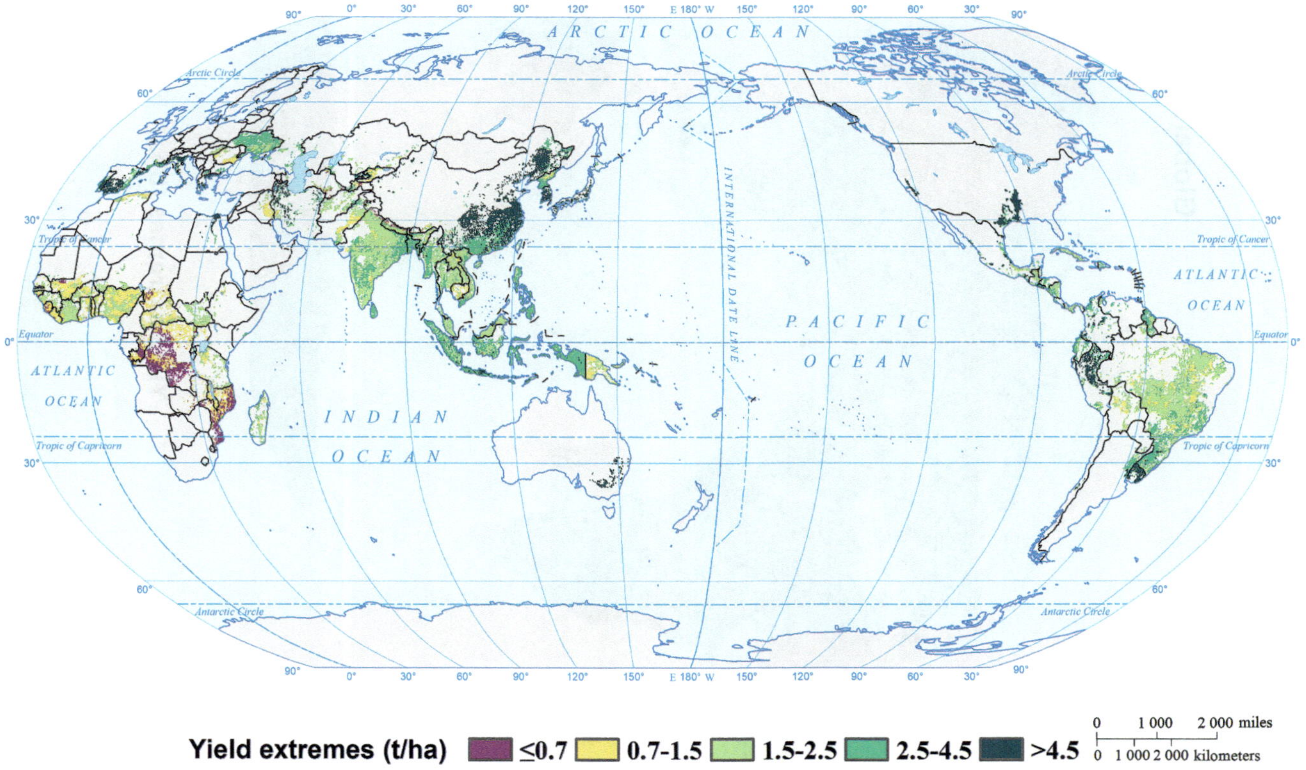

Yield extremes (t/ha) ≤0.7 0.7-1.5 1.5-2.5 2.5-4.5 >4.5

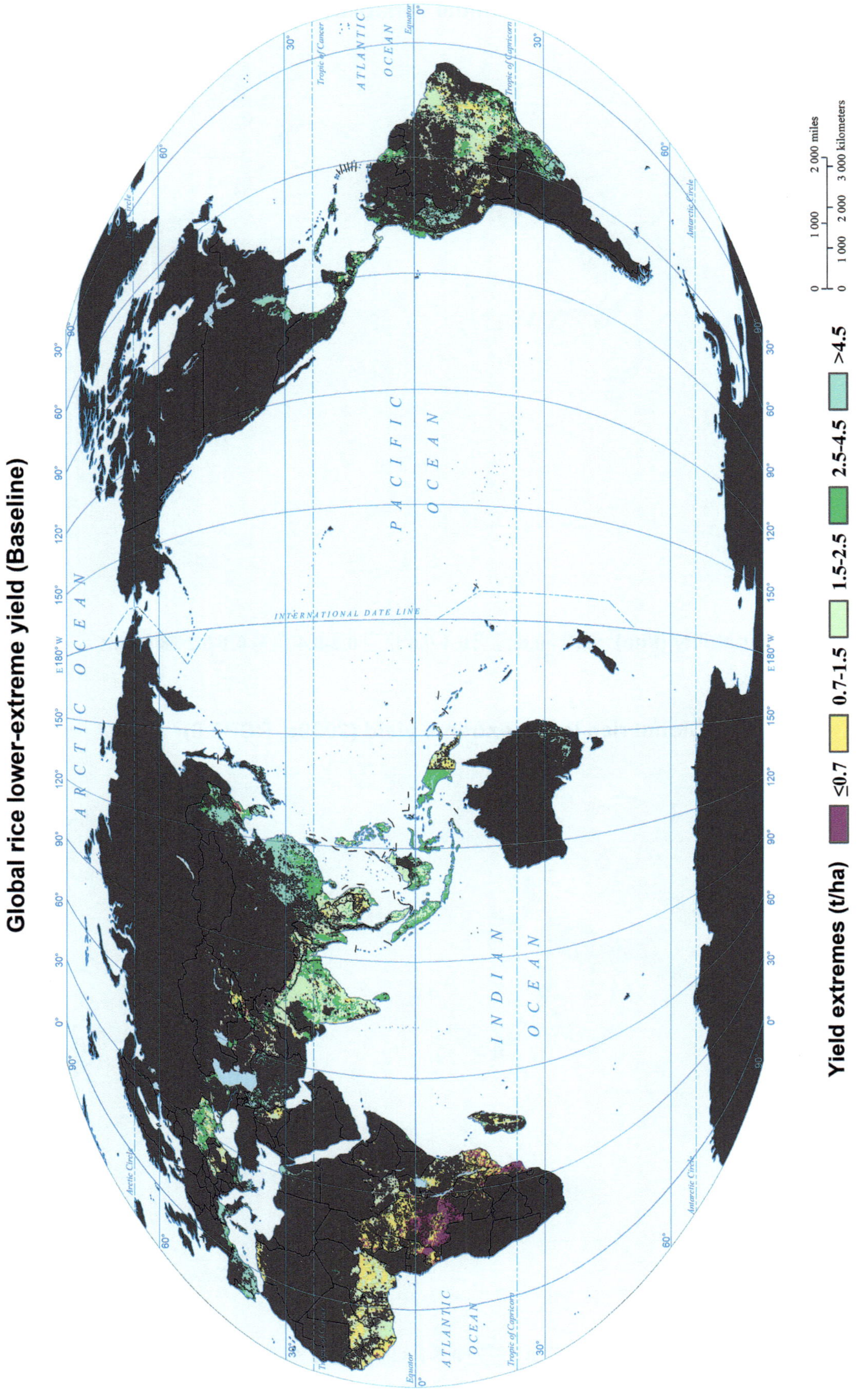

Global rice lower-extreme yield (Baseline)
ARCTIC OCEAN
PACIFIC OCEAN
INDIAN OCEAN
ATLANTIC OCEAN
INTERNATIONAL DATE LINE
Arctic Circle
Tropic of Cancer
Equator
Tropic of Capricorn
Antarctic Circle
Yield extremes (t/ha)
≤0.7
0.7-1.5
1.5-2.5
2.5-4.5
>4.5
0 1 000 2 000 miles
0 1 000 2 000 3 000 kilometers

Global rice lower-extreme yield (2030s, RCP4.5)

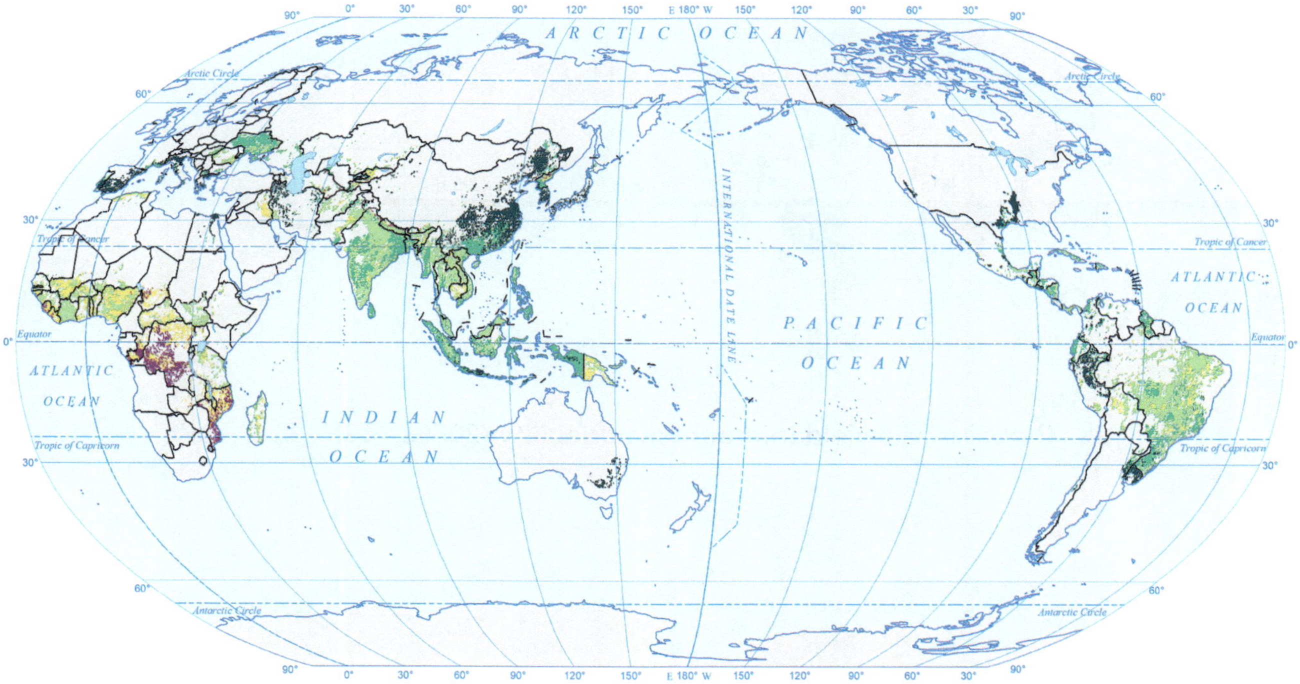

Yield extremes (t/ha) ≤0.7 0.7-1.5 1.5-2.5 2.5-4.5 >4.5

0 1 000 2 000 miles
0 1 000 2 000 kilometers

Global rice lower-extreme yield (2030s, RCP8.5)

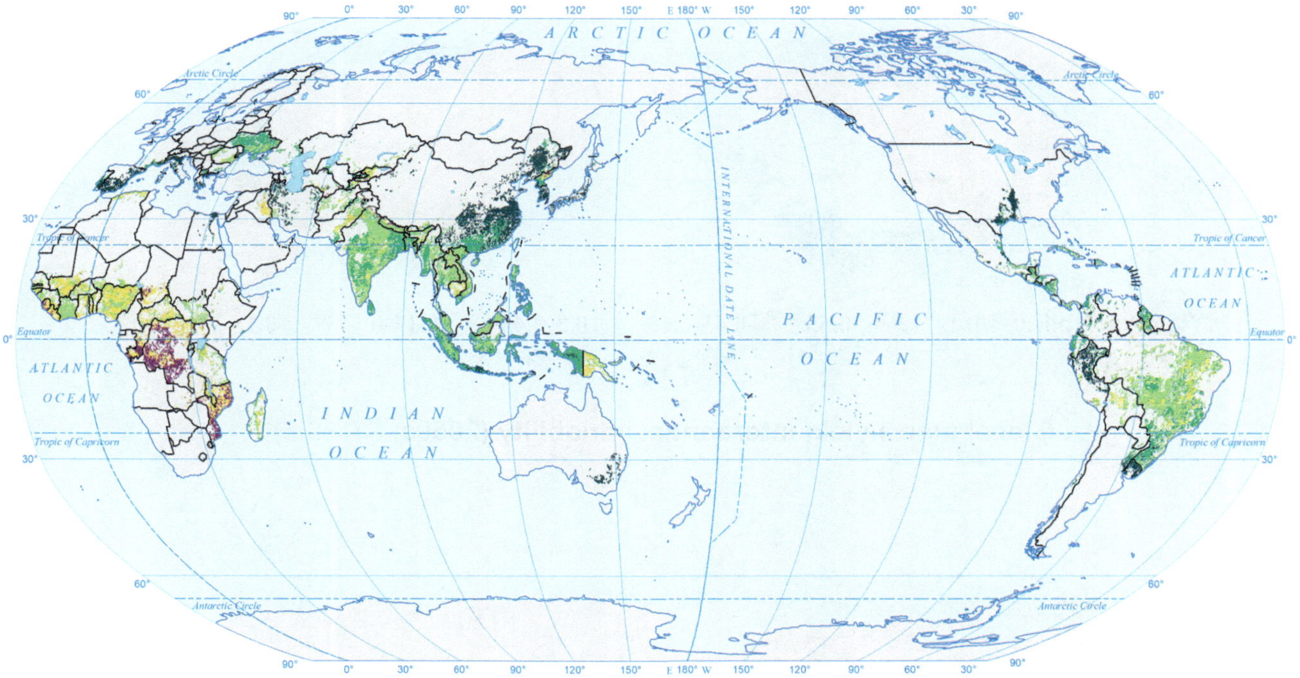

Yield extremes (t/ha) ≤0.7 0.7-1.5 1.5-2.5 2.5-4.5 >4.5

0 1 000 2 000 miles
0 1 000 2 000 kilometers

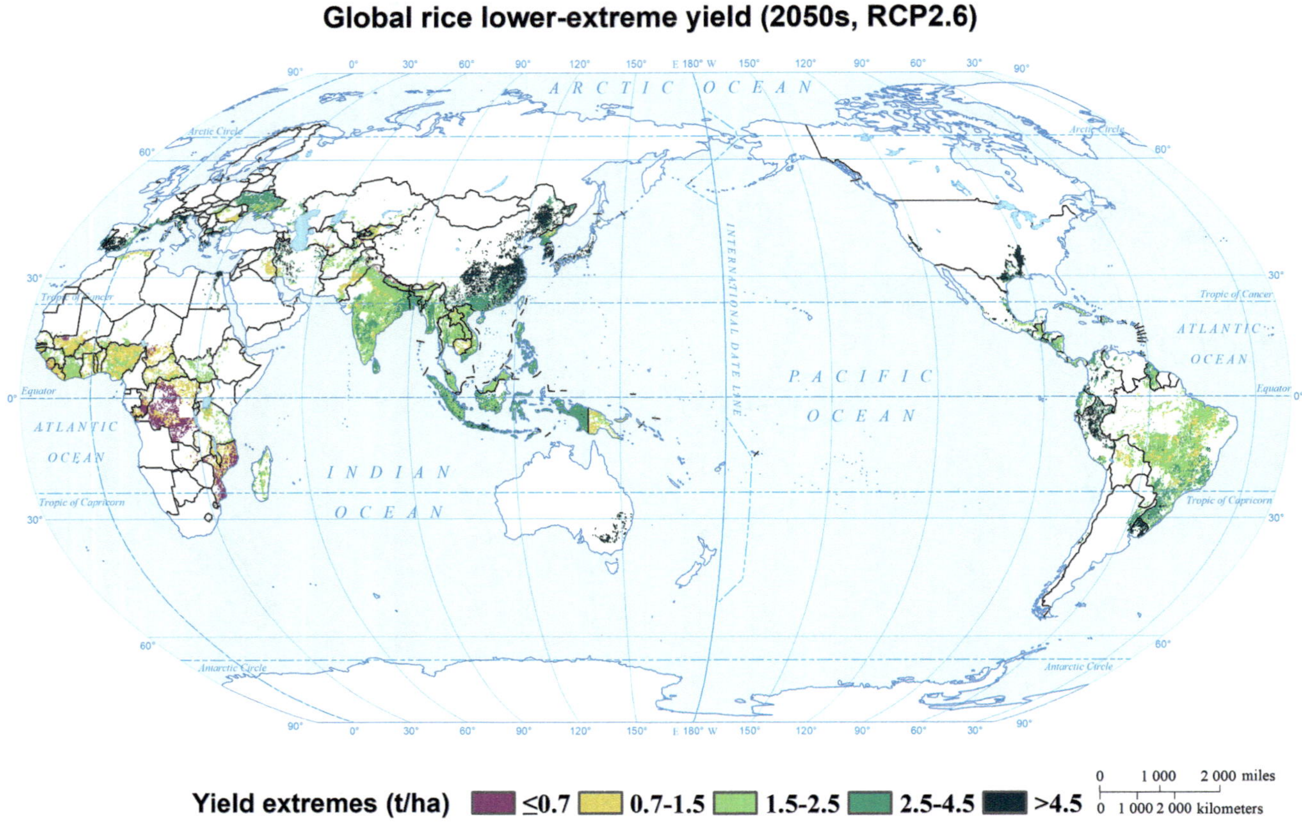
Global rice lower-extreme yield (2050s, RCP2.6)
ARCTIC OCEAN
PACIFIC OCEAN
INDIAN OCEAN
ATLANTIC OCEAN
INTERNATIONAL DATE LINE
Yield extremes (t/ha)
≤0.7
0.7-1.5
1.5-2.5
2.5-4.5
>4.5
0 1 000 2 000 miles
0 1 000 2 000 kilometers

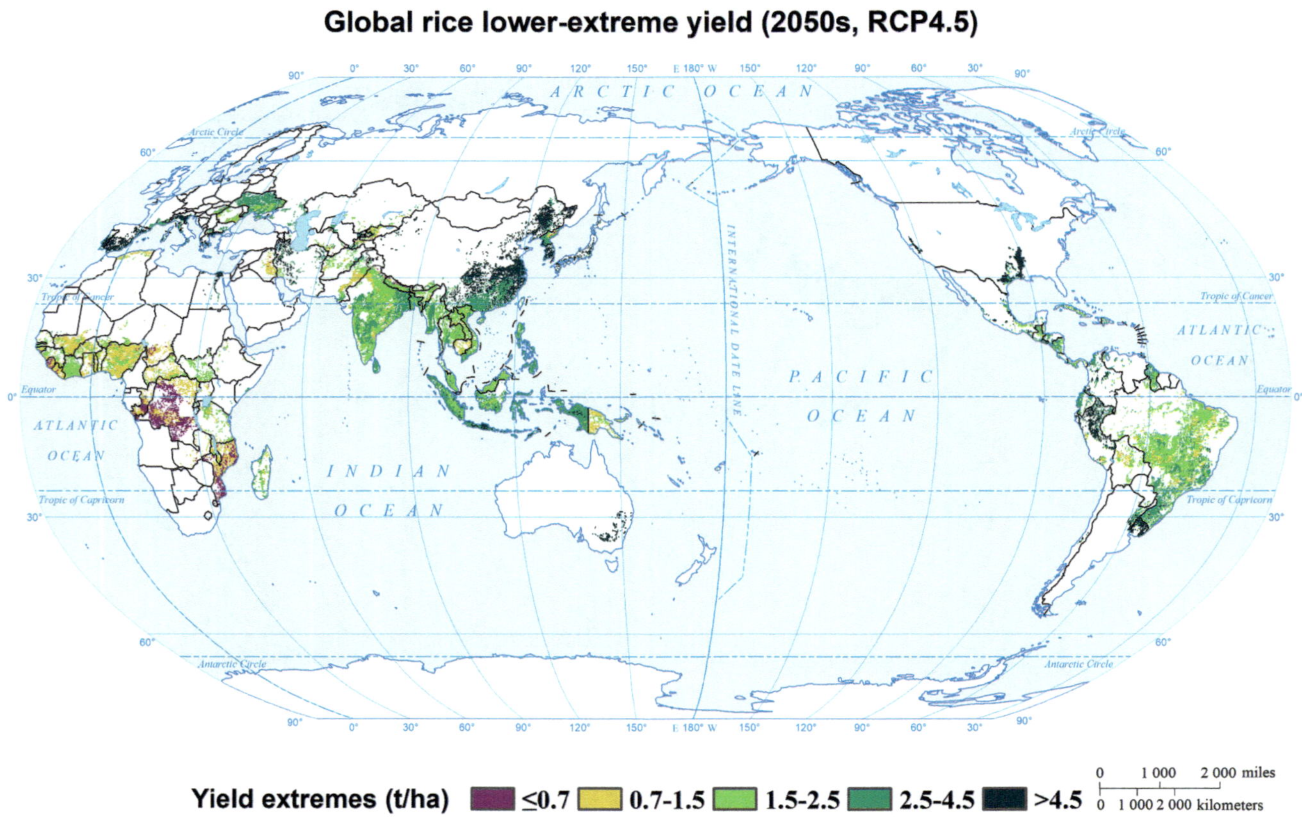
Global rice lower-extreme yield (2050s, RCP4.5)
ARCTIC OCEAN
PACIFIC OCEAN
INDIAN OCEAN
ATLANTIC OCEAN
INTERNATIONAL DATE LINE
Yield extremes (t/ha)
≤0.7
0.7-1.5
1.5-2.5
2.5-4.5
>4.5
0 1 000 2 000 miles
0 1 000 2 000 kilometers

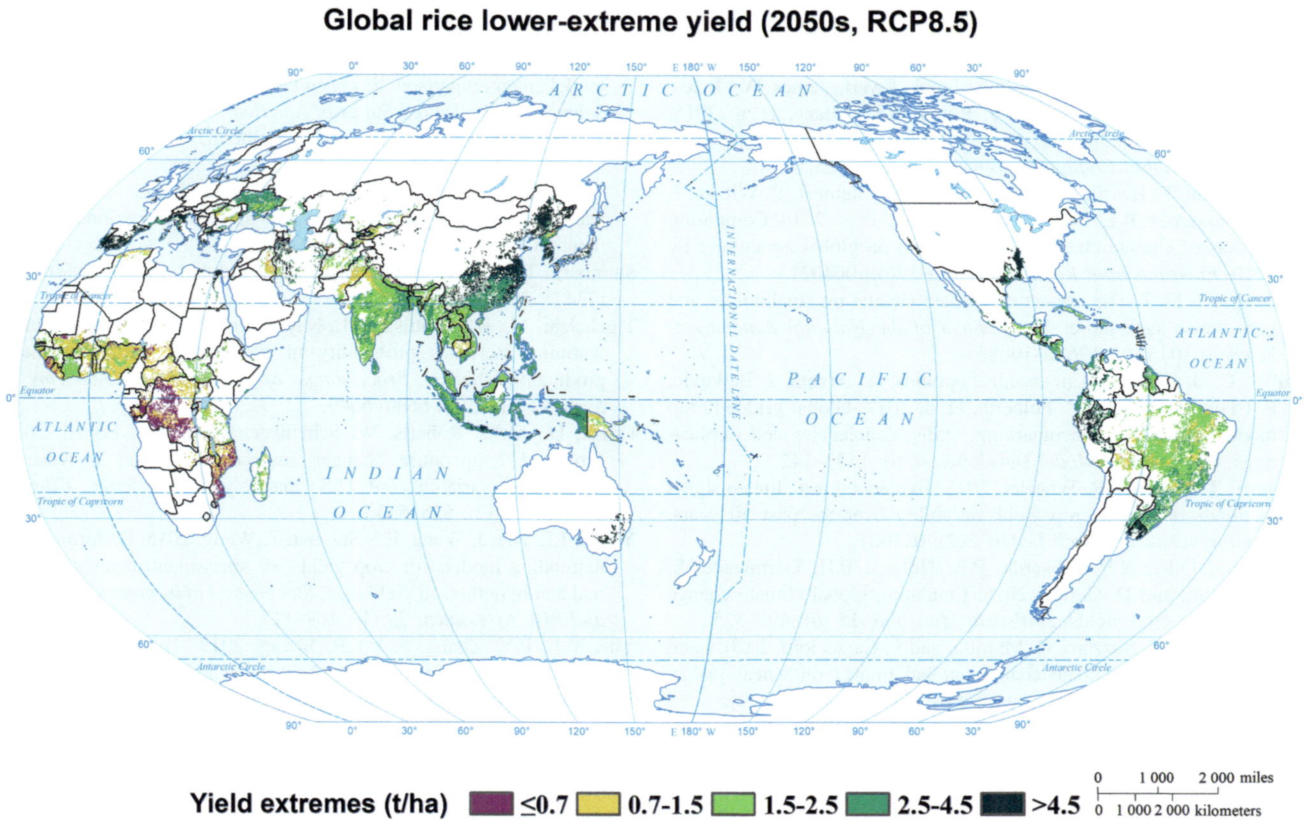

References

Asseng, S., F. Ewert, P. Martre, R.P. Rötter, D.B. Lobell, D. Cammarano, B.A. Kimball, M.J. Ottman, et al. 2015. Rising temperatures reduce global wheat production. *Nature Climate Change* 5 (2): 143–147.

Ben-Ari, T., J. Boé, P. Ciais, R. Lecerf, M. Van Der Velde, and D. Makowski. 2018. Causes and implications of the unforeseen 2016 extreme yield loss in the breadbasket of France. *Nature Communications* 9 (1): 1627.

Bobenrieth, E., B. Wright, and D. Zeng. 2013. Stocks-to-use ratios and prices as indicators of vulnerability to spikes in global cereal markets. *Agricultural Economics* 44 (s1): 43–52.

Coble, K.H., T.O. Knight, B.K. Goodwin, M.F. Miller, and R.M. Rejesus. 2010. A comprehensive review of the RMA APH and COMBO rating methodology. RMA Contract Report, Washington, DC. http://www.rma.usda.gov/pubs/2009/comprehensivereview.pdf.

FAO (Food and Agriculture Organization). 2019. *World food and agriculture statistical pocketbook*. http://www.fao.org/3/ca6463en/ca6463en.pdf.

Folberth, C., A. Baklanov, J. Balkovič, R. Skalský, N. Khabarov, and M. Obersteiner. 2019. Spatio-temporal downscaling of gridded crop model yield estimates based on machine learning. *Agricultural and Forest Meteorology* 264 (January): 1–15.

Frieler, K., B. Schauberger, A. Arneth, J. Balkovič, J. Chryssanthacopoulos, D. Deryng, J. Elliott, C. Folberth, et al. 2017. Understanding the weather signal in national crop-yield variability. *Earth's Future* 5 (6): 605–616.

Hawkins, E., T.E. Fricker, A.J. Challinor, C.A.T. Ferro, C. Kit Ho, and T.M. Osborne. 2013. Increasing influence of heat stress on French maize yields from the 1960s to the 2030s. *Global Change Biology* 19 (3): 937–947.

Holzkämper, A., P. Calanca, and J. Fuhrer. 2012. Statistical crop models: predicting the effects of temperature and precipitation changes. *Climate Research* 51 (1): 11–21.

Iizumi, T., and N. Ramankutty. 2016. Changes in yield variability of major crops for 1981–2010 explained by climate change. *Environmental Research Letters* 11(3): 034003.

IPCC (Intergovernmental Panel on Climate Change). 2017. Climate change and land: an IPCC special report on climate change, desertification, land degradation, sustainable land management, food security, and greenhouse gas fluxes in terrestrial ecosystems. In *Research handbook on climate change and agricultural law*, issued 2017.https://doi.org/10.4337/9781784710644.

Leng, G.Y. 2017. Recent changes in county-level corn yield variability in the United States from observations and crop models. *Science of the Total Environment* 607–608: 683–690.

Liu, B., S. Asseng, C. Müller, F. Ewert, J. Elliott, D.B. Lobell, P. Martre, A.C. Ruane, et al. 2016. Similar estimates of temperature impacts on global wheat yield by three independent methods. *Nature Climate Change* 6 (12): 1130–1136.

Liu, B., P. Martre, F. Ewert, J.R. Porter, A.J. Challinor, C. Müller, A.C. Ruane, K. Waha, et al. 2019. Global wheat production with 1.5 and 2.0°C above pre-industrial warming. *Global Change Biology* 25(4): 1428–1444.

Lobell, D.B. 2007. Changes in diurnal temperature range and national cereal yields. *Agricultural and Forest Meteorology* 145 (3–4): 229–238.

Lobell, D.B., and M.B. Burke. 2010. On the use of statistical models to predict crop yield responses to climate change. *Agricultural and Forest Meteorology* 150 (11): 1443–1452.

Lobell, D.B., M.J. Roberts, W. Schlenker, N. Braun, B.B. Little, R.M. Rejesus, and G.L. Hammer. 2014. Greater sensitivity to drought accompanies maize yield increase in the U.S. Midwest. *Science* 344 (6183): 516–519.

Lobell, D.B., W. Schlenker, and J. Costa-Roberts. 2011. Climate trends and global crop production since 1980. *Science* 333 (6042): 616–620.

Martre, P., D. Wallach, S. Asseng, Frank Ewert, James W. Jones, Reimund P. Rötter, Kenneth J. Boote, A.C. Ruane, et al. 2015. Multimodel ensembles of wheat growth: many models are better than one. *Global Change Biology* 21 (2): 911–925.

Meijl, H. van, P. Havlik, H. Lotze-Campen, E. Stehfest, P. Witzke, I. P. Domínguez, B.L. Bodirsky, M. van Dijk, et al. 2018. Comparing impacts of climate change and mitigation on global agriculture by 2050. *Environmental Research Letters* 13(6): 064021.

Morton, J.F. 2007. The impact of climate change on smallholder and subsistence agriculture. *Proceedings of the National Academy of Sciences* 104 (50): 19680–19685.

Müller, C., J. Elliott, J. Chryssanthacopoulos, A. Arneth, J. Balkovic, P. Ciais, D. Deryng, C. Folberth, et al. 2017. Global gridded crop model evaluation: benchmarking, skills, deficiencies and implications. *Geoscientific Model Development* 10: 1403–1422.

Osborne, T.M., and T.R. Wheeler. 2013. Evidence for a climate signal in trends of global crop yield variability over the past 50 years. *Environmental Research Letters* 8(2): 024001.

Oyebamiji, O.K., N.R. Edwards, P.B. Holden, P.H. Garthwaite, S. Schaphoff, and D. Gerten. 2015. Emulating global climate change impacts on crop yields. *Statistical Modelling* 15 (6): 499–525.

Raimondo, M., C. Nazzaro, G. Marotta, and F. Caracciolo. 2020. Land degradation and climate change: global impact on wheat yields. *Land Degradation & Development* 32 (1): 387–398.

Ray, D. K., Gerber, J. S., Macdonald, G. K. and West, P. C. 2015. Climate variation explains a third of global crop yield variability. *Nature Communication* 6 (1): 6989

Rosenzweig, C., J. Elliott, D. Deryng, A.C. Ruane, C. Müller, A. Arneth, K.J. Boote, C. Folberth, et al. 2014. Assessing agricultural risks of climate change in the 21st century in a global gridded crop model intercomparison. *Proceedings of the National Academy of Sciences of the United States of America* 111 (9): 3268–3273.

Ruane, A.C., J. Antle, J. Elliott, C. Folberth, G. Hoogenboom, D. Mason-D'Croz, C. Müller, C. Porter, et al. 2018. Biophysical and economic implications for agriculture of +1.5° and +2.0°C global warming using AgMIP coordinated global and regional assessments. *Climate Research* 76(1): 17–39.

Sternberg, T. 2011. Regional drought has a global impact. *Nature* 472 (7342): 169–169.

Tigchelaar, M., D.S. Battisti, R.L. Naylor, and D.K. Ray. 2018. Future warming increases probability of globally synchronized maize production shocks. *Proceedings of the National Academy of Sciences* 115 (26): 6644–6649.

Urban, D., M.J.J. Roberts, W. Schlenker, and D.B.B. Lobell. 2012. Projected temperature changes indicate significant increase in interannual variability of U.S. maize yields: a letter. *Climatic Change* 112(2): 525–533.

Ye, T., J.L. Nie, J. Wang, P.J. Shi, and Z. Wang. 2015. Performance of detrending models for crop yield risk assessment: evaluation with real and hypothetical yield data. *Stochastic Environmental Research and Risk Assessment* 29 (1): 109–117.

Yue, Y.J., P.Y. Zhang, and Y.R. Shang. 2019. The potential global distribution and dynamics of wheat under multiple climate change scenarios. *Science of the Total Environment* 688: 1308–1318.

Appendix A
Projection of Future Climate Change

Lianlian Xu and Aihui Wang

Introduction

The Coupled Model Intercomparison Project Phase 5 (CMIP5) multi-model ensembles under different emissions scenarios (Taylor et al. 2012; Dai 2013) are widely used to understand the past and predicit the future climate changes. However, the horizontal resolution of the general circulation models (GCMs) in the CMIP5 project are relatively coarse, which has limited the ability of the models to capture the spatially detailed information of (extreme) climate events at the regional or local scales (Ines and Hansen 2006; Xue et al. 2014). To overcome this limitation, statistical downscaling methods are widely used (Maurer and Hidalgo 2008; Li et al. 2010; Xue et al. 2014; Cannon et al. 2015; Chen et al. 2017).

Models and Method

The bias correction and spatial downscaling (BCSD) algorithm (Xu and Wang 2019) is used to correct and downscale the daily maximum temperature, daily minimum temperature, and daily precipitation from 13 GCMs in the CMIP5 project under the RCP2.6 scenario. Historical simulations (1961–005) and future projections (2006–065) under the RCP2.6 scenario in the CMIP5 project (Taylor et al. 2012) are adopted to be statistical downscaled in this study. Table A.1 provides the basic information about these GCMs.

In order to use the BCSD method, the Global Meteorological Forcing Dataset (GMFD, 0.25° × 0.25°) is taken as the observations. The GMFD is derived from merged reanalysis products of remote sensing and in situ observations (Sheffield et al. 2006), which was originally developed as the atmospheric forcing dataset for offline land surface models. After the three steps of BCSD, we obtain the global daily maximum temperature, daily minimum temperature, and daily precipitation at a spatial resolution of 0.25° × 0.25° from 13 GCMs under the RCP2.6 scenario.

References

Cannon, A.J., S.R. Sobie, and T.Q. Murdock. 2015. Bias correction of GCM precipitation by quantile mapping: How well do methods preserve changes in quantiles and extremes? *Journal of Climate* 28 (17): 6938–6959.

Chen, H.P., J.Q. Sun, and H.X. Li. 2017. Future changes in precipitation extremes over China using the NEX-GDDP high-resolution daily downscaled data-set. *Atmospheric and Oceanic Science Letters* 10 (6): 403–410.

Dai, A.G. 2013. Increasing drought under global warming in observations and models. *Nature Climate Change* 3: 52–58.

Ines, A.V.M., and J.W. Hansen. 2006. Bias correction of daily GCM rainfall for crop simulation studies. *Agricultural and Forest Meteorology* 138 (1): 44–53.

L. Xu · A. Wang
Nansen-Zhu International Research Centre, Institute of Atmospheric Physics, Chinese Academy of Sciences, Beijing, 100029, China

L. Xu
School of Atmospheric Sciences, Sun Yat-sen University, Zhuhai, 519082, China

P. Shi, *Atlas of Global Change Risk of Population and Economic Systems*, IHDP/Future Earth-Integrated Risk Governance Project Series, https://doi.org/10.1007/978-981-16-6691-9

Table A.1 List of the 13 general circulation models (GCMs) from the CMIP5 archive

Model Name	Horizontal Resolution in Degrees	Modeling Center
BNU-ESM	2.8°× 2.8°	Beijing Normal University
CanESM2	2.8°× 2.8°	Canadian Centre for Climate Modeling and Analysis
CNRM-CM5	1.4° × 1.4°	Centre National de Recherches Meteorologiques / Centre Europeen de Recherche et Formation Avancee en Calcul Scientifique
CSIRO-Mk3.6.0	1.875° × 1.875°	Commonwealth Scientific and Industrial Research Organization in collaboration with Queensland Climate Change Centre of Excellence
GFDL-CM3	2° × 2.5°	NOAA Geophysical Fluid Dynamics Laboratory
GFDL-ESM2G	2° × 2.5°	
GFDL-ESM2 M	2° × 2.5°	
IPSL-CM5A-MR	1.27° × 2.5°	Institute Pierre Simon Laplace
MIROC5	1.4° × 1.4°	Japan Agency for Marine-Earth Science and Technology, Atmosphere and Ocean Research Institute (The University of Tokyo), and National Institute for Environmental Studies
MIROC-ESM	2.8° × 2.8°	
MPI-ESM-LR	1.875 × 1.875°	Max-Planck-Institut für Meteorologie
MPI-ESM-MR	1.875° × 1.875°	
MRI-CGCM3	1.121° × 1.125°	Meteorological Research Institute

Li, H.B., J. Sheffield, and E.F. Wood. 2010. Bias correction of monthly precipitation and temperature fields from the IPCC AR4 models using equidistant quantile matching. *Journal of Geophysical Research: Atmospheres* 115(D10).

Maurer, E.P., and H.G. Hidalgo. 2008. Utility of daily vs. monthly large-scale climate data: An intercomparison of two statistical downscaling methods. *Hydrology and Earth System Science. Discussions* 12(2): 551–563.

Sheffield, J., G. Goteti, and E.F. Wood. 2006. Development of a 50-year high-resolution global dataset of meteorological forcings for land surface modeling. *Journal of Climate* 19 (13): 3088–3111.

Taylor, K.E., R.J. Stouffer, and G.A. Meehl. 2012. An overview of CMIP5 and the experiment design. *Bulletin of the American Meteorological Society* 93 (4): 485–498.

Xu, L., and A. Wang. 2019. Application of the bias correction and spatial downscaling algorithm on the temperature extremes from CMIP5 multimodel ensembles in China. *Earth and Space Science* 6: 2508–2524.

Xue, Y.K., Z. Janjic, J. Dudhia, R. Vasic, and F. De Sales. 2014. A review on regional dynamical downscaling in intraseasonl to seasonal simulation/prediction and major factors that affect downscaling ability. *Atmospheric Research* 147–148: 68–85.

Appendix B
Projection of Future Population and Economic System Changes

Fubao Sun, Jing'ai Wang, Yujie Liu, Wenxiang Wu, Huiyi Zhu, and Yaojie Yue

Introduction

The shared socioeconomic pathways (SSPs), which qualitatively and quantitatively describe broad patterns of possible global socioeconomic development with assumptions about climate change and policy responses under different challenges to mitigation and adaptation (O'Neill et al. 2014), are one of the core contents in the Intergovernmental Panel on Climate Change (IPCC) scientific assessment reports (IPCC 2014) and in the current literature (O'Neill et al. 2016). The increasing demand of the Scenario Model Intercomparison Project (ScenarioMIP) is calling for spatially explicit population, GDP, crop distribution, industrial value added, and road networks projections of high resolution for the future SSPs in both socioeconomic development and climate change adaption and mitigation research.

To date the global population, GDP, industrial values add, road networks projections for the five SSPs are mainly provided at the national and super-national scales from several global institutions, which have depicted a wide range of uncertainty within different organizations (Riahi et al. 2017; Xue et al. 2018) and limited the usage of integration with data from other disciplines. And the projections of gridded datasets at the global scale, to the best of our knowledge, are very limited so far.

This atlas has presented a set of spatially explicit global socioeconomic factors: population, GDP, crop distribution, industrial value added, and road networks projections that reflect substantial long-term changes of socioeconomic activities for both the historical period and future projections under different SSPs.

Data Availability

Some detailed information regarding this atlas is shown in the table below.

	Data and Method	Format	Advantages
Global population	Multi-source data from the World Bank, International Monetary Fund, United Nations, IIASA, and so on	geotif	Multi-source data based
Global GDP	Multi-source data as population;Official GDP at country and subnational levels; Chinese GDP updated under two-children policy.	geotif	Accuracy verified; High resolution; Two-children policy in China considered

(continued)

F. Sun (✉)
Key Laboratory of Water Cycle and Related Land Surface Processes, Institute of Geographic Sciences and Natural Resources Research, Chinese Academy of Sciences, Beijing, 100101, China
e-mail: sunfb@igsnrr.ac.cn

F. Sun
State Key Laboratory of Desert and Oasis Ecology, Xinjiang Institute of Ecology and Geography, Chinese Academy of Sciences, Urumqi, 830011, China

J. Wang · Y. Yue
Key Laboratory of Environmental Change and Natural Disaster of Ministry of Education, Faculty of Geographical Science, Beijing Normal University, Beijing, 100875, China

Y. Liu · W. Wu · H. Zhu
Key Laboratory of Land Surface Pattern and Simulation, Institute of Geographic Sciences and Natural Resources Research, Chinese Academy of Sciences, Beijing, 100101, China

P. Shi, *Atlas of Global Change Risk of Population and Economic Systems*, IHDP/Future Earth-Integrated Risk Governance Project Series, https://doi.org/10.1007/978-981-16-6691-9

	Data and Method	Format	Advantages
	NTL images and global gridded population of high resolution are used as base map in GDP disaggregation		
Global crop distribution	Global rice, wheat, and maize distribution; $\geq 0°C$ cumulative temperature, annual precipitation, annual average temperature, average temperature of the coldest month, pH, drainage, conductivity, exchangeable sodium percentage, soil property and soil depth; The Maxent model	geotif	Accuracy verified; High resolution
Global crop distribution	Value added in mining, manufacturing (also reported as a separate subgroup), construction, electricity, water, and gas; NTL images; Random forest method	geotif	Accuracy verified; High spatial resolution; The first global industrial value added dataset of the past and the future
Global road networks	Population, GDP per capita; Land use	geotif	High resolution

The global GDP disaggregation results for 2005 as historical period and for 2030–2100 as future projections for SSP1–5 at 10-year interval are provided (purchasing power parity in 2005 USD), acknowledging the two-children policy in China, with spatial resolutions of 30 arc-seconds (approximately 1 km at the equator) and 0.25 degrees. The global gridded GDP are provided in geotif format at https://doi.org/10.5281/zenodo.4350027. The GDP values are disaggregated within its administrative boundaries, and the Antarctica, oceans, as well as some desert or wilderness areas are filled with value 0.

The global industrial added value disaggregation results for 2010 as historical period and for 2030 and 2050 as future projections for SSP1–3 are provided (purchasing power parity in 2010 USD), with spatial resolutions of 0.5 degrees. The global and China 1 km gridded industrial added value is provided in geotif format at http://www.geodoi.ac.cn/WebCn/doi.aspx?Id=805 and https://www.scidb.cn/en/detail?dataSetId=633694460995174400&dataSetType=journal, respectively.

Further detailed information regarding the atlas is available from “Global dataset of gridded GDP scenarios”, which is provided by the Global Change Risk of Population and Economic Systems (GCR-PES): Mechanisms and Assessments Project, Beijing Normal University, Beijing, China (http://gcr.bnu.edu.cn/).

References

IPCC. 2014. *Climate change 2013: The physical science basis: Working Group I contribution to the Fifth assessment report of the Intergovernmental Panel on Climate Change*. Cambridge: U. K., Cambridge University Press.

O’Neill, B.C., E. Kriegler, K. Riahi, K.L. Ebi, S. Hallegatte, T.R. Carter, R. Mathur, and D.P. Van Vuuren. 2014. A new scenario framework for climate change research: The concept of shared socioeconomic pathways. *Climatic Change* 122: 387–400.

O’Neill, B.C., C. Tebaldi, D.P. Van Vuuren, V. Eyring, P. Friedlingstein, G. Hurtt, R. Knutti, E. Kriegler, et al. 2016. The scenario model intercomparison project (ScenarioMIP) for CMIP6. *Geoscientific Model Development* 9: 3461–3482.

Riahi, K., D.P. Van Vuuren, E. Kriegler, J. Edmonds, and B. C. O’neill, S. Fujimori, N. Bauer, K. Calvin, R. Dellink and O. Fricko. . 2017. The shared socioeconomic pathways and their energy, land use, and greenhouse gas emissions implications: An overview. *Global Environmental Change* 42: 153–168.

Xue, Q., W. Song, and H. Zhu. 2018. Global industrial added value 1 km square grid dataset. *Journal of Global Change Data & Discovery* 2 (1): 9–17.

Appendix C
Risk Assessment: Framework and Models

Tao Ye, Wei Xu, and Peijun Shi

Introduction

Global climate change featured by warming has created serious challenges to sustainable development and human security (IPCC 2014). It has become an important consensus of the international society to assess global change risk at the global scale and carry out tailored governance and risk-based adaptation (Aalst et al. 2014). Research initiatives from international organizations (World Economic Forum 2017), academic institutions, and state governments (DEFRA 2012) have covered global change risk assessment in some key fields, including people's health (Stephenson et al. 2013), population exposure and mortality (Hirabayashi et al. 2013), economic loss and assets exposures (Burke et al. 2015; Dietz et al. 2016), and food security (Rosenzweig et al. 2014). The assessment of global change risk is generally regarded as the evaluation of a co-evolutionary system consisting of changing climate and socioeconomic subsystems and their interactions (Winsemius et al. 2016). Correspondingly, such assessment predicts potential future losses using predefined quantitative models to integrate the combined dynamics of future climate change and socioeconomic development (Bouwer 2013; Liu et al. 2018; Wing et al. 2018). In the past decades, great efforts have been devoted to build climate projection (Eyring et al. 2016) and shared socioeconomic pathways (O'Neill et al. 2014) to produce datasets for common uses. Models used to describe the interactions, either statistical or process-based, are mostly problem-specific, and require substantially more efforts to catch up.

T. Ye
Institute of Disaster Risk Science, Faculty of Geographical Science, Beijing Normal University, Beijing, 100875, China

W. Xu
State Key Laboratory of Earth Surface Processes and Resource Ecology, Faculty of Geographical Science, Beijing Normal University, Beijing, 100875, China

P. Shi
Academy of Plateau Science and Sustainability, Qinghai Normal University, Xining, 810008, China
e-mail: spj@bnu.edu.cn

Conceptual Framework

The general conceptual framework to assess global change risk is to develop model systems that can integrate the combined dynamics of future climate change and socioeconomic development. Overall, it succeeded the hazard, vulnerability, and exposure framework of classic risk assessment models, but needs to accommodate the dynamics of climate, population, and economic systems (Fig. C.1).

Two types of major driving forces are considered in the models: climate change and socioeconomic system change. For climate change, three factors were considered: near-surface air temperature, precipitation, and wind speed. As suggested by the Intergovernmental Panel on Climate Change (IPCC 2012), climate change contains the changes in its mean, variability, and skewness. When focusing on the risk, the changes in variability and skewness alters climate extremes. In light of this, the changes in the mean, interannual variability, and extremes of each climate factor were considered as the driving forces from the hazard side of risk.

The changes of population and economic systems indicate the changes in both exposure and vulnerability. Exposed units considered in this atlas included population, staple grain production of paddy rice, wheat, and maize, and gross domestic product (GDP).

Risk assessment models were employed to describe the interactions between climate change hazards and the population and economic systems. The interaction-focused approach determines the models used correspondingly. For population affected and mortality and GDP losses, only

P. Shi, *Atlas of Global Change Risk of Population and Economic Systems*, IHDP/Future Earth-Integrated Risk Governance Project Series, https://doi.org/10.1007/978-981-16-6691-9

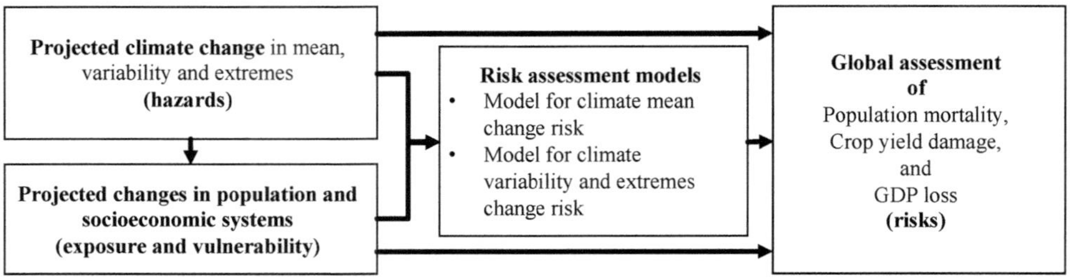

Fig. C.1 Conceptual framework of the global change risk assessment

Table C.1 Types of impact

Climate Change	Climate Mean Changes	Climate Variability Changes	Climate Extreme Changes
Population			
Affected/exposed	N/A	N/A	Extremely high temperature; rainstorms/droughts
Mortality	N/A	N/A	Extremely high temperature; floods
Economic system			
Crop exposed	N/A	N/A	Droughts/extremely high temperature
Crop yield loss	Mean yield changes	Changes in interannual variability of yield	Changes in lower extreme of yield
GDP exposed	N/A	N/A	Floods/droughts
GDP loss	N/A	N/A	Floods

climate extremes were considered, including extremely high temperature and rainstorms (and subsequent flood events). For these parts, classical natural disaster risk assessment models were employed by using a loss function (also know as quantitative vulnerability function) fitted from historical losses to link hazard and loss. For staple grain production, the impacts of changes in climate mean, variability, and extremes were all considered. Correspondingly, crop emulators derived from global gridded crop models were used to emulate crop yields under different climate change scenarios, from which the changes in mean yield, yield interannual variability, and lower extremes of yield were obtained (Table C.1).

Technical Specifications

Forcing and Scenarios

For climate change, Representative Concentration Pathways (RCPs) employed in the IPCC Coupled Model Intercomparison Project 5 (CMIP5) were used. To fully accommodate potential climate change, RCP2.6, RCP4.5, and RCP8.5 were actually used. Unless specifically indicated, climate forcing of NEX-GDDP released by the National Aeronautics and Space Administration (NASA) with 21 general circulation models (GCMs) and 0.25 degrees was used for RCP4.5 and RCP8.5. Thirteen GCMs in the CMIP5 project were downscaled using the bias correction and spatial downscaling (BCSD) method to supplement the NEX-GDDP dataset under the RCP2.6 scenario. More details about the downscaled data can be found in Appendix 1 (Projection of Future Climate Change).

For socioeconomic change, the Shared Socioeconomic Pathways (SSPs) setting the fundamental population and GDP growth paths in the future were considered. The SSP1, SSP2, and SSP3 were used in accordance with the RCP2.6, RCP4.5, and RCP8.5 scenarios. More details about the projection of global distribution of population, crop, GDP, industrial value added, and road networks are found in Appendix 2 (Projection of Future Population and Economic System Changes), and earlier chapters in Part II.

Risk Metrics

Annual average loss (AAL) was used as the key risk metric indicator. Simulated losses were averaged over each 20-year period, namely the baseline period (1986 − 2005), the 2030s (2016 − 2035), and the 2050s (2046 − 2065), following the CMIP5 framework. In the atlas, multi-model ensemble means and spreads of AALs were reported. The difference in cross-period AALs is the final measure of global change risk, the change in risk jointly driven by climate change and population and economic system changes.

References

Aalst, Maarten Van, Neil Adger, Douglas Arent, Jonathon Barnett, Richard Betts, Eren Bilir, Joern Birkmann, et al. 2014. "Climate Change 2014: Impacts, Adaptation, and Vulnerability." *Assessment Report 5*, no. October 2013: 1–76.

Bouwer, Laurens M. 2013. Projections of Future Extreme Weather Losses under Changes in Climate and Exposure. *Risk Analysis : An Official Publication of the Society for Risk Analysis* 33 (5): 915–930. https://doi.org/10.1111/j.1539-6924.2012.01880.x.

Burke, Marshall, Solomon M. Hsiang, and Edward Miguel. 2015. Global Non-Linear Effect of Temperature on Economic Production. *Nature* 527 (1): 235–239. https://doi.org/10.1038/nature15725.

DEFRA. 2012. *UK Climate Change Risk Assessment: Government Report.* https://www.gov.uk/government/uploads/system/uploads/attachment_data/file/69487/pb13698-climate-risk-assessment.pdf.

Dietz, Simon, Alex Bowen, Charlie Dixon, and Philip Gradwell. 2016. 'Climate Value at Risk' of Global Financial Assets. *Nature Climate Change* 6 (April): 1–5. https://doi.org/10.1038/nclimate2972.

Eyring, Veronika, Mattia Righi, Axel Lauer, Martin Evaldsson, Sabrina Wenzel, Colin Jones, Alessandro Anav, et al. 2016. ESMValTool (v1.0) – a Community Diagnostic and Performance Metrics Tool for Routine Evaluation of Earth System Models in CMIP. *Geoscientific Model Development* 9 (5): 1747–1802. https://doi.org/10.5194/gmd-9-1747-2016.

IPCC. 2012. *Managing the Risks of Extreme Events and Disasters to Advance Climate Change Adaptation.* Edited by Christopher B. Field, Vicente Barros, Thomas F. Stocker, and Qin Dahe. *Intergovernmental Panel on Climate Change Spacial Report on Managing the Risks of Extreme Events and Disasters to Advance Climate Change Adaptation.* A Special. Cambridge, United Kingdom and New York, NY, USA: Cambridge University Press. https://doi.org/10.1017/CBO9781139177245.

IPCC. 2014. "Climate Change 2014: Synthesis Report." *Contribution of Working Groups I, II and III to the Fifth Assessment Report of the Intergovernmental Panel on Climate Change. In: Core Writing Team, Pachauri RK, Meyer LA (Eds) IPCC, Geneva, Switzerland, 151 P.* Vol. 9781107025. https://doi.org/10.1017/CBO9781107415324.

O'Neill, Brian C., Elmar Kriegler, Kristie L. Ebi, Eric Kemp-Benedict, Keywan Riahi, Dale S. Rothman, Bas J. van Ruijven, et al. 2014. The Roads Ahead: Narratives for Shared Socioeconomic Pathways Describing World Futures in the 21st Century. *Global Environmental Change* 42 (January): 169–180. https://doi.org/10.1016/j.gloenvcha.2015.01.004.

Rosenzweig, Cynthia, Joshua Elliott, Delphine Deryng, Alex C. Ruane, Christoph Müller, Almut Arneth, Kenneth J. Boote, et al. 2014. Assessing Agricultural Risks of Climate Change in the 21st Century in a Global Gridded Crop Model Intercomparison. *Proceedings of the National Academy of Sciences of the United States of America* 111 (9): 3268–3273. https://doi.org/10.1073/pnas.1222463110.

Stephenson, Judith, Susan F. Crane, Caren Levy, and Mark Maslin. 2013. Population, Development, and Climate Change: Links and Effects on Human Health. *Lancet* 382 (9905): 1665–1673. https://doi.org/10.1016/S0140-6736(13)61460-9.

Wing, Oliver E J., Paul D. Bates, Andrew M. Smith, Christopher C. Sampson, Kris A. Johnson, Joseph Fargione, and Philip Morefield. 2018. Estimates of Present and Future Flood Risk in the Conterminous United States. *Environmental Research Letters* 13 (3). https://doi.org/10.1088/1748-9326/AAAC65.

Winsemius, Hessel C., Jeroen C. J. H. Aerts, Ludovicus P. H. van Beek, Marc F. P. Bierkens, Arno Bouwman, Brenden Jongman, Jaap C. J. Kwadijk, et al. 2016. Global Drivers of Future River Flood Risk. *Nature Climate Change* 6 (4): 381–385. https://doi.org/10.1038/nclimate2893.

World Economic Forum. 2017. The Global Risks Report 2017 12th Edition. *Insight Report.* https://doi.org/10.1017/CBO9781107415324.004.

Appendix D
Atlas Architecture and Design

Jing'ai Wang, Ying Wang, and Tian Liu

Design Concept

The maps of global change risk of population and economic systems are designed to express the spatial distribution of various elements. The main contents are the climate change, changes of exposures, and risks, and the core contents are the regional differences and changes of risks in different time periods and under different emissions and socioeconomic scenarios in the future. The maps visualize where the areas with the greatest changes in the extremes of basic meteorological elements will be, where or during what time period the risk will change the most, and where the highest risk zones will be found, which help the readers and users of the atlas to understand the spatial patterns of change in the future and make informed decisions. Every type of maps contains information on indicator (including mean, variability, and extremes), time period and scenario (including different time periods and emissions and socioeconomic scenarios), and attribute (different grades). The atlas is based on the three-dimensional structure (Fig. D.1) to present the contents, with the selected methods of presentation and color scheme and layout.

J. Wang · Y. Wang (✉) · T. Liu
Institute of Disaster Risk Science, Faculty of Geographical Science, Beijing Normal University, Beijing, 100,875, China
e-mail: wy@bnu.edu.cn

J. Wang · Y. Wang · T. Liu
School of Geography, Faculty of Geographical Science, Beijing Normal University, Beijing, 100,875, China

Cartographic Units

Grid is the fundamental unit for the assessment of basic climate factors, exposure, and risk of the population and economic systems as well as their cartographic presentation, which is of $0.25° \times 0.25°$.

Technical Flowchart

The mapping and compilation of this atlas include the following steps: preparation and design, map drafting, map plotting, and map generalization (Fig. D.2).

Cartographic Presentation

A variety of conventional cartographic presentation methods are used in this atlas to describe the change of basic climate factors, exposure, and risk (Table D.1), such as the quality-based method, line method, quantity-based method, etc.

Map groups can organize the maps of complex disaster processes in an intuitive way. Within a map group, each series contains seven combinations of time period and scenario—the baseline (2000s), 2030s-RCP2.6 (SSP1), 2030s-RCP4.5 (SSP2), 2030s-RCP8.5 (SSP3), 2050s-RCP2.6 (SSP1), 2050s-RCP4.5 (SSP2), and 2050s-RCP8.5 (SSP3). Examples of the map groups are provided in Table D.2 (Tables D.3 and D.4).

Map Color Design

The final color system of the maps in the atlas is presented in Table D.5.

P. Shi, *Atlas of Global Change Risk of Population and Economic Systems*, IHDP/Future Earth-Integrated Risk Governance Project Series, https://doi.org/10.1007/978-981-16-6691-9

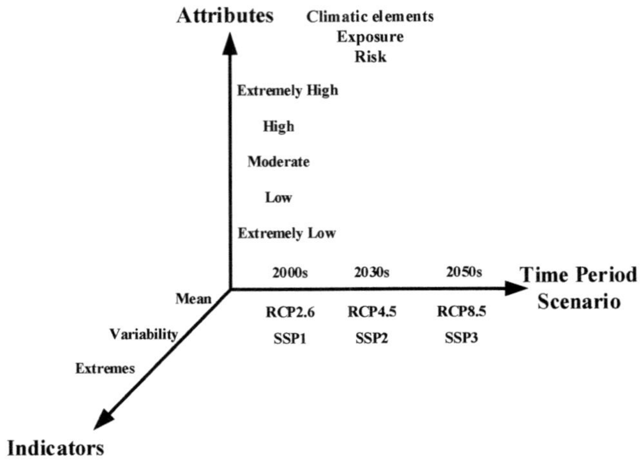

Fig. D.1 Design concept of the atlas

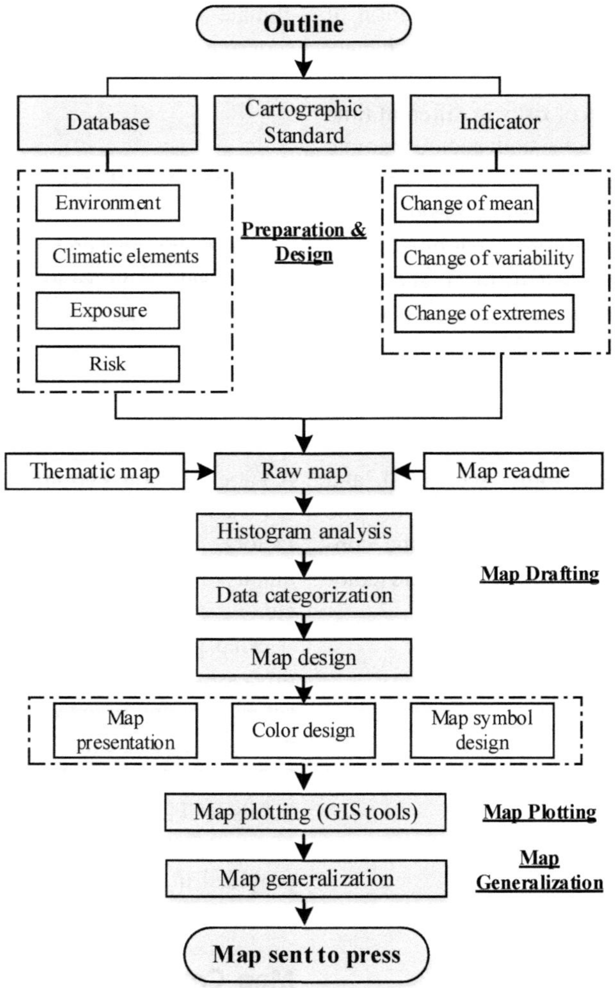

Fig. D.2 Technical flowchart for the mapping and compilation of the atlas

Table D.1 Map presentation methods

Maps	Presentation Method	Thematic Map Examples
Environment	Quality-based method	World political maps, global soil, global land cover, global climate zones
	Satellite image	Global satellite image
	Quantity-based method	Global digital elevation models, global terrain slope
	Line symbols	Global river systems
Climate change	Quantity-based method	All the maps
Change of exposure		All the maps
Change of risk		All the maps

Table D.2 Climate change map group

Temperature	Time period and scenario						
	Baseline	2030s, RCP2.6	2030s, RCP4.5	2030s, RCP8.5	2050s, RCP2.6	2050s, RCP4.5	2050s, RCP8.5
Change in mean	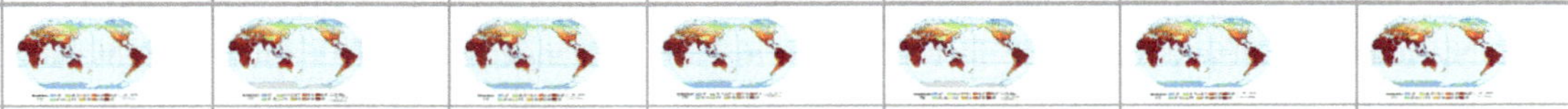						
Change in variability	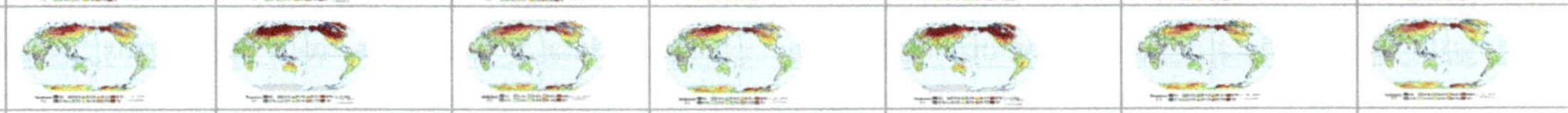						
Change in extremes	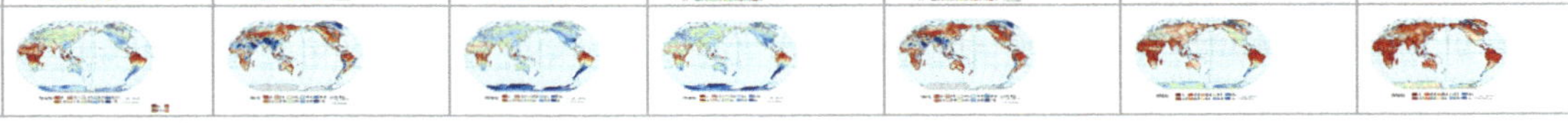						

Table D.3 Change of exposure map group

Exposure	Time period and scenario						
	Baseline	2030s, SSP1	2030s, SSP2	2030s, SSP3	2050s, SSP1	2050s, SSP2	2050s, SSP3
Population	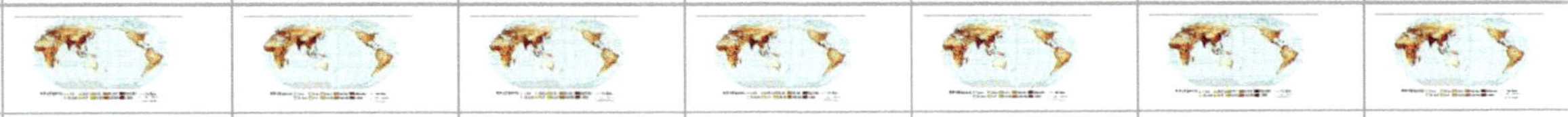						
GDP	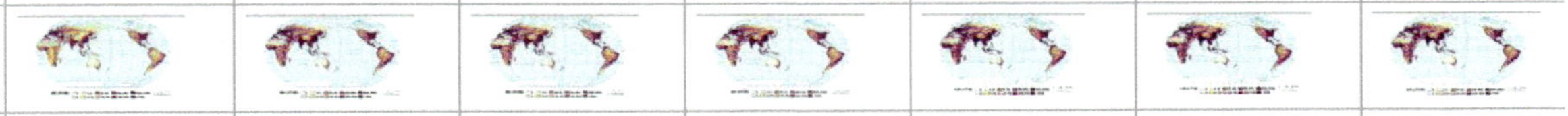						
Road system	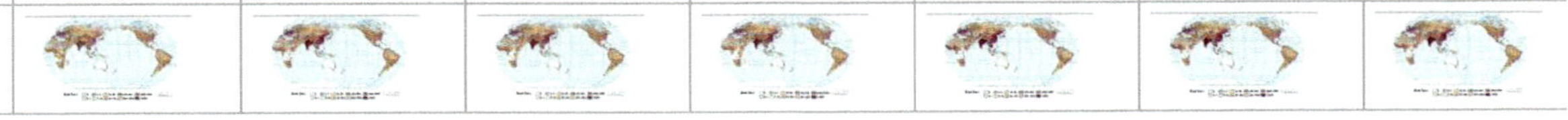						

In this atlas, the color design is demanding. In the part of climate change, the presentation of attributes adopts a ten-grade classification system. The color design principles are (1) emphasizing the areas with extreme values since these areas have the top level of risk for warning, and using light gray for areas with no data and dark gray for extremely low values in some maps (i.e., annual mean temperature variability); (2) highlighting the main colors (red, blue, brown) of the three basic climate factors (temperature, precipitation, wind); and (3) highlighting coastal boundaries by using thicker lines and darker color, for maps related to temperature and wind speed. In the rainfall maps, the coastal boundaries are depicted in black to distinguish them from the dominant color (blue).

In the part of exposures, the presentation of attributes also adopts a ten-grade classification system. The color design

Table D.4 Crop yield risk map group

Risk (mean yield)	Time period and scenario						
	Baseline	2030s, RCP2.6	2030s, RCP4.5	2030s, RCP8.5	2050s, RCP2.6	2050s, RCP4.5	2050s, RCP8.5
Wheat	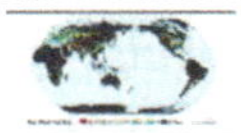	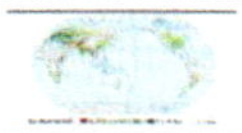		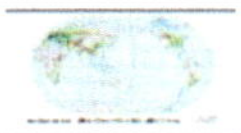	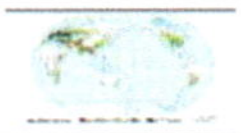	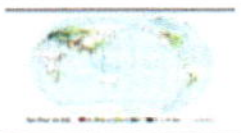	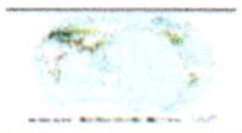
Maize		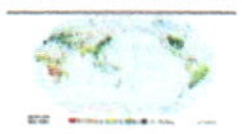	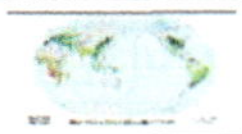		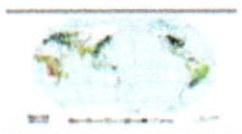	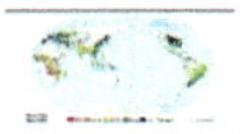	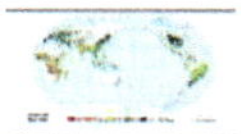
Rice	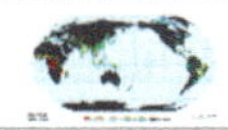						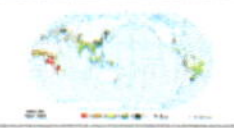

principles are (1) highlighting the main colors (orange, purple, green, brown) of each element (population, GDP, crops, and roads) in the distribution map of population and economic system elements; (2) emphasizing the colors of both the element itself and the hazard factor (rainstorm-blue, drought-brown, high temperature-red/purple); and (3) representing the 0 or no data areas with white or light gray, except for crop exposure to extremely high temperature (emphasized in black), and high values are emphasized in red or purple for warning in the exposure maps.

In the part of risks, the presentation of attributes adopts a 5/6-grade classification system. The color design principles are (1) using red/purple in all maps to emphasize the areas with high risk; (2) using white to represent areas with no data in the heatwave mortality risk, flood mortality risk, and flood GDP loss risk maps, and the color of the ocean is changed from dark blue to light blue to indicate different periods; (3) in the crop risk maps the color of areas with no data is changed from dark gray to light gray to indicate different periods.

All maps in the atlas follow the principle of applying bright and distinct colors to clearly show the spatial distribution of the attributes.

Table D.5 Color system of the maps

Color system of three meteorological elements					
	Base color		Base color		Base color
Temperature: red					
Mean	Red-blue	**Variability**	Red-green-grey	**Extremes**	Red-blue Red–green–blue Orange-red-roseo
Precipitation: blue					
Mean	Blue-green -orange	**Variability**	Cyan-yellow–red	**Extremes**	Hyacinth-red
Wind: brown					
Mean	Rufous-green -yellow	**Variability**	Brown	**Extremes**	Purple-light brown-yellow

Color system of exposure and hazard	
	Base color
Population: orange	Orange-crimson
GDP: purple	Light green-orange-purple
Crops: green	**Rice**: Orange-goose yellow-emerald green
	Wheat: Orange-goose yellow-grass green
	Maize: Orange-goose yellow-blue green
Industrial value added: fuchsia	Gray-orange-fuchsia
Road networks: brown	Lilac-brown-purple
Population exposure to heatwaves: orange	Orange-deep purple
Population exposure to rainstorms: blue	Blue-yellow–red
GDP exposure to drought: brown	Purple-green–brown
Crops exposure to extremely high temperature: red	**Wheat**: Grass green-yellow–red
	Maize: Blue green-yellow–red
	Rice: Emerald green-yellow-rose red

Color system of risk	
	Base color
Population risk of heatwaves: yellow	Purple-yellow
Population risk of flood: green	Red-green
GDP loss risk of flood: green	Purple-green

Crop yield risk: green					
Mean yield	Red-green	**Interannual variability of yield**	Rufous-green	**Lower extreme yield**	Purple-green

Cartographic Specifications

The national/regional boundary map of this atlas was same as the one in the *World Atlas of Natural Disaster Risk*. The designations employed and the presentation of material on the maps do not imply the expression of any opinion concerning the legal status of any country, territory, city, or area or of its authorities, or concerning the delimitation of its frontiers or boundaries. It uses the Equivalent Difference Latitude Parallel Polyconic Projection with Central Meridian of 150°E. We transformed the projection from the Equivalent Difference Latitude Parallel Polyconic Projection into the Robinson Projection and performed registration before using the boundaries in the maps of this atlas.

All maps in the atlas adopt the Robinson Project with Central Median of 160°E. According to the tasks and purposes of the atlas, we used the following scales for the full map of the world: 1:140,000,000 (Part I and baseline, single page) and 1:200,000,000 (Parts II, III, and IV, 1/2 page).

Appendix E
Integrated Assessment Model of Global Change Risk of Population and Economic Systems

Ning Li, Zhengtao Zhang, Jidong Wu, Saini Yang, and Hongjian Zhou

To integrate the population risk (injured and casualties, affected population) and the economic risk (production reduction of major crops, damage to road networks, GDP losses) due to global change, the integrated assessment model of population and economic system risks of global change is constructed and released.

This model integrates the theories and methods of global change risk assessment, and can quantitatively assess the risks of the population and economic systems caused by changes in the mean, variability, and extremes of climate elements (temperature, precipitation, wind). This model can meet the requirements of population and economic system risk assessment under different emissions scenarios (RCP2.6, RCP4.5, and RCP8.5, and SSP1, SSP2, and SSP3) at different spatial resolutions (50 km × 50 km for the global scale, and high resolution of 30 km × 30 km for hotspot areas), and in different time periods (the 2030s and 2050s). The global multi-regional assessment model and multi-scale assessment model are systematically integrated, and finally a global population and economic system risk assessment model is formed.

The model is mathematically reliable and produces useful results. It can effectively identify the quantitative relationships between the characteristics of different types of hazards and their intensity and the impact on the populations, the function of the transportation networks, the capital stock, and the industrial systems. It can also integrate different emissions scenarios and future near-term (the 2030s) and medium-term (the 2050s) risk changes, and help to understand the mechanism of climate change that may lead to risks.

Each module in this model describes the key steps in model construction and is based on a set of reasonable assumptions. Its results are replicable, and each model has been exemplified by actual cases (Fig. E.1). The platform is shown in Fig. E.2.

The three key modules and their component models are described as follows:

(1) Risk assessment module of population impact under future global change scenarios
- A model for evaluating the impact of future average temperature on China's population mortality. This model assesses the risk of population mortality under the RCP4.5 and RCP8.5 scenarios at a regional scale.
- A model for evaluating the impact of future drought on the exposure of the affected population in the Beijing, Tianjin, and Hebei region. The model evaluates the exposure of the population to droughts under different future scenarios.
- A model for evaluating the impact of future heatwaves on the global exposure of the affected population. The model considers the thermal stress of extreme temperature and humidity, and quantifies the risk of heatwaves affecting the population at the global and regional scales under different future scenarios.

N. Li (✉) · Z. Zhang · J. Wu · S. Yang
School of National Safety and Emergency Management, Beijing Normal University, Beijing, 100875, China
e-mail: ningli@bnu.edu.cn

H. Zhou
National Disaster Reduction Center of China, Ministry of Emergency Management of the People's Republic of China, Beijing, 100124, China

P. Shi, *Atlas of Global Change Risk of Population and Economic Systems*, IHDP/Future Earth-Integrated Risk Governance Project Series, https://doi.org/10.1007/978-981-16-6691-9

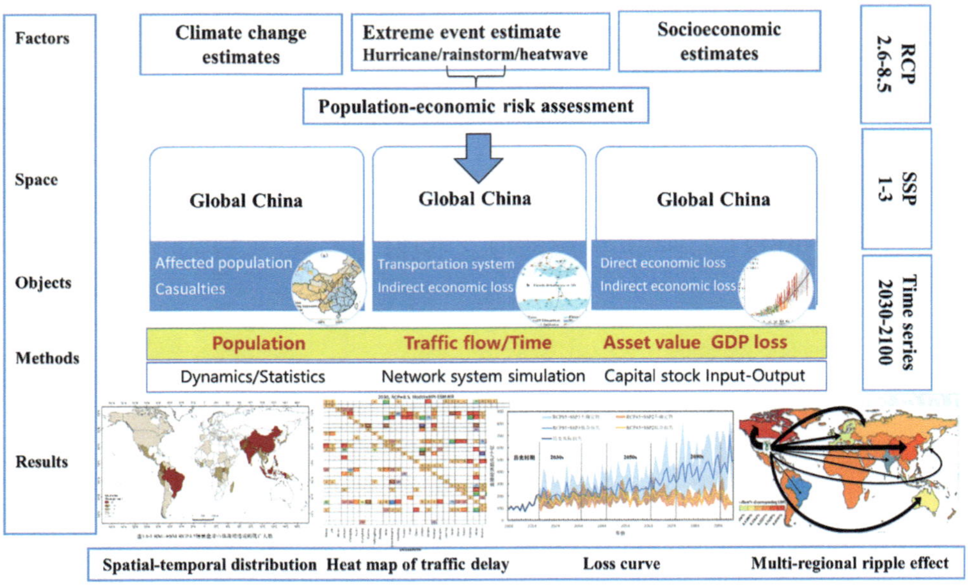

Fig. E.1 Composition of the integrated assessment model of global change risk of population and economic systems

Fig. E.2 Platform of the integrated assessment model of global change risk of population and economic systems

- A model for evaluating the impact of future extreme precipitation on the global affected population. This model quantifies the death risk of the population caused by rainstorms under different future scenarios.
- A model for evaluating the changes of population migration in China. The model assesses the migration status of the population.

(2) Risk assessment module of impact on China's economic system under future global change scenarios (focusing on hotspot areas)

- A comprehensive disaster impact assessment model of floods on the function of China's highway network system. The model evaluates the differences in the impact of floods on the function of highway network systems, changes in regional traffic flow, and different adaptation measures.
- An exposure assessment model of China's highways under the influence of extremely high temperature and extreme precipitation events. The model identifies the time and location at which the highways may suffer destructive damages, and assesses the degree of exposure to high temperature and precipitation events of highways in China in the future.
- A direct loss assessment model of typhoon events. This model uses the changes in the intensity and frequency of future typhoon hazards predicted by climate models to estimate the future spatiotemporal development trends of GDP and asset stock exposure of China to typhoons (based on SSPs data).
- An indirect economic impact assessment model of major crop yield reduction. The model assesses the economic ripple effects of agriculture output reduction on other sectors due to industrial linkages at the country scale.
- An indirect economic impact assessment model considering annual mean temperature rises at the global scale. The model assesses the indirect economic ripple effects among countries

(3) Assessment module of direct losses and indirect impacts of disaster events in China (hotspots)

- A direct loss assessment model of rainstorm and flood events. This model uses the nonlinear relationship between the direct economic losses of historical disasters and the three climate elements of risk to evaluate the direct losses of heavy rains and floods in China.
- An indirect economic impact assessment model of the disaster-affected labor forces. This model assesses the indirect economic impact of the decline in labor supply and its recovery due to a heavy rain in Wuhan City on Hubei Province and China.
- A functional impact assessment model of highway network system. This model evaluates the differences in regional traffic flow due to the floods in China, and changes and different adaptation measures.

Appendix F
Global Change Risk Map Platform of Population and Economic Systems

Peijun Shi, Wei Xu, Tao Ye, and Bo Chen

To further explore the spatial distribution and changes of various elements, the digital atlas platform—the Global Change Risk Map Platform of Population and Economic Systems—is designed.

The platform is the first to systematically present the global change risks of population and economic systems at a global scale. It is multi-period, multi-scenario, high-resolution, and interactive. The platform systematically presents the global changes in environmental elements, changes in climate elements (temperature, precipitation, wind), changes in population and economic systems (GDP, crops, industrial added value, transportation infrastructure), and global change risks in population and economic systems. The map data cover three periods—the baseline period (1985 − 2005), the 2030s (2016 − 2035), and the 2050s (2046 − 2065). For data in the 2030s and the 2050s, three Representative Concentration Pathway (RCP) scenarios (RCP2.6, RCP4.5, and RCP8.5), three Shared Socioeconomic Pathway (SSP) scenarios (SSP1, SSP2, and SSP3) and their combination scenarios are included. The platform includes about 1150 maps in total (Table F.1) and most maps have a spatial resolution of

P. Shi (✉) · W. Xu · T. Ye · B. Chen
Institute of Disaster Risk Science, Faculty of Geographical Science, Beijing Normal University, Beijing, 100,875, China
e-mail: spj@bnu.edu.cn

P. Shi · W. Xu · T. Ye · B. Chen
State Key Laboratory of Earth Surface Processes and Resource Ecology, Faculty of Geographical Science, Beijing Normal University, Beijing, 100,875, China

P. Shi
Academy of Plateau Science and Sustainability of the People's Government of Qinghai Province and Beijing Normal University, Xining, 810,008, China

P. Shi, *Atlas of Global Change Risk of Population and Economic Systems*, IHDP/Future Earth-Integrated Risk Governance Project Series, https://doi.org/10.1007/978-981-16-6691-9

Table F.1 Maps on the digital atlas planform of global change risk of population and economic systems

Content				Time Period	RCP/SSP	Value and Change	MME and MMS
Climate changes	Temperature	Mean	Annual Winter Summer	Baseline	–	Value	MME, MMS
				2030s 2050s	RCP2.6 RCP4.5 RCP8.5	Value, Change	MME, MMS
		Variability	Interannual Winter Summer	Baseline	—	Value	MME, MMS
				2030s 2050s	RCP2.6 RCP4.5 RCP8.5	Value, Change	MME, MMS
		Extremes	TXx TP10p TX90p	Baseline	–	Value	MME, MMS
				2030s 2050s	RCP2.6 RCP4.5 RCP8.5	Value, Change	MME, MMS
	Precipitation	Mean	Annual Winter Summer	Baseline	–	Value	MME, MMS
				2030s 2050s	RCP2.6 RCP4.5 RCP8.5	Value, Change	MME, MMS
		Variability	Interannual Winter Summer	Baseline	–	Value	MME, MMS
				2030s 2050s	RCP2.6 RCP4.5 RCP8.5	Value, Change	MME, MMS
		Extremes	R10mm, RX1day, RX5day	Baseline	–	Value	MME, MMS
				2030s 2050s	RCP2.6 RCP4.5 RCP8.5	Value, Change	MME, MMS
	Wind speed	Mean	Annual Winter Summer	Baseline	–	Value	MME, MMS
				2030s 2050s	RCP2.6 RCP4.5 RCP8.5	Value, Change	MME, MMS
		Variability	Interannual Winter Summer	Baseline	–	Value	MME, MMS
				2030s 2050s	RCP2.6 RCP4.5 RCP8.5	Value, Change	MME, MMS
		Extremes	Surface maximum wind speed	Baseline	–	Value	MME, MMS
				2030s 2050s	RCP2.6 RCP4.5 RCP8.5	Value, Change	MME, MMS

(continued)

Table F.1 (continued)

Content				Time Period	RCP/SSP	Value and Change	MME and MMS
Population and economic system changes	Global population			Baseline 2030s 2050s	SSP1 SSP2 SSP3	Value	–
	Global Gross Domestic Product (GDP)			Baseline 2030s 2050s	SSP1 SSP2 SSP3	Value	–
	Global crop distribution	Wheat Maize Rice		Baseline 2030s 2050s	RCP2.6 RCP4.5 RCP8.5	Value	–
	Global industrial value added			Baseline 2030s 2050s	SSP1 SSP2 SSP3	Value	–
	Global road system			Baseline 2030s 2050s	SSP1 SSP2 SSP3	Value	–
	Global population exposure to high temperature			Baseline	–	Value	MME, MMS
				2030s 2050s	RCP2.6-SSP1 RCP4.5-SSP2 RCP8.5-SSP3	Value, Change	MME, MMS
	Global population exposure to rainstorms			Baseline	–	Value	MME, MMS
				2030s 2050s	RCP2.6-SSP1 RCP4.5-SSP2 RCP8.5-SSP3	Value, Change	MME, MMS
	Global GDP exposure to drought	Normal Mild Moderate Severe Extreme		Baseline	–	Value	–
				2030s 2050s	RCP2.6-SSP1 RCP4.5-SSP2 RCP8.5-SSP3	Value	–
	Global crop exposure to extremely high temperature	Wheat Rice Maize		Baseline	–	Value	–
				2030s 2050s	RCP2.6 RCP4.5 RCP8.5	Value	–
Global change risks	Global morality risk to heatwave			Baseline	–	Value	MME, MMS
				2030s, 2050s	RCP2.6-SSP1 RCP4.5-SSP2 RCP8.5-SSP3	Value, Change	MME, MMS
	Global mortality risk to floods			Baseline	–	Value	MME, MMS
				2030s 2050s	RCP4.5-SSP2 RCP8.5-SSP3	Value, Change	MME, MMS
	Crop yield risk	Wheat Rice Maize	Yield mean Yield variability Yield extreme	Baseline	–	Value	MME, MMS
				2030s 2050s	RCP2.6 RCP4.5 RCP8.5	Value, Change	MME, MMS
	Global GDP loss risk to floods			Baseline	–	Value	MME, MMS
				2030s 2050s	RCP4.5-SSP2 RCP8.5-SSP3	Value, Change	MME, MMS

Note Value = the value for the current period (baseline, 2030s, or 2050s); change = the change between 2030s/2050s and the baseline period; MME = multi-model mean; MMS = uncertain of the multi-model; RCP = Representative Concentration Pathway; SSP = Shared Socioeconomic Pathway.

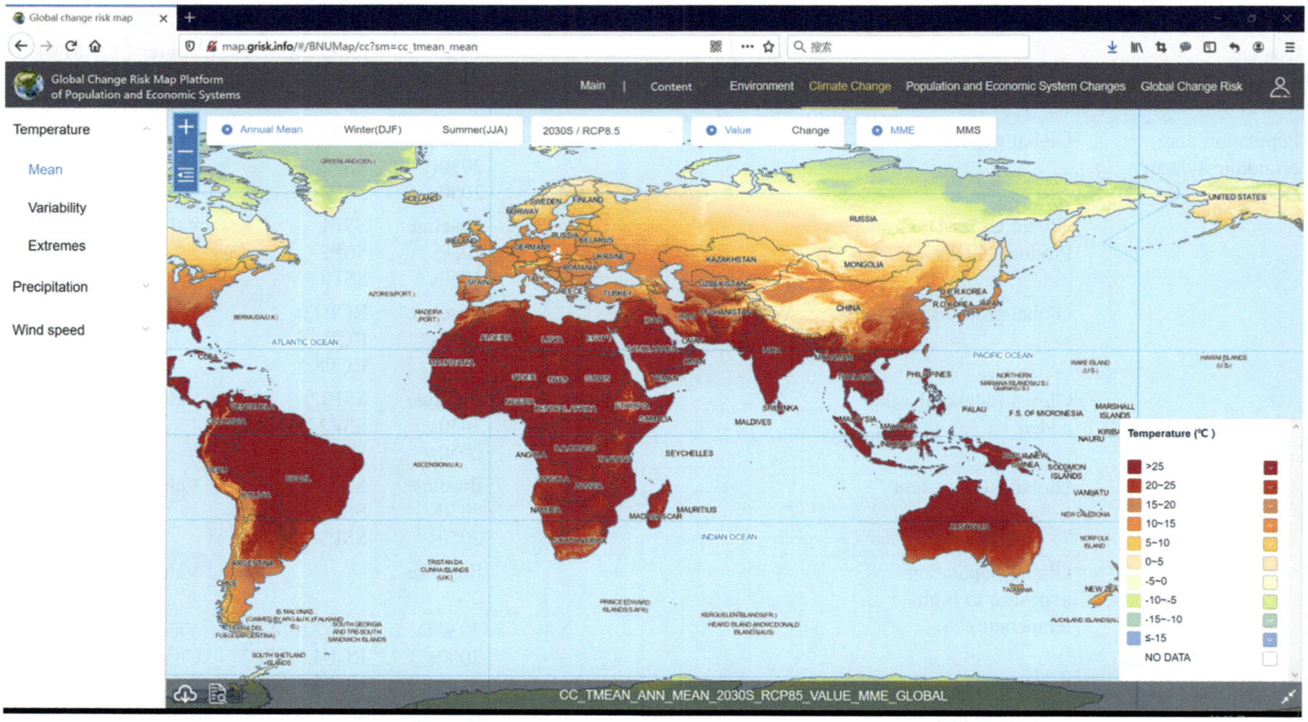

Fig. F.1 Global change risk map platform of population and economic systems

0.25° × 0.25°. All maps on the platform use the project of EPSG 4326 (the World Geodetic System 1984).

The platform enables map query, browsing and roaming, zoom in and out, map download, data display, data query, data download, and layer color manipulation (Fig. F.1). Compared with traditional paper maps, the platform is dynamic and interactive. More information about the platform can be found at http://www.grisk.info.

GPSR Compliance

The European Union's (EU) General Product Safety Regulation (GPSR) is a set of rules that requires consumer products to be safe and our obligations to ensure this.

If you have any concerns about our products, you can contact us on ProductSafety@springernature.com

In case Publisher is established outside the EU, the EU authorized representative is:

Springer Nature Customer Service Center GmbH
Europaplatz 3
69115 Heidelberg, Germany

Batch number: 10732815

Printed by Printforce, the Netherlands